76 Notes on Numerical Fluid Mechanics (NNFM)

Springer
Berlin
Heidelberg
New York
Barcelona
Hong Kong
London
Milan
Paris
Singapore
Tokyo

Aerodynamic Drag Reduction Technologies

Proceedings of the CEAS/DragNet European Drag Reduction Conference, 19–21 June 2000, Potsdam, Germany

Peter Thiede (Editor)

Springer

Prof. Dr.-Ing. Peter Thiede
EADS Airbus GmbH
28183 Bremen
e-mail: peter.thiede@airbus.dasa.de

Library of Congress Cataloging-in-Publication Data applied for

Aerodynamic Drag reduction technologies : proceedings of the CEAS, DragNet European Drag Reduction Conference 19 - 21 June 2000, Potsdam, Germany / Peter Thiede (ed.). – Berlin ; Heidelberg ; New York ; Barcelona ; Hong Kong ; London ; Milan ; Paris ; Singapore ; Tokyo : Springer, 2001
(Notes on numerical fluid mechanics ; Vol. 76)
ISBN 3-540-41911-X

ISSN 0179-9614
ISBN 3-540-41911-0 Springer-Verlag Berlin Heidelberg New York

Springer-Verlag Berlin Heidelberg New York
a member of BertelsmannSpringer Science+Business Media GmbH

http://www.springer.de

Printed in Germany

Typesetting: Camera ready by authors
Cover design: de'blik, Berlin

Printed on acid-free paper SPIN: 10835334 62/3020/M – 5 4 3 2 1 0

NNFM Editor Addresses

Prof. Dr. Ernst Heinrich Hirschel
(General editor)
Herzog-Heinrich-Weg 6
D-85604 Zorneding
Germany
E-mail: e.h.hirschel@t-online.de

Prof. Dr. Kozo Fujii
Space Transportation Research Division
The Institute of Space and Astronautical Science
3-1-1, Yoshinodai, Sagamihara, Kanagawa,
229-8510
Japan
E-mail: fujii@flab.eng.isas.ac.jp

Dr. Werner Haase
Höhenkirchener Str. 19d
D-85662 Hohenbrunn
Germany
E-mail: wug.haase@t-online.de

Prof. Dr. Bram van Leer
Department of Aerospace Engineering
The University of Michigan
Ann Arbor, MI 48109-2140
USA
E-mail: bram@engin.umich.edu

Prof. Dr. Michael A. Leschziner
Department of Engineering
Queen Mary & Westfield College (QMW)
University of London
Mile End Road
London E1 4NS
Great Britain
E-mail: mike.leschziner@qmw.ac.uk

Prof. Dr. Maurizio Pandolfi
Politecnico di Torino
Dipartimento di Ingegneria Aeronautica e
Spaziale
Corso Duca degli Abruzzi, 24
I - 10129 Torino
Italy
E-mail: pandolfi@polito.it

Prof. Dr. Jaques Periaux
Dassault Aviation
78, Quai Marcel Dassault
F-92552 St. Cloud Cedex
France
E-mail: periaux@rascasse.inria.fr

Prof. Dr. Arthur Rizzi
Department of Aeronautics
KTH Royal Institute of Technology
Teknikringen 8
S-10044 Stockholm
Sweden
E-mail: rizzi@aero.kth.se

Dr. Bernard Roux
IRPHE-IMT
Technopole de Chateau-Gombert
F-13451 Marseille Cedex 20
France
E-mail: broux@irphe.univ-mrs.fr

FOREWORD

This volume contains the proceedings of the CEAS/DragNet European Drag Reduction Conference held on 19-21 June 2000 in Potsdam, Germany. This conference, succeeding the First and Second European Forum on Laminar Flow Technology 1992 and 1996 respectively, was initiated by the European Drag Reduction Network (DragNet) and organised by DGLR (Deutsche Gesellschaft für Luft- und Raumfahrt - Lilienthal Oberth e.V.) under the auspices of CEAS (Confederation of European Aerospace Societies).

The development of aerodynamic drag reduction technologies is driven by predictions of remarkable fuel savings, promising substantial improvements not only of aircraft efficiency but also of environmental compatibility. However, considerable efforts on an European scale are needed in order to develop, qualify and demonstrate the means for their practical realisation.

The primary aim of this conference was to provide a comprehensive survey of the current status of research, development and application in all disciplines of aerodynamic drag reduction including laminar flow technology, adaptive wing concepts, turbulence and separation control, induced drag reduction and supersonic flow aspects. Besides aerodynamic topics the Call for Papers addressed also interdisciplinary aspects of design & system integration, structures, materials, manufacturing, operations and maintenance. The Programme Committee (PC), responsible for the scientific preparation of the conference, consisted of CEAS representatives and DragNet board members (see following page). At the PC meeting it was decided to limit the conference to two and a half days, to avoid parallel sessions, to arrange poster sessions and a round table discussion, and to organise a conference exhibition. Totally, 36 papers including 11 invited lectures as well as 10 posters were selected for presentation in 12 sessions.

The conference was opened at noon of the first day by representatives of CEAS and the German Government and by the PC chairman. The first afternoon session entitled *The Challenge of Drag Reduction* included three invited papers given by keynote speakers from the European Commission, industry and the research community. For the second session *Achievements of Technology Demonstration* three experts from industry (Germany, France, UK) were invited. The second day of the conference was devoted *to Laminar Flow Technology*, covering sessions on *HLF Design & Suction System Integration*, *HLF Operational Aspects*, *Transition Prediction*, *HLF Experimental Techniques* and a poster session, as well as to *Supersonic Flow Control Aspects* (one session). The third conference day addressed *Adaptive Wing Concepts*, *Turbulent Drag Reduction Methods*, *Separation Control & Induced Drag Reduction* (three paper sessions and a poster session). In addition, a session on *Future Prospects* including a round table discussion of representatives from the European Commission, industry and academia was organised at the last conference day. The conference was supplemented by an exhibition showing samples of drag reduction devices and aircraft models equipped with corresponding devices. The poster exhibition was completed by posters of DragNet member projects.

The conference brought together top level authorities from politics and RTD administration with engineers and scientists from industry and academia working in the field of aerodynamic drag reduction. It provided an outstanding opportunity to present and discuss the recent advances achieved in European initiatives (EC and GARTEUR activities), industrial projects and national programmes.

The review of the manuscripts published in this volume was undertaken by a Review Board (see following page) consisting of the session chairmen of the conference. Nevertheless, the authors are signing responsible for the content of their contributions. The manuscripts of most of the papers (36) and posters (10) presented during the conference are published in this volume in the same sequence as at the conference. Insofar, this volume gives an account of the complete conference.

On behalf of the Programme Committee, the editor would like to thank all colleagues for their efforts they put into the preparation and performance of the conference and exhibition in order to make them successful as well as into the publication of the conference proceedings. In this context, the involvement of the DragNet and Review Board members have to be particularly emphasized. Furthermore, thanks are due to DGLR for the layout and distribution of the conference announcement and programme, the local organisation of the conference and social events and for their engagement in the preparation of the conference proceedings. The performance of the conference and exhibition and the publication of the proceedings would not have been possible without financial support of the European Commission, which should be gratefully acknowledged. The editor is also grateful to Prof. E. H. Hirschel, the general editor of the "Notes on Numerical Fluid Mechanics", and to the Springer Verlag for making this publication possible.

Bremen, December 2000

The Editor
Peter Thiede

CONTENTS

Programme Committee

P. Thiede	DASA Airbus Bremen (Chairman)
A. Amendola	CIRA Capua
D. Arnal	ONERA Toulouse
A. Elsenaar	NLR Amsterdam
J. Fulker	DERA Bedford
D. Knörzer	EU Brussels
D. Lovell	DERA Farnborough
F. Monge	INTA Madrid
V. Schmitt	ONERA Châtillon
G. Schrauf	DASA Airbus Bremen
Y. Sedin	SAAB Linköping
E. Stanewsky	DLR Göttingen
J. Szodruch	DASA Airbus Hamburg
C. Warsop	BAE SYSTEMS Bristol
J. Wildi	SF Emmen

Review Board

D. Arnal	ONERA Toulouse
C. Atkin	DERA Farnborough
A. Elsenaar	NLR Amsterdam
J. Herrera	AI Toulouse
D. Knörzer	EU Brussels
K.E. Modin	SAAB Linköping
G. Redeker	DLR Braunschweig
J. Reneaux	ONERA Châtillon
V. Schmitt	ONERA Châtillon
N. Sellars	BAE SYSTEMS Bristol
J. Szodruch	DASA Airbus Hamburg
J.J. Thibert	ONERA Châtillon
P. Thiede	DASA Airbus Bremen

Session 1
The Challenge of Drag Reduction

Perspectives for the Future of Aeronautics Research

Dietrich Knörzer,
European Commission, Directorate General Research
Rue de la Loi 200, B-1049 Brussels, dietrich.knoerzer@cec.eu.int

Summary

Europe's modern societies are facing a number of challenges. In particular its industry has to master the increasing globalisation of the economy. Research and technological development - short R&TD play an important role to ensure the future compatitiveness on a global market especially in high-tech areas as aeronautics. Subsequently the role, the structure and the content of the effort in technology acquisition by the research framework programmes of the European Union will be highlighted. A special emphasis will be on the perspectives of specific aeronautics research within the technology programme "Competitive and Sustainable Growth".

The Challenges and Needs for Europe

Until the 80ies, the economy of most European countries was mainly nationally oriented with comfortable home markets for industries, which were often subsidised and well protected by trade regulations, as in many countries the aeronautics industry. For Europe this has changed - at latest since the Single European Market has been established at the begin of the nineties. The previously state-owned airlines have been more and more privatised and have to compete on a liberalised transport market. At the turn of the millennium, industry has to tackle the rapidly growing globalisation. Access to cheep labour, state subsidies and protected home markets do not anymore represent the useful means for Europe's industry to stay in business. Efficiency and the use of high-tech are tools to become competitive. Europe's society has to tackle all these challenges. Researchers and engineers have to contribute significantly by providing the needed new ideas, technologies and system solutions.

A New Strategic R&TD Approach

While the technology push was dominating the research activities in the past, today there is an increasing need for problem solving R&TD approaches. Innovation needs to comply with three main factors:

- The needs of the society,
- the demand of the markets,
- the technological advantage.

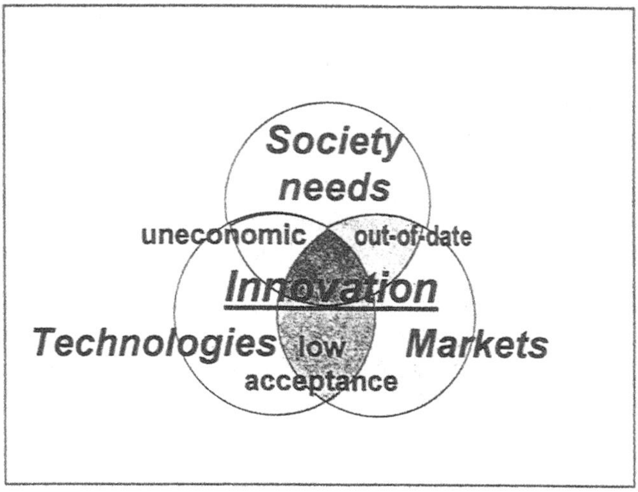

Figure 1: Factors of innovation

Only if all three are fulfilled, an innovative development can be launched in a specific industrial sector (see figure 1). Therefore the new strategic R&TD approach on European level has to comply with the following criteria and the needed scientific and technological resources have to be mobilised:

- Meeting the objectives of the society,
- ensuring economic development and scientific and technological prospects,
- creating a European added value.

Future Technology Needs of the Aeronautics Industry.

Within Europe the aerospace industry traditionally plays a key role in the development and use of advanced technologies. On the other side an efficient and environmentally sustainable air transport system is vital for Europe and its economy. Three main concerns are driving the technology needs of the aeronautics sector: efficiency, environmental sustainability and safety.

Efficiency

There is a need to increase the capacity and efficiency of both the transport system and of the aircraft, here specially by reducing the direct operation costs.

Environmental Sustainability

The increasing concerns about the environmental impact such as the global warming are becoming very demanding to the transport sectors. The future society will not easily accept any increase of pollution, emission and noise by the air transport system despite the increase of the traffic.
Interdisciplinary optimisation processes are needed for the new generation of aircraft bringing together the key disciplines such as fluid dynamics, materials, information technologies, electronics, and not at least the combustion process.

Safety

Despite a continuous growth of the air transport sectors, the society will not easily accept an increase of accidents proportional to the increase of air traffic. That means, measures need to be undertaken to improve the safety of air transport to cope with the growing traffic density.

Addressing the technology needs of the next generation aircraft, one can identify common needs:

- Reduction of production cost and time to the market by new design and manufacturing methods.

- Reduction of energy consumption and an increased aircraft efficiency leading to an overall emission reduction. By the use of new materials and improved structural concepts, a significant weight reduction can be achieved. Advanced aerodynamics contributes to a reduced energy consumption and external noise level.

- Improved operational reliability and safety
  The high frequent commuter links of the metropolitan areas require new concept and new technologies for a very high degree of reliability and safety. This includes aspects of maintainability.

However, the technology acquisition process until the first commercial application takes normally many years, starting from first focused fundamental research up to technology validation and demonstration. The Research Framework Programmes cover the entire process of technological developments.

Aeronautics in the Community Research Framework Programmes

Since 1984 the joint research activities on European level have been performed through research framework programmes. Their character and content has changed from technology push to problem-oriented technology acquisition, due to the changes of industrial needs in response to the market requirements. The budget has grown up to about 15.000 million Euro for the Fifth Research Framework Programme for a five years period covering all key technology areas.

Specific aeronautics research started in the Second Research Framework Programme by a small pilot phase of 35 million Euro for 1990/91. The community research activities in aeronautics has grown to 700 million Euro of EU funding that are spent in the Key Action *New perspectives in Aeronautics* of the Fifth Framework Programme. Figure 2 shows the significant development of the funding for specific aeronàutics research on Community level.

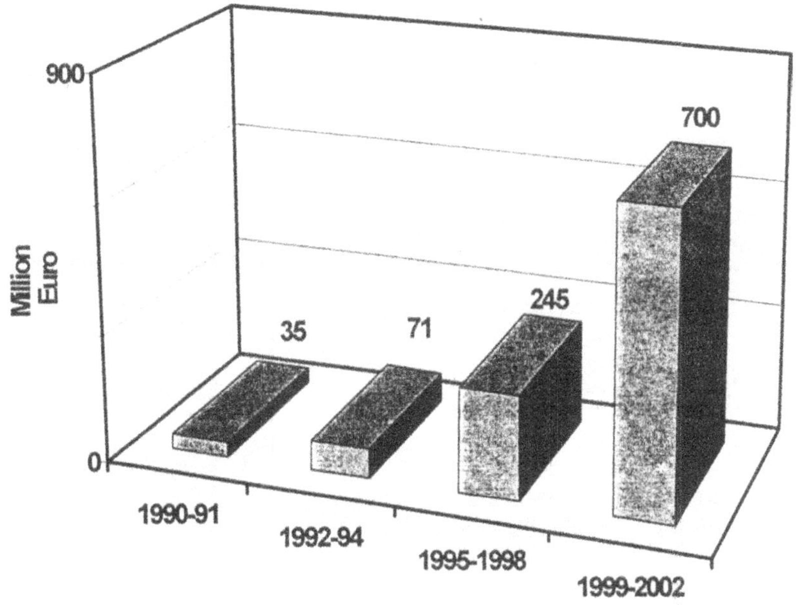

Figure 2: Development of the EU budget for specific aeronautics research of the Research Framework Programmes

Key Action: "New Perspectives in Aeronautics"

The Key Action Aeronautics is part of the Thematic Programme *Promoting Competitive and Sustainable Growth* that supports the technology acquisition contributing to industrial competitiveness and sustainable economic growth especially for the manufacturing and the transport industry. At a time of global challenges this is not anymore a matter of individual organisations or sectors alone. A multidisciplinary holistic R&TD approach is needed to encourage the development of the emerging technologies and their interrelations to meet future market demands and society needs.

The Key Action *New Perspectives in Aeronautics* has the overall objectives:

- Reducing aircraft development cost & time
  (35% in production costs; 20% in development time),

- Improving aircraft efficiency
  (20% less fuel consumption; more reliability),
- Improving environmental friendliness
  (20% less CO_2; 80% less Nox; 10dB less noise),
- Operational capacity and safety
  (increase airspace utilisation; 25% less maintenance costs; decreased accident rates).

Within the key action the critical technologies for civil aeronautics are developed along four technology avenues (see figure 3). A number of specifically identified *Technology Platforms* for integration and validation of advanced technologies addresses new technological concepts for tiltrotor, external aircraft noise, power optimised aircraft and modular avionics.

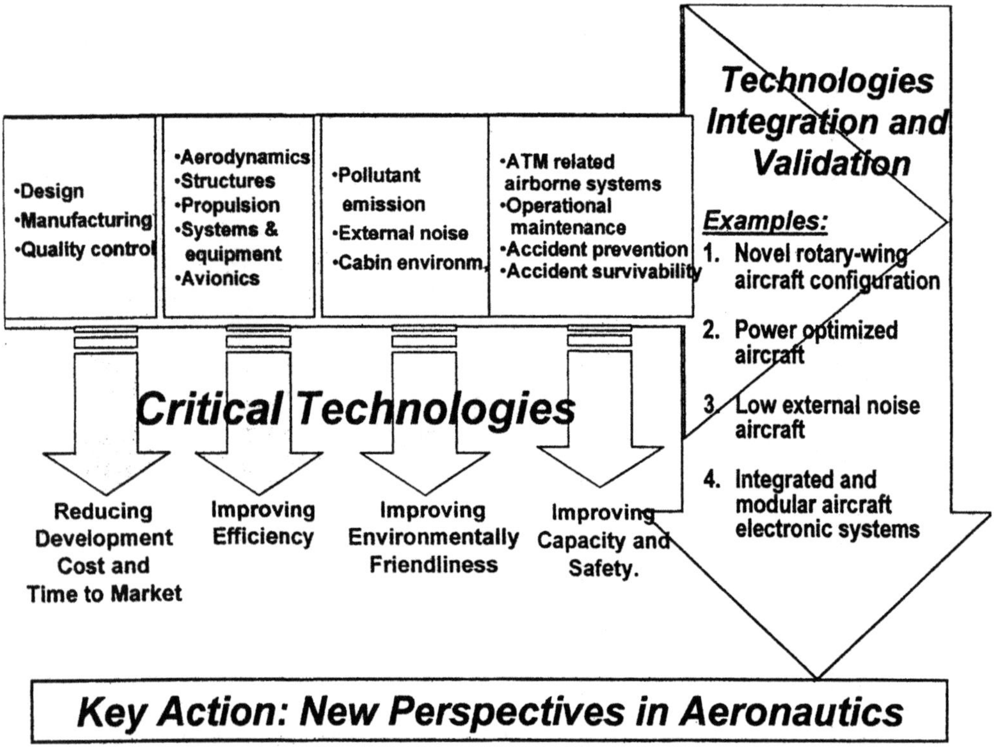

Figure 3: Main technology areas of the Key Action: *New Perspectives in Aeronautics*

Activities on Aircraft Efficiency on the Example of Aerodynamics

In the field of aerodynamics and aero-acoustic, numerous R&TD projects have been performed since the start of specific aeronautics research on EU level in 1989. They have contributed significantly to the development of the physical understanding of complex phenomena and of numerical design tools that enable industry to predict the flow around aircraft already with remarkable accuracy. A continuity of a joint European technology development has been ensured over a period of more than ten years. The total funding volume of aerodynamics related research projects on European level was more than 55 million Euro for the period 1990 to 2000.

Improving aircraft efficiency is one of the key elements for the competitiveness of Europe's aeronautics industry, therefore about 50% of the research activities aim for the following objectives:

- to reduce aerodynamic drag by 20% and improve lift-to-drag ratio,
- to reduce structural weight by 20%,
- to improve fuel economy by 20%,
- to reduce systems power take-up by 10% and weight by 20%.

In aerodynamics and related areas the main research tasks addressed are:

Aerodynamics

- drag reduction (laminar flow)
- boundary layer control
- adaptive wing
- high-lift aerodynamics
- CFD tools

Configurational and interdisciplinary aspects

- novel aircraft configurations
- airframe-engine integration
- dynamic aero-elastic phenomena

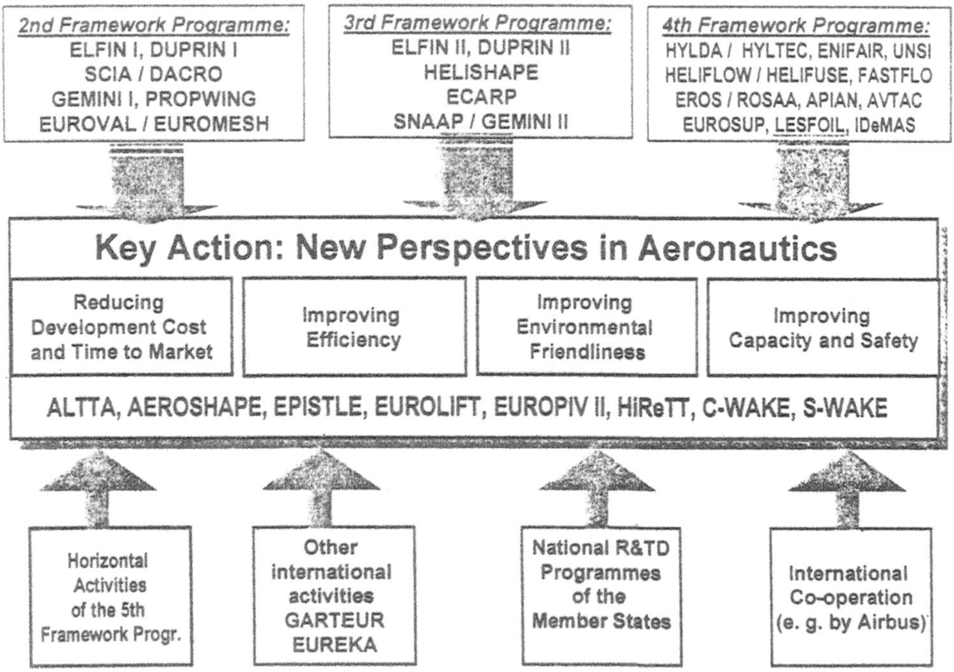

Figure 4: Maximal synergy by co-ordination and complementarity of research activities giving the example of European aerodynamics and aeroacoustics projects

The field of aerodynamics and aero-acoustics, the new R&TD activities have to rely on the technology basis created by the previous projects and have to be complementary to research performed on national and international level (see figure 4). Where appropriate, co-ordination will take place with other activities within the Research Framework Programme.

Outlook and Conclusions

The specific aeronautics reseearch has received a significant push within the Fifth Research Framework Programme. The technology acquisition through the Research Framework Programme of the EU has to be qualified by a strong European added value. It aims at research and innovation into new concepts of technologies for industrial products or for new production and service systems to boost sustainable competitiveness. The qualification of research activities for support by the European Union is based on five selection criteria:

- Ensuring scientific and technical excellence;
- Quality of approach, partnership and management;
- Stimulating European added value;
- Answering to society needs;
- Economic development as well as scientific and technological perspectives.

The policy of the European Union on R&TD is directed towards a European research area aiming to strengthen the scientific and technological bases of Europe's industry in an integrated way and encouraging to enhance the industrial competitiveness at international level by joint technology acquisition. Research upported financially by the European Union has to serve the needs of Europe's citizens in the first instance.

For aeronautics, the External Advisory Group of the Key Action Aeronautics has stated in its position paper of April 2000 [1]:

> *The European aeronautics industry is knowledge based.*
> *It faces three key high-level challenges:*
> - *Meeting the needs of an advanced society,*
> - *Creating competitive products for a global market,*
> - *Sustaining the flow of key technologies.*

Reference

[1] Aeronautics for Europe – A Partnership for Research and Technology and European Growth - A position paper produced by the External Advisory Group for Aeronautics; published by the European Commission, april 2000, EUR 19318, ISDN 92-828-8596-8

THE IMPORTANCE OF AERODYNAMICS IN THE DEVELOPMENT OF COMMERCIALLY SUCCESSFUL TRANSPORT AIRCRAFT

Wolfgang Schneider

EADS Airbus GmbH
P.O.Box 950109
D-21111 Hamburg

Summary

Aerodynamics - as an engineering discipline involved deeply in the aircraft development process - always have been and will continue to be essential for the commercial success of any aircraft programme. Past developments in computing methods and tools as well as in wind tunnel testing technologies have produced clear cost and performance benefits. For this reason it is absolutely justified to invest in further improvements concerning development of tools and methods including experimental technologies and facilities, taking advantage of the enormous leverage on both total programme costs and aircraft performance.

Introduction

This paper is intended to be an overview focusing on the role and importance of aerodynamics in the design process rather than a detailed presentation on specific aerodynamic technologies, which can be expected to be covered by the majority of presentations held at this conference. In order to successfully evaluate the importance of aerodynamics - as is the stated goal of the headline - it is necessary first to achieve an overview of today's process in the development of commercial aircraft. From this, the relative position and importance of the specific discipline aerodynamics can be evolved.
The basic reason for this is that the complete design process has changed over the time: what used to be a rather sequential process of design efforts of individual disciplines, has turned into a more or less integrated design effort that is arranged as much as possible in a concurrent way.
This is largely a consequence of the ever increasing demand for faster, better, cheaper design, which in turn triggers an ever increasing complexity in the design of modern high tech products. It is certainly an interesting question whether there would be alternative design processes that provide the necessary competitive edge. For the time being, it seems that nothing short of the perfect integration of Computer Aided Design, Engineering and Manufacturing (CAD / CAE / CAM) with all the relevant 'sub - Information Technologies' like Knowledge Based Engineering (KBE), Computational Fluid Dynamics (CFD) and all kinds of Computer Based Simulations (Virtual Reality) - to name a few - will define the ultimate competitive design process. And - this must be mentioned - the supply chain must be integrated in this effort. This trend seems only to be limited by the availability of engineers capable of fusing traditional engineering disciplines with the IT-world. Top talent is more and more attracted directly by the bonanzas of the IT-world.

Importance of Aerodynamics in the Design Cycle

Looking at the design of aeroplanes, it is a well-known fact that the impact of design decisions on overall costs reduces with increasing information about the aircraft, i.e. with progress during the design process. In other words, those decisions which are made early in the design process, have the highest impact on the technical and financial success of the aircraft (see **Fig. 1**).

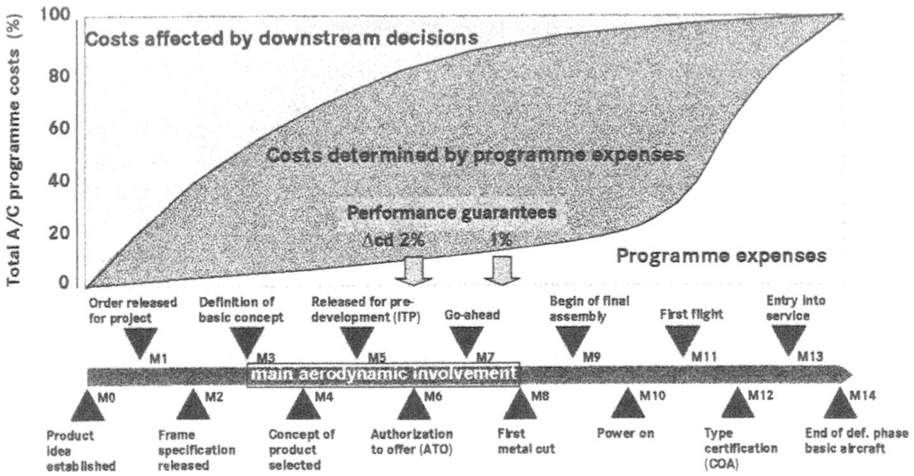

Fig. 1 Typical a/c programme cost versus milestones

Aerodynamic design is one of those disciplines involved in the very early stages of aircraft design and must consequently provide mature technologies and use methods which assure sufficiently exact prediction of aerodynamic efficiency. A level of confidence of 2% remaining uncertainty at ATO (Authorisation to Offer), or 1% before first metal is cut, must be guaranteed. On the other hand, amount and quality of available data is very limited in the early design phases. In order to cope with these opposite trends, the product development cycle was reconsidered.

Fig. 2 gives an indication of increased pre-development work, which must be heavily complemented by technology preparation and validation. Although initially a higher cash flow before Go-Ahead is required, the maturity of the design at that milestone is increased. The intended consequence is that actual development cost and time summed up until Entry Into Service (EIS) can be reduced significantly. The main message here: boost the spending for technology and pre-development activities in order to reduce the total development costs.

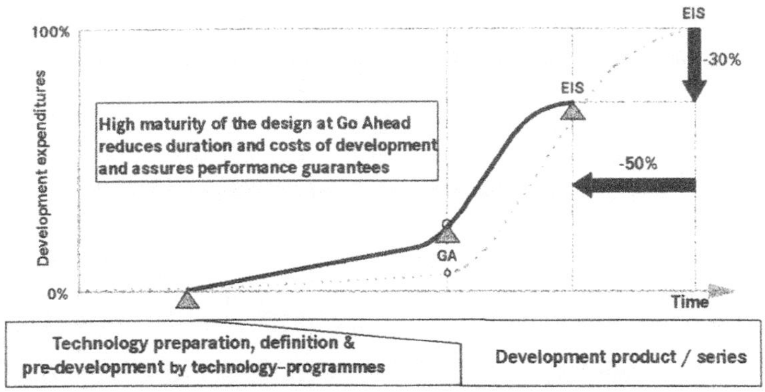

Fig. 2 Faster, better, cheaper approach

Cost – Benefit of Aerodynamics

Analysing the costs of aircraft programmes one can identify: Non-Recurring Costs (NRC) - based on Non-Specific Design Work (NSDW) and Specific Design Work (SDW) plus more than a dozen of other Cost Chapters - and Recurring Cost (RC). The cost chapters NSDW, NRC and RC relate to each other like 1 : 10 : 120 (see **Fig. 3**). This relation indicates that with a rather small budget in the early NSDW phases, the majority of the total programme costs are affected. Thus, it is evident that the definition of the basic concept of an aircraft already determines about 50 - 60% of the total programme costs.

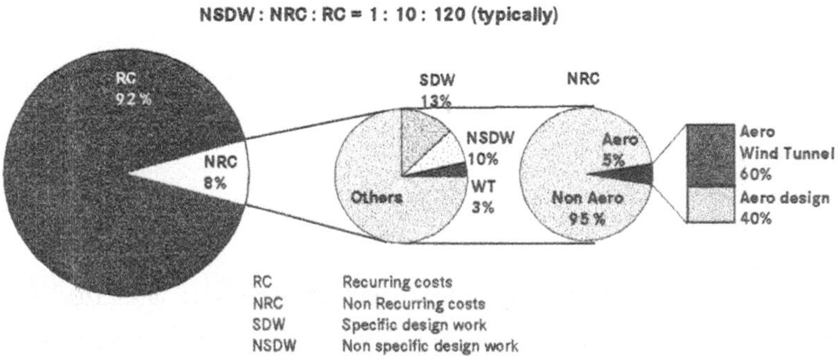

Fig. 3 Costs of aerodynamic development

In order to quantify the costs of aerodynamic development, the NRC (basically the development costs) can be analysed and the aerodynamic design efforts and the wind tunnel activities separated. Doing this, the total aerodynamic contribution results in about 5% of NRC, where aerodynamic design contributes about 2% and the wind tunnel testing approximately 3%.

In spite of the relative small share of aerodynamics at the development costs of a new aircraft, it must be pointed out that total development costs are approaching prohibitive proportions (e.g. more than 10 bn U.S. $ in case of the A3XX, money which needs to be raised from the capital market by the manufacturer, with the prospect of achieving break-even years later). That's why there seems to be no alternative to the faster, cheaper, better approach.

The development of an aircraft during all phases is nowadays an extremely interdisciplinary activity. Although we have separated aerodynamics for the purpose of cost transparency, it is evident that they only can contribute to reductions in costs and development-time in a highly interdisciplinary environment as sketched in **Fig. 4**. Here the typical data flow to and from Flight Physics/ Aerodynamics is depicted. This is of course an iterative process of the same kind as the one within the aerodynamic group itself.

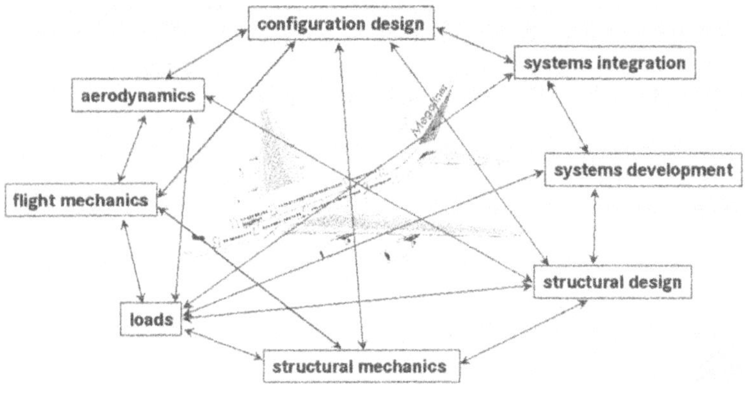

Fig. 4 Role of aerodynamics

However, within aerodynamics further enhancement and validation of numerical tools, as well as continued development of experimental technologies and facilities is paying. More precise (and more costly) computational methods will lead to fewer, albeit more precise (and more expensive) experimental testing. As an example, cryogenic test technology, which is admittedly expensive, will allow the investigation at real world Reynolds and Mach number effects and deliver precise performance data. Resulting higher upfront cost should be easily recoverable by reduced efforts in flight-testing and costly hardware modifications late in the development process.

Aerodynamic flight efficiency (in terms of M*L/D) has been increased by more than 30% relative to the A300 basis (see **Fig. 5**).

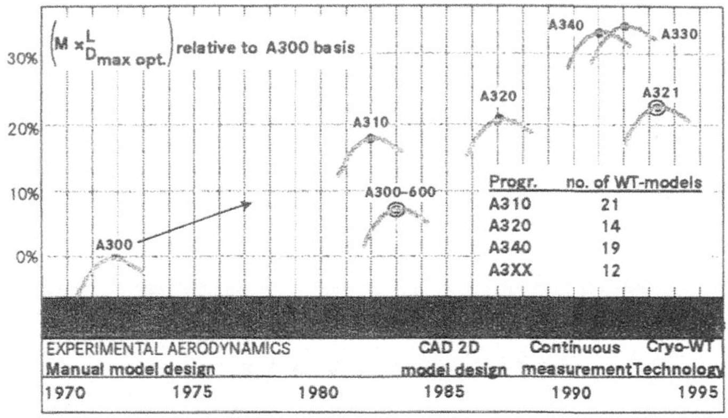

Fig. 5 Progress in aerodynamic efficiency

In order to get a better understanding of the importance of aerodynamic refinement we shall consider the following numbers: a 1% reduction in aerodynamic drag for an A340 aircraft operating in the long range mode will result in savings of about 400,000 liters of fuel per year and aircraft, which in turn will save the environment roughly 5000 kg of noxious emissions (see **Fig. 6**).

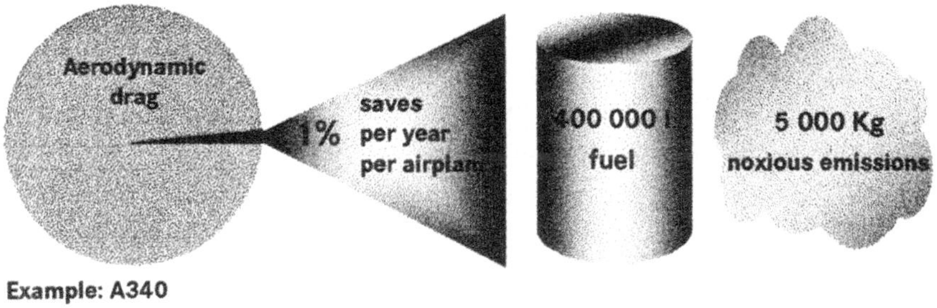

Fig. 6 Drag reduction and environmental impact

The savings in Direct Operating Costs (DOC) for an airline can amount to millions of U.S. $ over a typical period of operations. It needs to be kept in mind that net profit for an airline is the (positive) difference between two large numbers: revenue and costs. It can be stated that in the recent past, a margin of about 5% has been indicative for a well managed airline. In the economic conditions of 1996 fuel accounted for about 30% of the A340 DOCs. A 15% increase in fuel cost would require a 10% increase in revenue to balance the equation and keep the same positive 5% margin. In case revenue can't be increased, profit of the airline would be erased to zero. Conversely these numbers indicate that a 15% reduction in fuel costs - e.g. achieved by burning less fuel, either by better engines or an aircraft with less drag - would double the profit for the airline. This clearly indicates the potential benefit of improvements in aerodynamics.

Future Challenges

Traditionally the challenge has been to make an aircraft ever more efficient in terms of L/D ratio - at a given reasonable manufacturing effort. A combination of superior low speed/high lift- and cruise performance as well as benign handling qualities of the aircraft over the whole flight envelope has been the justified goal for past decades. This will continue, but with the additional challenge of simultaneously designing for low aerodynamic noise.

In a similar way there will be the additional challenge to design for low vorticity in the wake of a starting or approaching aircraft. Especially in view of the new generation of very large aircraft it will be mandatory to achieve a breakthrough in low vortex energy design, if the gain in capacity due to the increased size of the aircraft is to be exploited. Vortex considerations determine the distance and sequence of approaching aircraft. The current rule is based on empirical data using as main parameter the weight class of an aircraft, with the highest weight aircraft requiring the widest spacing of approaching traffic. So, unless the physics are fully understood and engineers are able to design for low vortex intensity, the air transport industry will not be able to significantly improve on capacity bottlenecks at major congested airports.

Other challenges that are visible today, are also increasingly of multi- or interdisciplinary nature like e.g. aeroelasticity, the interaction of fluid dynamics and structures. This shows that the main challenges cannot be mastered by the work of a single discipline anymore, but instead it is the multi- and interdisciplinary effort which is required.

One of the remaining challenges in the sense of aerodynamic performance improvement will be the introduction of laminar flow in large commercial transport aircraft. Laminar flow has been the subject of research and technology work for decades, but so far has found actual application only in niche markets e.g. in glider planes or in a couple of general aviation or business aircraft. Here 'natural' laminar flow is achieved by specific shaping of the airfoil and it is possible because of the lack of disturbing leading edge slats with associated gaps and irregularities. The application for large commercial transport planes, with corresponding large Reynolds numbers, will require 'hybrid' laminar flow, which is a combination of airfoil shaping and a powered suction systems which control the boundary layer for sustained laminarity of the flow.

Principles are well known and research work culminated in recent full-scale laminar fin flight tests of an A320 that have confirmed real world laminar flow performance (see **Fig. 7**). It is time now to state that laminarisation is not any more an aerodynamic problem in first place, but rather it is an enormous challenge for structures and systems design. Making laminar flow a successful operational reality will certainly require an intense and costly interdisciplinary research and development effort. However net drag reduction could considerably contribute to achieve overall reductions in emissions that might be requested by regulatory authorities in response to concerns that continuous growth of air traffic (with current emission rates) will lead to an adverse impact on global climate.

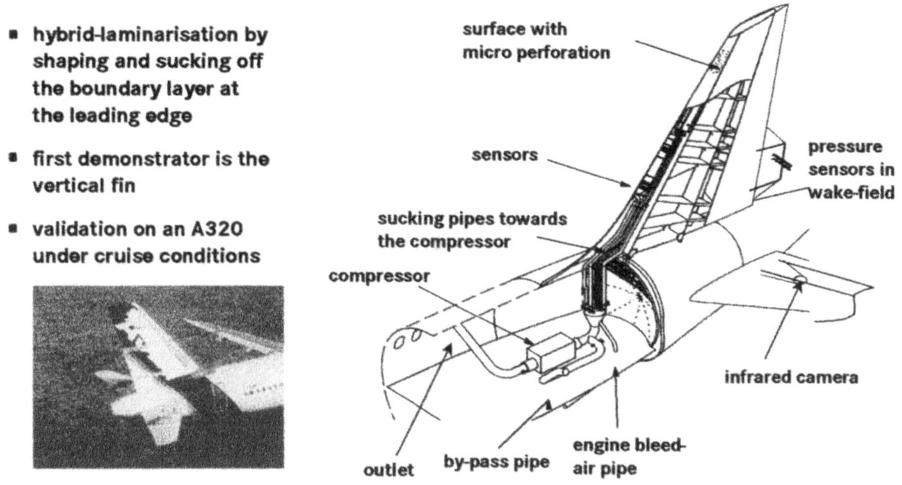

Fig. 7 Future challenges

Conclusion

The development of an aircraft during all phases is nowadays an extremely interdisciplinary activity. In spite of the relative small share of aerodynamics of the development costs for a new aircraft, total development costs are approaching prohibitive proportions and consequently there is continuous pressure - for each specific discipline - to be more efficient at reduced costs and shorter cycle times. Therefore, permanent efforts to optimise expenses for aerodynamic investigations are of high importance. Many impressive improvements in wind tunnel model design and test practices have been achieved. Furthermore the precision of flow calculation methods is progressing continuously, resulting in reduced expenses in wind tunnel and flight testing.

As shown above, costs reduction is not only a topic for the development and manufacturing process but also of vital importance for the operator. Therefore, research and technology must further concentrate on aircraft performance improvements, however, harmonised with manufacturing requirements and operational aspects.

Further reductions in fuel consumption can be envisaged originating from advanced technologies in the fields of aerodynamics, structure and materials, systems and engine design. Despite the breakdown in single disciplines, it is in the end the interdisciplinary, concurrent adaptation of all those technologies which might contribute to an anticipated 50% fuel savings indicated as possible over the next 20 years (see **Fig. 8**).

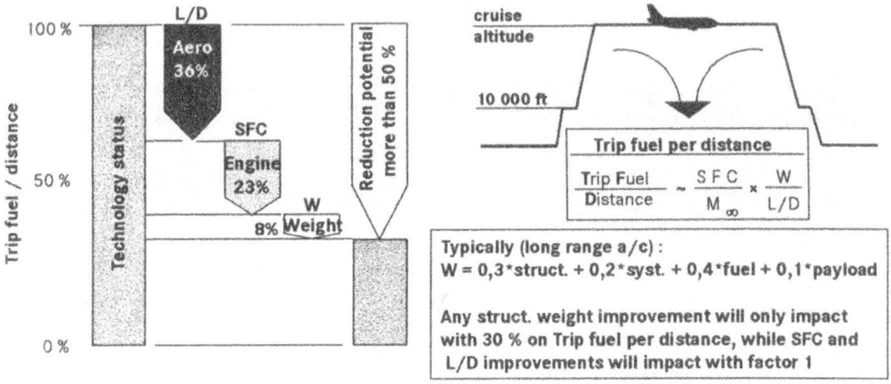

Fig. 8 Potential for reduction of fuel consumption over next 20 years

Improvements in aerodynamics can provide a significant contribution to reduce the overall impact of air travel on the environment as well as significantly improve the bottom line for the operator.

Drag Reduction : a Major Task for Research[1]

J.-P. Marec,
General Scientific Director, ONERA, B.P. 72, 92322 Châtillon, France

Summary

The paper recalls how research programs can be ranked by priority in light of a necessarily restricted budget, through a judicious mix of three-type ingredients : (1) current and future market needs as identified by industry, users and governmental authorities ; (2) scientific and technical progress ; (3) demands of our customers for higher performance, greater safety, lower cost and environmental friendliness.
Through this process, drag reduction has been identified as a major task for research in view of its potentialities for fuel consumption reduction, and consequently DOC (Direct Operating Cost) and environmental effect reduction. However, potentiality does not necessarily mean operational applicability and a complete cost-benefit analysis has to be performed in all cases.

Starting from the classical drag breakdown for a typical transport aircraft, the different drag reduction technologies are considered in turn : natural or controlled (HLFC) laminarity ; reduction of turbulent skin friction drag, induced drag, wave drag …
In each case, a short historical review is given, with examples of research results and emphasis on european cooperation. The different research phases are identified : basic (numerical/experimental) research, evaluation in large wind tunnels, in-flight validation (demonstrators). Applications concern mainly Airbus-type aircraft, but also business jets.
Perspectives offered by some possible technological breakthroughs are mentioned. EREA (European Research Establishment Association) is considered as a major european tool to face the new challenges.

Introduction

When I was invited to say a few words in this first session of the Dragnet Conference, I answered positively with great pleasure.
First because it is hosted by DGLR and organised jointly by Dragnet and CEAS which can be considered as the European ICAS, in which I am presently fully involved.
Then, of course, because of the subject of my talk. Since you are attending this conference, I presume I shall not have to make great effort to convince you that "Drag reduction" is indeed "a major task for research". Most of you are experts in the field. I shall not dare enter into scientific and technological details that will be presented during the Conference. Let me just recall a few obvious facts that will serve as a convenient introduction to the following sessions and presentations.

---

[1] Invited paper presented at the CEAS/DragNet European Drag Reduction Conference 2000, Potsdam, Germany, 19-21 June 2000.

Research strategy

And since I have to emphasise the research aspect, let me first recall (Fig. 1) the classical scheme of research strategy used by all players of the aerospace sector : national governmental authorities, international bodies such as the European Union, industry, Research Establishments such as, for example, ONERA.

Being given a limited budget for research, one has to carefully set priorities to define research efforts, plans and programmes, by a well-known double approach : "bottom-up" and "top-down". Among all the possibilities offered by the scientific and technical advances, one has to select the development of the key-technologies which are best suited to satisfy the requirements for the development of competitive future industry products and also the demands of operational users. In this multi-criteria optimisation, the objective is not only to increase the performance of the product, but also to increase safety and – to a greater extent today – reduce cost and environmental impact. In this respect, drag reduction of civil transport airplanes directly concerns performance, but indirectly, of course, cost and, to a certain extent, environment. Cost reduction concerns not only airlines – and consequently users – but also industry, both in a very competitive market.

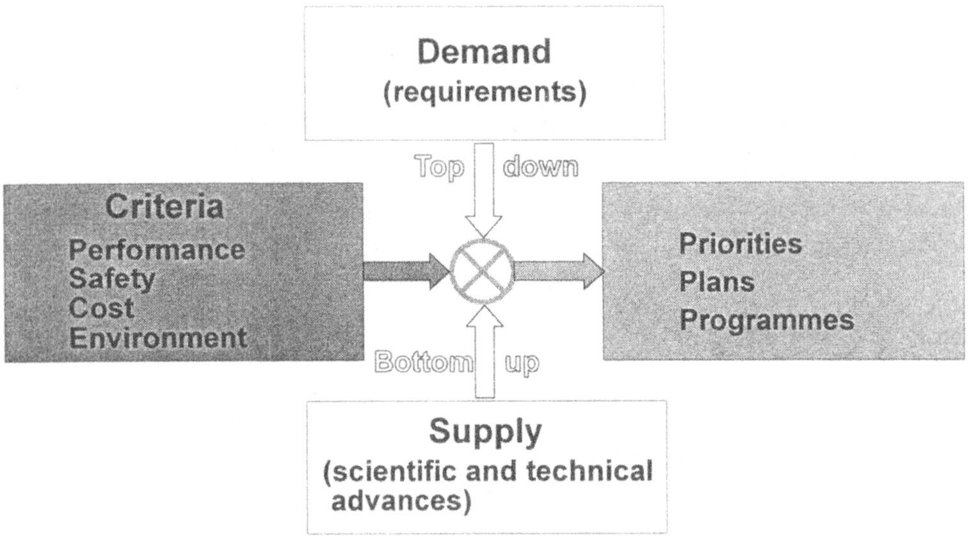

Fig. 1 – Research strategy.

Direct Operating Cost

Figure 2 recalls the DOC breakdown for a typical long range transport aircraft. DOC is of utmost importance for the airlines. Drag reduction directly impacts on fuel consumption which represents in this case about 22 % of the DOC. However, one has to carefully check that the corresponding technology implementation does not have induced detrimental effect on aircraft weight, aircraft price because of increased complexity, and maintenance cost because of new operational constraints.

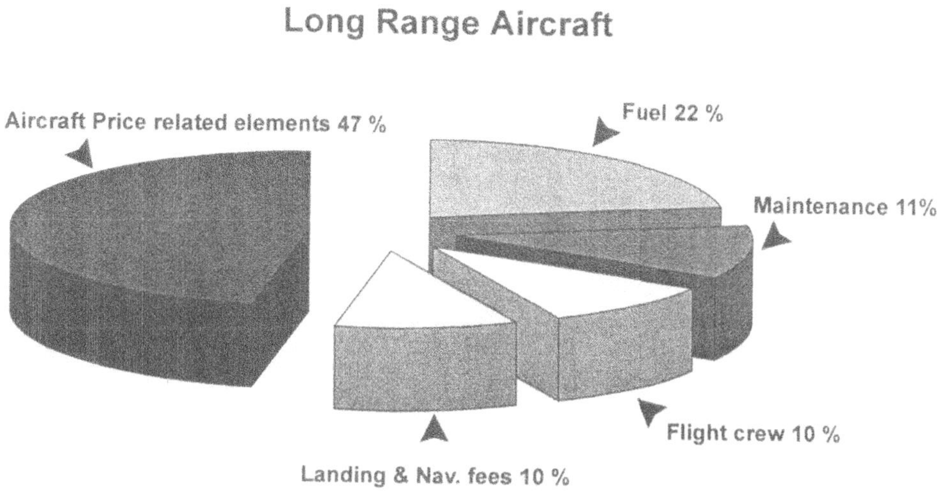

Fig. 2 – DOC breakdown.

I am sure that the research scientists attending this conference are already convinced that before making any conclusion about the interest of an apparently promising concept, a complete cost/benefit analysis has to be performed, and this cannot be done without the help of industry.
However, research remains of utmost interest at the beginning of the process and, at least apparently, the financial support which can be devoted to these research actions is far from negligible, in view of the expected benefits.

Aircraft sale price

Figure 3 shows, in addition to the previous DOC breakdown, the aircraft sale price breakdown. It is borrowed from the well known EUROMART study and the figures may have somewhat changed now, but not the principle of the demonstration.
It appears that the research, testing and project design part represents only 5 % of the aircraft sale price and hence 5 % x 40 %, i.e. 2 % of the DOC. Hence, in order to obtain a 10 % fuel saving, i.e. a 10 % x 20 % = 2 % reduction of DOC, it is worth to almost double the total research effort.
These fuel savings are not only linked to the improvement of engines, but also, to drag reduction, in a global efficiency improvement approach. This justifies the important research effort carried out by industry and by the Research Establishments, and the relatively large amount of money dedicated to this efficiency topic by the European Union in the successive aeronautical cooperative programs.

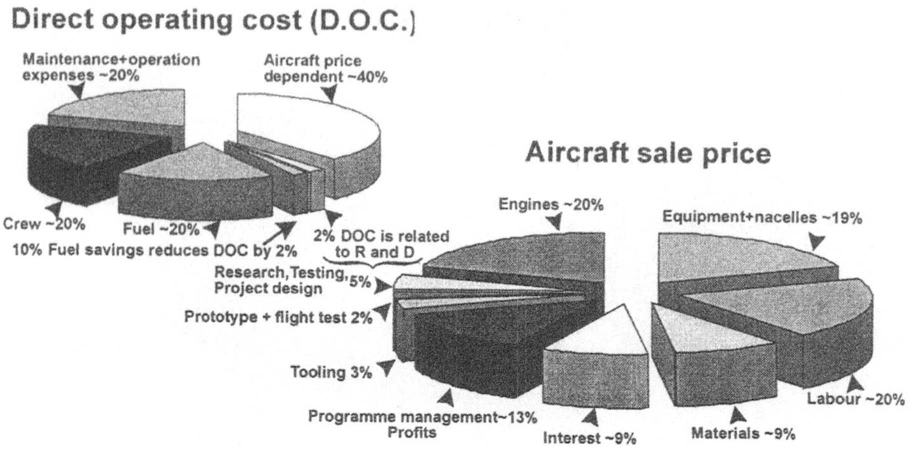

Fig. 3 – Effect of technology improvements on aircraft sale price.

Drag breakdown

Drag reduction is a great challenge. But there is certainly room for improvement. Figure 4 recalls the drag breakdown of a typical civil transport aircraft. It shows that skin friction drag and lift induced drag represent together more than 80 % of the total drag and may offer the highest potential for drag reduction. The other components only represent about 20 %, but cannot be neglected.

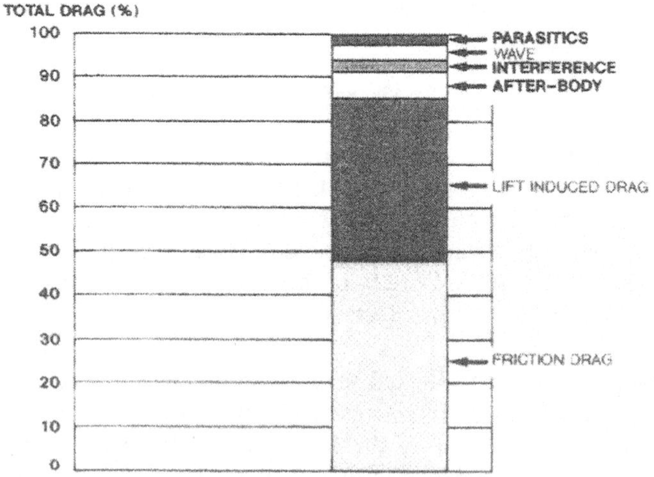

Fig. 4 – Drag breakdown of a typical transport aircraft.

Methodology

What I would like to recall rapidly now is how the challenge of drag reduction has been faced in the past 15 years and what remains to be done. In this short historical presentation, I would like to demonstrate that for each concept, the same methodology is applicable (Fig. 5) :

- the action starts with theoretical, numerical and experimental basic research ;
- it is followed by wind tunnel validation, and then by flight test demonstration ;
- finally, it ends by industry complete cost/benefit analysis. For several of the examples cited here, this last step is now on the way.

This historical survey, elaborated with the help of Volker Schmitt from ONERA, is far from being exhaustive. I shall focus on the main efforts made in France and in Europe. Because of the severe Boeing/Airbus competition, very little information is available from the US.

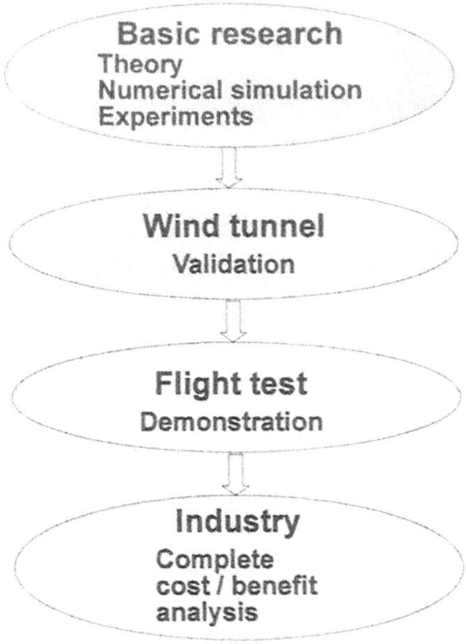

Fig. 5 – Methodology.

Skin friction drag

We have seen that skin friction drag (Fig. 6) represents about 50 % of total drag and hence offers a strong potential of improvement.

One solution is to try to achieve a significant laminar flow extent. It is the NLF (Natural Laminar Flow concept), which is applicable to small size aircraft, such as regional aircraft and business jets.
The investigation of this concept led to important basic research (BR) efforts in France, at ONERA, and in Europe, mainly at DLR, on boundary layer stability and transition. Numerical tools have been used to optimise the shape of airfoils as well as complete wings. This will be completed by the European Union EUROTRANS project.

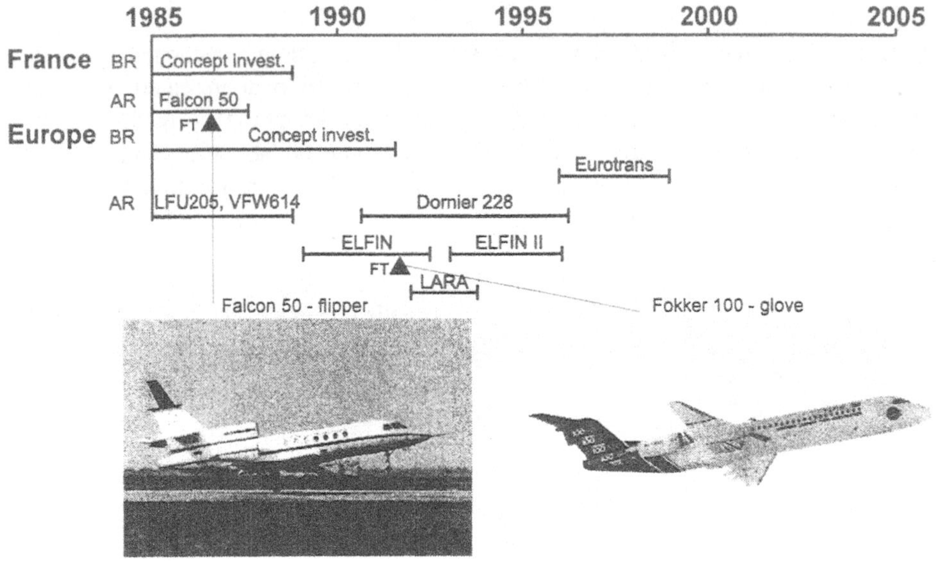

Fig. 6 – Skin friction reduction through natural laminar flow (NLF).

In applied research (AR), let us recall :

- in France, the Falcon 50 flipper in-flight experiment conducted by Dassault Aviation with participation of ONERA, as shown on the bottom-left picture ;
- the DLR in-flight investigations on several planes (LFU 205, VFW 614 and Dornier 228) ;
- and, of course, the NLF part of the European Union projects ELFIN (European Laminar Flow Investigation), and also LARA (Laminar flow Research Action) on wing and nacelle drag reduction. Let us mention, in particular, the flight test demonstration of a laminar glove on a Fokker 100 in the frame of the ELFIN project, as shown on the bottom-right picture.

The fact that the research effort has not been continued in recent years may be explained by the present industrial context concerning regional aircraft.

For Airbus-type civil transport aircraft, the corresponding Reynolds numbers and sweep angles are such that the NLF concept cannot provide any delay of the transition process and hence one has to call for other concepts such as the HLF (Hybrid Laminar Flow) using suction (Fig. 7).
For the HLF concept, basic research efforts were made in particular at ONERA and DLR again on boundary layer stability, transition process but also on leading edge contamination, Gaster bump concept and suction.
In France, important applied research in-flight demonstrations were conducted on a Falcon 50 and then on a Falcon 900 (see bottom-middle picture).
In Europe, let us recall the HLF part of the ELFIN and LARA European Union projects and other more technology related EU projects : HYLDA (Hybrid Laminar Flow Demonstration on Aircraft) ; HYLTEC (Hybrid Laminar Flow Technology on Transport Aircraft) and ALTTA (Application of hybrid Laminar flow Technology on Transport Aircraft). The bottom-left picture shows the ELFIN first wind tunnel test in ONERA Modane S1 wind tunnel.

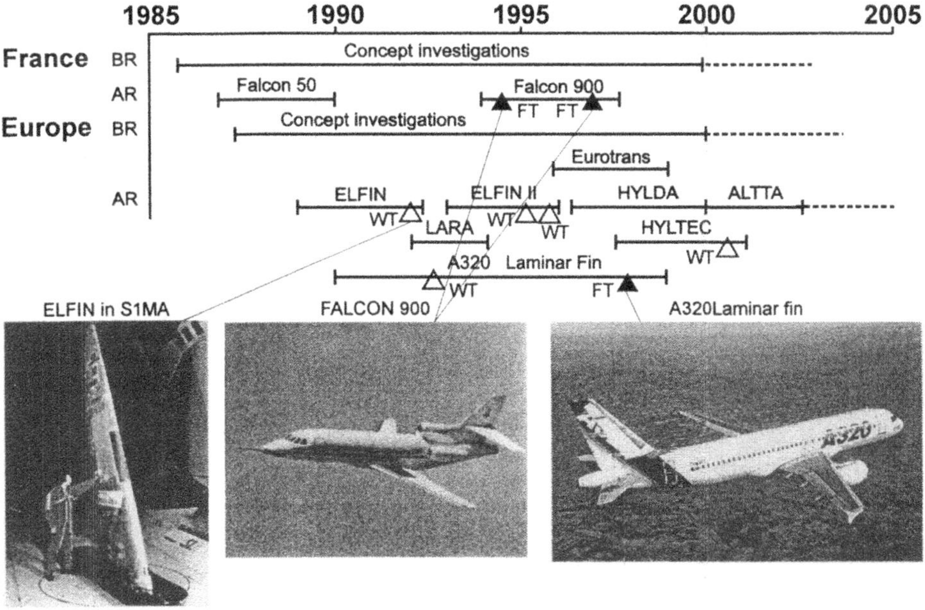

Fig. 7 – Skin friction reduction through hybrid laminar flow (HLF).

One has also to mention the Airbus Industry investigation on the A320 laminar fin, in cooperation with ONERA and then DLR. Former basic research of ONERA showed a potential reduction of 38 % on the skin friction drag of the fin (close to the total drag of the fin) and 1.3 % on fuel consumption. Important wind tunnel tests were conducted in S1MA in 1993 and 1994 and, finally, a flight test demonstration was performed in 1998 by Daimler-Chrysler Aerospace Airbus, with all Airbus partners, in association with ONERA and DLR (see bottom-right picture).

The HLF concept has been proven to be very efficient, at least from the aerodynamic point of view. The global interest of such a concept has still to be demonstrated and also there is room for further research to simplify the suction device and to look for other concepts such as active control technology.

Once the turbulent boundary layer has developed, one has to face the reduction of the turbulent skin friction drag (Fig. 8), and this can be done through use of turbulent boundary layer manipulators.

Basic research on LEBU, and riblets has been conducted at ONERA, and on riblets at DLR. The bottom-left picture shows a riblet test performed in the T2 wind tunnel of ONERA/CERT in Toulouse for a generic cylindrical shape. A more realistic wind tunnel test was conducted in S1MA wind tunnel on a A320 model (see bottom-middle picture).

In Europe, a flight test demonstration was done in 1988, conducted by Daimler Chrysler Aerospace Airbus, with Airbus partners, 3M France and ONERA, where 700 m^2 of the skin of an A320 were covered with riblets (see bottom-right picture). A gain of 1 to 1.5 % on the fuel consumption was demonstrated.

The investigation of the operational aspects is still being conducted on an A340 airplane of Cathay Pacific Airlines.

Apart from the continuation of this demonstration, much remains to be done in basic research : refined optimisation of the shape of riblets, use of other concepts to be imagined.

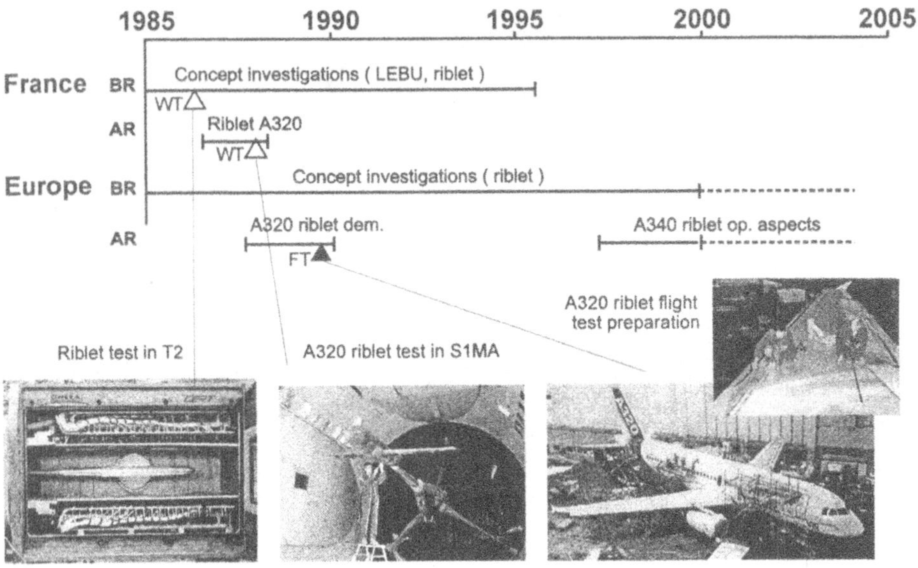

Fig. 8 – Turbulent skin friction drag reduction.

Induced drag

Induced drag (Fig. 9) is one of the major contributors (about 35 %). However for optimised aspect ratio wings, the reduction potential is rather small.

Applied research effort has been concentrated on wing-tip devices to increase the effective span of the wing. Airbus has tested wing-tip fences (see bottom-left picture) in flight on an A320, with a total drag reduction of about 1 %. This concept has been implemented on the product.

Wing-tip turbines have three functions : not only induced drag reduction, but also vortex power recovery and vortex alleviation.

The last two are not the subject of this conference, but should not be neglected in a global approach of the concept. The concept has been thoroughly studied at ONERA with the support of Airbus Industrie. Wind tunnel test were performed in the Modane ONERA wind tunnels S2 (for five-hole probe survey), and S1 as shown on the bottom-middle picture. Reductions of 5 % of induced drag and 1.5 % of total drag have been obtained. An A320 wing-tip turbine mock-up (see bottom-right picture) has been developed in cooperation by Airbus partners, Sundstrand and ONERA (for the aerodynamic design) but not yet tested.

Integrated winglet concept, with continuous evolution of the wing-tip shape has been studied by ONERA. New concepts such as spiroïds are also envisaged.

Wave drag

Wave drag represents only about 3 % of the total drag but should not be negleted.

Basic research has been performed at ONERA on Trailing Edge Optimisation (TEO) and the bump concept as shown on Fig. 10, with application to Single Aisle Aircraft (SAA).

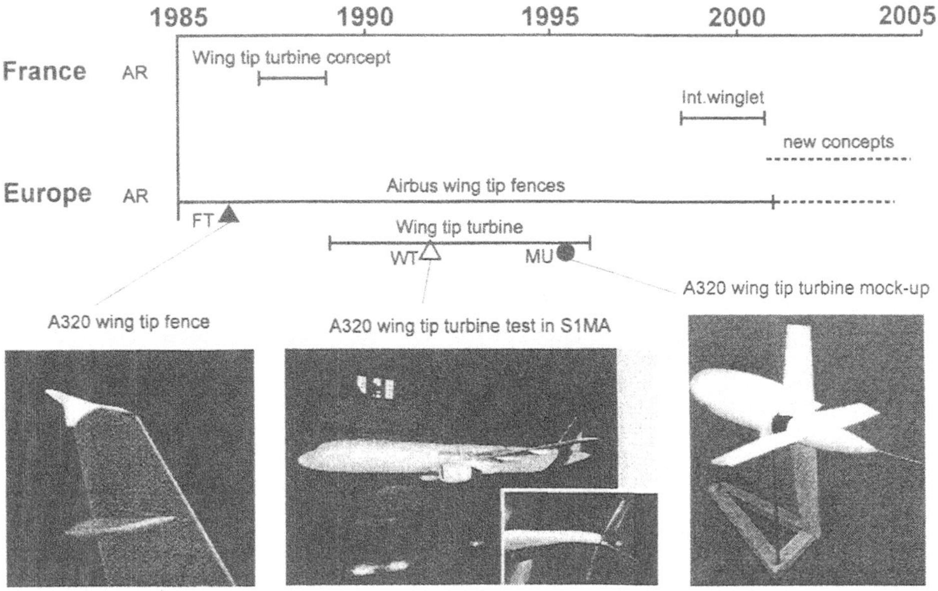

Fig. 9 – Induced drag reduction.

In Europe, an important basic research effort has been conducted in the frame of the European Union projects EUROSHOCK I and II. Applied research on TEO has been supported by Airbus Industry, with contributions of ONERA, DLR and DRA, in 2D and then in 3D, where a reduction of 2 % of the total drag has been demonstrated. The bump concept for Large Transport Aircraft (LTA) will also be checked in the ONERA S2 wind tunnel in Modane in cooperation with DASA and DLR.

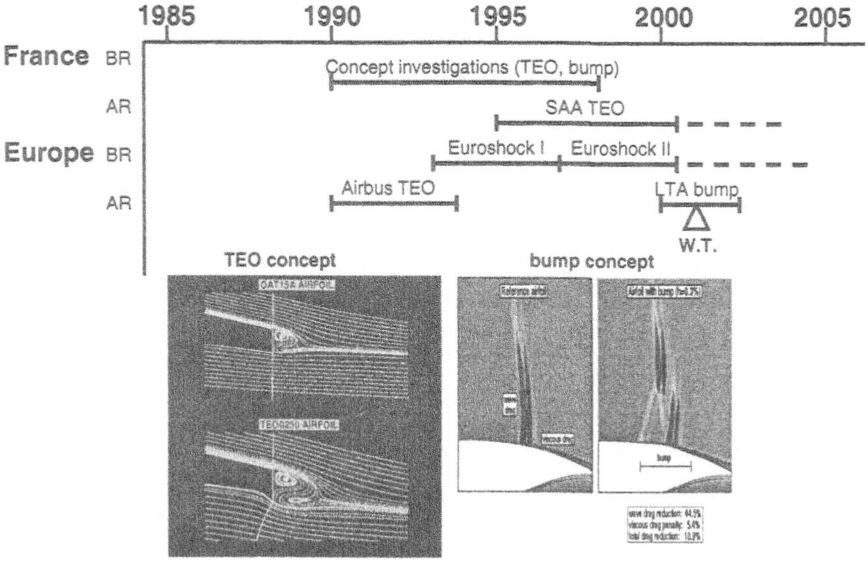

Fig. 10 – Wave drag reduction.

International cooperation

Research on drag reduction involves costly phases, such as the use of demonstrators, which require international cooperation. Examples have been given of past cooperation, in particular under the aegis of Airbus Industrie and of the European Union. Research Establishments have an increasing role to play.
Already ONERA and DLR have developed very strong links. Their research on helicopters is now completely integrated. There is a single common research program to satisfy the needs of Eurocopter. We are now trying to do the same for civil transport aircraft with respect to Airbus Industrie and later AIC (Airbus Integrated Company). Let us recall that ONERA and DLR have now a common yearly Symposium ODAS (ONERA-DLR Aerospace Symposium). The second one took place in Berlin last week. ONERA and DLR have also a common scientific journal AST (Aerospace Science and Technology) published by Elsevier, in which other RE participate.

But this only involves two research establishments. The true challenge is to join all European forces. And in this respect I would like to say a few words about EREA, which is certainly well known by all of you, but which has recently experienced important changes you may not be aware of.
EREA is the Association of the seven European Research Establishments in Aeronautics, recalled on Fig. 11. In May 1999, it became a legal entity, with an extended perimeter. It now deals not only with civil, but also with military and space-related aeronautics.
The EREA objectives aim at making EREA one of the major players in the aeronautical sector.
The organisation is now simplified. Temporary joint activities are conducted as projects, in the frame of networks. More permanent joint activities are coordinated by groups, such as the ARG (Aeronautical Research Group), that most of you know, since it insures the interface between EREA and the European Union. Answers of the RE to the EU calls for proposals are now made in a concerted way, and this is true in particular for drag reduction projects.
EREA vision of a joint future has not changed since the 1993 joint Position Paper. For example the objective of the creation of research capacity for long term new topics led to the elaboration of a LTTP (Long Term Technology Plan), where drag reduction was recognised as a major key-technology.

EREA

Association of
European
Research
Establishments in
Aeronautics

CIRA	Centro Italiano Ricerche Aerospaziali	Italy
DERA	Defence Evaluation and Research Agency	United Kingdom
DLR	Deutsches Zentrum fur Luft- und Raumfahrt	Germany
FFA	Flygtekniska Försöksanstalten	Sweden
INTA	Instituto Nacional de Técnica Aeroespacial	Spain
NLR	Nationaal Lucht- en Ruimtevaartlaboratorium	Netherlands
ONERA	Office National d'Études et de Recherches Aérospatiales	France

Fig. 11 – EREA establishments.

Conclusion

To conclude, I hope that this short presentation has met its objective : to recall the challenge represented by drag reduction. Much progress has already been made, but there is still room for improvement, and drag reduction remains a major task for research.
The charts of the historical survey are open after year 2000, either to complete the study of on-going concepts with phases of the recalled methodology not yet performed (see dotted lines), or to develop new concepts. Some of them have been mentioned. The active flow control, often linked to the development of smart materials and structures, appears to be one of the most promising.
International cooperation has to be strengthened. The EREA stepwise approach which from an association would lead to partnership and perhaps a true Union is also a challenge. Let us hope together that taking at least the first steps of this approach will be beneficial to the particular challenge of drag reduction.

Session 2
Achievements of Technology Demonstration

Airbus A 320 HLF Fin Flight Tests

R. Henke
EADS Airbus GmbH
D-28183 Bremen
Germany

During the presentation of this paper in the 1st European DragNet Conference in Potsdamm 2000, the video about this programme has been presented, called "A Nose Ahead".

Introduction

Today it is often discussed whether the performance of a modern transport aircraft in the usual wing-body configuration can be further increased at all, or whether its curve over time has reached an asymptotic stadium.

At least two main topics remain in aircraft design which will result in more than just a marginal bit of performance increase but in a quantum leap: on one hand side there is laminar flow, and on the other hand there is turbulent flow control, sometimes linked to adaptive wing concepts, e.g. circulation control by variable camber. Each of both topics has a potential of some 15% performance increase which can be gained within the next 10 – 15 years. Finally, both research activities will meet because laminarity is of course part of flow control in general.

Looking at the key buying factors which lead an airline to buy a specific aircraft, one can see that performance is still a major topic, also related to DOC where performance contributes via fuel consumption, and that environmental issues which again are linked to performance will have an increasing importance. Therefore, it is very well worth to work hard on performance increase of transport aircraft.

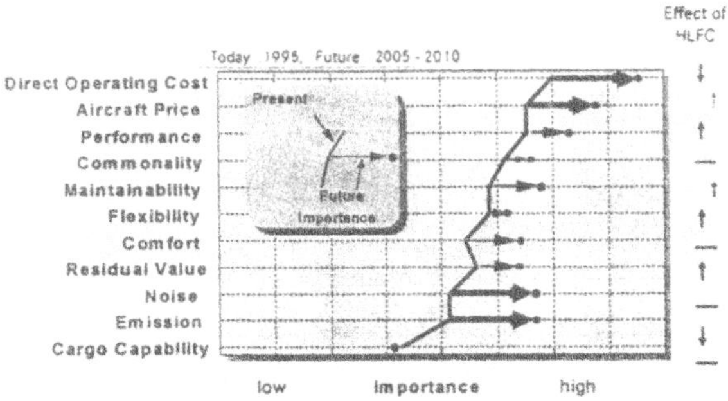

Fig. 1: Key Buying Factors

Both topics on flow control have something in common: they have their roots in aerodynamics, but they are highly interdisciplinary. Only if aerodynamicists, system and structure specialists, flight test specialists and in general future project specialists join their efforts towards a truly interdisciplinary, not just multidisciplinary design, the large potentials visible today can be gained.

The "A 320 HLF Fin Programme"

For laminarity, there are three areas of realisation to be looked at:

- Natural laminar flow NLF, i.e. keeping the flow laminar just by having the right contour,
- Hybrid laminar flow HLF, i.e., adding an active device to the right NLF shape, which will be necessary for large and fast transport aircraft, and finally
- Laminar flow control LFC in general where one would have active devices all over the surface to be laminarised.

The knowledge about the boundaries and conditions of each area is vital for an aircraft manufacturer because any such boundary is valid for his competitor as well.

The active devices mentioned above could be sound, heating, vibrating and others. It has been found out that suction is the most promising way to influence the boundary layer especially from the application point of view, so transport aircraft will be laminarised by a combination of suction at the leading edge and a suitable contour shape, resulting in HLF.

Under these assumptions, the Airbus partners have set up a global laminar strategy called Laminar Technology LaTec: In a three-step-approach, HLF should be made available for all new Airbus including investigation on retrofit solutions. Step 1 would be a laminarised fin, in a first phase as technology validation, and in a second phase by operational flight tests. Step 2 would be a laminar wing glove, and step 3 a full laminar wing. Hence, the laminarisation of the vertical stabiliser is just the first phase of the first step of this strategy, and the Airbus partners agreed in using this first phase for learning about the physics of laminar flow, i.e., by varying the main parameters: Mach and Reynolds number. The types of laminarisation, the LaTec strategy and the influence of Reynolds number boundaries have been presented before, e.g. in /1/.

The objectives of the "A 320 HLF Fin" programme were:

- To laminarise the fin of an Airbus A 320 so that about half of the fin has laminar flow under cruise condition
- To vary Mach and altitude in flight in a way that HLF flow can be studied a wide range of Reynolds numbers
- To use this test for rating HLF flow in general, which asks for a test programme including tests on allowable manufacture tolerances, power supply, under yaw angles to simulate lift etc.

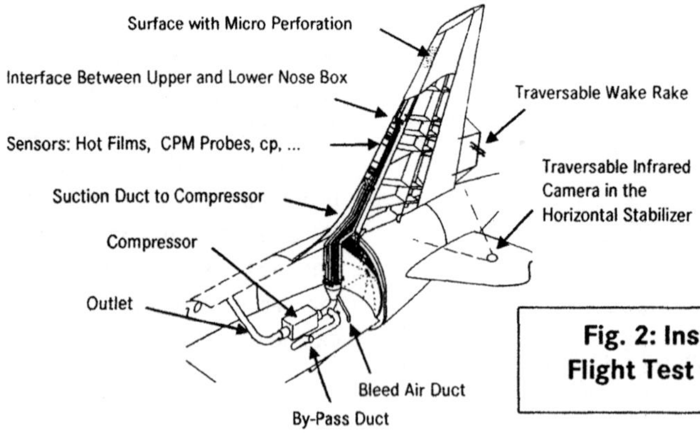

This led to a setup for the flight test as shown in the sketch.

Fig. 2: Installation of Flight Test Components

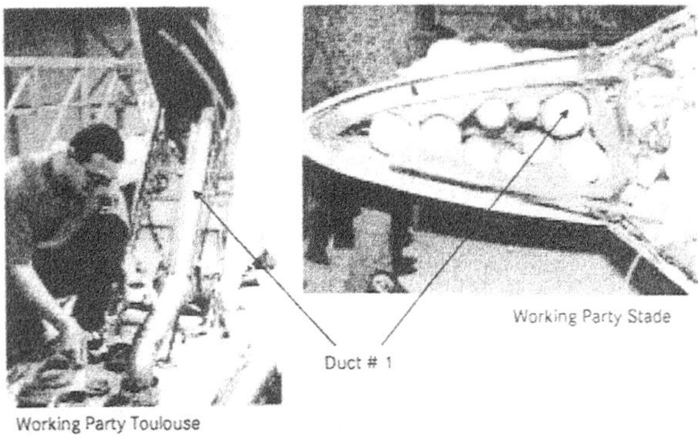

Fig. 3: Suction Nose and Duct Installation

Fig. 4: Suction System Installation in Cabin

The major steps within the programme were:

- Application and qualification of advanced measurement techniques, including a first flight test, the so-called Turbulent Reference Flight Test TFT where the whole measurement chain has been tested before being applied to the laminar fin itself.
- Development of the suction system, including another large test phase in an altitude chamber for checkout reasons as well as for operational training.
- Design and manufacture of the suction nose, including many test pieces and careful checkout procedures concerning mass flow rates, pressure losses, suction distribution etc.
- Installation of engineering teams for flight test purposes in order to organise the complete technology flight test consisting of two so-called working parties, suction system, structural and test integration, data acquisition and reduction, etc., and finally the successful flight test itself in autumn 1998.

Evaluation and Results

After completion of the flight test phase, all data had to be pre-processed, calibrated, condensed, qualified, and distributed to all partners: the infrared images, the pressure distributions, the time curves of hot films etc., being a process which took some time which is why not many of the flight test points are evaluated yet.

Many steps form the complete evaluation process chain:

From the infrared images, the exact transition position has to be determined along the span on both sides of the fin, in case of flight tests with artificial imperfections on the fin in an even closer detail. Together with the pressure distribution and unsteady signals from the hot films

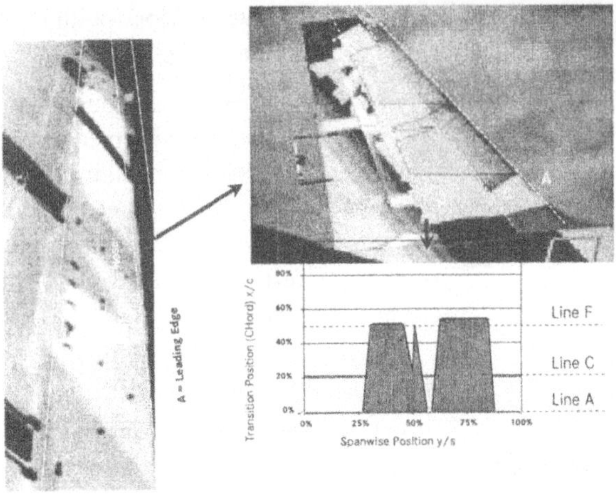

Fig. 5: Evaluation of Transition Position

and piezoelectric sensors and combined with boundary layer calculation, these data result in the detailed knowledge on the boundary layer at one single flight test point. On top, the data from the wake rake, the suction data from the perforated surface down to pressure and temperature values in the different tubes, and the fan data, especially the amount of power supply have to be added. The last part of the picture comes from the test protocols containing information e.g. about surface disturbances, whether these are artificial or accidental, all together in order to correlate boundary layer data with system and surface data.

This very complex evaluation process which on top has to be shared and agreed between all partners is still running, no final results can be presented. But from the evaluation so far, even from on-line analysis during the flight tests, the following statements can be made already:

- Suction directly at the leading edge may not be necessary in case of a functioning device against Attachment Line Transition ALT such as a Gaster Bump.
- Surface imperfections are less critical than expected after several wind tunnel tests.
- The suction power needed was below the expected maximum value.
- There is a self-cleaning effect in flight especially when flying through clouds, but water ingress must be avoided, e.g., by blowing instead of sucking during climb/descend.
- A surface or hole quality degradation effect was not experienced.
- Even with the existing fin which was partially outside the specified values by means of surface tolerances the expected amount of laminarity was reached.
- By measuring the fuel flow at constant aircraft conditions with and without suction, the expected drag reduction was proved.

Overall, the results of the HLF Fin Flight Test evaluated so far are most promising. No severe show stopper for HLFC was found.

Apart from the direct evaluation for rating single data points, all results must be used to gain knowledge on general HLF application:

One must know the correlation between e.g. system complexity, i.e. penalty, and transition position, i.e. benefit in order to balance efforts for the system with fuel saving.

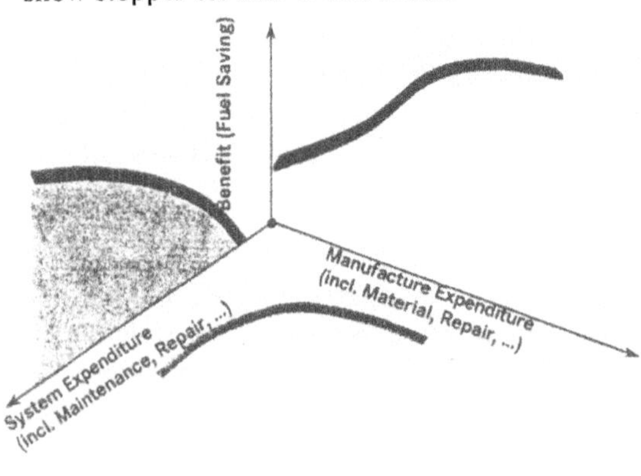

Fig. 6: Benefit / Penalty Analysis

Even today it seems clear that not the best aerodynamic performance, i.e. not the maximum laminar area is the solution to application but the right balance between all factors. The same is true for the field of structure / manufacture, and this field again is linked to systems because e.g. a very smooth surface may allow to decrease suction power and vice versa.

Hence, system layout, surface characteristics and, by this, manufacture layout are linked together, resulting in the benefits of laminar flow. And finally, each of the curves sketched above is time dependent as new fuel consumption taxes, new materials, new power supply methods etc. will change the picture. This sort of knowledge about the main properties related to laminar flow is needed if one wants to apply HLF on an aircraft to be sold to a world wide operating customer.

Next Steps

After this successful flight test, which was the first one world wide with an HLF fin on a transport aircraft, and after agreeing the two assumptions that HLF flow still has a major potential in performance increase, and that no show stopper was found in HLF flow validation, there are two chances on how to continue: One could

- look at the other parts of the aircraft now, especially the wing, but also the horizontal stabiliser and the nacelle and by intensive research push HLF on these parts to the status the fin has reached already, or
- continue with the development of an industrial solution for the fin and, by this, pave the way for HLF application and acceptance for the other parts to be addressed in the future.

Both choices have their pro's and con's as one can easily imagine and because both options are known for many years this led already to the choice of a laminar fin programme itself.

Under this logic it has been agreed between many European partners to continue the way to HLF application on the fin.

Therefore, the next logic step is an operational flight test preferably on an Airbus in service. In order to prepare this flight test, a series solution must be developed and flight tested before. So there are three steps left towards an HLF fin being ready for series application:

1. Development of an aircraft series solution in suction structure, suction system and whole system integration
2. Flight testing this solution in order to prove its viability and to get customer acceptance
3. Operational in-service flight tests on an airline aircraft for about a year as the final proof of concept

The first topic is partially covered in the European programme "Application of Laminar flow Technology on Transport Aircraft ALTTA". In that programme, mainly structural aspects are covered, leading to an HLF fin mock-up with a new structure. Furthermore, inputs will come from flight tests on a Saab 340 and on a Dornier 228 within the EC programme "Hybrid Laminar Flow Technology HYLTEC", where advanced system and de-/anti – contamination concepts are under investigation. Further work in national programmes is necessary to fulfil step 1 defined above. E.g., the integration of the flow control system in cockpit systems has not been addressed anywhere.

The flight tests for the second step is envisaged to take place again in an EC programme for a flight test campaign comparable to the one described in this report.

Meanwhile, research should take place on an HLF wing, HTP and nacelle in national and European approaches in a small scale e.g. at research establishments in order to prepare for a larger programme to come after the HLF fin has proven its reliability for industrial use.

Summary and Outlook

The following table summarises the state-of-the-art in technology development for an HLF fin:

Topic	Available / Status	Needed
Multidisciplinary Design	O.K. for specimen and models	Design based on total aircraft
Part Manufacture	Concepts, tests on specimen	Concept rating: design to cost
System Control	Sophisticated but isolated solutions for tests	Aircraft system integration
Handling (Repair, Maintenance, ...)	First ideas	Customer oriented solution
Reliability	First ideas, some F/T data	Lifetime performance guarantees

So in general, after some 10 years of lab tests and after the A 320 HLF Fin Flight Test, now the aircraft in a customers hand must be addressed.

Looking at the main topics of the A 320 HLF Fin flight tests, the outcome can be summarised as follows:

- Laminar Flow is still the most promising technology to increase aircraft performance. After some 10 years of EC programmes, a solid base of knowledge has been established in Europe.
- The principle concept for an HLF fin has been proven numerically, with a wind tunnel model and in the A 320 HLF Fin Flight Test. This flight test has shown that many problems which were expected based on wind tunnel experience are less critical in flight.
- As a next step, a series concept must be developed. This will partially be done in the EC CTP "ALTTA". But there is still more work to be done.
- Following a successful ALTTA programme, the European effort must then focus on a flight test with such a system in order to prove reliability, get customer acceptance and gain industrial experience in commercial, operational and certification aspects.
- The final step towards an HLF fin would be operational flight tests on an airline aircraft in service.
- BUT: Even though an HLF Fin already offers a reasonable industrial benefit, its development was always meant to be the first step, the main HLFC benefit has still to be gained. At latest after having realised an industrial solution for an HLF fin, the other parts of the aircraft, especially the wing must be addressed.

Reference

/1/ R. Henke et al.: The "A 320 HLF Fin Programme": Objectives and Challenges
CEAS 2nd European Forum on Laminar Flow Technology, Bordeaux, June 1996

LESSONS LEARNED FROM DASSAULT'S FALCON 900 HLF DEMONSTRATOR

Jacques FITON

DASSAULT AVIATION - FALCON DIVISION
BP 24 – 33701 Mérignac Cédex – France

Extended Summary

Reducing aircraft drag is an economic challenge not only for airlines for which the fuel consumption reduction is of the foremost importance for D.O.C. reduction, but also for smaller aircraft like business jets where the improved aerodynamic performances make it possible to build smaller, lighter and therefore less expensive aircraft. It is therefore one of the major keys for the improved competitiveness of the business jets.

DASSAULT AVIATION initiated the first experiment on LAMINAR FLOW on the FALCON 50 in 1986 with the financial support of D.G.A.C. and S.P.A.E. In 1997, the FALCON 900 N°1, which was certified in early 1995, has terminated its operational demonstration after having cruised more than two years with hybrid laminar flow over its wings, and was flight tested again in order to draw up a full assessment of the experiment.
This paper presents the tests, needed and performed for laminar flow validation at operational level. After a quick overview of the tests performed on **FALCON 50** (natural and hybrid laminar flow), a detailed description of the **FALCON 900** is made.

First, due to the high complexity of the phenomena involved in the transition mechanism, one of the main conditions for the use of laminarity on the industrial level was the validation of prediction tools, not only in laboratory (wind tunnel testing) but also by full scale flight testing. Tests in **natural laminar flow** performed on the FALCON 50 (fin) between 1986 and 1987 have generated a significant amount of data for validation of transition prediction tools developed at **ONERA-CERT**.

Secondly, even if the means for transition delaying were relatively well known in theory, the use of HLF concept needed the development of specific technologies, which then had to be validated by flight testing. Demonstrations in Hybrid Laminar Flow performed on the FALCON 50 right inboard wing, and terminated in 1990, have required the design and the manufacturing of a leading edge, incorporating an anti-icing and cleaning system as well as a device for leading edge contamination prevention. Tests also made it possible to determine the various parameters that might affect the system performances and showed that suction rates to develop controlled laminarity were relatively small in comparison with the expected gains.

Finally, applying laminar flow at production level has required the demonstration that the tools, methods and technologies were robust enough at operational level in order to validate laminar flow for the design of a future business aircraft . This is why the Falcon **LAM**inar operation **(FLAM)** was initiated on the two FALCON 900 inboard wings, selected to support these experiments. In addition to its ambitious objectives in terms of laminarity (broad range of CL and flight conditions), the purpose of this demonstration was to design, manufacture and certify an aircraft in **hybrid laminar flow** using industrial methods with consideration of weight and cost constraints and then put the aircraft into service to analyse the behaviour of the devices and look for any operational problem there may be.

This paper gives an overview of methods used for designing and manufacturing the modified aircraft as well as the different requirements (operational and certification) which had been fulfilled. After its certification in February 1995, the FALCON 900 has been in operation at DASSAULT FALCON SERVICE during more than two years and has accumulated more than 1500 flight hours, including 1000 hours spent under hybrid laminar flow at various climatic conditions. At the end of its in-service operation, the FALCON 900 was flight tested again. A number of lessons learned for the use of laminar flow on business jets are presented, especially concerning:

- maintainability and reliability of the different devices,

- appropriate modelisation (geometry and CFD), taking into account the manufacturing tolerances, to use for designing laminar aircraft.

FLIGHT TESTING OF A HLF WING WITH SUCTION, ICE-PROTECTION AND ANTI-CONTAMINATION SYSTEMS.

Bryan E. Humphreys* and Karl-Heinz Horstman‡

*Aerospace Systems and Technologies Ltd.,
Consett, County Durham, England. DH8 6SR
E-mail: beh@aerosystech.com

‡DLR – Deutsches Zentrum fur Luft und Raumfahrt
Postfach 3267, D-38108
Braunschweig, Germany.
E-mail k.h.horstmann@dlr.de

Key Words: Laminar Flow, Hybrid Laminar Flow, Systems, De-icing, Decontamination, CEAS/DragNet, Potsdam 2000.

1. SUMMARY

Much effort has been put into the development of laminar flow technology, leading to improved understanding and prediction of the aerodynamic phenomena involved, but relatively little flight experience has been gained with the hardware and systems that will be required on a laminar aircraft. The joint European program HYLTEC plays a significant role in adding to the knowledge and experience of materials and systems for HLFC by conducting two complementary flight tests. One test, using a SAAB 2000 aircraft will gain airline service experience of a range of porous surfaces both with and without insect contamination protection, whilst the other test on a Do 228 will explore the operating performance of various combinations of suction, ice protection and insect contamination protection systems. This paper reviews the concept, design and construction of the Leading Edge for the HYLTEC Do228 systems flight test experiment.

The layout of the leading edge required careful consideration because the intent is to evaluate two alternate approaches to the contamination control problem. Financial considerations limited the experiment to one wing of the aircraft and the need to avoid the region in the propeller slipstream further limited the span available for test regions. In addition, part of the experiment will use a Kruger high lift device and it was necessary to ensure that this would not introduce unacceptable asymmetric effects or impair control surface efficiency. The two combinations of systems chosen for the test were thermal anti-icing (TAI), used in conjunction with a Kruger flap for insect contamination protection and liquid for both ice and insect contamination protection. In both cases these systems are combined with suction systems for HLFC. Furthermore, the TAI application utilizes an approach that has never been used in Europe.

In addition to the test zones it was necessary to provide a reference region where the accretion of insect debris on a non-protected surface could be recorded and for safety reasons both this and the leading edge in the propeller slipstream had to be protected against icing.

The introduction of TAI posed constraints on materials and assembly methods, together with duct sizing requirements that had to be optimised for both TAI and suction. In addition, start-up and failure cases for TAI imposed the need for alternate exhaust routes and protection against excessive temperatures and pressures.

Despite assistance from the aircraft manufacturer, certification required modeling of the wing and leading edge. The HLF Leading Edge was inevitably stiffer than that normally fitted to the aircraft, requiring careful treatment of fastener loads due to wing bending and the Kruger flap added loads that were never expected by the manufacturer. However, due to a stiff wing and conservative original design the test installation could be accommodated.

Partners in this experiment are AS&T, DASA, DLR, and Nord Micro with technical assistance from Sonaca and the University of Limerick.

2. OBJECTIVE OF TEST

The objective of this test is to investigate and to demonstrate, under flight conditions, solutions for ice protection and anti-contamination systems and their use in combination with a suction system for boundary layer control.

3 DESCRIPTION

The work within this project can be split into activities on the following topics:

1. Conceptual Arrangement
2. Energy Sources
3. Systems
4. Structure
5. Instrumentation
6. Certification
7. Flight testing

At the time of writing, the concept and design stages of this project have been completed. Manufacture of the test leading edges is part completed as is the provision of test equipment, energy sources and other modifications to the aircraft.

4 CONCEPTUAL ARRANGEMENT

Figure 1 illustrates the test aircraft and was taken during a previous, unrelated, laminar flow campaign. However, for this test, the same pod mounted infra-red boundary layer imaging system will be used and a laminar flow glove will be installed similar to the one that can be seen on the right wing.
The experiment will include three different test zones as illustrated in figure 2. These are built into a new leading edge that is being manufactured from three separate nose boxes together with a short transition section. The innermost box is in the propeller slipstream and does not form part of the test but is equipped with a TKS liquid ice protection system to ensure safe operation in icing conditions. The next nose box is provided with a Kruger flap for contamination protection and a hot air system for anti-icing. The Kruger flap can not be retracted but is removable so that tests can be flown with the leading edge representing high and low speed configurations. It is also possible to adjust the position of the Kruger flap for optimisation of

performance if required.

The thermal anti-ice system blows hot air into the five forward suction plenum chambers and differs from normal practice by exhausting the air through the porous surface. Special codes were developed by SONACA for prediction of this process.

Figure 1: Do228-100 Showing Pod Mounted Infra-red Camera for Boundary Layer Imaging.

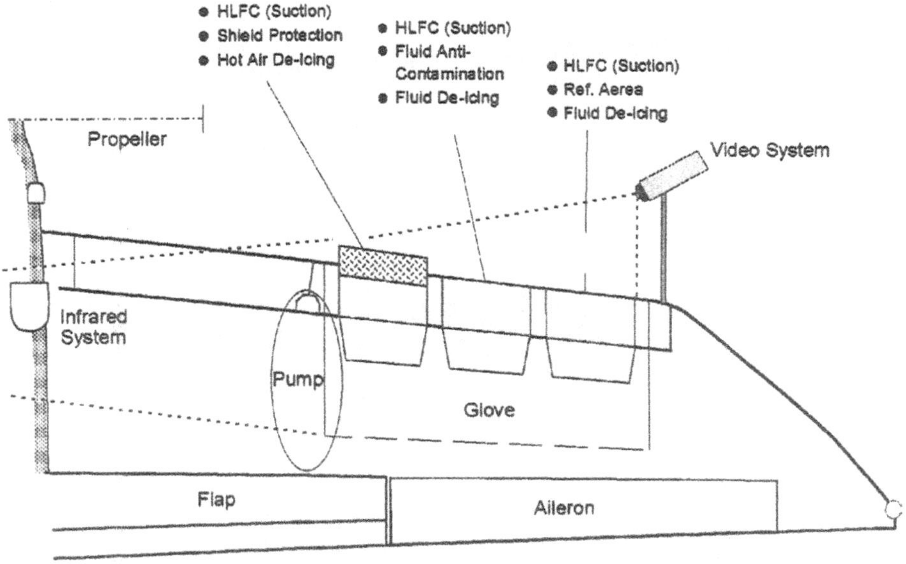

Figure 2: Conceptual Layout of Right Wing of Do228 for HLF Systems Flight Test.

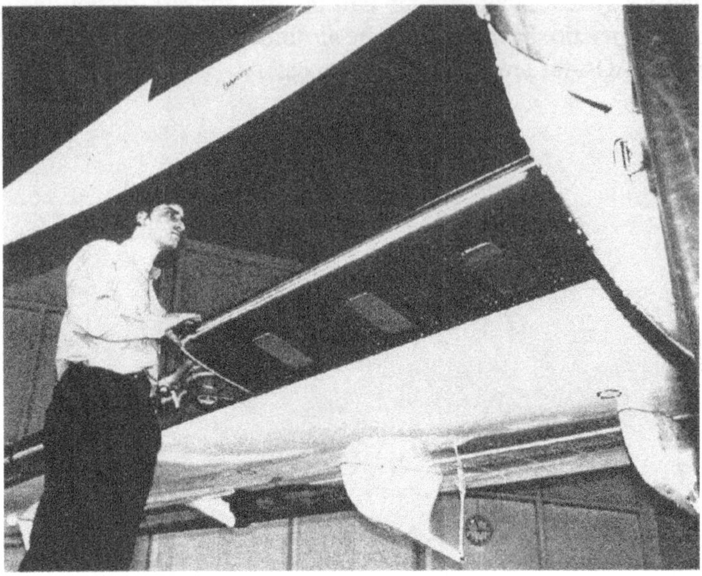

Figure 3: Inner (TKS De-iced) Nose Box During Trial Installation

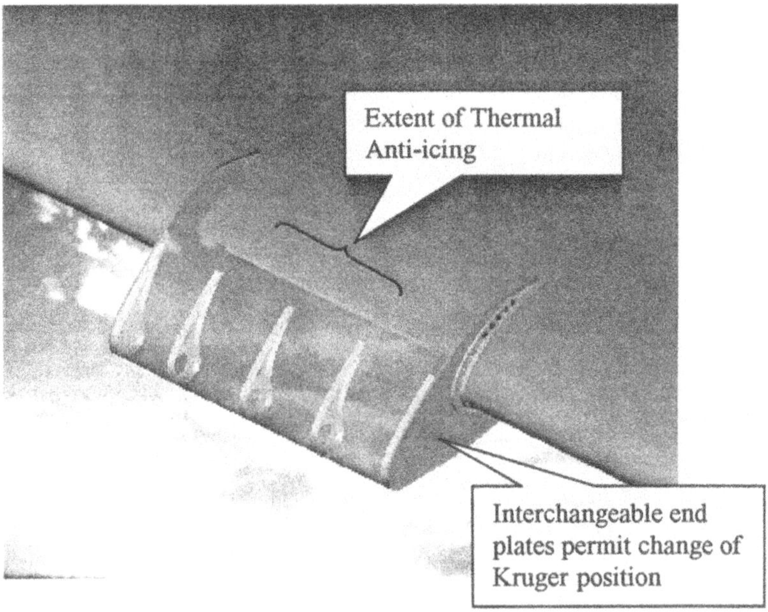

Figure 4: Kruger Shield and TAI Zone

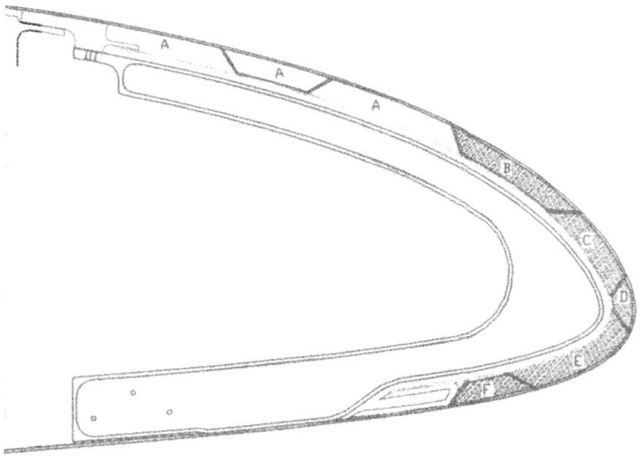

Figure 5: Cross Section of Leading Edge Showing Plenum Chambers Used for TAI

The outer nose box contains two similar zones; each equipped with a fluid foam system for ice protection and anti-contamination purposes. These zones can be operated independently so that, during insect contamination test flights, the outer panel protection system can be de-activated to serve as a reference panel in order to quantify the effectiveness of the anti-contamination measures. During icing flights both outer panels will act as ice protection test panels. A reference region is not desirable during icing tests because this part of the span is ahead of the aileron and icing could impair control characteristics.

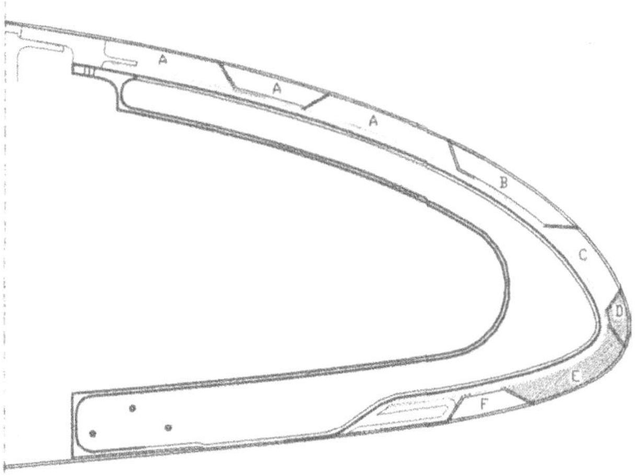

Figure 6: Cross Section of Leading Edge Showing Plenum Chambers Used for Liquid Foam Exudation

The wing profile will be modified slightly to allow incorporation of a fibreglass fairing on the upper surface. The main purpose of this fairing is to provide the thermal insulation necessary for infra-red boundary layer visualisation. However, it is also a requirement for the test that the wing would be naturally turbulent, with a laminar boundary layer existing only when the leading edge systems were working effectively, and this is achieved with the modified profile.

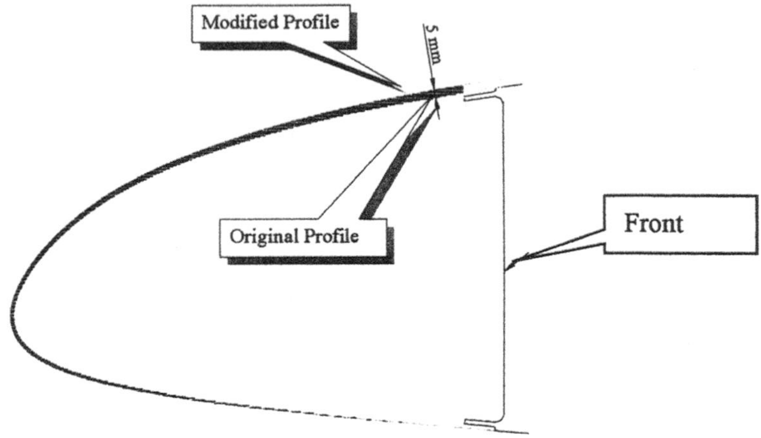

Figure 7: Aerofoil Profile Modification

5. ENERGY SOURCES

Energy sources are required for boundary layer suction, thermal anti-icing, liquid foaming, instrumentation and control systems. Power available is limited to bleed air from the right engine and 28 volts d.c. electrical power.

Regulation of bleed air temperature and pressure is required and this is done in two stages to produce hot air, at nominally 185°C, for thermal anti-icing and cool air for foam generation. However, the most difficult facility to provide was the suction needed for boundary layer control. Bleed air, from the right hand engine, was eventually selected as the energy source to produce suction. This bleed air powers a jet pump producing a suction mass flow of 100 g/sec.

The suction pump presented a significant challenge because no "of the shelf" equipment was available. A number of alternate solutions were examined including aircraft pumps intended for other purposes, industrial units and automotive turbochargers. None of these presented an acceptable solution being either too expensive or too heavy or, in the case of electrically powered units, exceeding the spare power available on the aircraft. Finally an ejector (or jet pump) was chosen because this was affordable and sufficient bleed air power was available. However, limited design information was available and this unit was developed jointly by DLR and Nord Micro. With careful design and manufacture acceptable pump efficiency was achieved. Using a bleed air mass flow of 160 g/sec at 3 bar, a suction mass flow of 100 g/sec is obtained at 15-kPa pressure difference between suction surface pressure and pump outlet pressure. Figure 8 illustrates the unit that was developed. This pump will be mounted in a pod below the wing, close to the inboard end of the Test Region.

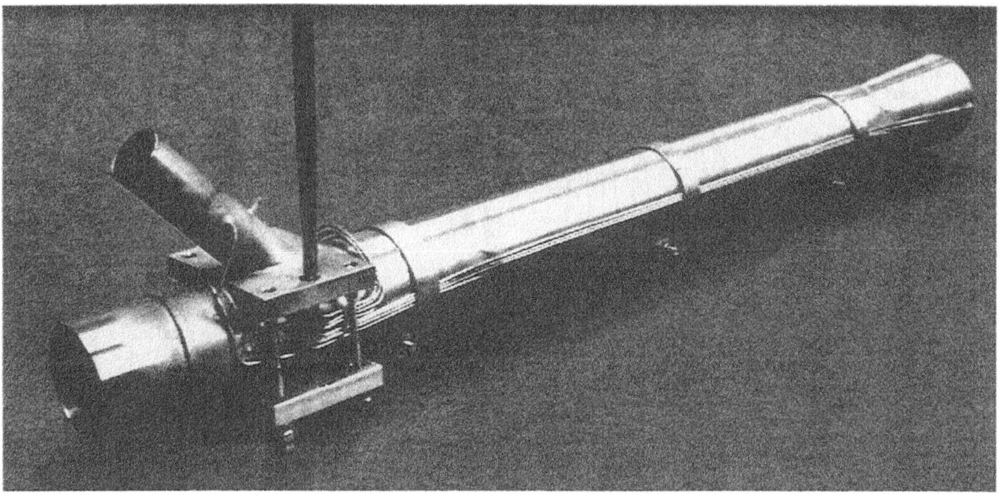

Figure 8: Jet pump for boundary layer suction

6. SYSTEMS

The supply and measurement system is rather complex. In addition to the boundary layer suction system provided for the three test panels, a bleed air supply system for inner panel thermal anti-icing, a foam supply system for the two outer panels and a fluid supply system for the inner, non-test, part of the leading edge have to be installed and controlled in use.

Figure 13 shows the pneumatic and liquid systems for the three test zones. The main energy source is bleed air from the right hand engine. Suction flow distribution is achieved by 6 computer-controlled butterfly valves for the 15 suction chambers of the three test panels. The bleed air is further used for hot air anti-icing of the inner test panel and for the foam production for the outer two zones. Overall, 35 valves, 135 pressure tappings and 34 temperature sensors are managed by a PC based computer system.

The systems for a flight experiment such as this are more complex than would be used for production systems because it is desirable to make the arrangement as flexible as possible, giving a range of operating flow and pressures, rather than the single point operation that would be more suitable for production aircraft. In addition, extensive instrumentation is required for the flight test. Although it would have been possible to install all of the equipment within the leading edge some components, such as the jet pump, are installed in a pod below the wing to improve overall accessibility.

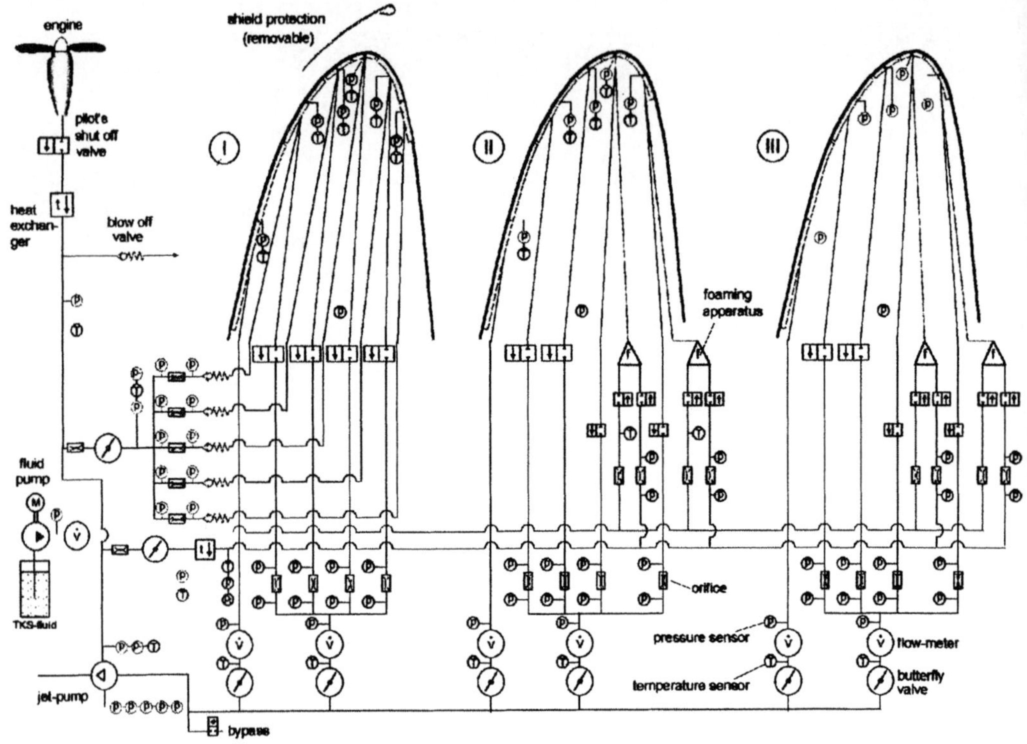

Figure 9: Pneumatic and Liquid supply system for the three test panels of the Do228

Thermal Anti-icing

Thermal anti-icing makes use of the porous surface by blowing hot air into the boundary layer. It is predicted that each injection point will protect about 1.5 metres of span. The span of the test region is less than this, but by offsetting the injection point, a representative arrangement can be tested. At start up, air can not exhaust through the skin if the surface is ice covered. In this event a pressure relief valve at the end of the chamber allows spanwise flow until the ice is melted.

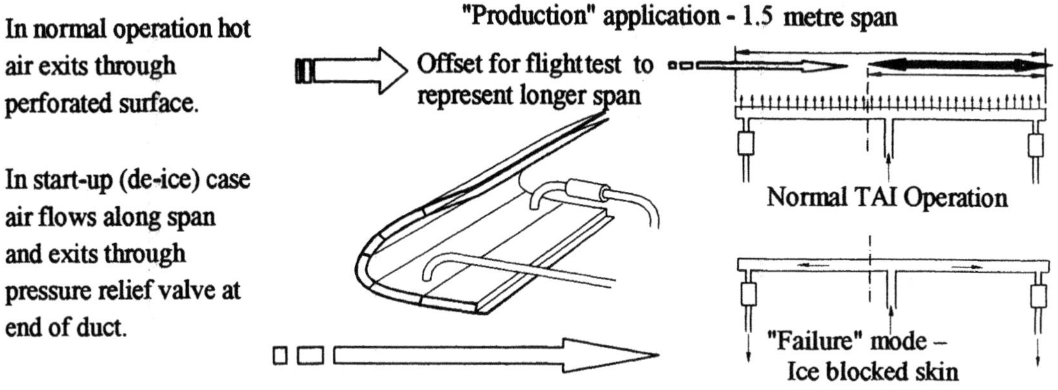

Figure 10: Details of Thermal Anti-icing

Liquid (Foam)

Liquid transpiration through a porous zone within the leading edge is commonly used for ice protection on smaller aircraft and has been used experimentally for insect contamination avoidance, on laminar flow aerofoils, with good results. However, for applications where attachment line suction is required, it is not practical to exude liquid directly through the plenum chamber skin. There are two reasons for this: Firstly, the skin porosity required for boundary layer suction in incompatible with that required to control the dispersal of the relatively low flow rate of liquid that required for ice and insect protection. Secondly, the volume of the plenum chamber is unnecessarily large for liquid dispersal, this implies a large mass of liquid to fill the plenum at start-up and the need to remove this liquid before suction can be established. Apart from the obvious weight and liquid usage disadvantages, undesirable time delays would also result at liquid start-up and shut down.

These incompatibilities and disadvantages can be overcome if the liquid can be transported through the plenum chamber and skin in a foamed state. Only limited data and experience is available with this concept and this experiment is the first time that it has been tested in flight.

A number of methods of foam generation have been examined and the most suitable for this application has found to be the use of a porous Septum (divider). High-pressure air is expanded through the septum into liquid covering the downstream surface. For this application a simple gravity dependent device is used, but for more demanding applications centrifugal techniques could be employed.

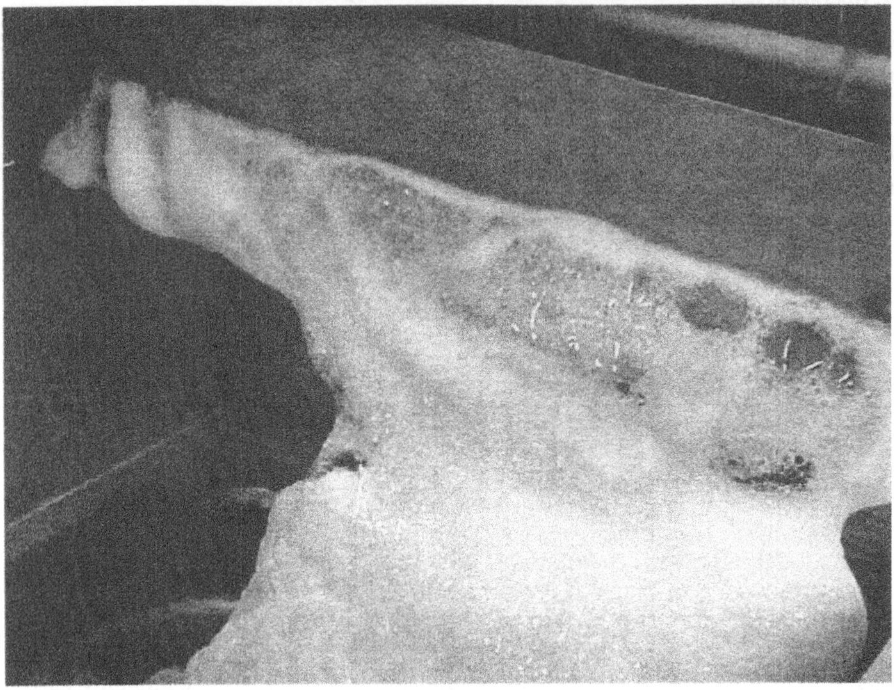

Figure 11: Foam exudation from a suction duct.

7. STRUCTURE

Leading Edge Nose Boxes

The original leading edge consists of two nose boxes of nominally equal length. These will be replaced for the test by three separate nose boxes and a short transition section. The test leading edge follows the design of the original leading edge as far as possible, using skins supported by ribs at about 200-mm spacing. A single row of screws, top and bottom, attach the aft edge of the leading edge skin to the front spar.

The innermost box is in the propeller slipstream and does not form part of the test but is equipped with a TKS liquid ice protection system to ensure safe operation in icing conditions. The next nose box is provided with a Kruger flap for contamination protection and a hot air system for anti-icing (TAI). The Kruger flap can not be retracted but is removable so that tests can be flown with the leading edge representing high and low speed configurations. It is also possible to adjust the position of the Kruger flap for optimisation of performance if required. The outer nose box contains two similar zones; each equipped with a fluid foam system for ice protection and anti-contamination purposes.

All leading edge skins are manufactured from titanium, made porous by laser perforation where required. Ribs and most other internal structural items are aluminium.

Relatively high temperatures result from the use of thermal anti-icing, restricting the use of bonding techniques. It is also desirable that the skin close to the leading edge has the minimum amount of non-porous regions, since these would impair ice protection and liquid decontamination system performance, in addition to their potential effect on laminar flow.

Taking these factors into account, a hybrid design for the suction surface was developed using welded and mechanically fastened joints. This is illustrated in figure 11

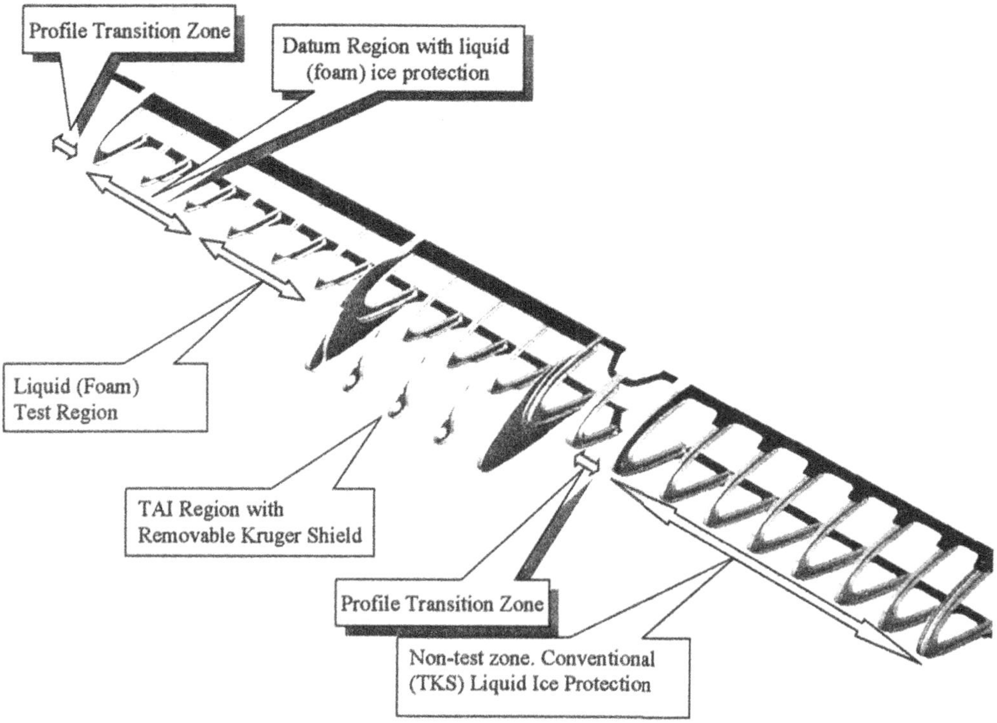

Figure 12: HLF Leading Edge Structure

In addition to providing the necessary profile blend between the original wing section and the modified test section, the transition skin is removable for access. Further access is provided by a number of spring-loaded doors fitted in the lower surface of the leading edge, which also vent the leading edge in the event of accidental pressurisation due to a pipe or component failure. Cooling air is also provided along the leading edge using a NACA intake at the root with an exhaust at the tip.

Glove

The wing upper surface will be fitted with a smooth fibreglass fairing varying from 5mm thickness at the front spar to zero at the aft end of the main wing box. This fairing will extend along the span of the test section only and taper to blend with the basic wing section at each end. Figure 14 illustrates the profile modification in front of the wing spar, the 5mm offset is maintained over most of the wing box chord, tapering at the aft end to blend with the original profile at the rear spar.

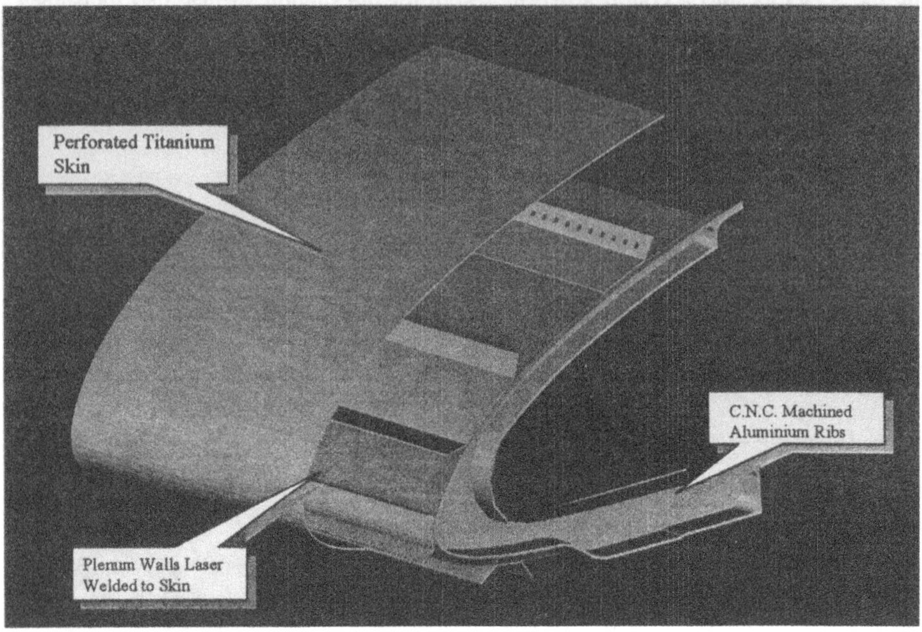

Figure 13: Suction Surface Construction

8. INSTRUMENTATION

Two observation systems will be used. An infrared camera, mounted two metres above the fuselage on a "quadpod", (figure 1) will display the extent of the laminar boundary layer enabling the influence of contamination to be observed. Whilst visible contamination and icing behaviour of all panels will be observed using a video system. A row of pressure tappings will be installed between the two outer panels to measure the pressure distribution in the chordwise direction. Pressure tappings are also included in each plenum chamber to provide information on system operation and further pressure tappings monitor flow and pipeline system performance.

9. CERTIFICATION

This flight test will be made under experimental certification rules, requiring that all aspects must be shown to be safe for flight.

Leading edge to wing fastener loads have been subject to particular examination because there a number of factors that will cause an increase in the load carried by the fasteners:

(i) The titanium skins of the test leading edge are stiffer than those of the normal aluminium leading edge. This means that higher loads will be induced in the test leading edge and the fasteners due to wing bending. This applies along the entire leading edge but tends to be greater towards the inboard end.

(ii) Thermal expansion will give rise to loads in the fasteners of the hot air anti-iced section.

(iii) Loads from the Kruger Flap are carried via the leading edge and so will cause a change in fastener loads in the hot air anti-iced section.

Invaluable assistance has been provided by the aircraft manufacturer who was able to advise the group that the existing fasteners are able to carry twice the loads applied by the normal leading edge. A finite element model of the wing and leading edge was used to evaluate the change in loads due to wing bending and thermal effects. This showed that the fasteners would be subjected to about 1.5 times the original loading due to these effects. Loads arising from the Kruger are very dependent on speed and g loading. At high speed, 1g conditions the Kruger induces a reduction in leading edge fastener loads, but at lower speeds and higher g levels fastener loads are increased. Reducing the manoeuvring speed limit when the Kruger is fitted will be required to cover this.

Spring loaded access panels are used to provide venting in the event of a failure of pipes or components carrying pressurised air. These panels also provide easy rapid access to the internal components and systems.

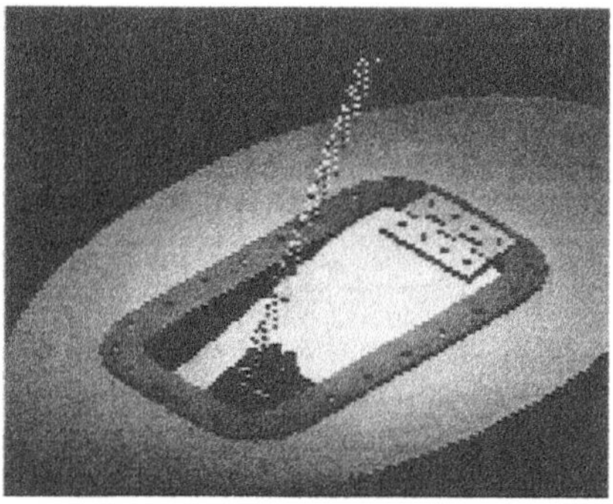

Figure 14: Spring Loaded Access/Blow-off Panels

10. FLIGHT TESTING

Because the aircraft is being used for other programmes in addition to this experiment, the time schedule for the flight tests is dependent on availability of the aircraft as well as equipment. It is planned that the leading edge nose boxes, the data acquisition system and the control systems will be available at the end of August '00. The aircraft also becomes available at about this time so installation will commence, leading up to a period of system ground testing in mid October. A short period of certification flight-testing will follow before the aircraft is ready to commence testing of the HLF systems.

10 hours of icing flight tests are planned using a second Do228 tanker aircraft to produce artificial icing clouds. This testing will take place during winter '00/'01. Following this, 20 hours of low level flight in natural insect collection conditions are planned during spring and early summer '01, this having been found from previous experience to be the period when the greatest airborne insect population exists in Northern Germany.

Session 3
HLF Design & Suction System Integration

New Aerodynamic Approach to Suction System Design

Chris Atkin, Aerodynamics Department,
Küchemann Building, DERA Farnborough, Hants, GU14 0LX, UK.

Summary

A new approach to the aerodynamic design of Hybrid Laminar Flow Control suction systems is presented. The definition of suction chamber layout and pressures has been closely coupled with the boundary layer and stability analysis methodology to provide a numerical tool to help in the design of a suction system. The new approach also provides a direct link between the cost functions of suction system mass and power with the aerodynamic drag benefit, yielding a more streamlined design procedure. Practical constraints appear at an early stage in the process rather than late in the day after much effort has been expended. To demonstrate the power of the technique, the advantages and penalties associated with two different chamber layouts are discussed. Further research is required into the control of crossflow instability and the over-suction phenomenon before the method can be fully exploited.

Introduction

The past two decades have seen a revival of interest in Hybrid Laminar Flow Control (HLFC) for the reduction of drag/fuel burn of transport aircraft, largely because of increased concerns about the environmental impact of commercial air traffic at high altitude. As confidence has grown that the technical problems do, in fact, have solutions, the question of commercial viability has arisen. Recent progress means that this issue can now be addressed with some confidence. Another important issue is that the introduction of a new aircraft boasting a new technology such as HLFC may be a risk too far for the civil aerospace industry. One of the issues tackled by the EU 4th Framework HYLTEC project is the possibility of retro-fitting HLFC technology to a mature aircraft product. The assessment of the potential of a retrofit solution is being undertaken within task 2 of the HYLTEC project. The Airbus Industrie A310 was selected as the baseline aircraft for this task.

The aerodynamic design of HLFC systems focuses on two issues: where to apply suction, and how much suction to apply. In earlier HLFC programmes, the suction region was limited to the area forward of the front spar, so as to minimise impact on wing structure and fuel volume, and suction rates had to be flexible to aid in the learning process. With the maturing of HLFC technology, the answers and indeed the questions have become more sophisticated. The concern is now directly with the design of the plenum chambers: where to place them, how many to have and how large, and what the chamber pressures should be. The goal is not necessarily to maximise laminar flow, but to optimise *performance* including aerodynamic, system and structural penalties as well as simple profile drag reduction.

The objective of the present work is to re-organise the aerodynamic design process to reflect modern design issues and to facilitate the integration of aerodynamics into a multi-disciplinary design procedure. For brevity, the results shown in this paper focus on one of the HYLTEC design points, that of the A310 wing at a Mach number of 0.8, a mean chord Reynolds number of 30 million, and a sectional lift coefficient towards the upper end of the operational range. The form of the sectional pressure distribution for this case is shown in Figure 1(a). Clearly, the aircraft designer would cover a range of operating points, but the application of the approach to this single case will serve to demonstrate the basic principles.

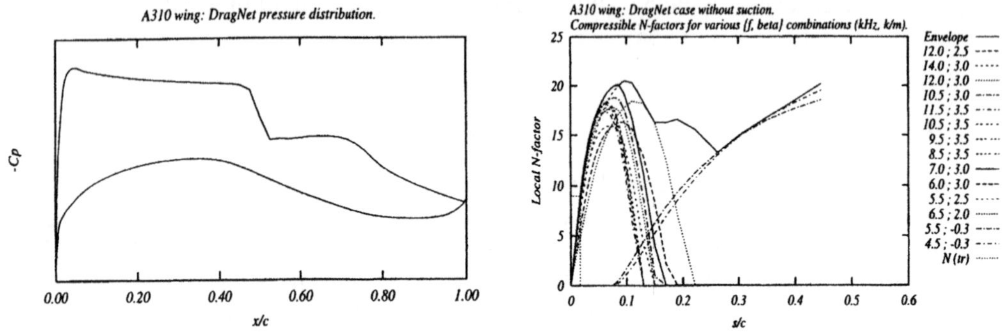

Figure 1 (a) Pressure distribution for A310 test case and (b) corresponding N-factors (no suction).

The basic tools of aerodynamic HLFC analysis are the swept-tapered laminar boundary layer and the e^N transition prediction methods. An example of the output from these tools is illustrated in Figure 1(b). The amplitude of all boundary layer instabilities of crossflow (CF) or Tollmien-Schlichting (TS) type is expressed in terms of N-factors, one curve for each possible mode of instability. Figure 1(b) shows a selection of most-amplified modes and the envelope-of-envelopes curve showing the variation of maximum N-factor with chordwise position. The DERA stability method [1] uses the constant-spanwise-wavenumber integration strategy and the envelope-of-envelopes analysis; no filtering of modes takes place and a single N-factor criterion is applied for all modes. Of course there is nothing to stop the automatic technique which follows from being coupled with any other e^N strategy.

For this single N-factor strategy a correlated value of N = 9 at transition can be inferred from the literature [2] (coincidental, perhaps, with the classically quoted value for 2D flows). Applying this criterion, it can be seen that, for this test case, transition without HLFC would be expected to occur at about 2% chord *s/c*. The N-factors exhibit a peak near the leading edge where crossflow instability causes transition in the absence of any turbulent contamination of the leading edge flow. The crossflow instabilities are subsequently damped downstream of the suction peak where Tollmien-Schlichting instabilities take over. No N-factors are seen beyond 45% chord where the shock wave would cause laminar separation.

Review of classical aerodynamic HLFC system design

Laminar flow control originated with theoretical aerodynamics and the flow through a real porous wing surface is still usually modelled with an analytical velocity distribution. This distribution observes certain practical constraints, for example that suction be applied only on the wing upper surface and that the suction system cannot extend into the main wing box aft of 20% chord. Since it is observed that suction is more efficient towards the leading edge, where instability first occurs, the distribution is trapezoidal in shape. Furthermore, by constraining the shape the suction distribution can be characterised by a single parameter (e.g. maximum suction velocity). This simplifies the optimisation of such a suction system.

The crudest approach is to apply sufficient suction to remove all instability over the porous region, thereby delaying the whole transition process by at least the length of the suction region, or in the present case pushing transition as far aft as 35% chord, Figure 2(a). The corresponding velocity distribution, Figure 2(b), is expressed as notional local hole velocities. A more sophisticated approach is to vary the suction rate and to examine the movement of transition: as the mean suction velocity increases, the transition mechanism changes suddenly from CF-induced to TS-induced, much further aft on the wing. This leads to a recommended suction rate which is just sufficient to push transition aft to the mid-chord region. Figure 3 shows how, with a transition N-factor of 9, this can be achieved with much less suction (about half) than that required for the complete stabilisation of CF modes.

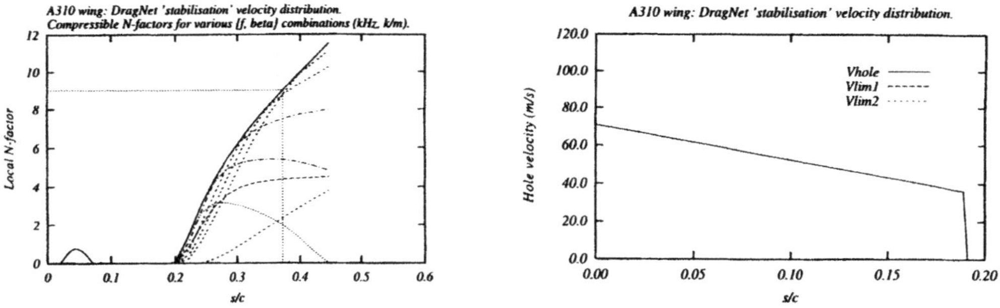

Figure 2 (a) Complete stabilisation of CF modes and (b) corresponding velocity distribution.

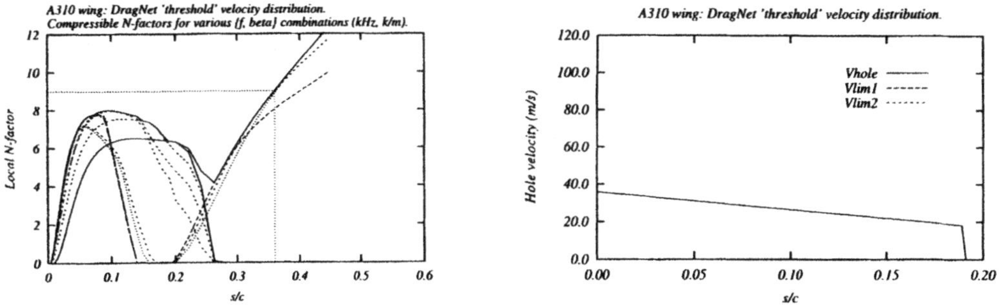

Figure 3 (a) 'Avoidance' of CF transition and (b) corresponding velocity distribution.

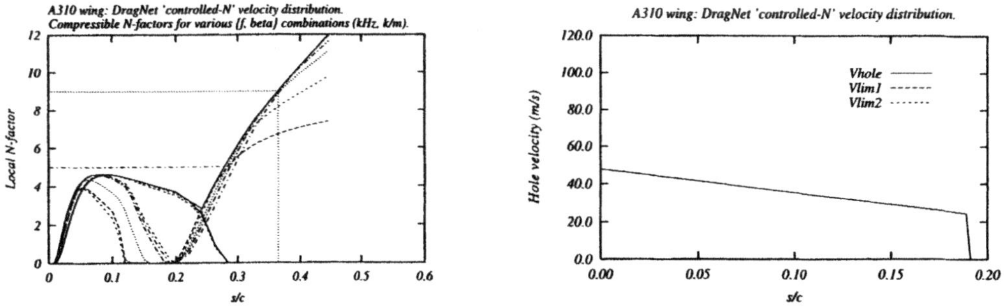

Figure 4 (a) N-factor control of CF modes and (b) corresponding velocity distribution.

There are risks with this approach: the e^N method bundles all the non-linear effects known to occur in the latter stages of transition into the critical N-factor. However, the threshold design approach allows instability growth right up to non-linear amplitudes some way ahead of the predicted transition location. In practice the flow is likely to be dominated by non-linear effects from that point onward, invalidating the subsequent predictions of the linear model. A third approach has therefore been proposed where the crossflow N-factor region is controlled so that instabilities are contained within the boundaries of linear behaviour. For example, limiting the N-factor to be 5 or lower represents a factor of 50 between the control amplitude and the 'critical' amplitude. This approach is shown in Figure 4. The hole velocities are about 40% higher than for the threshold approach but 30% lower than for complete stabilisation. The question remains as to what is an acceptable N-factor margin to avoid non-linear effects.

Practical realisation of surface suction

Wall transpiration is actually implemented by sucking air through a laser-drilled skin. The velocity through each drilled hole is determined by the pressure drop across the skin, the geometry of the hole and the flow conditions at the mouth of the hole. The exact relationship used in this work was derived by ONERA [3]. We average out the separate hole flows into a mean transpiration velocity using the hole area ratio but the exact validity of this averaging has never been thoroughly investigated. We know that high individual hole velocities generate secondary flows in the boundary of a vortical nature which cannot be modelled by the linear stability tools used for HLFC design. These secondary flows are avoided by using over-suction criteria. Two recent experimental investigations into over-suction were carried out by Reneaux & Blanchard (R&B) [4] at ONERA, who derived the expression

$$\frac{V_{h,crit}}{U_e} = \frac{2}{\pi R_{\delta^*}}\left(\frac{\delta^*}{\delta}\right)\left[P_0^2\left(\frac{\phi}{\delta^*}\right)^{-1} - 2P_0P_1 + P_1^2\left(\frac{\phi}{\delta^*}\right)\right], \tag{1}$$

and by Ellis & Poll (E&P) [3] at Manchester University, whose results can be expressed by

$$\frac{V_{h,crit}}{U_e} \sim \left(\frac{\phi}{\delta^*}\right)^{-7/4}\left(\frac{L}{D}\right)^{-1/4}. \tag{2}$$

V_h and U_e are the hole and edge flow velocities respectively, ϕ and δ^* are the hole diameter and boundary layer displacement thickness, R_{δ^*} is the Reynolds number based on δ^*. L and D are the suction length and hole spacing respectively: L/D represents the number of rows of suction holes The P_n terms are constants obtained from a line-fit. One difficulty in reconciling the results of E&P with those of R&B is the lack of an explicit dependence on R_{δ^*} in equation (2) compared with equation (1). Another obvious difference is the functional dependence of V_h on ϕ/δ^*, arising principally from the linear and logarithmic figures used by the two research groups, although it may be connected with the differing R_{δ^*} dependence. It does, however, appear that a power law might fit the data of R&B better than the linear relationship, especially for $\phi/\delta^* < 1$ which would be typical of flight conditions. The effect of suction length, observed by both research groups, is only quantified by E&P. The differences between the two investigations can be resolved only by further experiments which should also cover the influence of the local boundary layer profile shape, including three-dimensionality. Concluding this too brief review on over-suction, a composite relation, more general than equations (1) and (2) but calibrated against them, was used in the current work:

$$\frac{V_{h,crit}}{U_e} \sim \frac{1}{R_{\delta^*}}\left(\frac{\phi}{\delta^*}\right)^{-7/4}\left(\frac{L}{D}\right)^{-1/4} \tag{3}$$

Practical suction distributions

The simplest approach to chamber layout is to try to match the analytical transpiration velocity distributions derived using the aerodynamic analysis described above. This is clearly an easier job than tackling chamber layout *ab initio* because the aerodynamic parametric study then involves only one variable, the average suction rate. Figure 5(a) illustrates such a 6-chamber layout designed to reproduce the velocity distribution of Figure 4(b): the actual resultant distribution is shown in Figure 5(b). These figures also show the over-suction limits *Plim1/Vlim1* corresponding to equation (3) above, and *Plim2/Vlim2* corresponding to a hole Mach number limit of 0.5. The first limit only seems to be significant near the leading edge where the boundary layer is at its thinnest.

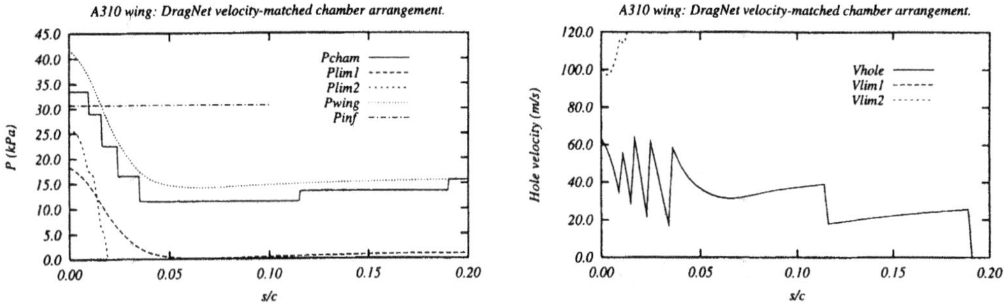

Figure 5 (a) Chamber layout and (b) corresponding velocity distribution devised to match Figure 4.

The resultant velocity distribution shown in Figure 5(b) is very jagged near the leading edge due to the finite chamber lengths and the external pressure gradient. Although it is recognisably close to the distribution in Figure 4(b), the aerodynamic constraints met by the analytical suction distribution are not reliably met by the chamber design. In this case the N-factors exceed the suction-zone limit by about 20%. However the approach can always be improved in this respect by using a larger number of smaller chambers. But is this is actually necessary?

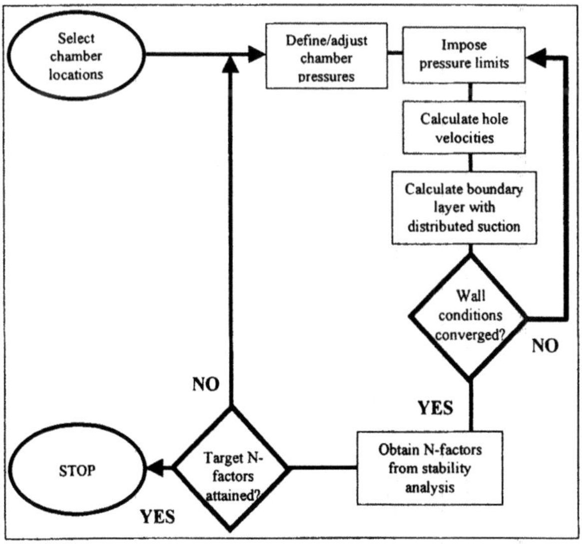

Figure 6: flow diagram of integrated chamber layout and aerodynamic analysis process.

New approach

The new approach simply involves the integration of the various steps described above, with the important difference that the chamber layout is proposed first and used directly to generate input for the boundary layer and stability analysis: no analytical velocity distributions are required. The N-factors from the stability analysis are then used to control the chamber pressures until the aerodynamic laminar flow constraints are achieved. A schematic of the process is shown in Figure 6. The process starts with the specification of chamber layout alone. The scheme initially sets chamber pressures for minimum mass flow, determined by the no-outflow criterion. An initial control loop checks lower pressure limits for each chamber on the basis of boundary layer output before proceeding to the stability analysis phase. The most complicated part of the process is the control of chamber pressures on the basis of N-factor output, and the success of the method relies in no small part upon the good qualities of the boundary layer and stability methods used at DERA. Precise resolution of small changes in suction rates is required, as is a smooth response to these changes of the maximum N-factor. The process must also work without any user intervention.

The N-factor control scheme is illustrated in Figure 7 which shows a the development of a typical N-factor curve over a series of discrete suction bands. The effect of each chamber is assessed over two regions: the suction region, *'a'* in the figure, and any gap *'b'* before the start of the next chamber. N-factors are measured relative to the start of the suction-controlled region, labelled *'u'* on the figure this being simplified greatly by the lack of significant upstream-influence of boundary layer control. Consideration is also given to N-factor values at the downstream end of the control region, *'d'* in the figure, since these may place a burden on the following control region if they are close to the N-factor limit. Target N-factors are then derived for the control region, and a revised chamber pressure is prescribed based on these targets. Newton's method is used where possible; interval search where not. In certain circumstances maximum N-factors in a control region may be independent of the chamber pressures and this must be recognised by the control scheme.

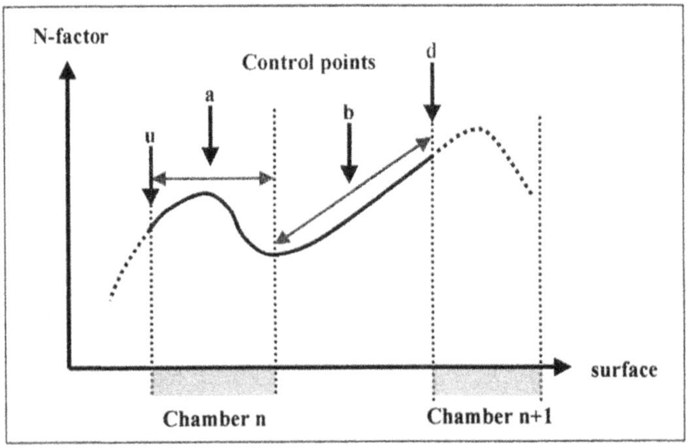

Figure 7: illustration of N-factor control regions.

For multiple chamber arrangements, the upstream chambers are allowed to settle down before the control loop is applied to the downstream chambers.

The end result of the process is a chamber pressure specification which observes the N-factor constraints for each control region: usually this represents just one of a number of possible solutions to the control problem (for a fixed chamber layout), but it is easy to adjust chamber pressures manually to investigate other solutions. This usually involves increasing the suction levels over the upstream chambers. The output from the process is chamber pressures and mass flow rates (dependent on the spanwise extent of the chambers) which can be used as the basis for a system design. In the HYLTEC project this approach is being used to investigate HLFC performance issues across all the issues of profile drag, pump power, system weight and cost.

The numerical features of the process are as follows. An interpolation scheme is applied to the basic mean flow specification to include explicitly the chamber start and end points. These are resolved over four intervals of 0.05% chord each. The final distribution of points yields about 100 boundary layer stations for analysis, although the iterative stability calculations are restricted to the currently active control regions, saving unnecessary analysis. Nonetheless at each station the stability of some 400 modes is analysed, and a complete control loop might involve the calculation of between 5000 and 30 000 eigenvalues. The whole process takes from one to four hours on a Pentium 2 PC at 350 MHz depending on the number of chambers involved. Clearly there are enormous opportunities for the replacement of the full stability analysis method with a robust, validated database-type method.

Application of the new process

The opportunities offered to the designer by such a tool are demonstrated by the following examples. The first is a seven chamber layout shown in Figure 8. This can be compared with the manually-designed example presented in Figure 5.

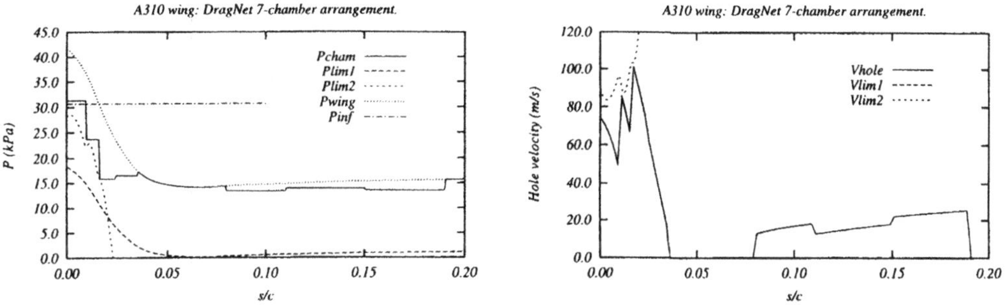

Figure 8 (a) Seven-chamber layout and (b) velocity distribution satisfying the control requirements.

Here the attachment line chamber pressure has been relaxed to achieve a target R_θ along the attachment line: the second and third chambers have been set to give maximum possible suction rates, while the fourth is free to respond to the N-factor distribution. The fifth chamber has been removed, while the sixth has been subdivided into three to introduce some flexibility into the control of the Tollmien-Schlichting modes. Of these three, the middle chamber is at the maximum allowable pressure (with a safety .margin against outflow) while the outer two respond to the N-factor distribution. The saving in mass flow compared to Figure 5 is 19%. The steps leading to this choice of distribution started with simple N-factor control over chambers 2, 3 and 5: the results show that suction over chamber 5 is quite inefficient, and that control is achieved more economically by removing this chamber, which pushes most of the suction into chamber 4 . Shifting suction even further upstream by manually increasing the suction in chambers 2 and 3 reduces the required mass flow rate even further.

Figure 9 illustrates a simplified chamber layout with only three chambers which also satisfies the control requirements. Here the control of crossflow modes is achieved with a single chamber in addition to the one on the attachment line. The middle of the three chambers downstream of the suction peak has been removed. As well as the system simplification, a further 4% reduction (compared to Figure 5) in mass flow is achieved overall (although the TS-control mass flow has increased). Note, however, that the pressure losses have increased.

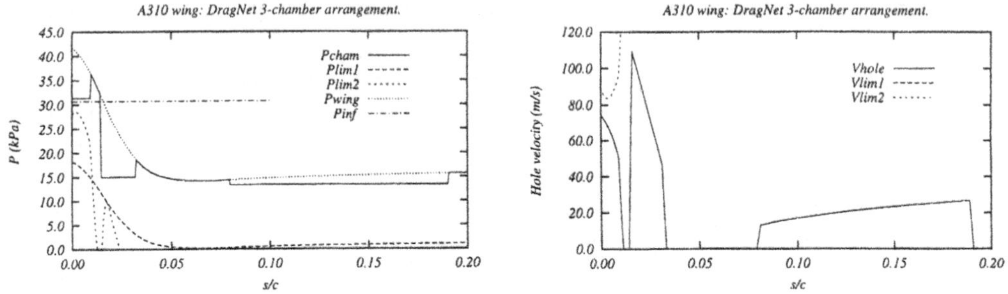

Figure 9 (a) Three-chamber layout and (b) velocity distribution satisfying the control requirements.

The N-factor plots for these two arrangements both look very similar to Figure 4(a) because the chamber pressures have been adjusted iteratively to satisfy the same N-factor control criterion.

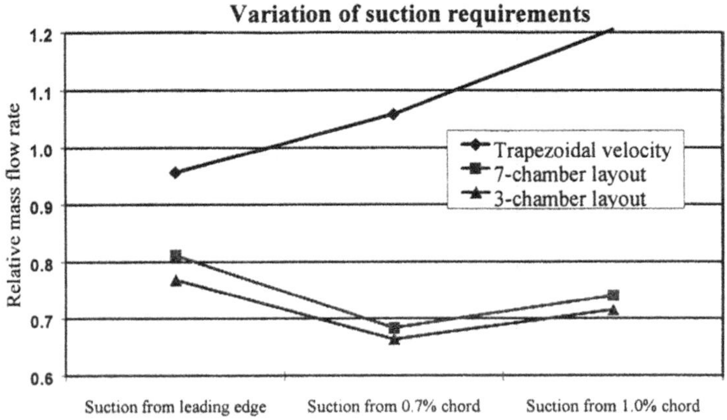

Figure 10: Effect of moving suction away from the leading edge.

The impact of these changes on the system specification, including power requirements, weight and cost, is one the issues being investigated in the HYLTEC project. From the systems point of view the analysis is greatly facilitated by the immediate availability of mass flow and pressure loss information from the aerodynamic design. In all cases it is demonstrable that significant improvements, in terms of reduced mass flow, can be made over the use of a simple trapezoidal transpiration distribution. Any design studies undertaken using a velocity-distribution analysis would be completely obscured by the uncertainties involved in implementing the chamber layout. The present approach allows the implementation to be controlled and compared while the aerodynamic parameters are varied.

Both of the chamber layouts described above feature gaps in the suction distribution which have been introduced on the basis of suction effectiveness. In practical cases there may be a requirement for gaps in the suction distribution to accommodate other systems (e.g. de-icing) near the leading edge. Figure 10 illustrates the effect of moving suction aft from the attachment line for the velocity-distribution, 7-chamber and 3-chamber approaches. Clearly, in mass flow terms, there is a significant benefit in not sucking right at the attachment line (if an alternative contamination avoidance system is used); but there is a point beyond which total suction effort increases for a given configuration. For obvious reasons, the simple velocity-distribution approach fails to capture these effects.

Conclusions and recommendations

An automatic tool has been developed to satisfy N-factor requirements which also allows the designer to constrain chambers en route to an optimised layout. The method relies on the existence of a robust, well-resolved and automatic stability analysis method.

The new approach has highlighted some of the deficiencies of the trapezoidal-velocity-distribution approach. The importance of modelling the control system is perhaps a useful message to those who are developing analytical suction optimisation tools for HLFC design.

A new N-factor control philosophy has been proposed which attempts to balance the technical risk of an HLFC system against commercial gain. The philosophy is based on the likely onset of non-linear behaviour, and the consequent failure of the e^N model. The philosophy must be properly validated and/or modified with further research, particularly in the area of non-local and non-linear analysis of crossflow instability, since it dictates about 50% of the mass flow requirements - at least for the test case presented here.

A review of recent over-suction studies by two different research groups has shown that the two sets of findings differed significantly in terms of important parameters. A hybrid criterion, not yet validated, has been used for the present work: further work in this area is essential to the development of a simple chamber layout and plumbing system. Future work should also focus on the highly swept flow close to the attachment line where the over-suction problem appears to have the greatest impact on high-Reynolds-number HLFC design.

References

1 Atkin C J, Poll D I A. *Correlation between linear stability analysis and crossflow transition near an attachment line.* Transitional Boundary Layers in Aeronautics, Henkes and van Ingen eds., North-Holland, 1996. pp. 155-164.

2 G. Schrauf, J. Perraud, F. Lam, H.W. Stock, D. Vitiello, A. Abbas, *Transition prediction with linear stability theory – lessons learned from the ELFIN F100 flight demonstrator,* 2nd European Forum on Laminar Flow Technology (1996), 8.58-8.71.

3 Preist J, Paluch B. *Design specification and inspection of perforated panels for HLFC suction systems.* Paper 6.3, 2nd European Forum on Laminar Flow Control, Bordeaux (June 1996).

4 Reneaux J, Blanchard A. *The design and testing of an airfoil with hybrid laminar flow control.* 1st European Forum on Laminar Flow Control, Hamburg (March 1992).

5 Ellis J E, Poll D I A. *Laminar and laminarizing boundary layers by suction through perforated plates.* 2nd European Forum on Laminar Flow Technology, Bordeaux (June 1996).

Retrofit Studies based on Airbus A310 for HLFC Application on Aircrafts

Heribert Bieler *, Pete Swan †
* DaimlerChrysler Aerospace Airbus GmbH, 28183 Bremen, Germany
† BAE SYSTEMS Airbus UK Ltd., New Filton House, Filton Bristol BS99 7AR, England

Summary

About 50 % of the drag the aircraft experiences during flight occur due to the friction between the aircraft skin and the air. Laminar flow technologies, if applied to wings, nacelles, empennage reduce the surface friction significantly.

HYLTEC (HYbrid Laminar flow TECnology) is a follow-on activity to previous projects funded by the EC within the programme "Industrials and Materials Technologies" [1].

In order to improve the competitiveness of new and derivative products we consider the potential for HLFC-retrofit (Hybrid Laminar Flow Control - retrofit) already in the initial design phase.

The following results are a 1st pass approach to the requirements for the practical implementation and exploitation of HLFC on existing aircrafts resp. provisions for future ones. Several iterations in aerodynamics, systems and structure converged to the case of suction on the slat at moderate levels and (even with less constraints) suction at the nose of the empennage.

1 Introduction

Laminar flow technologies, if applied to wings, nacelles, empennage reduce the surface skin friction significantly. While fuel prices have been relatively low for a long time, energy taxes - in particular for burning fossil fuel - and, starting in a few decades, shortage of fossil resources will increase the price for fuel.

It is therefore a strategic requirement for aircraft manufacturers to advance fuel saving technologies like laminar flow in time.

Hybrid Laminar Flow Control (HLFC) is a combination of suction around the nose (wings, nacelles etc.) and shaping in the box region. By these means the flow close to the surface (boundary layer flow) is stabilized and the point of transition to turbulent flow is shifted downstream. While the application of HLFC on aircraft is a medium to long term topic, the products of aeronautical industry normally have a rather long commercial life-time (about 30 to 40 years). It is therefore attractive for the industry to investigate the retrofit topic, by which aircraft presently in airline service might be equipped (retrofitted) with HLFC when the market asks for it, or the design of an aircraft which takes a later HLFC retrofit into account right from the beginning of the design process.

This dual approach offers the airline to make full use of HLFC even during the life-time of a presently turbulent aircraft.

In order to make best use of the available budget, a well known real configuration (with known difficulties) has been sought. For such a baseline vehicle all HLFC relevant areas have been addressed in order to illustrate the design constraints for HLFC on wings and empennage [2]. The study is focused on modern commercial aircraft.

Pro's and con's of the different Airbus families:

Single aisle:

This class is economically attractive for the manufacturer since a large number of aircraft (which reduces the business risk for the producer) exist. However, the benefit is due to rather short missions limited - assuming that the cruise part of the mission , where the nose suction is switched on - has to be a significant portion of the whole flight path.

Twin aisle:

This Airbus is technically a suitable candidate. Here we have a good balance between risk and benefit for both: manufacturer and customer. This is also a market segment for which an update of products is needed within foreseeable future. Because of the age of the aircraft - however- a retrofit for the actual A300 or A310 is unlikely. But - due to limited budget and time - only a trend analysis can be performed in such a project, this has not been regarded as a handicap. Because of the nature of the study, the conclusions drawn here are regarded to be valid for any widebody aircraft in the 250 seater segment.

Long Range:

Here the operator enjoys the largest operational benefit, but the risk for the introduction of such a technique is rather high. These risks come from high Reynolds number in combination with high wing sweep angles , which promote the amplification of disturbances in the boundary layer flow and the challenge of 4 engines (for the A340), which poses severe questions in terms of 3D effects and aeroelastic effects.

Future Configuarions:

Aircraft designs presently in consideration at future project offices have the disadvantage, that the benefit / risk assessment might be difficult to quantify, since the turbulent reference is less well known. This accounts also for detailed data (geometrie etc.).

It was decided to investigate the A310, which is even more appropriate for such a study than later aircraft due to structural advantages over the single aisle type and more or less 2D flow outside of the engine, which facilitates the use of quick and robust aero codes.

2 Aerodynamic Specification Estimates

Within aerodynamics, systems and structure multidisciplinary work has been done to cover all important aspects of HLFC retrofit. The configuration is the A310 and Fig. 1 shows at which portions of wings and empennage suction has been investigated. Since only trend analysis is intended for a variety of parameters, quick and robust aerodynamic codes have been selected. These are capable to treat the laminar and turbulent flow (viscous and inviscid) within a so-called conical approach. Complicated 3D effects in the vicinity of engines, tip and wing root regions have been excluded from the study. The percentage numbers in Fig. 1 are the regions where the isobar concept is satisfactorily fulfilled.

To prove the possibility of laminar flow via HLFC on a A310 wing, several partners checked the laminar extent for the most important lift coeffients during cruise (Fig. 2). On the upper side the laminar region is shock limited up to about 35 % to 40 % and on the lower surface up to 40 % laminar flow has been predicted [3]

Since the objective of this study is to understand the impact of all key parameters in aircraft design onto HLFC retrofit, simplifying assumptions and rapid techniques were selected in order to deliver the information required for a global assessment of the benefits and penalties of retrofit.

Fig. 3 shows as an example that within aerodynamics the boundary layer stability analysis (linear theory) resulted in certain mass flow requirements. The suction velocity (averaged over the whole chamber area as shown in Fig. 3) represents the input for the porous panel design. These mass flows can be realized with suitable internal pressure settings over separately controlled chambers. The upper figure contains the external and (in a stepwise fashion) the internal pressure, which has to be lower than the external to realize the suction flow. This pressure drop across the panel has been modelled empirically by using the ONERA conical hole flow approach [4] As a compromise between complexity and space and system constraints, 7 chambers have been proposed for the wing case to provide systems and structure departments with aerodynamic input for their iterations. Some remarks on system aspects (indicated in Fig. 3 right hand side) are part of chapter 3.

3 Interdisciplinary Work

As it usually the case for investigations aiming to improve the aerodymanic efficiency of aircraft the aerodynamic requirement have to be formulated first and then (sometimes also in parallel) other departments start their work in order to realize the aerodynamic designs or to propose compromises to fulfill all constraints in a balanced way. Fig. 4 gives an overview of influential parameters for these studies.

Within the parameter space we have 2 principal directions of investigations:

1. all relevant cruise conditions like several Mach numbers, lift coeffients etc. have to be addressed. In addition off-design conditions, operational issues (failure of HLFC system, limited laminarity due to surface damages, insect contamination etc.) have to be checked.

2. in order to arrive at an overall benefit, several iterations between aero, systems and structure are necessary. As an example the mass flow requirements have to be fulfilled without an excessive penalty on systems complexity and weight increase due to duct and pump installations.

Finaly the proposed solution has to be realistic from a manufacturing point of view, since tighter tolerances for laminar flow surfaces are valid, which increase the cost of production and consequently the price for the aircraft. Some system aspects are illustrated in Fig. 3 which contains some key topics at the interface between aerodynamics and systems. The result is a ducting layout for the 7 chamber proposal for the wing with suitable control devices (valves etc.) and suction pump (e.g. electric fan).

As an example for the structural work Fig. 5 indicates the importance of a theoretical joint analysis at the junction slat to wing box. Laminar flow requires smooth surfaces and empirical predictions show that the present manufacturing standards for the A310 wing do not allow laminar flow at such a step. A design improvement shown in the figure is therefore needed, which might be realized via stop devices to ensure a sufficient smooth gap and step at the slat end. More detailed information may be found in [5].

In such a overview paper we can point only to some highlights of the retrofit study. The major loops which resulted in a convergence of results and requirements are given in Fig. 6. More detailed internal minor loops have also been done within the project period. A systematic compilation of all results for the system topics is contained in [6].

It came out in general that the wing poses the most severe problems in terms of allowable number of chambers, mass flow levels and space considerations (esp. in relation with high lift devices).

The fin and horizontal tail plane are less critical and technologically less challenging, mainly due to absence of movable parts at the nose.

4 Conclusions

The retrofit task in HYLTEC addressed all key topics of HLFC constrained design for aerodynamics, systems and structure.

The A310 with known properties has been selected as a baseline vehicle for all investigations, which means - since trend analysis has been done - the typical transonic cruise of a 250 seater has been treated.

Finally, the studies converged - due to fruitful and efficient cooperation among all partners - to the case of suction on the slat at moderate suction levels. The empennage constraints were less critical, in order to realize laminar flow.

Within the phase 1 of the study only minumum changes of the baseline aircraft were permissible. These results were shown here.

In phase 2 (limited in scope however), the design for retrofit will be addressed. These studies will include more changes of the baseline configuration nameley variations of sweep, noise radii, use of Krueger devices etc.

The final activity will be feeded into a global assessment task, which addresses overall benefits and penalties of HLFC retrofit.

5 Acknowledgement

This work was supported by the Commission of the European Communities Industrial and Material Programs (BRITE/EURAM IMT) within the project HYLTEC (EC Contract BRPR-CT97-0606). The authors appreciate the support by all HYLTEC partners for this paper and would like to thank the scientific officer from the EC for his assistance.

References

[1] Bieler, H. : *HYLTEC - Hybrid Laminar Flow Technology.* Project programme EC-Contract BRPR-CT97-0606.

[2] Swan, P. : *Design restrictions. Retrofit and Design for Retrofit.* HYLTEC TR-03. 1998.

[3] Stuke, H., Schröder, W., Reneaux, J. and Ruiz, C. : *Transition prediction for a A310 aircraft with HLF retrofitted wing, stabilizer and fin-noses* HYLTEC TR-01. 1998.

[4] Preist J. and Paluch B. *Design specification and inspection of perforated panels for HLFC suction system.* Paper presented at 2nd European Laminar Forum on Laminar Flow Technology, Bordeaux;1996.

[5] Overbergh, C.: *Joining Techniques for HLFC Suction Panels.* HYLTEC-TR 28. 2000.

[6] Pfennig, J. : *Task 2.2.3: Development of System Parameters. Part A: Retrofit Design and System Layout* HYLTEC TR-21. 2000.

6 Figures

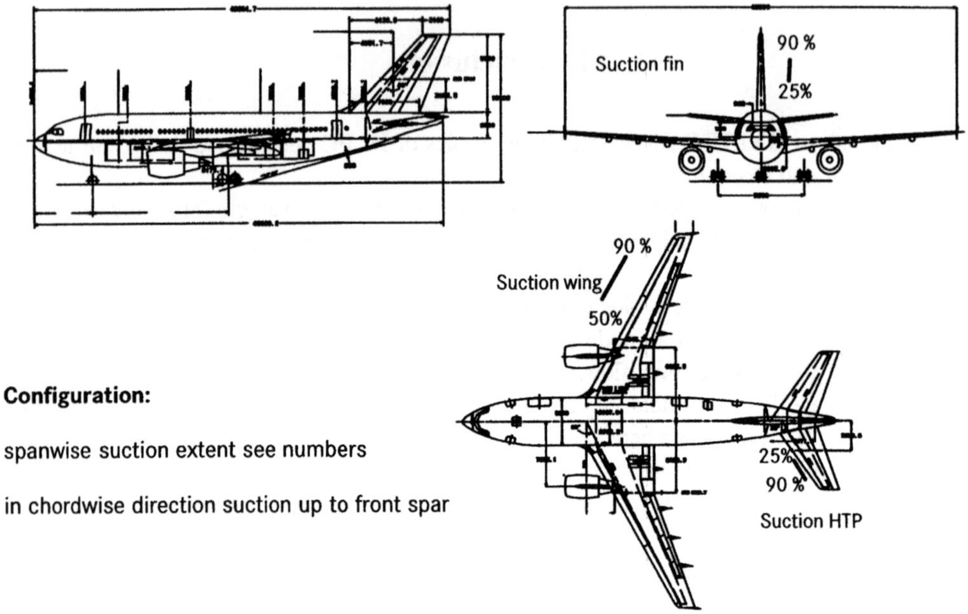

Figure 1 A310 General Arrangement; suction up to x/c = 0.18

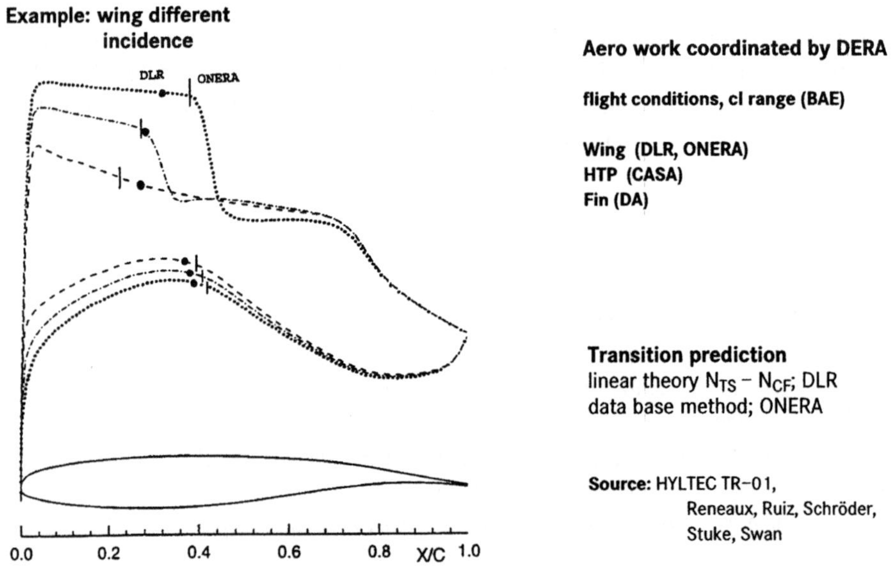

Figure 2 Transition Prediction - A310 wing at cruise

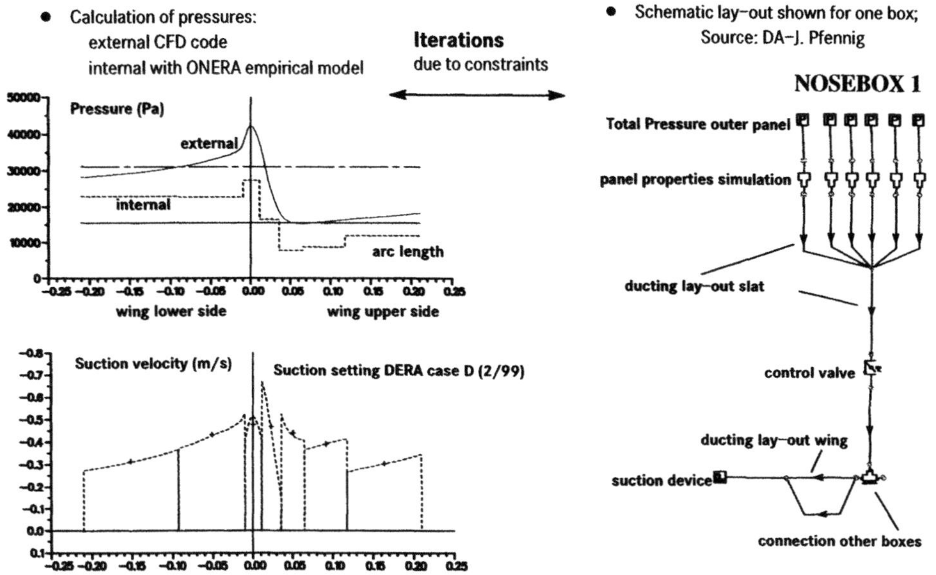

Figure 3 Interface Aerodynamics - Systems; starting case suction upper and lower surface

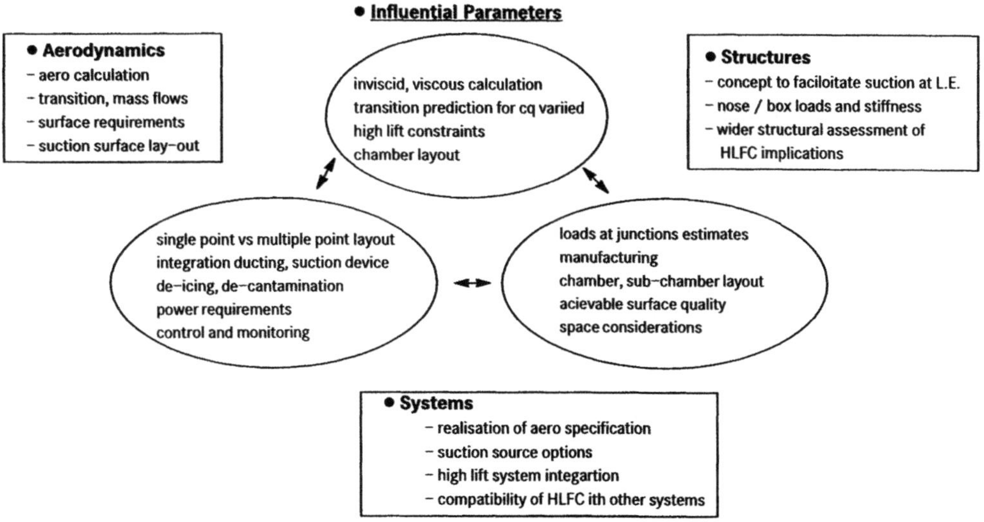

Figure 4 Scope of Studies

Influential parameters:

- Wing flexibility
- Stiffness of porous outer panel
- Front spar properties
- Fatique
- Erosion
- Maintainability

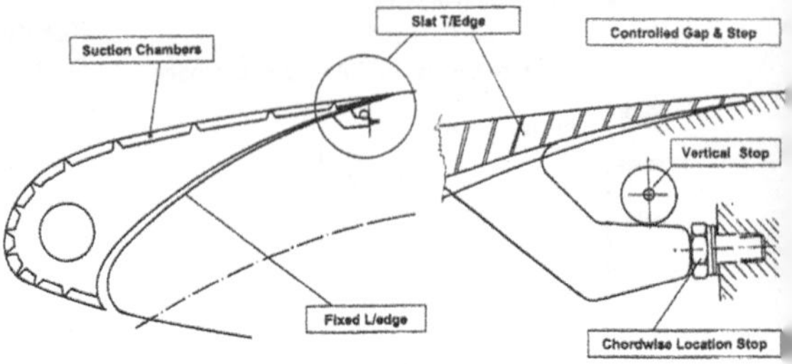

Joint analysis:
between Slat and wing box

Design improvement:
Stop devices for step gap control

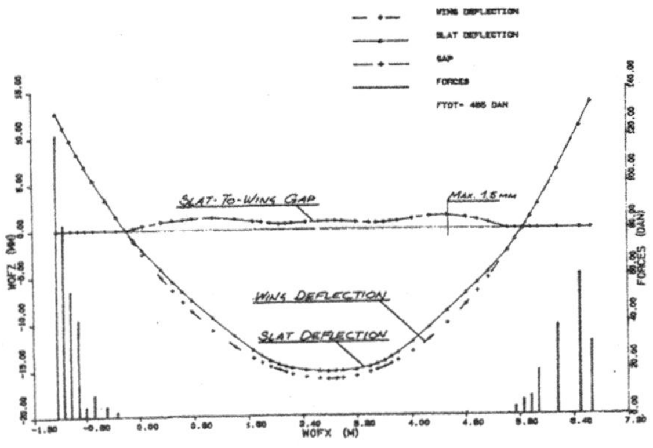

EXAMPLE OF GAP DISTRIBUTION ALONG THE SPAN :

Cruise condition 2

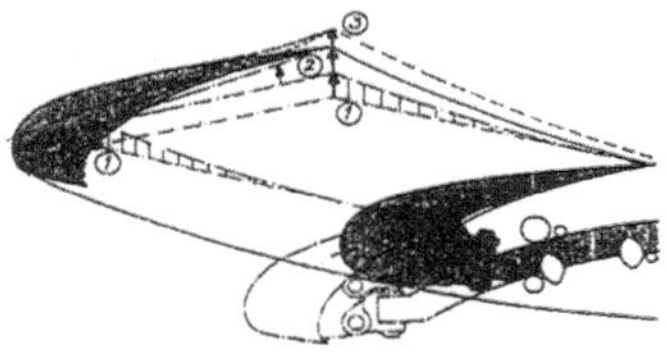

Source:

SONACA, C. Overbergh

Figure 5 Structural Improvement at Slat; case for suction on slat only

Overview of results (retrofit):

- **1. loop**
  - aero spec based on turbulent operations (BAE SYSTEMS)
  - from– from 01/99 onwards according to DERA plan aero specification for systems, structures
  - mass flows for wing, fin (ONERA, DERA)
  - HTP mass flows (CASA)
  - based on given suction data 1. chamber layout wing, fin (DA); for HTP (CASA)
- **2. loop**
  - space allocation study (AM) showed 1.6 kg/s is not realistic (mid 99)
  - new limit of 0.7 kg/s estimated
  - group decided to limit suction on the slat
  - for most critical case (wing) 2. chamber layout, mass flows (DERA) (end 99)
  - mass flows allow duct integration in wing (fin, HTP less critical)
  - examination of high lift situation (BAE and FFA)
- in parallel structural, system work was done
  - e.g. structural study at slat wing box junction (SONACA) etc.

Figure 6 Convergence of Iterations

Light transmission control technique and correlation with pressure loss characteristics of perforated panels for Hybrid Laminar Flow Applications

B. Paluch
ONERA - Dept. DMSE/RCS - 5 Boulevard Paul Painlevé – 59045 Lille - France

Summary

In order to quickly establish a cartography of the perforation quality on significant panel sizes designed for HLF applications, the control method which was developed consists in measuring the light transmission through the holes of perforated surface. The principle of control lies on equivalence between the quantity of light transmitted through the holes and the air flow passing through these same holes. According to this principle, the cartography of the light transmission will provide an image of what should be the corresponding homogeneity of the velocity suction field for iso-pressure loss conditions. After having shown that it is possible, by choosing adapted cell surface and sampling step, to minimise the aliasing effect, the evaluation of the method sensitivity highlight the influence of the perforation characteristics on the light transmission measurement. A correlation established, on a real panel, between the light transmission measurement and the suction velocities showed the validity of the principle of equivalence.

1. Introduction

The control of perforated panels designed for hybrid laminarity applications is of great importance. At present, this control is carried out in two different ways:

- the measurement of the hole diameters D and the estimation of their statistical distribution parameters : mean <D> and standard deviation s(D),
- the measurement of the pressure loss characteristics such $\Delta P = f(V_p)$

However these two methods are not very efficient to locate quickly the areas where the perforations could be of more or less good quality, because the measurements are only performed in a given number of positions, usually equally spaced on the perforated surface. Moreover they induce times of control the more significant as the number of holes or numbers of pressure loss curves to be measured is large. For several years, the control of a great number of perforated panels shows that local fluctuations of the suction velocity V_p at iso-pressure loss ΔP can sometimes be very large. These fluctuations are mainly due to the presence of local perforation inhomogeneities that the two preceding methods are unable to highlight without carrying out their exhaustive application on the whole perforated surface. To avoid these disadvantages and to quickly establish a cartography of the perforation quality on significant areas (higher than the square decimetre) the control method we propose in this paper consists in measuring the transmission of light through the holes of the perforated surface.

2. Measurement principle and apparatus

The principle of control lies on equivalence between the quantity of light transmitted through the holes and the air flow passing through these same holes. These two quantities are, for a given unit area S_u, locally proportional to the perforated surface. They depend at the same time on the mean hole diameter < D > and of their proportion in S_u. With regard to the quantity of transmitted light and the air flow, the proportionality is not completely respected, in a strict sense, because of the phenomena of different physical nature, specific to the two concerned variables, that are:

- the light diffraction in the immediate vicinity of the edge of each hole
- the incidence of air viscosity (there is a boundary layer on the internal walls of each hole), of flow contraction at the entry of holes, and some possible compressibility effect (at high V_p), on the air flow passing through the holes

Nevertheless for a certain range of low level light power and low suction velocities one can consider that these two phenomena deteriorate only little the proportionality, and consequently the principle of equivalence can be considered as respected. Measurement, in each point of perforated surface, of the local percentages of light transmission is then able to provide us a cartography of the homogeneity of the perforation, and thus an image of what should be the corresponding homogeneity of the suction velocity field at iso-ΔP. On a practical way, the measurement of the light transmission forces us to put, on one side of the panel a source of light, and the photoelectric cell on the opposite side. Two processes are then possible:

- The surface of the light source is fixed and of significant size, and the photocell moves inside the field of this source so as to draw up the cartography. The major disadvantage of this method lies in the difficulty to manufacture a light source of big size (typically more than one dm^2) and of uniform intensity in space.
- The surface of the light source is reduced (but higher than that of the cell) and moves at the same time as the photoelectric cell. One then eliminates the preceding problem while being sure that the intensity of the light beam projected on the panel remains constant during the control process

We chose to employ the second way because of its simplicity of implementation and its reliability. The measurement device consists of a fork (*figure 1*) on which are fixed the light source (halogen lamp) and the photoelectric cell (photodiode with an integrated amplifier). Panel scanning is ensured by step by step motors with a positioning precision of 0.05mm. The perforated surface is scanned by a succession of lines spaced by a step p_s. On each line the fork moves at constant speed, and during this movement the signal coming from the cell is sampled at constant intervals p_s. At the end of each line the data are transferred to the microcomputer charged to draw up the cartography of the light transmission.

3. Photo cell size, sampling frequency and aliasing effect

The cell surface S_c and the sampling pitch p_s are the most significant factors likely to deteriorate the measuring accuracy of light transmission, because an aliasing effect can occur during the scanning. In theory such a measurement has to give coherent results (constant response amplitude of the cell) since the number of holes seen by the cell, for each sampling step p_s, must remain constant. To be more precise the sum of surfaces of the luminous spots projected by each hole on the cell which must remain constant during the scanning. If the holes were distributed according to square mesh of regular drilling pitches $p_x = p_y$, and that S_c was equal to a multiple of p_x, such as $S_c = m\, p_x^2$, there would be no aliasing effect since $p_s = m\, p_x$. Actually this purely theoretical configuration does not meet on real panels because :

- on a same line of drilling the holes are not completely regularly spaced and a slight pitch fluctuation between lines of drilling can occur (the pitch standard deviations $s(p_x)$ and $s(p_x)$ are very small but not zero).
- Hole rows are shifted between consecutive lines of drilling.

As a consequence of an industrial drilling process these slight imperfections are inevitable, because the panels are usually drilled at a constant scrolling speed with a frequency from 20 to 30 holes a second. Moreover during the perforated surface scanning, the cell can not move completely in parallel or perpendicular to the lines of drilling, due to the panel positioning precision compared to the scanning direction. It would be possible to measure the light transmission hole by hole, but the presence of these imperfections would make this type of method very difficult, even impossible, to perform. To highlight clearly the aliasing effect we established an algorithm allowing to calculate the light projected by each hole on the cell. On a real panel, the aliasing effect would be partly masked by the variation of the light flow induced by the hole diameter scatter. Consequently we calculate the cell response by carrying out a numerical scanning on a network of holes of constant diameter, reproducing the three types of imperfections with a drilling step of p_x=p_y=0.6 mm, in order to evaluate the aliasing effect. Since the cell is square, the light flow l_c measured by the cell have initially be calculate when it scans with a sampling step p_s equal with $p_s = \sqrt{S_c}$, either in parallel or perpendicularly to drilling lines and for different value of S_c (see *figure 2*). The response l_c is not constant in function of cell displacement, and the fluctuation is the more significant as S_c is small. Whatever the scanning direction $<l_c>$ remains constant in function of cell size but is obviously proportional to S_c, whereas $s(l_c)$ only increases slightly according to S_c. To quantify the aliasing effect in function of S_c the best way is to calculate the ratio $s(l_c)/ <l_c>$. The aliasing effect disappears either for $s(l_c)$ tending towards zero, or for $<l_c>$ tending towards the infinite. This is only possible that for a cell size higher than the perforated surface, which removes obviously any interest with this type of method since one wants, on the contrary, to highlight the presence of possible local drilling inhomogeneities. *Figure 3* shows sees that $s(l_c)/ <l_c>$ decreases very strongly for S_c ranging between 1 and 5 mm^2, and then more slightly. In theory one will be able to minimise the aliasing effect only with the detriment of increasing S_c. However it remains possible to reduce again this effect by using the slight signal periodicity property of the cell when p_s=p_x (dashed curve). One then notices that the ratio $s(l_c)/<l_c>$ strongly decreases, with a lowest value for a scanning direction perpendicular to the drilling lines. The choice of S_c thus results from a compromise between two opposed tendencies, namely the aliasing effect minimisation (by increasing S_c) and the minimisation of the cell averaging response effect (by decreasing S_c). In order to satisfy this compromise we fixed S_c to 5 mm^2 with a sampling step p_s=p_x for all the panels which we controlled with this type of method.

4. Evaluation of the control method sensitivity

On real panels the quantity of light passing through the N_c number of holes seen by the cell will depend on the local hole diameter statistical distribution and on the percentage P_b of the number of blocked holes. These two factors will contribute to the variation of the light flow passing through the holes. And, according to the principle of equivalence, the measurement of the transmission of light must be locally proportional to V_p at iso-ΔP. Let us examine the incidence of the first factor on the variation of the signal delivered by the cell. Let $<D>$ and $s(D)$ being the mean and standard deviation of the hole of the whole perforated surface, and $<d>$ and s(d) the mean and standard deviation on the diameter of the N_c holes. If $r_d = <d>/<D>$, the variation Δl_c of the cell signal is related directly to this ratio, since:

$$\Delta l_c = \left(N_c \pi <D>^2 /4 - N_c \pi <d>^2 /4\right) / \left(N_c \pi <D>^2 /4\right) = 1 - r_d^2. \qquad (1)$$

According to (1) a variation of ± 20% on $<D>$ induces a variation of a little more than the double (45 % approximately) on l_c. One could think that this variation does not depend, a priori, of the number of holes seen by the cell. In reality the variation of $<D>$ will be the more significant as S_c will be low. Knowing $<d>$ and s(d), it is possible to calculate the variation interval of r_d. Each measurement made by the cell is equivalent to sample N_c holes among the total population of holes of parameters ($<D>$, s(D)). While considering a sufficiently large number of samples (greater than 100),by using for example a Monte-Carlo's type method, one can estimate s(d) precisely and calculate $r_{d\,min}$ and $r_{d\,MAX}$ from $<d>_{min}$ and $<d>_{MAX}$, such as they delimit a probability interval of about 99 %, *i.e.*:

$$r_d \in \left[r_{d\,min}, r_{d\,MAX}\right] = \left[<d>_{min} / <D>, <d>_{MAX} / <D>\right] = \left[1 \pm 3 s(d) / <D>\right]. \qquad (2)$$

So, considering a real population of holes with, for example, the following characteristics: $<D>$=60 μm, s(D)=8 μm, we plotted on *figure 4* the ranges of Δl_c and $<d>$ versus S_c, where it is noting that s(d) decrease when S_c increases. That then enables us to calculate the variation on $<d>$ in function of S_c as well as the associated Δl_c variation. Consequently the smaller the cell size, the most sensitive the luminosity measurement to the variation of $<d>$. Δl_c is about 90 % for S_c = 1 mm^2 and approximately of 40 % for S_c = 5 mm^2. Beyond this value Δl_c tends towards an asymptote, indicating that a cell of 5 mm^2 seems to constitute a critical threshold of sensitivity. We find approximately the same sensitivity about the percentage of blocked holes P_b on Δl_c in function of S_c. The probability that the cell sees N_b blocked holes ($0 \le N_b \le N_c$) follows a binomial statistical law, such as:

$$P(n_b = k) = C_{N_c}^k P_b^k (1 - P_b)^{N_c - k}. \qquad (3)$$

If it is supposed, in order to dissociate the respective influences of P_b and $<D>$, that all the holes have a constant diameter, the reduction Δl_c of luminosity induced by the own presence of blocked holes is then equal to $\Delta l_c = 1 - r_b$ with $r_b = n_b/N_c$. The values of the maximum reductions of luminosity for increasing values of P_b in function of S_c are plotted on *figure 5.* The light transmission is strongly affected below S_c=3 mm^2. The incidence of the number of blocked holes on the luminosity variation measured by the cell is thus of the same order of magnitude as that induced by the variation of the local mean hole diameter $<d>$, and the real cell response is thus a function of these two factors, without it being however possible to distinguish their respective roles. For realistic values of the hole diameter statistical parameters and moderate percentages of blocked holes (P_b<5 %, case of usual perforated panels), the combination of these two factors should nevertheless give a significant variation of the local luminosity, conferring on this type of method a great sensitivity to the measurement of local perforation quality. Taking into account the physical phenomena involved in the light transmission through the holes, the local luminosity measurement can only be relative. It is preferable to standardise the measurements by the maximum value of the measured luminosity, *i.e.* $l_i = l_i / l_{i\,MAX}$ with $0 \le l_i \le 1$. Let us notice that a constant luminosity is synonymous with an homogeneous quality of perforation. Now let us calculate (without aliasing effect) the variation of relative luminosity for a hole population characteristics ($<D>$, s(D), P_b). For each point of measurement the local luminosity is given by:

$$l_i = \sum_{i=1}^{N_c - n_b} \pi d_i^2 / 4 . \quad (4)$$

The values of l_{min}, <l> and s(l) were calculated for 1000 samples (or measurement points) with <D>=60 μm, p_x=0.6 mm, s(D)=8 μm and for various values of P_b. These values have been plotted on *figure 6*, where it is noted that in the absence of blocked holes the value of l_{min} is already equal to 0.56. Dispersion on the mean hole diameter <d> has a dominating influence on the local variation of luminosity. The line (obtained by linear regression) shows that l_{min} decreases when P_b increases. Moreover, this graph can be divided into two zones showing the respective effects of s(D) and P_b. As we underlined previously, the influence of the second factor is less significant than that of the first, and becomes sensitive only from P_b>10 %. <l> decreases slowly in function of P_b, contrary to s(l) which grows much more quickly according to the same parameter. In other words, the percentage of blocked holes affects particularly the scatter on the local light transmission measurements. It is necessary to bear in mind that this is only a numerical model, in which the distribution of the hole diameter over the panel is assumed as random. Although it highlights, in a rough way, the influence of the various parameters (<D>, s(D), P_b, p_x) on the luminosity distribution (<l>, s(l)) this model cannot, in any case, describe local inhomogeneity of the hole distribution over the panel, as is it going to be shown.

5. Correlation with pressure loss characteristics

A perforated panel drilled for one of a study carried out within the framework of HLF programme has been controlled using the method we have just detailed. Its characteristics are the following ones: < D>=68.83 μm, s(D)=14.71 μm, P_b=2%, p_x=0.36 mm and the size of the perforated zone is of 280x180 mm. We are going to check the validity of the principle of equivalence by proceeding to a comparison between the average luminosities and the corresponding values of V_p at iso-ΔP. The pressure loss curves have been measured at 25 equidistant positions with a suction head of 20 mm diameter. This panel has been selected because, on the one hand of the significant value of s(D), and on the other hand because of its rather scattered pressure loss characteristics, which will facilitate the correlation between l_i and V_p at iso-ΔP. From a practical point of view, the control time is about 30 mn/dm^2 (it is obviously shorter when p_x increases). The luminosity cartography raised on this panel is shown on *figure* 7, with normalised values 20% ≤ l_i ≤ 100 %. On this cartography one can note that the luminosities are not distributed in an homogeneous way. One clearly distinguishes alternations of zones of high and low light transmission ratios l_i, highlighting areas of perforation inhomogeneity more or less marked. The percentage of blocked holes being low, these inhomogeneities are to be charged directly to local variations of <d>. The calculation carried out according to the method of chapter 3 gives 30% ≤ l_i ≤100 %, which is very close to reality. The numerical model developed previously being based on the assumption of a random distribution of mean hole diameters is not able, now, to reproduce the local inhomogeneities encountered in the spatial luminosity distribution. It is of course what confers all the interest to the inspection method we developed. In order to establish a correlation between l_i and V_p, we determine the V_p from the pressure loss curves for ΔP=20 000 Pa. We then calculated the average luminosity L of the area contained inside the suction head such as :

$$L = 1/n \sum_{i=1,n} l_i \tag{5}$$

n being the number of points contained in the circle of diameter D_a. The values of V_p in function of L are plotted on *figure 8*, on which we plotted the regression line, with a correlation coefficient equal to 0.758. This value gives a first confirmation about the validity of the principle of equivalence. In any rigour, the points should be located on the regression line, but one does not to forget that because of the phenomena invoiced in chapter 2 there is no exact proportionality between V_p and L. This is also probably due to a slight aliasing effect. It is thus normal that the couples of points (V_p, L) are affected by scatter, but we have to notice that they are included in the confidence intervals plotted for risks α= 1, 5 and 10%. In order to make sure that the light transmission measurement technique reflects the local hole diameter distribution, we selected 8 zones (of size 5 X 5 mm) in each one of which we measured the diameters of 30 randomly sampled holes. In these same zones we calculated L using (5) and plotted the couples ($< d >$, L) on *figure 9*. The linear regression line shows the excellent correlation (coefficient of 0.956) between these two quantities, confirming the preceding assumption well.

6. Conclusion

The light transmission control method of perforated panels allows to draw up a cartography relative with perforation quality on significant perforated areas, in a fast, precise and reliable way, since it exist a strong correlation between L and V_p. The control of other panels carried out in the framework of various programmes [1] shown the potentiality of this method, which one of the most interesting lies, according to the principle of equivalence, in the possibility to locate very quickly some critical area where the suction velocity gradient at iso-ΔP can be so significant that it could not be acceptable for a given HLF application.

Reference

[1] Paluch B., *Contribution to HYLDA subtask 2.3 : analysis of an Nd-YaG drilling process experiment plan*, CEE HYLDA report TR 30, June 1999

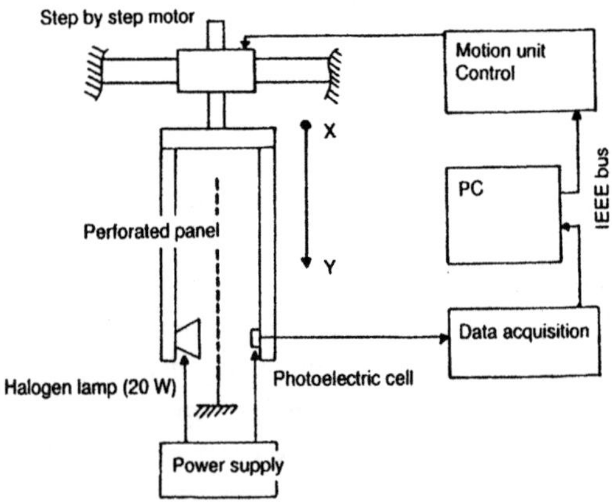

Fig.1 : measurement apparatus

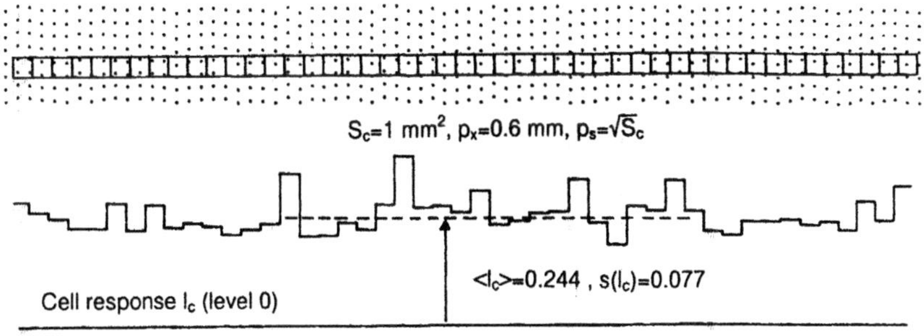

Fig. 2 : Influence of the aliasing effect on the cell response l_i.

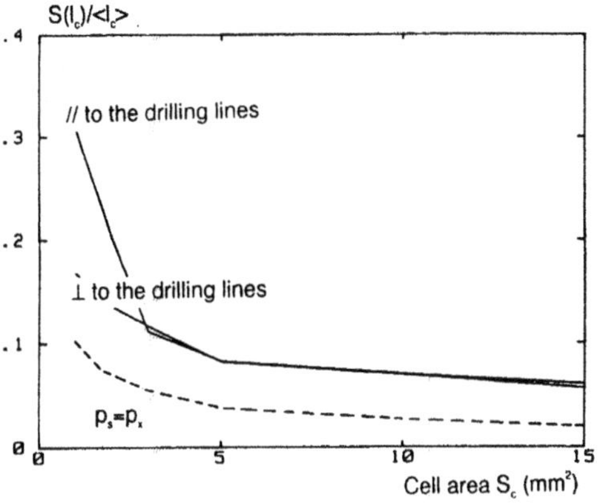

Fig.3 : Aliasing effect in function of cell size

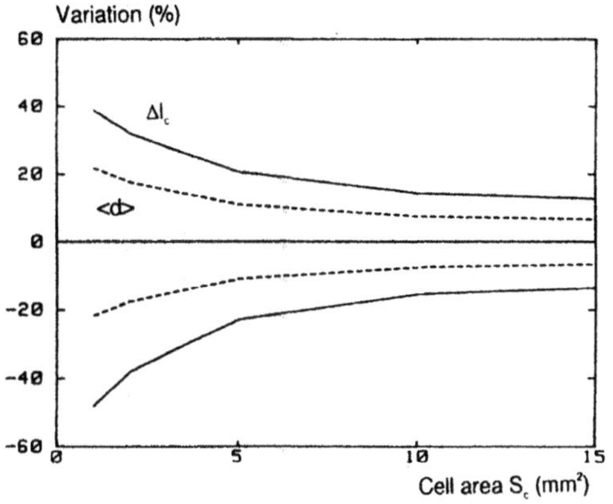

Fig.4 : Variation of Δlc and <d> in function of cell size

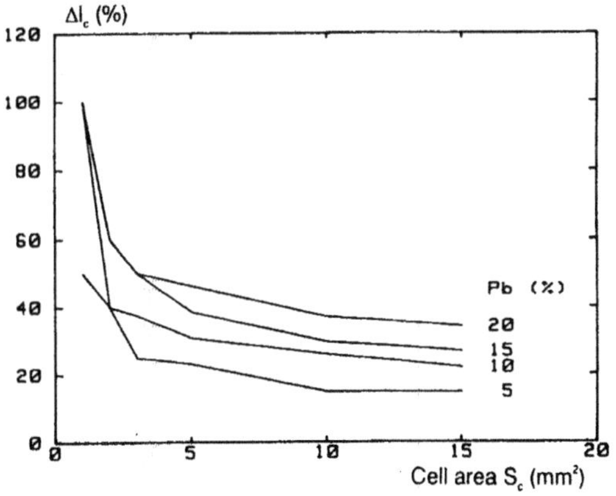

Fig. 5 : Influence of cell size on ΔL_c for different values of P_b

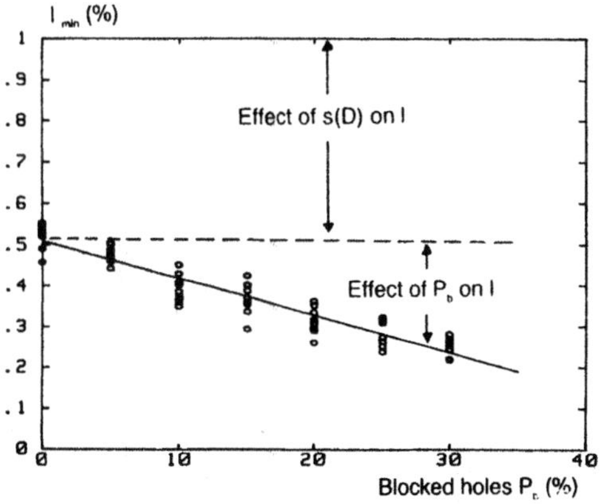

Fig. 6 : Variation of l_{min} in function of P_b

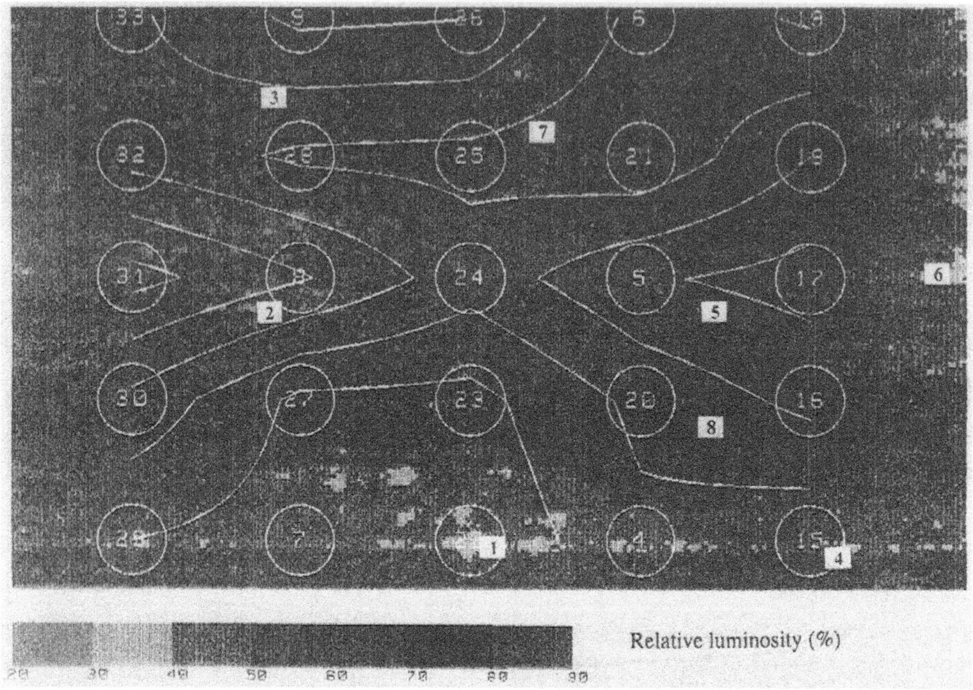

Fig. 7 : Cartography of normalised light transmission l_i on a real perforated panel (the circles and the small white areas indicate the positions of the suction head used to measure the ΔP and the positions where the local mean hole diameter were measured respectively)

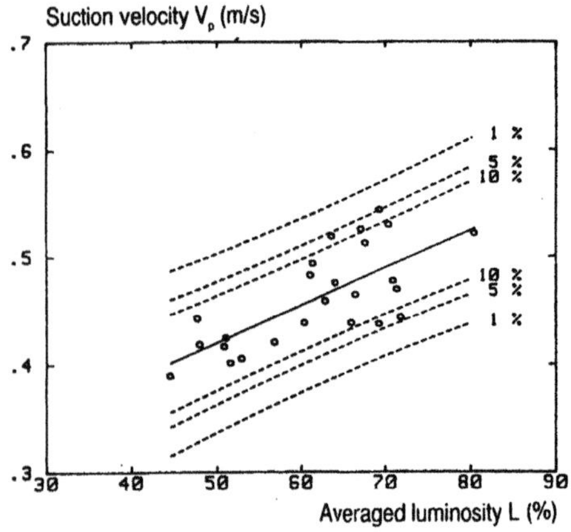

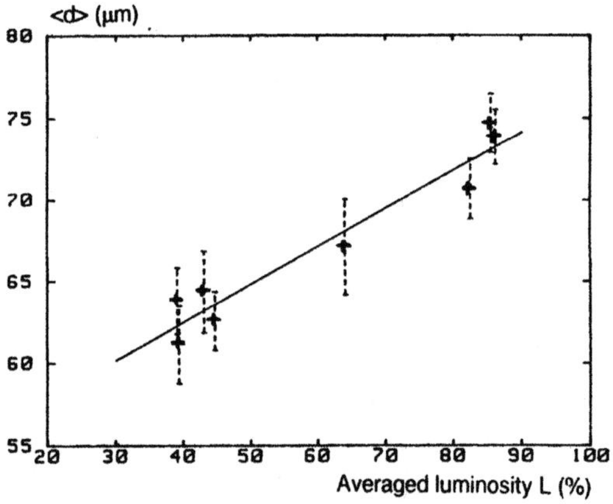

Fig. 8 : Plot of V_p vs L (from *fig. 7*)

Fig. 9 : Plot of <d> vs L (from *fig. 8*) dashed lines show confidence intervals

DNS Study of Suction through Arrays of Holes in a 3-D Boundary-Layer Flow

Ralf Messing, Markus Kloker
Institut für Aerodynamik und Gasdynamik, Universität Stuttgart
Pfaffenwaldring 21, D-70550 Stuttgart, Germany

1 Summary

Spatial direct numerical simulations based on the full incompressible unsteady Navier-Stokes equations are employed to explore the effects of arrays of suction holes on an accelerated 3-D laminar boundary layer. For a hole array with 35 hole rows it turns out that the hole-row order typically has no influence on the amplitudes of the 3-D disturbance modes at the end of the hole array. Due to an averaging effect caused by the changing flow direction no permanent amplification of upstream induced vortex structures is discernible. Localized suction induces disturbance mode combinations consisting of crossflow vortex type eigenmodes (CFV modes) subject to possible primary downstream growth, and other (suction vortex, SV) modes. However, continued discrete suction seems to suppress the CFV-type eigenmodes and favor non-or only weakly growing SV modes.

2 Introduction

The capability of boundary-layer suction in delaying 3-D boundary-layer transition has recently been proven by free-flight tests with an Airbus A320 using a modified swept vertical fin where suction has been applied up to 20% in chordwise direction. The introduction of hybrid laminar flow technology in operational service of an aircraft has yet not taken place due to significant uncertainties remaining for a reliable design of suction devices. The microscale flow phenomena in the vicinity of suction holes are barely known and are the major subject of the present numerical investigations providing finally a better understanding of the integral effects of suction.

Typically suction is applied in the front region of the back-swept wing where the flow is accelerated and crossflow instability is dominant. According to linear stability theory stationary crossflow modes can be amplified in 3-D boundary layers but it is not at all clear to what extent these unstable modes are generated by suction through holes. In previous investigations the influence of the hole diameter d, spanwise hole spacing s_z and the volume rate per hole $\dot{V}$ has been studied for the case of a single row of holes. With the assumption of homogenous discrete suction (i.e. d, s_z, $\dot{V}$ are constant) which is valid for the present investigations the smallest spanwise disturbance wave number is fixed by the spanwise hole spacing ($\gamma_0 = 2\pi/s_z$). The generation of higher harmonics

depends on the ratio d/s_z. The smaller the ratio d/s_z, the broader is the range of relevant discrete spanwise wave numbers which are excited. A major result of these studies is that for subcritical suction a transient region exists extending approximately 70 δ_1 downstream of the hole row wherein all modes are damped [3]. The observed damping even of modes with wave numbers that should be amplified according to Linear Stability Theory (CFV modes) cannot be attributed to suction as the suction rate was very weak and the mean flow alteration is negligible. In Fig. 1 the streamwise development of stationary disturbance modes is shown. The hole row is located at $x - x_0$=0.083. The mode (0,1) attenuates up to $x - x_0$=0.42 according to 72 δ_1 past the hole row. Only after a distance of 140 δ_1 the mode (0,1) evolves as predicted by Linear Stability Theory (LST)(Fig. 1, curves on top). Obviously, the SV mode (0,1) directly excited by suction is not an eigenmode (with wall-normal disturbance velocity $v'(y=0)=0$) but after some transient region changes over to it.
Further simulations with nine hole rows show that the disturbance development depends on the streamwise hole order [4]. In case of a staggered hole order (spanwise offset of streamwise successive hole rows is $s_z/2$) the mode (0,1) is amplified wheras in case of a non-staggered hole order the same mode does not grow leading to lower amplitudes dowstream of the hole array. For the staggered hole order the growth is caused by a permanent amplification of upstream induced suction vortices at each subsequent downstream hole row. As the propagation direction of the suction-induced SVs is right the mean of the direction of the wall streamline and the potential streamline and as the local sweep angle changes only slightly within the considered domain the staggered order is unfavourable for the considered flow region (details of the base flow will be given in section 3).
After the study of a single row of holes and an array with a few hole rows where the properties of the base flow barely alter, mainly two questions arise if we extend our investigations to hole arrays with more hole rows. Does an averaging effect occur as for the amplification and cancellation of vortices if the properties of the base flow, especially the local sweep angle, change non-negligeably along the hole array? Does a hole array generate significant CFV modes that grow eventually as predicted by LST?

3 Numerical Method and Base Flow

The thouroughly tested combined compact Finite Difference / spanwise Fourier spectral method solves the full incompressible unsteady Navier-Stokes-equations in vorticity-velocity formulation within the integration domain shown in Fig. 2. As we limit the present investigations to simulations of spanwise equally spaced holes the use of periodic boundary conditions in z-direction with the periodicity length s_z is appropriate. Steady suction through holes with diameter $d = 2 \cdot r$ is modeled by prescribing a wall-normal disturbance velocity distribution $v'(r) = -v_p cos^3(\pi r/d)$ with $r = \sqrt{(x-x_L)^2 + (z-z_L)^2}$, $0 \le r \le d/2$, x_L, z_L - hole center coordinates, v_p - peak suction velocity. All velocities are normalized by the reference velocity $\overline{U}_\infty$=159 m/s, and spatial variables by a reference length $\overline{L}$=9.434 mm such that the global Reynolds number is Re=$\overline{U}_\infty \overline{L}/\overline{\nu}$=$10^5$. Further details of the numerical method are reported in [3],[2].
The base flow has been chosen to match all essential properties of a three-dimensional boundary-layer flow in the front region of a swept wing. We use a Falkner-Skan-Cooke (FSC)-type boundary layer with a Hartree parameter β_H=0.4 and a local sweep angle

$\overline{\varphi_e}$=35^0. Other important base flow parameters are shown in Fig. 3. Within the integration domain the freestream velocity increases by 25% leading to a decrease of the local sweep angle from $\overline{\varphi_e} = 35^0$ at the inflow boundary to 29^0 at the outflow. The fundamental spanwise wave number is γ_0=115 corresponding to a spanwise hole spacing $\overline{s_z}$=500μm. Thus (0,1) represents a stationary mode with β=0 and γ=115. The spanwise discretization is done with KMAX=21 positive Fourier modes, the step sizes in streamwise and wall-normal direction are Δx=0.4188 $* 10^{-3}$ and Δy=0.45 $* 10^{-3}$ with a wall zone where Δy is halved for 32 intervals.

4 Results for arrays of holes with 35 hole rows

To examine the influence of the hole order two simulations have been performed. The array has 35 hole rows with constant spanwise and streamwise hole-grid spacing $\overline{s}_z$=$\overline{s}_x$=500μm=10$\overline{d}$. The hole diameter d based on the cos^3-distribution of v' corresponds to about d_{eff}=0.65d for a parabolic Hagen-Poiseuille distribution. Thus the effective porosity of the suction panel is (0.0078 $* 0.65^2$)%. The suction rate $c_q=\overline{v}_{av}/\overline{Q}_\infty$ with $\overline{v}_{av}$ as the average suction velocity over the whole suction panel is 0.5 $* 10^{-4}$ yielding a maximum suction velocity of $\overline{v}_p/\overline{Q}_\infty$=0.029 ($\overline{Q}_\infty$=194$m/s$, $\overline{Q}_\infty=\overline{U}_\infty/cos\overline{\varphi_e}$) for the used suction-hole velocity distribution. The two cases differ only in the order of the holes. For Case 1 the holes are ordered in a regularly staggered arrangement (constant successive offset of $s_z/2$) whereas the holes in Case 2 are brought into line with the wall streamline (Fig. 4 and 5).
As expected the amplitude of mode (0,1) increases within the first (12) hole rows in the staggered case (Fig. 4) but further downstream the growth stops and the amplitude even slightly decays. In Case 2 mode (0,1) grows at the beginning and oscillates then around a constant amplitude level. At the end of the hole array mode (0,1) has nearly the same magnitude in both cases indicating the disturbance averaging effect by the out-of-propagation-line hole order. The absent growth may be also attributed to the pure suction effect, mode (0,0). However, in the considered cases, (0,0) is too small to be accounted for. As observed for a single row of holes a transient region downstream of the hole array exists where the mode (0,1) is damped. For comparison the amplitude development for the single row of holes located at $x - x_0$=0.084 is plotted in Fig. 4 and Fig. 5. Most interestingly, on the array mode (0,1) does effectively not grow as long as suction is active although the transient downstream region of several upstream situated hole rows is exceeded on the array. The growth as predicted by LST first sets in after the hole array and the transient region downstream of the hole array. In Case 1, at $x - x_0$=2.05 the CFV mode abruptly dominates, as is visible by the slope change of the amplitude curve, whereas in Case 2 a more gradual changeover from the SV to the CFV mode takes place. This behaviour suggests that suction at the wall induces a combination of several vortex modes for a fixed spanwise wave number. To substantiate this finding eigenfunctions of the disturbance velocities as obtained from spatial LST and disturbance profiles from DNS are compared at several chordwise positions for Case 1 (Fig. 6). The analysis with LST is done for the superposition of the base flow and the spanwise averaged disturbance flow (base flow plus mode (0,0)) to capture the influence of suction on the mean flow profiles. The comparison clearly shows that suction excites a combination of modes whose amplitude distributions differ substantially from the distributions of the pure CFV mode as obtained from LST. The $u'(y)$-, $w'(y)$-profiles from DNS have two local maxima, one

close to the wall, and the second near the eigenfunction maximum of the CFV mode. Tracing downstream this second peak gets more pronounced. It clearly can be assigned to the CFV mode which is also triggered by suction. But obviously the development of this possibly growing mode is hindered by the other induced mode resulting eventually in no growth of mode (0,1) over the hole array. As soon as the boundary layer is no longer subject to suction the CFV mode evolves and prevails. At $x-x_0$=2.51 the eigenfunctions and the disturbance profiles from DNS agree very well.

5 Conclusion

Direct numerical simulations of hole arrays with 35 hole rows with spanwise equally spaced holes have been performed to investigate the disturbance evolution induced by suction panels with different hole orders. The streamwise changing flow direction leads to an averaging effect in terms of damping and amplification of induced vortex modes resulting in a comparable disturbance level downstream of the hole arrays. This is confirmed by further simulations not shown here. Thus the hole order has no relevance for practical applications. Assuming a homogenous arrangement of suction holes no growth of induced vortex modes can be observed in the considered base flow as long as suction is active. For a fixed spanwise wave number, suction through holes generates a combination of vortex modes where the crossflow vortex eigenmode part seems to be depressed by regularly appearing „spots" of non-zero wall-normal velocity at the wall. Thus, considering steady modes, the principal effect of subcritical suction through an array of holes in a 3-D boundary layer is not only induced „passive" stabilization of the mean flow, but also by direct influence on the CFV modes. This holds as long as the spanwise hole spacing coincides roughly with the spanwise wavelength of the most unstable mode according to LST. This is not the case in the respective DLR-Göttingen suction experiments [1]. We want to point out that this conclusion has so far only be verified for the base flow as described in section 3 at weak suction rates within the linear regime of disturbance development.

Acknowledgement

The financial support of this research work by the Deutsche Forschungsgemeinschaft (DFG) under contract number KL 890/3-1 is gratefully acknowledged.

References

[1] Abegg, C.; Bippes, H.; Janke, E.: Stabilization of boundary-layer flows subject to crossflow instability with the aid of suction. In Fasel, H.; Saric, W. (eds.): *Laminar-Turbulent Transition* Proc. IUTAM Symposium Sedona, Az./USA (1999), Springer-Verlag, 2000.

[2] Bonfigli, G.; Kloker, M.: Spatial Navier-Stokes Simulation of Crossflow-Induced Transition in a Three-Dimensional Boundary Layer. In Nitsche, W.; Heinemann, H.-J.; Hilbig, R., (eds.): *New Results in Numerical and Experimental Fluid Dynamics II.*. Proc. 11. AG STAB/DGLR Symposium (1998), NNFM 72, Vieweg Verlag, Braunschweig, 1999.

[3] Messing, R.; Kloker, M.; Wagner, S.: Direct Numerical Simulation of Suction through Discrete Holes in a Three-Dimensional Boundary Layer. In Nitsche, W.; Heinemann, H.-J.; Hilbig, R., (eds.): *New Results in Numerical and Experimental Fluid Dynamics II.*. Proc. 11. AG STAB/DGLR Symposium (1998), NNFM 72, Vieweg Verlag, Braunschweig, 1999.

[4] Messing, R., Kloker, M.: Effects of suction through arrays of holes on a 3-D boundary layer investigated by spatial direct numerical simulation. In Fasel, H.; Saric, W. (eds.): *Laminar-Turbulent Transition* Proc. IUTAM Symposium Sedona, Az./USA (1999), Springer-Verlag, 2000.

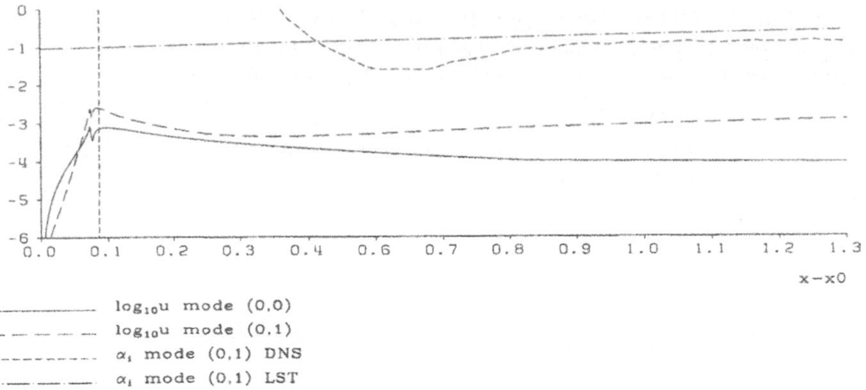

Figure 1 u'-amplitudes (y-max) and amplification rate $\alpha_i = -d/dx(lnA)$ for a single row of holes, holes located at $x - x_0 = 0.083$

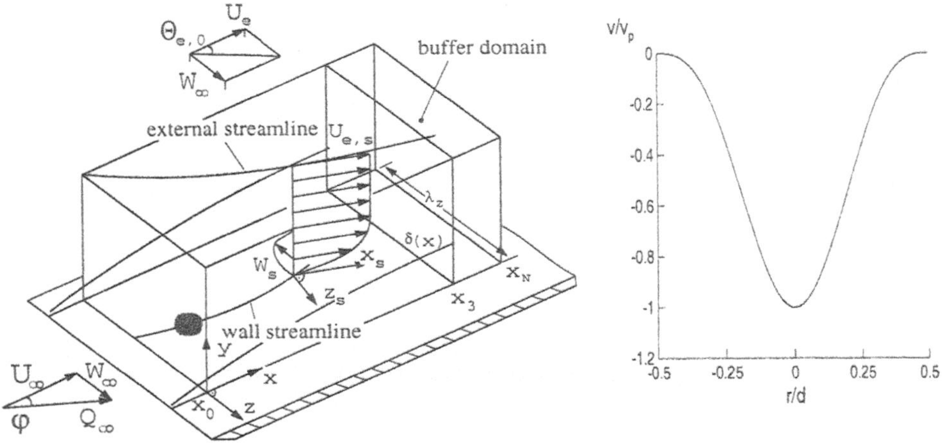

Figure 2 Integration domain and v'-distribution at the wall for a hole

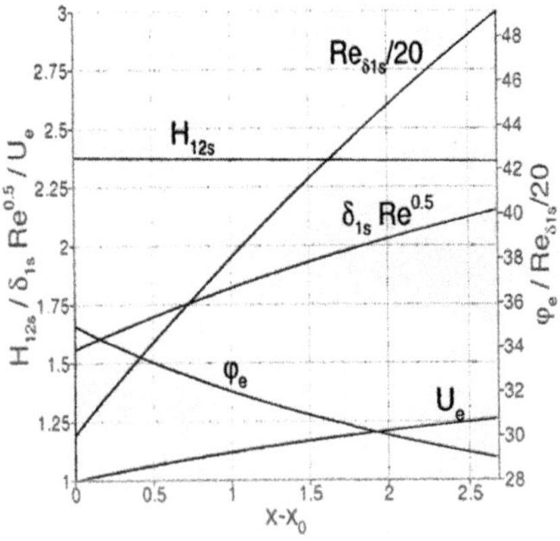

Figure 3 Base flow parameters
Suffix s: Values in a coordinate system aligned with the local external streamline direction (cf. Fig. 2)

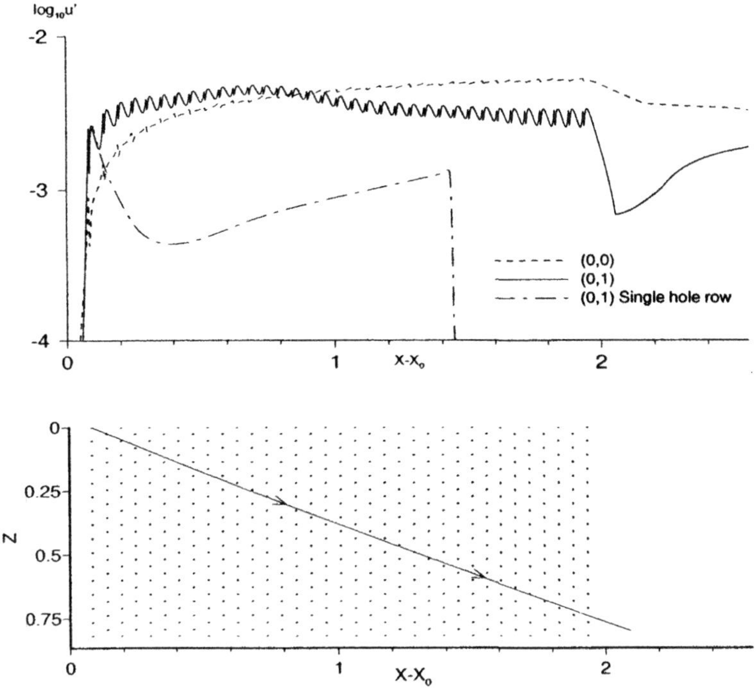

Figure 4 Top: u'-amplitudes (y-max) versus x for Case 1
Bottom: hole order (offset $s_z/2$), and a wall streamline

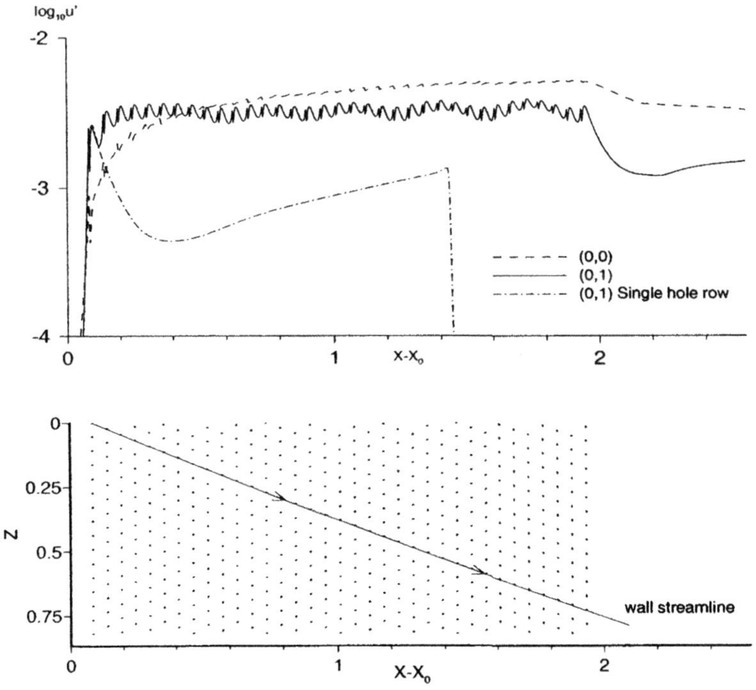

Figure 5 Top: u'-amplitudes (y-max) versus x for Case 2
Bottom: hole order (offset aligned along wall streamline)

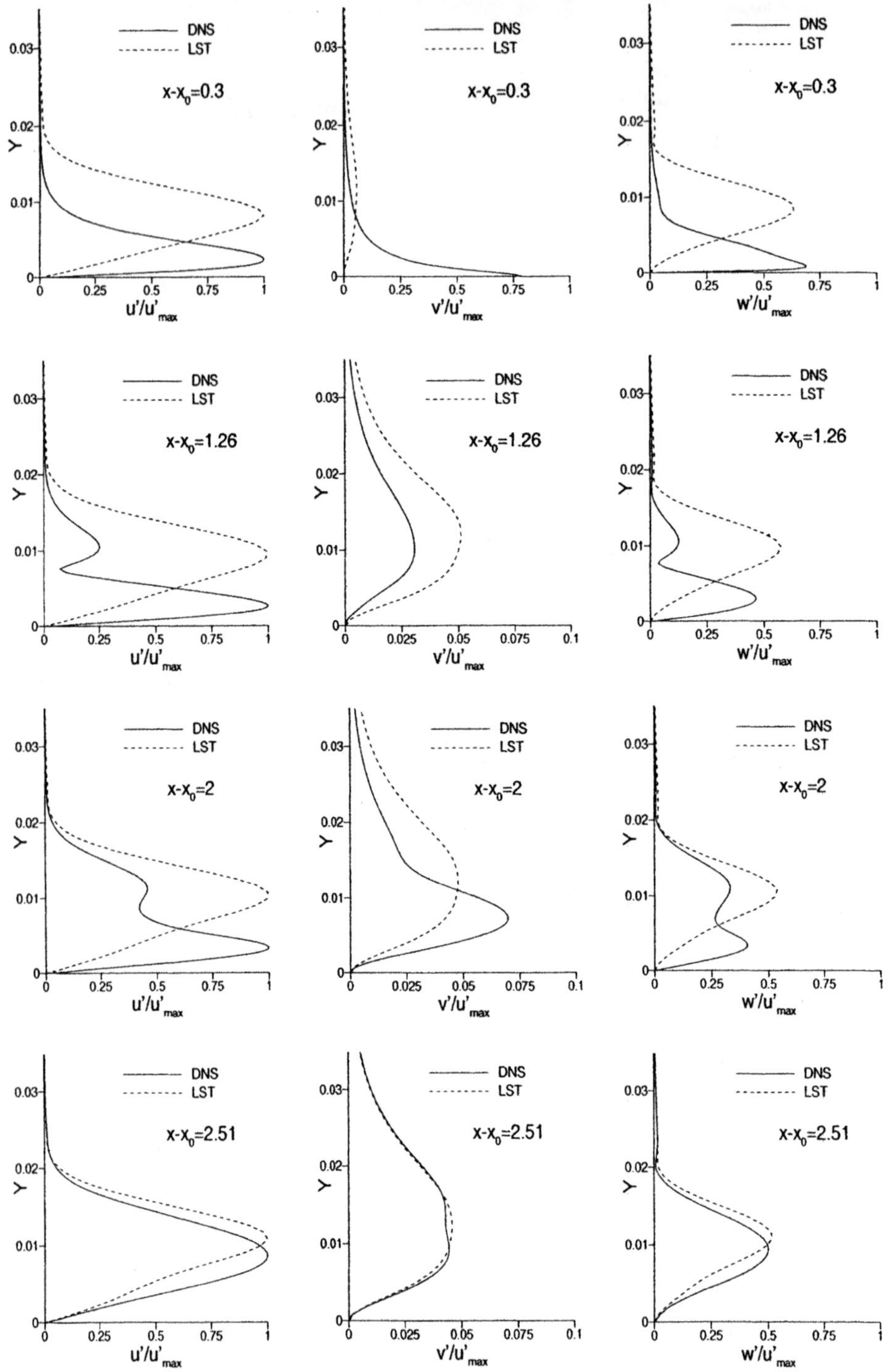

Figure 6 Disturbance amplitude profiles for Case 1 (cf. Fig. 4) and eigenfunctions of LST-analysis of base flow plus mode(0,0) for β=0 and γ=115
$x - x_0$=0.3: within 5th hole row
$x - x_0$=1.26: after 22 hole rows
$x - x_0$=2: shortly after suction array

Session 4
HLF Operational Aspects

Saab 2000 In-Service Test of Porous Surfaces for HLFC

Bryan E. Humphreys* and Ernst J. Totland†

*Aerospace Systems and Technologies Ltd.,
Consett, County Durham, England. DH8 6SR
E-mail: beh@aerosystech.com

†SAAB Aerospace
SE-58188 Linköping, Sweden
E-mail: ernst.totland@saab.se

1. SUMMARY

The paper describes a flight test, carried out within the European HYLTEC (**HY**brid **L**aminar flow **TEC**hnology) programme, with the objective of investigating and gaining experience on the contamination and durability aspects of porous surfaces for Hybrid Laminar Flow.

The test is aimed at gathering data and experience relating to materials for porous suction surfaces and the way that these will behave under in-service conditions. Two small span panels have been installed in the leading edge of a SAAB 2000 aircraft, one panel with a fully automatic contamination control system. No active suction system has been installed, but both panels utilise the pressure difference on the wing nose for suction and blowing.

The experiment started in July 1999 and is still on-going at the time of writing this paper, which describes the preparation, experience and results, available at the time of compilation.

2. BACKGROUND

Much effort has been put into the development of laminar flow technology, leading to improved understanding and prediction of the aerodynamic phenomena involved, but relatively little flight experience has been gained on in-service aspects of Laminar Flow.

Considering that a new aircraft has to be designed for a service life of 20 years or more, it is important to know the long-term effects of daily airline service and its impact on the laminar flow concept. The behaviour of porous suction panels exposed to ageing and contamination is of prime interest as both functionality and maintenance costs are affected. In addition, there is very little practical experience regarding the integration of the systems that will be required for normal service use.

The collaborative European program HYLTEC plays a significant role in adding to the knowledge and experience of materials and systems for HLFC. The test described here, using a SAAB 2000 aircraft operated by SAS, is aimed at gaining airline service experience relating to a range of materials for porous suction surfaces, both with and without insect contamination protection. The test is specifically aimed at gathering data and in-service experience relating to the effects of contaminants encountered in daily operation and the effectiveness of one type of anti-contamination system.

Although the SAAB 2000 class of aircraft is far from the long range jet transport aircraft being the prime target for HLFC, it was considered to be a useful platform for this test as climb and descent speeds do not differ significantly from the long range jet. A main benefit of using the SAAB 2000, is the high degree of utilisation, i.e. a large number of flights a day. A factor was also the fact that the aircraft is owned and maintained by Saab Aircraft.

The experiment commenced July 1999 and is incomplete at the time of writing this paper, which describes the preparation, experience and results, obtained at the time of compilation (originally written May 2000, data updated Oct 2000). Ultimately the test will give at least 18 months of experience in typical airline operating conditions.

This work was carried out within the HYLTEC programme and the writers would like to acknowledge the help and support of the European Commission, BAe Systems Power & Control and the University of Limerick. The paper was written as a collaborative venture between Aerospace Systems and Technologies (AS&T) and SAAB.

3. DESCRIPTION OF TEST

Two small span test regions, incorporating a number of porous surfaces have been installed in the leading edge of a SAAB 2000 aircraft, illustrated in Figure 1, that operates in airline service mainly in Northern Europe. No active suction system has been installed, but both test panels utilise the pressure differences around the wing nose for suction and blowing. One of the panels has been fitted with a fully automatic Liquid Contamination Control System (LCCS); the other region is "passive".

Figure 1: SAS Operated SAAB 2000 Aircraft showing Test Specimens and their locations

Surface contamination and changes in porosity are being monitored and durability aspects such as erosion, corrosion and damage are targeted. Current utilisation is approximately 250 flights a month covering Sweden, Finland, the Baltic States and Germany.

The test is addressing several key issues, such as choice of material for the perforated surfaces, effect of liquid protection system, weather effects, operational and certification issues. Spanning at least an 18-month period, including two summer seasons, the panels will be exposed to a variety of meteorological and entomological conditions as well as exposure to normal cleaning and de-icing procedures.

A monitoring procedure has been implemented involving regular inspections, porosity checks, collecting records of routes and basic weather conditions.

AS&T and SAAB are jointly responsible for the design, construction and installation of systems and test specimens as well as test monitoring. The University of Limerick is assisting with the analysis of data.

4. OPERATING FLIGHT PROFILE FOR SAS SAAB 2000

The SAAB 2000 is clearly a much smaller aircraft than the long range jet transports aircraft that will potentially derive the most benefit from HLFC. Although cruise speed and flight length differs significantly from these larger aircraft, the speeds and times for climb and, especially, descent differ less. However, most atmospheric contamination (rain, ice, snow, hail, insects, dust etc.) occurs below cruise levels so that the SAAB 2000 provides a useful test vehicle.

Based on data for the actual routes operated, the operating conditions for the test aircraft are as follows:

Route distance:		Flight time:	
Max:	461 nm	Max:	2:13
Min:	139 nm	Min:	0:27
Mean:	273 nm	Mean:	0:54 (54 minutes)

Climb speed:

The crew has three speeds to choose from (depending on flight condition):

CAS=190 kts

CAS=220 kts

CAS=240 kts   M<0.5 at alt<10 kft

Cruise speed:

The cruise speed depends on the flight condition (altitude, load, and distance).
Mach No is typically 0.62. At the high speed/low weight end of the range,
CAS varies from 350 to 370 kts reducing to 285 – 315 kts for long range.
Typical cruising altitude is 28,000 – 31,000 ft

Descent:

On descent, the CAS above 11,000 ft is typically 265 kts, reducing to 205 to 245 kts below 11,000 ft where the pilot selects the speed depending on flight and traffic conditions.

5. DESIGN AND ENGINEERING

Both wings of the SAAB 2000 aircraft designated for the test have non de-iced panels in the leading edges at mid-span (fig 1). These panels extend along the span approximately 300mm and terminate at the front spar. This location originally housed and provided access to a flux gate, however this is not used for this purpose on the aircraft available for the test.

The test panels consist of a specimen holder with embedded specimens. Each specimen holder is aluminium alloy, and replaces the original leading edge panel. It is designed to have at least the same strength (for bird impact) as the original part. Each specimen holder has recesses that accommodate specimens of laminar flow materials. The specimens can be removed and replaced without removing the specimen holder.

Laser perforation of materials for suction surfaces typically produces a tapered hole as shown in Figure 2. Most experimental work up to this date has orientated the material such that the small diameter end of the holes is on the outer surface, on the premise that particles drawn in by suction airflow will pass through the enlarging diameter, rather than wedge as might happen if the taper was in the opposite direction. However, there are certain aerodynamic and manufacturing reasons why it might be advantageous to orientate the holes in the reverse direction (i.e. with the large diameter on the outside surface). In view of this, this experiment seeks to determine if the direction of taper does indeed have any influence on susceptibility to blockage.

Figure 2: Orientation of Laser Perforated holes.

The starboard test panel (fig 3 - a) consists of a laser perforated Titanium skin. Approximately 70% of the skin surface is perforated with the hole taper in the normal direction for boundary layer suction, the remainder has a reversed hole taper. This specimen is protected from contamination during the take-off phase (0 to 1500 ft) and landing phase (1500 to 0 ft) using a Liquid Contamination Control System (LCCS). The system is activated at a throttle lever position above Ground Idle and at a radar altimeter height below 1500 ft to ensure that a liquid film is maintained above 40 knots.

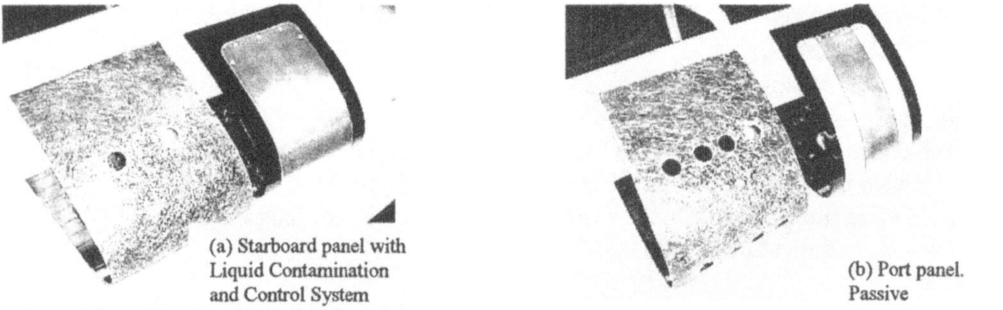

(a) Starboard panel with Liquid Contamination and Control System

(b) Port panel. Passive

Fig 3: Test panels and fibreglass flowmeter-positioning templates

The LCCS is based on liquid ice protection system principles and uses standard components as far as possible. The LCCS reservoir, pump, pressure switch and filter are installed in the non-pressurised forward section of the starboard nacelle.

The port test panel (fig 3 - b) consists of a structure supporting three test zones. Three perforated materials are being tested during the first phase of the flight test program: Un-clad chromic acid anodised aluminium; Titanium and Un-clad hard anodised aluminium. The port test panel is inactive (i.e. does not have any form of contamination protection).

6. MANUFACTURING AND CERTIFICATION ISSUES

Conducting a test of this nature on a revenue earning aircraft naturally imposes strict constraints. There are regulations to comply with and business interests of the operator to consider. Of vital importance was assurance that there would be no adverse effects on safety, neither aerodynamically or on aircraft systems. It was also important to consider airline and passenger reactions. Cost and disturbance to the operator had to be minimised. This necessitated a fully automatic fluid system that would not require the attention of the crew at any time. This was essential in order to avoid the prohibitive costs that would occur if crew activities were involved (due to training), and in order to avoid additional certification costs due to flight manual amendments etc.

Due to the intensive utilisation of the aircraft only weekly inspections are possible on the test panels and LCCS. As a result, sufficient fluid for at least one week's consumption had to be carried on-board; this requirement is comfortably met by the 7-litre reservoir provided. It also meant that visual inspections of the test specimens could not be performed as often as desired. The eventual format of the weekly inspection was restricted to visual inspection the test specimens, checking fluid system event-recorder and time registration as well as extracting a list of routes and flight time from the "digital" logbook of the aircraft. The weekly inspections were supplemented at approximately monthly intervals by a visit from a SAAB engineer to carry out additional inspections and make porosity checks.

The manufacturing issues mainly related to the test specimen on the starboard wing, i.e. the "active" panel. Because of the relatively simple nature of the port panel, it was a matter of simply exchanging the original panel with the "passive", multi-sectional test panel. In the "active" panel case, however, a complex fluid system had to be built into the aircraft wing and nacelle affecting structure, electrical system, etc. Several disciplines were involved: mechanical systems, structures, electrical/avionics, production, maintenance etc. Structural integrity had to be ensured and since the aircraft has fully electronic flight instruments, stringent electrical interference requirements had to be met. Only flight qualified components were used. In some instances, where the required certification was not available, special qualifying tests had to be performed.

Having agreed the design principles with SAAB, AS&T was responsible for the detail design and manufacture of all test components. As far as possible, the fluid system components were adapted from normal production items, but the specimen holders required special attention. In addition to providing a suitable mounting for the test surfaces each specimen holder had to provide a duct for airflow behind the perforated skin and, most importantly, had to provide at least the same resistance to bird strikes as the original panels.

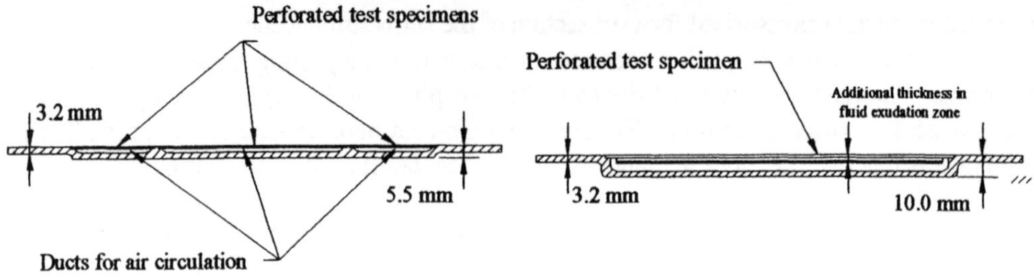

(a) "Inactive" Specimen Holder for Port Wing (b) "Active" Specimen Holder for Starboard Wing

Figure 4: Details of Specimen Holders

Typical cross sections are shown in figure 4. In the case of the "passive" panel, the holders were approximately twice the thickness of the original panels (i.e. 6mm instead of 3mm) and were made by CNC machining a flat plate which was then formed to shape. For the starboard wing the specimen holder had to be thicker in order to provide the required air passage behind the additional thickness of the liquid exudation zone.

7. SERVICE EXPERIENCE TO DATE

The SAAB 2000 flight experience experiment has now been flying since July 11 '99 and accumulated more than 3382 flights and 2958 flight hours (status per end Sept 2000). Utilisation is about 200 flights per month with some seasonal variation (fig 5).

The flights have covered a large part of northern Europe and exposed the aircraft to widely varying meteorological and entomological conditions. So far the porous panels have endured very well showing no visible sign of erosion, or damage due to ground handling. In fact, the aluminium panels are performing better than expected although the un-clad hard anodised aluminium panel is showing evidence of pitting due to corrosion, while the un-clad chromic acid anodised aluminium panel is not. This is believed to be caused by poor surface treatment or damage during installation, rather than environmental influence. Further analysis is required when the panel has been removed from the aircraft.

The fluid system has performed as planned and has been found to protect the leading edge well against insect debris and ice. So far the system has performed 7000 operations, amounting to an operating period of 270 hours. A period of intermittent LCCS operation occurred during spring '00 due to a failing electrical relay, but apart from this period, insect remains have never been found on the starboard panel which has always been thoroughly wet with decontamination fluid when inspected. Also, the "active" test panel has always been ice-free on inspection even when the adjoining rubber boots had collected ice at the leading edge. The port panel however, has naturally shown insect hits and ice, but usually the insect debris has been smeared out having little or no perceptible thickness, in contrast the adjoining rubber boots have shown numerous insect hits of a height that would be expected to cause boundary layer transition. It might be expected that the smeared out debris would cause blocked holes in the suction surfaces, but this has not shown up in the porosity measurements.

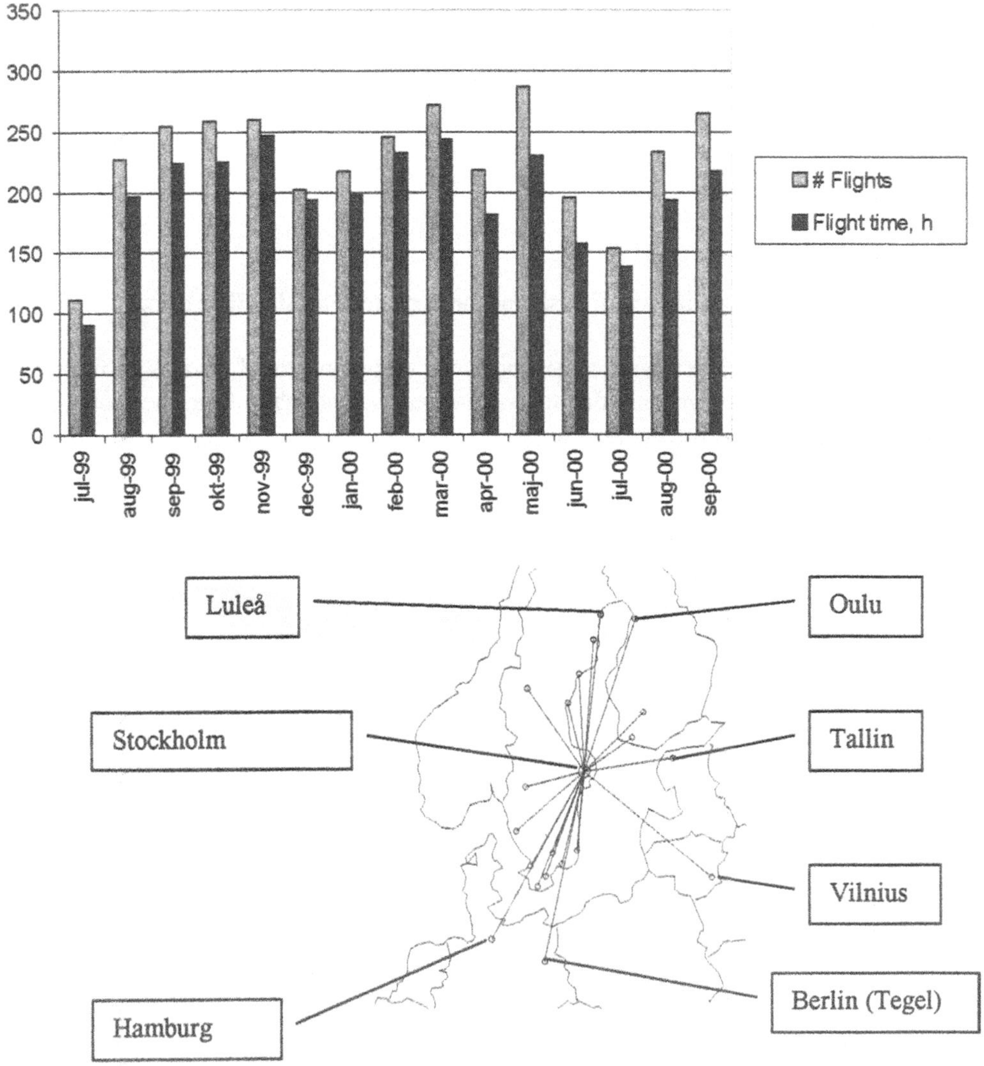

Figure 5: Flight Experience and Routes Flown

porous surface. Templates (see figures 3 and 6) are used for these measurements to ensure consistent positioning. The flow measurements are rather difficult to perform due to difficulties in obtaining a good seal between probe and panel surface, something that shows up as scatter in the results. Seasonal and day to day (dry/wet) conditions also appear to have an influence. However, some general trends are beginning to appear. The results of the first fourteen months of testing are shown in figures 7 and 8. These figures show measured data for each of the test positions.

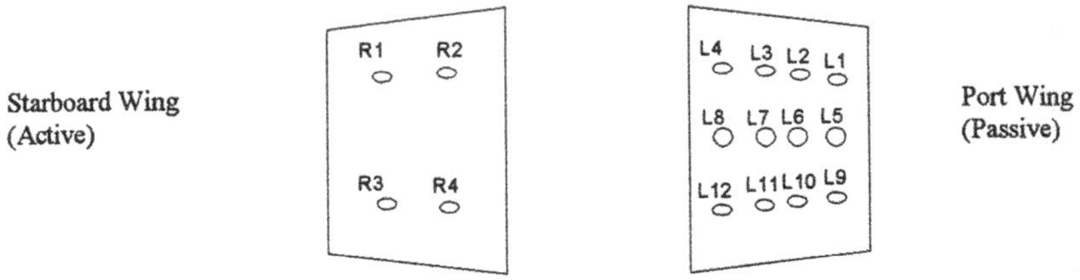

Figure 6: Identification of Positions for Repeated Porosity Flow Checks

In the case of the starboard panel all test positions show some trend towards an increase in airflow porosity, which is probably due to gradual removal of debris from the laser drilling process, from within the holes, by rain erosion.

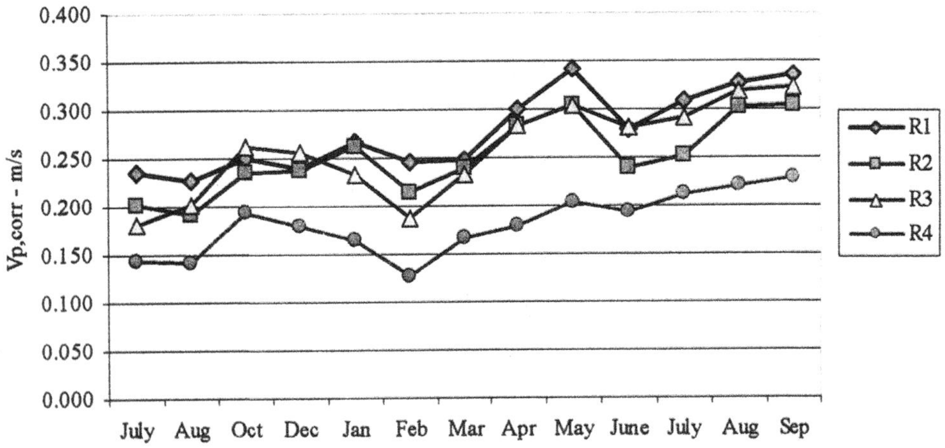

Figure 7: Suction Velocity at 5 kPa vs. Time, for "Active" Starboard Test Panel (LCCS)

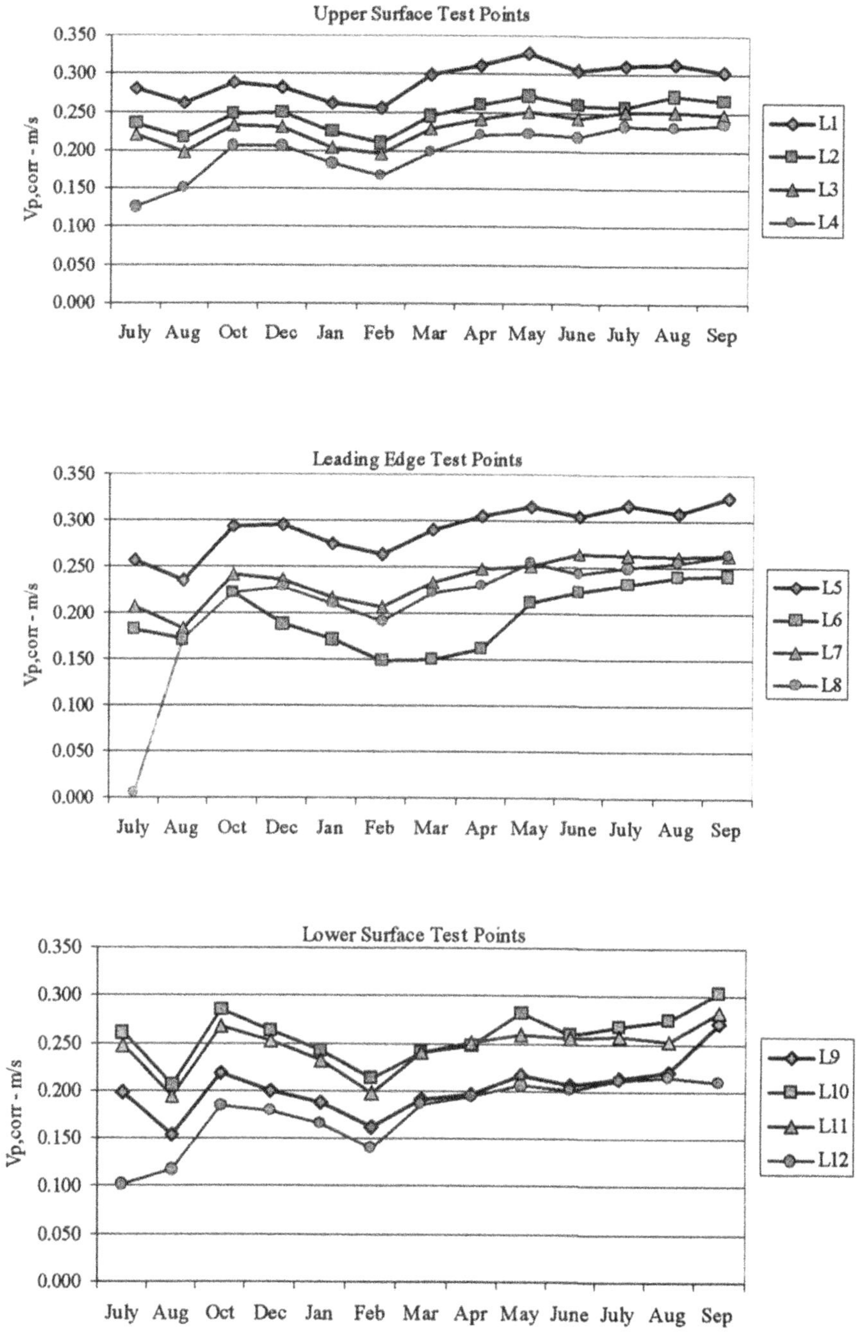

Figure 8: Suction Velocity at 5 kPa vs. Time Port Test Panel

It is evident from figure 8 that the Port panel shows more variation in trends. Most test points either show no change, or a slight increase in porosity, but test point L6, and to a lesser extent L 10, show a tendency for the porosity to decrease. Note that the region L8 was known to be blocked during installation so that the first data point is not valid. However, it is interesting to note that, within the first month, erosion had cleaned this region and that thereafter this test point followed the same variations as the adjacent test point (L7).

Finally, basic flight weather (METARS) is collected twice a day (weekdays) from the LFV (Swedish Air Authority) website. It planned to correlate general meteorological conditions with recorded data at the end of the test.

No panel maintenance has been necessary so far. However, a malfunction of the automatic fluid system caused a one-week interruption, but did not effect the normal operation of the aircraft. Normal maintenance consists only of the planned refill of fluid and data collection. Although anti-contamination fluid is always evident on the wing behind the "active" test panel after landing, this has not been a maintenance issue. It might be an environmental issue, though, in a full-scale application if present de-icing fluids are to be used.

8. CONCLUDING REMARKS

The test program will continue throughout the year ultimately giving some 18 months of experience in typical airline operating conditions. A more thorough investigation of the flight-tested panels will be performed after removal from the aircraft, hopefully giving more detailed information about test surfaces behaviour. So far the porous surfaces have performed very well showing little effect of ageing, except for the pit corrosion on the hard anodised aluminium panel. The flow measurements show only small effects on porosity, but scatter makes definite conclusions difficult at this stage. The overall trend is towards less blockage of holes.

FLIGHT OPERATIONAL ASSESSMENT OF HYBRID LAMINAR FLOW CONTROL (HLFC) AIRCRAFT

T.M. Young[a*], J.P. Fielding[b]

[a] Department of Mechanical and Aeronautical Engineering, University of Limerick, Ireland.
[b] College of Aeronautics, Cranfield University, Cranfield, Bedfordshire MK43OAL, UK.
[*] Corresponding author: Fax: +353-61-202944, Email: trevor.young@ul.ie

SUMMARY

Hybrid Laminar Flow Control (HLFC) offers the potential for significant fuel burn reductions. However the system is likely to have a lower operational reliability than other aircraft systems. The "failure modes" which will result in a loss of laminar flow, are mechanical (i.e. system failure) and environmental (i.e. rain, ice or insect contamination). The typical contingency fuel taken onboard to accommodate unplanned contingencies is 3 to 5% of the trip fuel. This is inadequate to cover a complete loss of laminar flow for an extended period during the cruise. The probability of an in-flight diversion decreases as the planned contingency fuel is increased; however this leads to a reduced fleet efficiency due to increased take off weights. A computer performance model of a twin engine aircraft in the class of the Boeing 757, has been used to study the change in block fuel for alternative fuel planning assumptions based on a loss in laminar flow due to cloud encounters in the cruise.

NOMENCLATURE

C_D	Drag Coefficient
FL	Flight Level
HLFC	Hybrid Laminar Flow Control
ICAO	International Civil Aviation Organisation
JAR	Joint Airworthiness Requirements
NASA	National Aeronautics and Space Administration
SFC	Specific Fuel Consumption
TIC	Time-in-Cloud

INTRODUCTION

Hybrid Laminar Flow Control (HLFC) is an active drag reduction technique that permits extended laminar flow on an aircraft surface at chord Reynolds numbers normally associated with turbulent flow. The delay in transition of the boundary layer is usually achieved by the application of suction over the first 10 to 20% of the chord. The resulting drag reduction on the aircraft has been estimated to be in the region of 10%. The flight operational problem explored in this study is that the contingency (en-route reserve) fuel, generally taken to be between 3% and 5% of the trip fuel [1], is inadequate to cover the increase in fuel burn due to a total loss of laminar flow. Clearly a decision to take a 10% contingency fuel to cover HLFC system failure mitigates some of the advantage of the inclusion of the system.

HLFC SYSTEM FAILURE

Five independent events have been identified which will result in a partial or a total loss of laminarity for part of, or for the entire mission (table 1).

(1) HLFC System (Mechanical) Failure

It is expected that system failure rates will be comparable with other aircraft systems. Initial comparisons have been made with air conditioning systems. Mechanical system failure would result in a partial or a complete loss of laminar flow for the remainder of the mission.

(2) Damage to the Perforated Surface

The most likely causes of in-flight damage are hailstone or bird impact. Current requirements are based on retaining structural integrity following an impact; however of concern to a laminar flow surface is a dent, which could cause transition. A single impact could result in a turbulent wedge aft of the impact site. Appropriate structural design of the laminar flow surface is required to reduce the likelihood of such events to an acceptably low level.

Table 1 *Events which will impact fuel usage on HLFC aircraft*

Description	Mission Phase	Consequence	Mitigation	Factors
System failure	Take off Climb Cruise Initial descent	Partial or complete loss for remainder of mission	- System design - Maintenance	- System reliability - Design issue
Damage to laminar flow surface (eg. hail impact)	Take off Climb Cruise Initial descent	Partial or complete loss for remainder of mission	- Route planning - Pilot avoidance - Surface design	- Weather - Design issue
Insect contamination	Take off Initial climb	Partial or complete loss for remainder of mission	Cleaning by: - On-board system - Rain / Ice	- Weather - Season - Location - Design issue
Airborne particles	Take off Climb Cruise Initial descent	Partial or complete loss for remainder of mission	Cleaning by: - On-board system - Rain / Ice	- Volcanic - Weather - Operational issue
Ice, rain and snow	Top of climb Cruise Initial Descent	Complete loss of laminar flow for finite time	- Route planning - Pilot avoidance	- Weather - Operational issue

(3) Insect Contamination

Data on insect aerial distributions indicate that insect populations have a definite height distribution [2, 3]. Whilst there is no definitive agreement on the exact height at which insect impacts are no longer likely to occur, the data indicates that encountering insects at heights above 500ft is remote, although it should not be discounted. Insect protection systems fall into two categories - mechanical protection (e.g. Krüger flaps, shields or covers) and fluid protection (normally glycol based solutions, which would also be responsible for de-icing / anti-icing).

(4) Airborne Particles

Suction surface hole blockage due to dust particles is unlikely to be a major problem because the largest dust particles are less than the typical hole diameter used for the suction surface [4]. Potential difficulties could be encountered in dust storms or after volcanic eruptions. It was suggested that the HLFC system could be inactivated or the flow reversed in these situations.

(5) Ice, Rain and Snow

Ice build-up and snow will result in a loss of laminar flow by blocking the suction holes. The adaptation of existing hot air or fluid de-icing / anti-icing systems for HLFC surfaces is required. Rain will also result in blockage of the suction holes and the water will have to be purged before suction can commence. Ice crystals in clouds between 25000 and 40000ft, are of a sufficient size and result in sufficient particle flux (when the aircraft flies through the cloud), to cause transition [5].

IMPACT ON FLIGHT OPERATIONS

The design requirement for a HLFC system in terms of the items described above should be such that these failures have a probability of occurrence of the order of 10^{-2} to 10^{-3} per flight hour (or less). Such failures could occur several times in the life of the aircraft and the consequences of these failures should be minor, having importance only to fleet service management. The reliability of the system must be sufficiently high and the probability of en-route-damage must be adequately low. *Operational procedures* to ensure this for the environmental factors of ice, rain and snow have not been fully investigated. The HLFC system is only likely to be used in the last part of the climb, during cruise and for the initial part of the descent. The frequency of rain at these heights is not that great and can thus be largely discounted. Ice crystals in cirrus clouds pose a unique problem for HLFC aircraft, as they are not an issue for current turbulent designs. This is further explored in this paper.

The most complete set of cloud encounter data available to date was obtained during the NASA Global Atmospheric Sampling Program, conducted from March 1975 to June 1979, using four in-service Boeing 747 aircraft. Distinct seasonal variations were reported, with the incidence of cloud encounters being greatest in the winter. Jasperson *et al.* [6] reported on the probability of cloud encounter on seven routes at two flight level (FL) bands. The results indicated that the probability of encountering clouds for more than 10% of a flight was less than 34%. The probability of flying in cloud for more than 50% of the cruise, was estimated to be between 0 and 1.4% for the routes studied. Also of interest was the route average *Time-in-Cloud (TIC)*. Jasperson *et al.* [6] presented this for the US East Coast to NW Europe route. They showed that the gamma probability density distribution provided an acceptable correlation with the measured data. For 50% of the flights, the average TIC was between 0 and 5%. The percentage falls off rapidly with increased average TIC, as shown in figure 1.

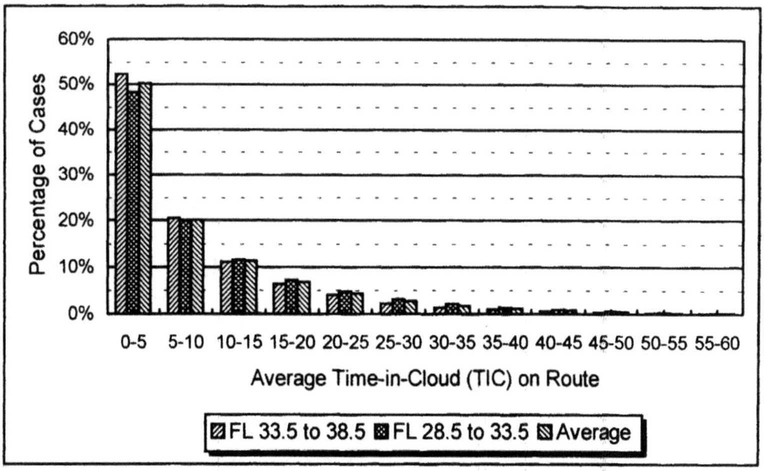

Fig. 1 *Average Time-in-Cloud (TIC) on route US East Coast to NW Europe based on gamma probability distribution (From data presented by Jasperson, et al. [6])*

FUEL PLANNING

The ICAO provides recommendations for the international flight operations in ICAO Annex 6-4.3.6.3. When an alternate aerodrome is specified on the flight plan, the requirements are to fly to the destination aerodrome, execute an approach and a missed approach, and thereafter:

(1) To fly to the alternate aerodrome; and then
(2) To fly for 30 minutes at holding speed at 1500 feet above the alternate aerodrome under standard temperature conditions, and approach and land; and
(3) To have an additional amount of fuel sufficient to provide for the increased consumption on the occurrence of any of the potential contingencies [7].

At the planning stage, not all factors which could have an influence on the fuel consumption to the destination aerodrome, could be foreseen. According to JAR-OPS 1 Subpart D [1] *contingency fuel* is carried to compensate for:

(1) Deviations of an individual aeroplane from the expected fuel consumption data;
(2) Deviations from forecast meteorological conditions;
(3) Deviations from planned routings and/or cruising levels/altitudes.

The *contingency fuel* may be either 5% of the planned trip fuel, or 3% of the planned trip fuel provided that an en-route alternate aerodrome is available. Furthermore, *additional fuel* (i.e. in addition to the contingency fuel) is required under certain circumstances. This covers the possible failure of a power unit or loss of pressurisation (and is based on the assumption that such a failure occurs at the most critical point along the route) and permits the aeroplane to divert to an adequate aerodrome. Such additional fuel is only required if the on-board, fuel calculated as described above, is not sufficient for such an event [1].

The aircraft's flight profile may be divided into several parts for the purpose of flight planning (figure 2). The fuel, time and distance covered for each segment must be determined.

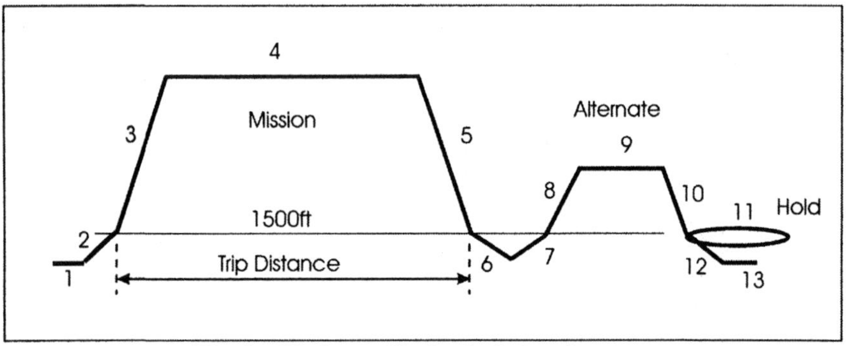

Fig. 2 *Flight profile for fuel planning*

Index

1	Engine start-up, taxy and line-up	**8**	Climb to alternate cruise altitude
2	Take-off and climb to 1500ft	**9**	Alternate cruise
3	Climb to cruise altitude	**10**	Descent to 1500ft
4	Cruise	**11**	Hold for 30 minutes
5	Descent to 1500ft	**12**	Approach and land
6	Approach	**13**	Taxy and shutdown
7	Overshoot, climb to 1500ft		

FUEL CALCULATION METHODOLOGY

To study the impact of HLFC system on the block fuel, a computer performance model of a twin-engine aircraft in the class of the Boeing 757 was developed. Data from a Performance

Engineers Manual [8] was used. The pertinent data included the drag polars, corrected fuel flow and thrust in climb, cruise and idle conditions. These data were used to compute the fuel and time required for specified mission distances. The calculations were performed according to the International Reserves flight profile described earlier. Numerical techniques such as the Integrated Range and Integrated Hold methods have been used, rather than analytical expressions such as the Breguet formula.

After verification of the basic performance model, it was modified to take into account the HLFC system. Three input fields in the program were created to allow the user to account for: (1) drag reduction; (2) system weight increase; (3) SFC increase. As the cruise had been divided into ten segments, it was possible to apply a drag reduction and SFC increase to individual portions of the cruise.

(1) Drag Reduction

The selection of this particular class of aircraft allowed direct comparisons to be made with the study performed by Boeing on the impact of HLFC on the 757-200 aircraft [9]. The Boeing study focused on HLFC for the upper and lower wing, however data was provided on the change in profile drag due to HLFC on the empennage. By removing the estimated influence of the lower wing surface and adding the impact on the nacelles, an estimate of the drag reduction for the desired configuration was obtained (table 2). In the Boeing report [9] the total drag for the baseline 757-200 at 37000ft at Mach 0.80 for a lift coefficient of 0.50, was given as 0.03079. This compared well to the performance model developed by the authors. A profile drag coefficient reduction due to HLFC of 0.00425 was used for the study.

Table 2 *Drag coefficient reduction due to HLFC*

	Baseline 757-200*	Boeing study*		Current study	
		HLFC System	HLFC 757	HLFC System	HLFC Aircraft
	C_D	ΔC_D	C_D	ΔC_D	C_D
Wing	0.00694	0.00335	0.00359	0.00225	0.00469
Fuselage	0.00653	0.00000	0.00653	0.00000	0.00653
Vertical tail	0.00128	0.00050	0.00078	0.00050	0.00078
Horizontal tail	0.00181	0.00080	0.00101	0.00080	0.00101
Nacelles	0.00164	0.00000	0.00164	0.00070	0.00094
Struts & flap tracks	0.00055	0.00000	0.00055	0.00000	0.00055
Total profile drag	**0.01875**	**0.00465**	**0.01410**	**0.00425**	**0.01450**

**Data from [9]*

(2) Weight of HLFC System

The weight of the HLFC system was based on the work of Wilson [10], in which he investigated the impact of HLFC on five different classes of aircraft. A mean ratio of 0.89% of HLFC system weight to MTOW was determined. For the aircraft under consideration, with a MTOW of 115900kg, the HLFC system weight was estimated to be 1032kg.

(3) Specific Fuel Consumption Penalty

The Specific Fuel Consumption (SFC) increase due to the operation of the suction pumps was based on an estimate of the required suction. The total suction power off-take from the engines was estimated to be 266kW for the installation of HLFC on the upper wing surface

(124kW), the fin (53kW), horizontal tailplane (44kW) and nacelles (45kW). These values were based on estimates taken from Wilson's study [10] and from the Boeing study [9]. By comparison the Boeing estimate of power extraction was 206kW for the upper and lower wing surfaces [9]. The SFC penalty was based on a conversion of 0.70% SFC increase per 100kW power off-take. This gave a SFC penalty for the aircraft of 1.86%.

Ground Rules for the Studies

(1) The mission range was 2900nm. The route was assumed to have a suitable en-route alternative aerodrome. Failure of the HLFC was considered to have economic rather than safety implications. The payload was 25670kg.

(2) The International Reserves policy for fuel planning was used. The alternate leg was 200nm and the hold time was 30 minutes. In all cases the planned contingency fuel was 3% of the planned trip fuel. The trip fuel allowed for a loss of laminar flow ranging from 0% (most optimistic) to 100% (most pessimistic).

(3) For the HLFC aircraft the Operating Empty Weight was increased by 1032kg (over the turbulent baseline aircraft), the SFC was increased by 1.86% and the drag coefficient was reduced by 0.00425. The HLFC system was assumed to operate in the cruise only.

(4) The following conservative assumption was made regarding the cloud encounter model. The upper value of the TIC band (see figure 1) was used for each case (e.g. 5% TIC was taken to correspond to 50% of the cases, rather than a value mid-way in the band).

(5) It was assumed that laminar flow would be lost whenever the aircraft was in cloud (and immediately regained upon leaving the cloud).

Study 1

The objective of the study was to investigate the impact on the block fuel, of the loss of laminar flow due to cloud encounters in the cruise. The fuel planning was performed with an assumed loss of laminar flow in the cruise (25% for example). If the en-route conditions are exactly the same as those taken into account in the planning of the flight and the aircraft is forced to divert to an alternate aerodrome (again as per the planned mission), then the fuel on landing at the alternate aerodrome will exactly equal the calculated contingency fuel. The fuel planning model described above was used to calculate the Brake Release Weight and block fuel for an assumed 25% TIC during the cruise. The program was then re-run several times with the same Brake Release Weight (and total fuel), but with different average TIC. The *block fuel* and the fuel on *landing at the alternate* aerodrome were calculated (table 4). If the TIC is less than that assumed for the fuel planning, then the aircraft will burn less fuel and the fuel on landing at the alternate aerodrome will be greater than the contingency fuel. Figure 3 is a plot of fuel on *landing at the alternate* aerodrome and the planned contingency fuel, versus the percent loss of laminar flow. (It is evident from the figure that the two fuel quantities are equal for the 25% point.) For the cases where the TIC exceeded 25% of the cruise, it was required to use part of the contingency fuel to get to the alternate aerodrome; however the contingency fuel was adequate to allow for a loss of laminar flow for 55% of the cruise.

Study 2

The second study investigated the fuel savings for a fleet of HLFC aircraft over the equivalent turbulent aircraft (assuming that the cloud encounter model conformed with that described earlier). Table 3 contains block fuel results for the analysis which assumed 25% TIC for the fuel planning. Results for the equivalent turbulent baseline are also shown. The total fuel consumed by the fleet was determined by summing the product of the block fuel for a particular TIC by the corresponding number of flights that the fleet will encounter that TIC (as per the model given in figure 1). The resultant fuel usage for a fleet of HLFC aircraft was 8.6 %

less than that of the equivalent turbulent baseline. The results for various fuel planning "policies" based on the assumed loss of laminar flow in cruise are shown in figure 4.

Table 3 *Output summary for 25% planned loss of laminar flow during cruise (mass in kg)*

Output Summary	Turbulent Baseline	Average TIC (Time-in-Cloud) on route					
		5%	15%	25%	35%	45%	55%
Brake Release Weight	115650	114650	114650	114650	114650	114650	114650
Operating Empty Weight	58380	59410	59410	59410	59410	59410	59410
Total Fuel (at the ramp)	32010	29970	29970	29970	29970	29970	29970
Block Fuel	27180	24740	24950	25160	25380	25600	25830
Planned Contingency Fuel	798	737	737	737	737	737	737
Fuel on Landing at Alternate	1758	1144	940	737	527	310	87

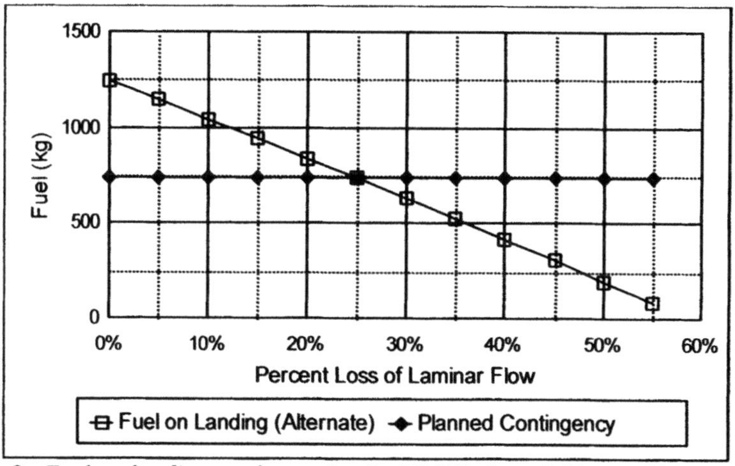

Fig. 3 *Fuel on landing at alternative for HLFC aircraft (Range 2900nm)*

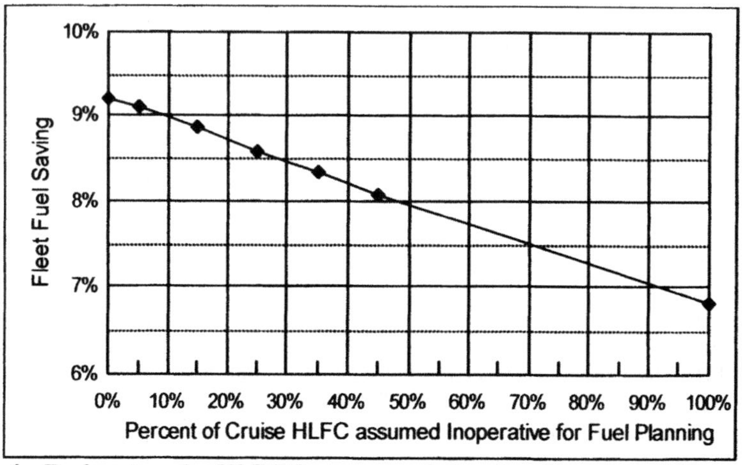

Fig. 4 *Fuel savings for HLFC fleet over turbulent baseline aircraft (Range 2900nm*

DISCUSSION

The consequence of one of the failure events outlined earlier is that the aircraft will burn more fuel than planned. The contingency fuel (en-route reserve) can accommodate a minor loss of laminarity over a small part of the wing or a total loss of laminarity for a limited time period. If the HLFC system is assumed to function for 100% of the cruise and credit is taken for this in the fuel planning, then an extensive loss of laminarity for a significant time during cruise will "deplete" the contingency fuel and the pilot will be forced to divert. The consequence of a diversion is extremely unpopular with any airline. A refuelling stop is expensive and results in customer dissatisfaction. Where to set the limit of the en-route reserve so as to reduce the probability of diversion to an acceptable level is the critical question. If it is set too high then too much fuel is carried and if it is too low, the probability of a diversion increases.

The savings in block fuel for the HLFC fleet over the turbulent baseline aircraft (for the assumed cloud encounter model described earlier) is shown in figure 4. The lowest value of 6.8% corresponds to a fuel calculation that assumed no credit for the HLFC system in the fuel planning. As the planned fuel is reduced (by taking credit for the HLFC in the fuel planning) the fuel saving for the fleet increases. A point is reached where the contingency fuel is inadequate to cater for the extreme case of 55% TIC on the route. For the cloud encounter model used in this study, this occurred when the fuel planning assumed a loss of laminar flow of less than 25% of the cruise.

CONCLUSIONS

- Based on the adopted models for aircraft performance and probable cloud encounter, the study showed that if the fuel planning assumed 25% TIC during the cruise, then:
  (1) Less than 7% of flights would encounter more cloud than this.
  (2) In the extreme case of 55% TIC during the cruise, the contingency fuel (taken as 3% of the trip fuel) would be sufficient for the aircraft to complete the mission (including the alternate leg and hold).
- The fuel savings obtainable from HLFC must be considered for a *fleet* over an extended period of time, rather than for a single flight, due to the variability of cloud encounter.
- For a fuel planning "policy" that assumed 25% TIC (based on the cloud encounter model used), a fuel saving for the fleet of 8.6% was calculated over the turbulent baseline.

REFERENCES

[1] JAR-OPS 1, *Joint Airworthiness Requirements OPS 1,* European Joint Aviation Authorities.

[2] Coleman, W.S., "Roughness due to Insects", *Boundary Layer and Flow Control,* Vol. II, Editor: Lachmann, G.V., Pergammon Press, 1961.

[3] Croom, C.C. and Holmes, B.J., "Insect Contamination Protection For Laminar Flow Surfaces", Langley Symposium on Aerodynamics, Vol. 1, Dec. 1986.

[4] Humphreys, B., 1992, "Contamination Avoidance For Laminar Flow Surfaces", First European Forum on Laminar Flow Technology, Hamburg, March 1992.

[5] Hall, G.R., "On the Mechanics of Transition Produced by Particles Passing Through an Initially Laminar Boundary Layer and Estimated Effect on the LFC Performance of the X-21 Aircraft", Northrop Corp., 1964.

[6] Jasperson, W.H., Nastrom, G.D., Davis, R.E., and Holderman, J.D., "GASP Cloud Encounter Statistics: Implications for Laminar Flow Control Flight", Journal of Aircraft, November 1984.

[7] Boeing, "Performance Engineer General Course Notes - Volumes No. 1 and 2", 1996.

[8] Boeing, "Performance Engineers Manual 7G7", undated.

[9] Boeing, "Hybrid Laminar Flow Control Study: Final Technical Report", NASA CR-165930, 1982.

[10] Wilson, R.A.L., "The Introduction of Laminar Flow to the Design and Optimisation of Transport Aircraft", PhD thesis, Cranfield University, 1997.

Application of HLF Technology to Civil Nacelle

Pascal Meyer, Ing Dipl, CEng MRAeS
Installation Technology, Rolls-Royce plc, PO Box 31, Derby DE24 8BJ, United Kingdom

Summary

The paper summarises the work carried out in the Nacelle task of the HYLDA (HYbrid Laminar flow Demonstration on Aircraft) project 1996-1999, funded by the European Community. A nacelle *demonstrator* has been defined for the CFM56-5C engine installed on the A340. The Hybrid Laminar Flow nacelle features a translating forebody replacing both the intake and the fan cowl doors and also an upgraded thrust reverser. The major conclusion of the multidisciplinary work on this specific installation is that the technology is available to proceed to a flight demonstrator programme. Nevertheless constraints arising mainly from the objective of having laminar flow over half of the nacelle surface, led to an unattractive product in terms of net performance benefits. The recently launched ALTTA project is based on a trade study for optimum benefit and focuses on the top risks, mostly non-aerodynamic, analysed in HYLDA.

Introduction, HYLDA project.

Hybrid Laminar Flow (HLF), a technology based on suction of the boundary layer to extend the laminar flow region and hence reduce fuel burn through a reduction of drag, is currently receiving considerable attention. In 1993, [1] &[2], a HLF nacelle, designed for transition at 50% chord, was successfully tested at Mach 0.6 and altitudes up to 8000m in an over wing installation on the ATTAS VFW614 aircraft. In 1994, [3] GE, HLF technology was applied to a CF6-50C2 nacelle on an A300/B2 aircraft and a 1% fuel burn saving was validated. [4] ELFIN II and LARA nacelle studies, completed in Europe, have illustrated the feasibility of high speed HLF designs at both cruise and off design conditions and culminated in a large scale, high speed wind tunnel test of an isolated, aerodynamically optimised, nacelle design in 1996.

The herein-developed HYLDA (HYbrid Laminar flow Demonstration on Aircraft) project 1996-1999, being carried out in the European 'Industrial and Materials Technologies' research and technological development programme, had the aim to define a technical specification for a HLF nacelle on a transport aircraft. This project brought together some fifteen European partners, including all those who participated in the ELFIN II and LARA projects.

The vehicle chosen for this project is the nacelle of the CFM56-5C engine that is installed on the Airbus A340. Throughout the project, two HLF nacelle designs were considered in order to satisfy the nacelle manufacturers' requirements of assessing commercial viability. Both designs

were based on the translating forebody concept: a *demonstrator nacelle* which replaces the existing intake and fan cowl doors, puts an over-structure on the TRU and retains the production nozzle, and a notional *production nacelle* where these constraints are removed. This underlying notional production design serves two purposes: it allows a comparison to be made between the compromised demonstrator design and a production solution after flight test, and allows an assessment of the cost, benefits and risks of a production installation.
The HLF nacelles have been defined in terms of aerodynamics, structure and systems.
A Risk Assessment, comprising a risk analysis and a cost and benefit analysis, has been carried out.

Aerodynamics

HLF technology applied on nacelle enables extended laminarity and reduced drag, and provides freedom to design nose region to meet low speed off design requirements.

Within HYLDA the aero lines have been defined using [5] HTC described technique. The leading edge radius of all nacelles yields a velocity peak with a subsequent decelerated region. Instead of a convex pressure recovery after velocity peak, as with the conventional turbulent nacelle, a concave pressure recovery is used for the decelerated region. The concave pressure recovery has two advantages: to prevent shock formation and to prevent laminar separation even without surface suction. Suction is, however, applied to this decelerated region to maintain the laminar flow. A favourable pressure gradient is then used to maintain continuous laminar flow afterward. A concave pressure recovery then follows, aft of the second pressure minimum, to prevent turbulent flow separation. As a result, the local radius to maximum diameter increases, and the new maximum diameter moves downstream.

The effect of applying this technique to the CFM56-5C lines is shown on Figure 1 below.

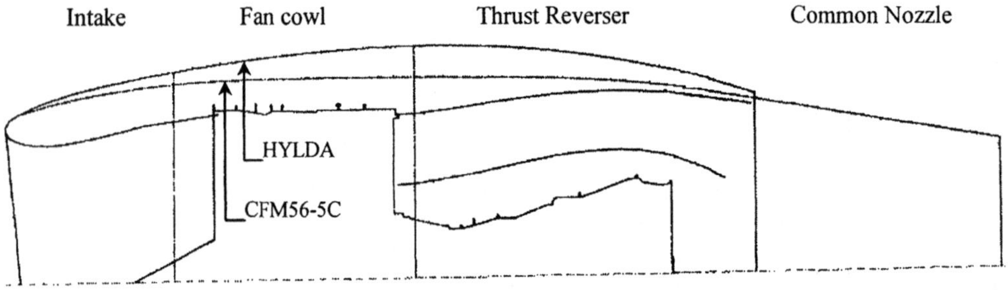

Figure 1: Typical section showing the HYLDA loft lines over the original CFM56-5C lines

Computational Fluid Dynamic (CFD) flow calculations and boundary layer calculations have been performed. As shown in Figure 2, the objective of 50% laminar flow extent has almost been reached resulting in a 35% nacelle friction drag reduction.
An installed nacelle high-speed wind tunnel campaign, Figure 3, has been conducted and highlighted a risk of turbulent flow separation in the rear part of the nacelle.

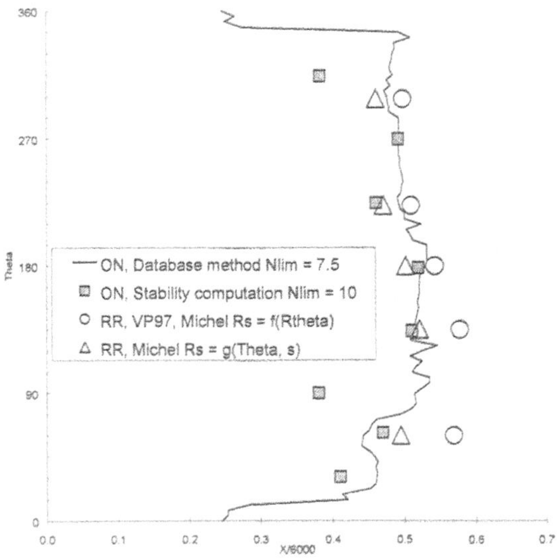

Figure 2: Transition location %chord at θ position
Mach 0.82; 2.5deg incidence; Spillage 75% cruise
Suction 5% to 20% of the chord at V=0.08m/s

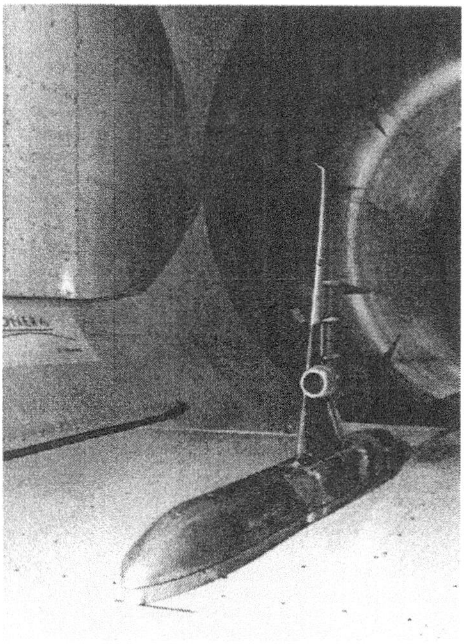

Figure 3:Model set-up of the HYLDA Nacelle in the ONERA S1MA high-speed wind-tunnel

Structure

The required features of a nacelle, such as joints between adjacent components, steps and/or gaps at these joints and shape deviation due to manufacturing tolerance, compromise the benefits of laminar flow drag reduction. The specification of the HYLDA nacelle mechanical design has been tailored to achieve the high level of aerodynamic performance. Figure 4 shows the selected novel translating forebody concept, integrating the intake and fan cowl doors functions. The design details the actuation and the HLF system integrated in the CFM56 nacelle. The choice of the nacelle concept was made at the beginning of the project. Intensive work has been carried out on this nacelle concept and no alternative has been studied.
It was concluded that the door type thrust reverser was the most appropriate thrust reverser for HLF. This is derived from improved aerodynamic line smoothness, good external skin stiffness and practical structure for the installation of the suction pump exhaust pipe routing through the thrust reverser.

In Figure 1, the four divisions of the nacelle (i.e. Intake, Fan cowl, Thrust Reverser and Common Nozzle Assembly) are shown. As seen, the thrust reverser region is severely affected by the change of the aero lines; its volume has nearly doubled. As a result, the weight of the nacelle structure has increased considerably.

The characterisation of candidate lightweight composite materials for the perforated suction surface has been achieved through laboratory studies including mechanical testing.

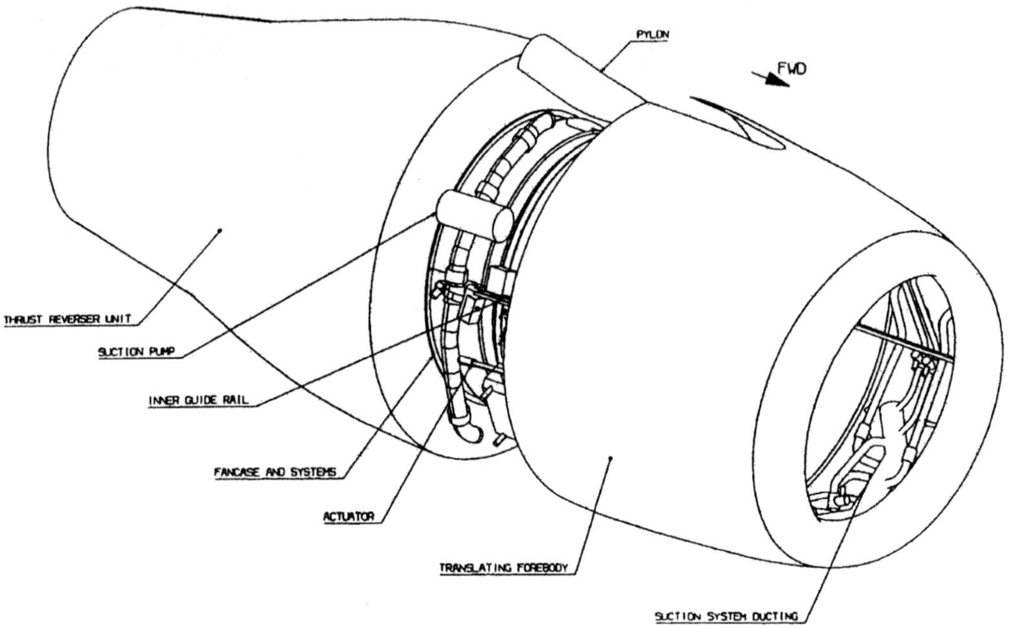

Figure 4: Isometric view showing the translating forebody nacelle

Systems

[6] Fokker 100 and [7] Falcon 900 tests highlighted the need for an anti-contamination system on HLF profiles. [8] U. Bristol studies have shown that the use of over-suction to suppress disturbances would require non-available sophisticated systems. A fluidic anti-contamination system has been tested, see [9], on a Dornier Do 228 NLF wing glove reducing the number of insects adhering to the surface to 10% of the number obtained in the reference region where no anti-contamination system was installed. Difficulties remained to achieve a complete wetting of the surface and a fast and complete relaminarization after switching the system from the active to the passive phase.

Within HYLDA, a nacelle system design integrating boundary layer suction, anti-contamination and anti-icing systems has been delivered. Requirements for these different systems were assessed by individual partners and discussed with all members.

It was assumed that the flight test *demonstrator* would feature a perforated titanium suction surface and a sixteen compartment separately measured and controlled suction system, whereas the *production nacelle* would use a perforated composite (likely to be a thermoplastic to obtain adequate rain erosion resistance) and a simplified suction system.

A three-stage axial compressor prototype has been performance tested. A suction optimisation algorithm, to minimise pump energy, has been demonstrated in a wind tunnel.

Risk Assessment

Towards the end of the project a ***Risk Analysis*** has been conducted.
First, hazards have been identified. Then, the level of uncertainty has been assessed. Finally for each hazard the HYLDA partners have examined and agreed ways in which either the likelihood or the consequences of that hazard could be reduced to an acceptable level. This process involved all the partners and experience from previous projects has been reviewed and taken into account (e.g. ELFIN, LARA, RR/DLR VFW614…).

As discussed previously a *HLF demonstrator nacelle* and a *HLF production nacelle* were defined. 46 Top risks (from 130 risks total) were identified for the *HLF production nacelle*. These top risks are largely affecting manufacturing capability and certification, and fall into four groups as shown on the graph below. The number of Top risks is shown together with the proportion of risks that requires the [10] ALTTA new project for mitigation. The acronym "HYLDA+", used in the graph, states for a hypothetical project that will have to deal with the remaining unassessed risks and with in-flight demonstration.

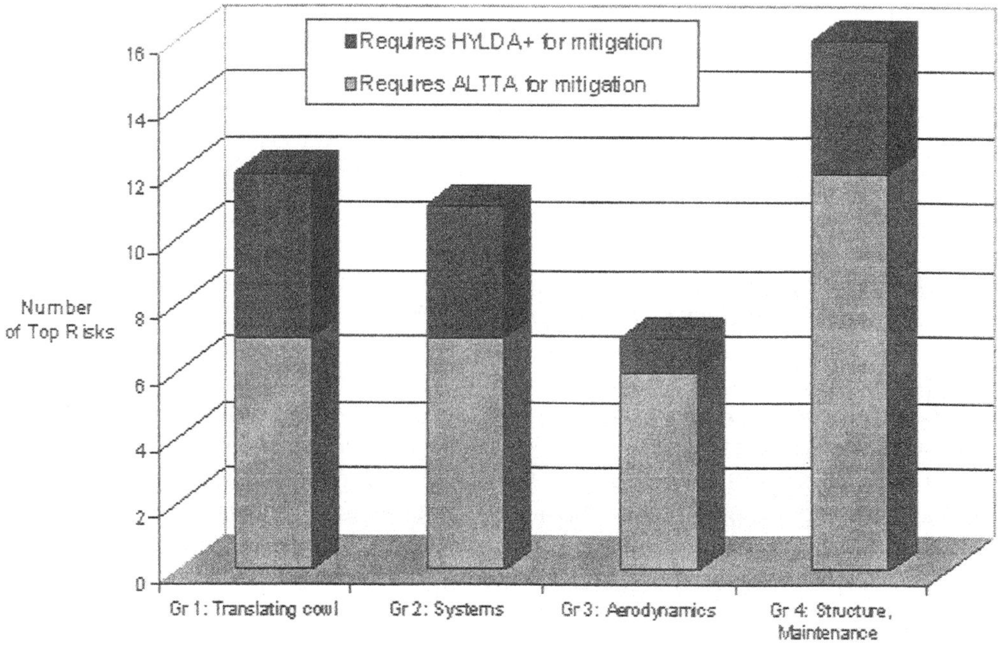

For the HYLDA *demonstrator nacelle* a total of 84 risks was identified. These risks fell into all groups of the *production nacelle* risks. Availability of the flight test aircraft and permission to fly issues are additional risks specific to the *demonstrator nacelle*.

Certification and critical operational issues have been identified and the associated risks have been monitored in the risk assessment study. The current regulation does not particularly consider systems such as suction system, anti-contamination system and translating forebody. Current requirements are applicable but not sufficient to certify an HLF nacelle.

Costs and Benefits of the *HLF production nacelle* have been assessed and the current stack up shows a net benefit of 0.56% sfc (specific fuel consumption). However, this does not assume any reduction in fuel carried (expected to be the case for first application). Once reliability is demonstrated it would be expected that fuel loading could be reduced on subsequent applications and this would increase the benefits.

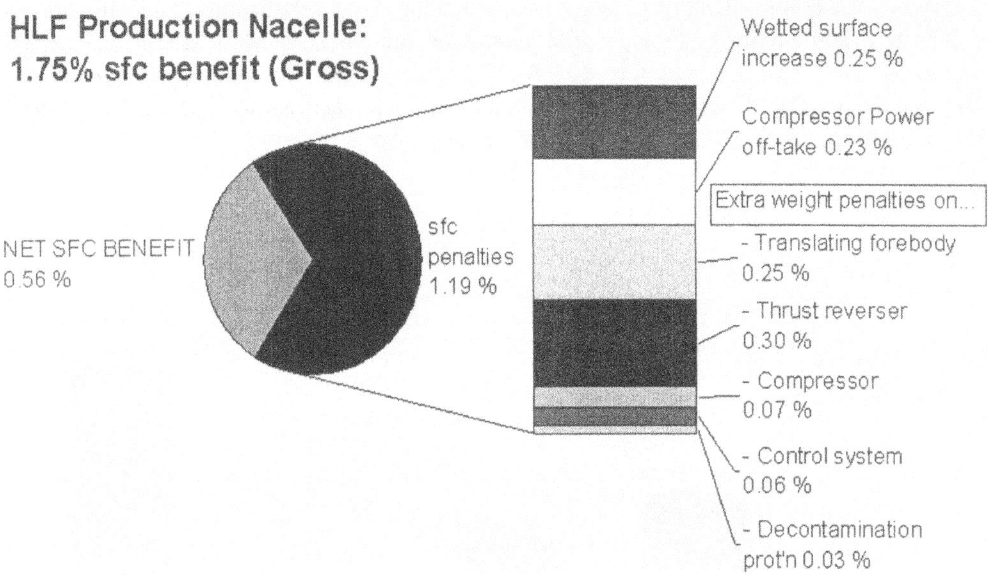

The *HLF production nacelle* is significantly heavier than the current A340/CFM56 nacelle. The extra weight, alone, halves the total performance benefit.

No cost/benefit analysis has been attempted for the *demonstrator nacelle* due to the detrimental effect on fuel burn from the extra weight of the titanium perforated surface, the thrust reverser over structure, the oversized pump and valves and the extra instrumentation.
[11] and [12] are studies on the impact, in terms of DOC (Direct Operating Cost), of HLF technology on a large subsonic aircraft. It has been found that greatest savings are achieved when the aircraft configuration is fully re-optimised. Taking into account airline considerations, significant DOC benefits are required for HLF technology to be acceptable.

Technical feasibility

The objective of the HYLDA Nacelle task was to achieve a state of technology readiness for partners to commit to a flight test demonstrator programme. From the Technical Feasibility final meeting it appears that the technology is available to proceed to a flight project utilising the *demonstrator nacelle* concept. This would demonstrate the suction system and ice & contamination protection system integration and the aerodynamics of a HLF nacelle on a high-speed modern aircraft. During this meeting it has been stated that the technology to test an "Industrial" flight demonstrator using composite perforated panels and production manufacturing methods is not sufficiently mature at this time, and that an interim project is needed to address the outstanding risks.

Conclusion, ALTTA Project.

The principal conclusion of the HYLDA project is that the technology is available for flight test of a HLF nacelle demonstrator on an Airbus A340. Multidisciplinary work on the CFM56-5C nacelle installation has been carried out. The shape of the nacelle has been defined in order to maximise the extent of laminar flow. The developed aerodynamic lines give transition at about 50% of the chord resulting in a 35% nacelle friction drag reduction. The overall surface quality of the nacelle being a key factor for laminar flow success, the new translating forebody nacelle concept has been studied. Laboratory experiments have been conducted on lightweight composite materials. A generic database has been produced for system design covering suction, insect contamination control and ice protection.
In order to assess viability of such a HLF nacelle, both a cost/benefit analysis and a risk analysis were conducted. A low, 0.56% sfc, net performance gain has been recorded. This is mainly due to the increase in weight brought by the change of the aerodynamic lines. Risks associated with the HLF demonstration on nacelles have been assessed and mitigation has been proposed. A majority of these risks concerns the structure, material and maintainability of the nacelle and the HLF system.

Within the [10] ALTTA (Application of Laminar flow Technology on Transport Aircraft) project 2000-2002, a group of HYLDA partners, reinforced by a nacelle manufacturer, is conducting a trade study to maximise sfc/DOC benefits and is focusing on the HYLDA identified risks that do not require a flight test sequence. It is expected that balancing the reduction of the extent of laminar flow with the use of a simplified lighter structure, less constrained by the aerodynamics requirement, will lead to a better solution in terms of sfc/DOC.

In the World of Aeronautics there is continual change. Airbus has recently increased its involvement in new nacelle definition and manufacture. This will probably enhance HLF nacelle work from the Airbus Partner's Strategy described in [13] and [14]. The HLF nacelle performance gain and the associated reliability/maintenance requirements for the airline operators, emphasise that the complete aircraft should be considered for HLF demonstration.
New aircraft designs like the 'More Electric Aircraft/Engine' and the 'delta aircraft with suction' could well boost the HLF introduction on nacelles. The [15] More Electric Engine, incorporating new advances in motors/generators, active magnetic bearings, power electronics and other electrical technologies, promises to be lighter, more efficient, more reliable and less costly than current designs. The elimination of fan case mounted accessories would enable an easier installation of an HLF system and the absence of fan cowl doors/access doors would relax the aerodynamic constraints. The [16] 'delta with suction aircraft' quotes a 50% reduction in profile drag…

Acknowledgements

The author would like to thank the European Commission, represented by Dr Knoerzer, for the backing it has given to the HYLDA project. The following partners have made possible this further step on Hybrid Laminar Flow technology applied on nacelles: AM-Airbus, AS&T, BAE SYSTEMS P&C, DLR, Hispano Suiza Aérostructures, Hurel Dubois, Nord-Micro, ONERA, Rolls-Royce, Bombardier Shorts, SNECMA and the universities of Cranfield, Limerick and Southampton. Special thanks to the Co-ordinator G. Schrauf (DA) and the Task leaders, E. Maingre/C. Plaisance (SN), A. Blair/P. Thompson (SB) and A. Hurez (AM-A) for the permission to publish the herein contained information.
The author accepts responsibility for the ideas expressed in this paper which do not necessarily reflect the policies of the sponsoring organisations.

References

[1] B. Barry, S. Parke, N. Bown, H. Riedel, M. Sitzmann, "The Flight Testing of Natural and Hybrid Laminar Flow Nacelles", ASME Paper 94-GT-408.
[2] A. Mullender, H. Riedel, "A Laminar Flow Nacelle Flight Test Programme", 2nd European Forum on Laminar Flow Technology, AAAF, 1996.
[3] F. Tegarden, "Hybrid Laminar Flow Nacelles-A Test for the Future" General Electric, 1994.
[4] A. Mullender, J-L Lecordix, E. Lecossais, J-L Godard, M. Hepperle, "The ELFIN II and LARA HLF Nacelles: Design, Manufacture and Test", Forum on Laminar Flow Technology, AAAF, 1996.
[5] Y. Wie, F. Collier, R. Wagner, "Application of Laminar Flow Control to High-Bypass-Ratio Turbofan Engine Nacelles", High Technology Corporation - NASA Langley, ASE 912114.
[6] N. Voogt, "Flight Testing of a Fokker 100 Test Aircraft with Laminar Flow Glove", 2nd European Forum on Laminar Flow Technology, AAAF, 1996.
[7] J. Maestrati, C. Bulgubure, "Laminar Flow for Business Jets: Falcon 900 HLFC Demonstrator", AAAF, 1996.
[8] R. Eustace, R. Barrett, "The use of suction to suppress disturbances in laminar flow caused by insect and other surface debris", IMechE Vol 213 Part G, G02299, 1999.
[9] K. Horstmann, H. Koerner, B. Wagner, D. Welte, "Natural Laminar Flight Test Investigation for Commuter Aircraft Application", 2nd European Forum on Laminar Flow Technology, AAAF, 1996.
[10] G. Schrauf, "Application of Laminar flow Technology on Transport Aircraft", CEAS Drag Reduction Conference 2000
[11] D. Sawyers, R. Wilson, "Assessment of the Impact of Hybrid Laminar Flow on a Large Subsonic Aircraft", 2nd European Forum on Laminar Flow Technology, AAAF, 1996.
[12] R. Wilson, R. Jones, "Operational and Certification Considerations for Subsonic Transport Aircraft with Hybrid Laminar Flow Control", AAAF, 1996.
[13] R. Henke, P. Capbern, A. Davies, R. Hinsinger, J. Santana, "The A320 Fin-Programme: Objectives and Challenges", 2nd European Forum on Laminar Flow Technology, AAAF, 1996.
[14] J. Roeder, "Laminar Flow Application - Past Realities and future prospects", Forum, 3AF, 1996.
[15] M. Provost, "The More Electrical Aircraft and Beyond", IMechE Seminar, 17 May 2000.
[16] R. Denning, J. Allen, F. Armstrong, "Future large aircraft design - the delta with suction", RAeS paper No. 2212, 1997.

Session 5
Supersonic Flow Control Aspects

EUROPEAN RESEARCH TO REDUCE DRAG FOR SUPERSONIC TRANSPORT AIRCRAFT

D A Lovell*

Defence Evaluation and Research Agency, Farnborough, Hampshire GU14 0LX, UK

Summary

Research to reduce the drag of supersonic transport aircraft at low speed and cruise (transonic and supersonic), and to assess relevant aerodynamic design methods, is described. The evaluation of the aerodynamic design of three wing configurations by wind-tunnel model testing is described. Results are presented for independent CFD analyses of the optimised wing configurations, and a comparative analysis of results from design, model test and CFD evaluation is presented. It is concluded that aerodynamic design by direct shape optimisation, using Euler multiblock CFD methods, with geometric and aerodynamic constraints, is capable of producing aerodynamic configurations of high efficiency. Target values for L/D at cruise and at low speed were achieved and excellent agreement was obtained with the predicted performance. Major limitations were exposed in current computational and design methods for low speed.

Introduction

Europe has a firm base of experience, gained in the design and operation of first-generation supersonic transport aircraft, on which to approach the technologies required for a second-generation aircraft. A substantial potential market exists if environmental concerns can be met and fare levels can be maintained close to those for subsonic transport aircraft, by the application of advanced technology. Major improvements are required in materials, structures, aerodynamics, propulsion (including emissions), and aircraft systems. For aerodynamic technology, significant advances are necessary in wing design, propulsion-system installation and configuration integration. In particular a substantial reduction in the level of the drag of a new supersonic transport aircraft is essential to achieve these goals.

The paper describes the final results from a 3-year research project 50% funded by the Commission of the European Union, and member states, on the reduction of drag (at take off and landing, cruise over land and cruise over sea), and on the assessment of aerodynamic design methods. In the three-year project "EUROSUP" 10 industrial and research establishment partners contributed to a 1.5 million ECU computational and experimental programme co-ordinated by the UK Defence Evaluation and Research Agency.

Specific objectives for the EUROSUP project were:
(1) the reduction of aircraft drag at cruise conditions over land (transonic) and over sea (supersonic), to reduce the fuel required and hence increase the payload and cost effectiveness of the aircraft.

---

* Technical Manager, Applied Aerodynamics

(2) the reduction of aircraft drag at low speed, including take-off and landing conditions, to minimise the engine thrust required and hence the noise and emissions generated by the aircraft

On the basis of these requirements quantitative targets were set for L/D performance at the three design points to provide a focus to assess the capability for analysis and design of current computational methods in the supersonic transport application:
(1) lift/drag ~ 10 at supersonic cruise (M = 2.0)
(2) lift/drag ~ 15 at transonic speeds (M = 0.95)
(3) lift/drag ~ 8 at low speed (M ~ 0.3)
These targets for aerodynamic performance represent improvements of between 20% and 30% relative to that achieved for the first generation of supersonic transport aircraft.

Of equal importance to industry is the rapid completion of a design cycle. Within the design cycle rapid aerodynamic design methods are required to optimise the geometric shape for maximum aerodynamic performance, subject to packaging and off-design performance constraints. Thus another important objective for the EUROSUP project was to evaluate the relative merits of alternative aerodynamic shape design methods with respect to speed of execution, ease of use and the ability to incorporate updates to design constraints.

The EUROSUP consortium consists of airframe companies from France (Aerospatiale), Germany (DASA), Italy (Alenia), Sweden (Saab) and UK (BAE Systems Airbus), and research establishments from France (ONERA), Germany (DLR), Italy (CIRA), the Netherlands (NLR) and UK (DERA). In 1998 papers were presented on the aerodynamic design tasks in the project[1], and on the overall status of the project as at April 1998[2],[3]. In this paper the remaining research tasks are described: experimental evaluation of the aerodynamic designs, computational fluid dynamics (CFD) analyses of the aerodynamic designs, and overall assessment of the aerodynamic performance and design methods.

Research Programme

To achieve the objectives stated above a programme of computational and experimental research was followed. Figure 1 shows the broad breakdown of the research activities among the EUROSUP partners. Each task has a leader and contributors drawn from the consortium.

Several CFD methods, covering a range of fidelity in flow modelling, were first evaluated for their accuracy and speed of prediction against an existing set of wind-tunnel data for a generic configuration for a supersonic transport aircraft. The outcome of this work fed into the analysis work in subsequent tasks. Two linked wing-design tasks were then completed in parallel; for transonic/supersonic cruise design and for low-speed high-lift design. Several CFD-based methods were used for these tasks, again covering a range of speed and accuracy of flow modelling. These design tasks defined three configurations of a single wing shape for a wind-tunnel model. The model was tested at low-speed, transonic and supersonic conditions in July 1998. The experimental data was analysed to assess the design methods, and to compare the measured aerodynamic performance against the levels predicted by the CFD analyses (which were completed in parallel with the experimental work), and the initial aerodynamic targets for L/D. Figure 2 shows the time-scales and interrelation of the research tasks.

Evaluation of CFD methods

It was concluded[4] that linear methods are adequate for supersonic design for the initial design phases. They offer acceptable prediction accuracy for drag and lift, while their high computational efficiency permits the analysis of a much larger number of configurations than Euler and Navier-Stokes codes. It was also concluded that supersonic and two-point supersonic / transonic design could be carried out with optimisation methods based on Euler solutions on less fine grids.

The use of medium to coarse grids in three-dimensional shape optimisation, for reasons of cost and speed, is entirely justified in supersonic flow. Grid requirements for accurate aerodynamic performance prediction are more severe in transonic flow. The reliability of computations on a coarse grid in the optimisation process becomes questionable. Thus it is essential to use fine grids for the analysis by Euler codes of optimised shapes obtained at transonic and transonic / supersonic dual-point conditions. Computations using Navier-Stokes flow analysis codes are essential to complement Euler calculations as viscous effects are unlikely to be negligible in transonic flow. At low speed, fine-grid Euler computations were able to capture the major features of the flow.

A study of drag sensitivity to wing shape changes carried out by Alenia[5] concluded that:
(1) optimisation of the inner wing has the greatest effect on supersonic L/D
(2) the lift-induced drag is very sensitive to the shape of the inboard wing sections; reduced incidence of the airfoil at the wing-fuselage junction increases aerodynamic efficiency (L/D)
(3) longitudinal shifting of the wing along the fuselage, or a change of the outer wing sweep had a significant effect on wave drag

Transonic/Supersonic wing design

New design methods based on the use of CFD codes in combination with numerical optimisation techniques have been demonstrated in the EUROSUP research. Wing shapes were defined having significantly higher values of L/D than those achievable with linear theory methods, while considering multiple design points (low-speed and transonic cruise in addition to supersonic cruise). Because of the strong interaction between low-speed and high-speed design (for example the degree of bluntness employed on the wing leading edge) it was essential to maintain close co-ordination of these tasks in the project.

The European Supersonic Civil Transport (ESCT) configuration[6] formed the basis for the aerodynamic design. The geometry of this research configuration is shown in figure 3. To meet the requirements for manufacture of a wind tunnel model the fuselage was simplified to have a circular cross section, however the cambered nose and fuselage cross-sectional area distribution were not modified. This fuselage geometry and the wing planform remained fixed throughout the design work. The untwisted, uncambered wing was used as the datum against which the performance of the new wing designs was assessed. Design variables were the wing twist, camber and thickness distributions, and the overall configuration incidence at each design point. Geometric design constraints were defined[6] for the undercarriage bay, spar depths, wing thickness, and wing/cabin floor relationship (figure 4). Overall lift coefficient was constrained to the design point value. For the later stages of design[7] local Mach number was constrained on the upper surface of the wing.

Four methods were used for the transonic / supersonic design with the common aim of minimising drag at the high-speed design points. Rolston et al[1] have described this work in detail. A linear theory method for manual design was used by Alenia for single-point optimisation of wing twist and camber at the supersonic design point (the wing thickness distribution was fixed to satisfy the geometric constraints). The Alenia design[8] exceeded the transonic L/D target and came within 4% of the supersonic target (point B in figure 7). In a parallel activity ONERA used an Euler CFD code for single-point optimisation[9] at the supersonic design point. Wing thickness was first optimised for an untwisted and uncambered wing, and this thickness distribution was then used for wing twist and camber optimisation. This produced improvements in L/D of 1% transonically and 0.5% supersonically (point C in fig 7), relative to the Alenia design, but it featured a round leading-edge shape which would be likely to give improved low speed performance.

Some minor refinements were made to the ONERA single-point supersonic design by DERA, who then completed transonic design[7]. For this work it was essential to resolve the wing flow features on a grid coarse enough for design by optimisation. Figure 5 shows that the SAUNA[10] CFD code captures the main features of pressure distributions with a coarse (85000 cells) grid. From initial shape optimisation it was concluded that a fixed geometry could not meet all the design objectives. The basic wing shape was therefore frozen as the single-point supersonic design and DERA defined the optimum deflection angles for segmented leading- and trailing-edge flaps to maximise transonic L/D, with the constraints applied on upper surface pressure distributions to ensure attached flow (point D2 on fig 7). It was concluded[11] that the transonic / supersonic design problem was closely constrained, so the scope for geometric shape modification was restricted very considerably by geometric constraints.

Low-speed design

Variable leading-edge geometry was examined[12] to reduce drag to meet the airport noise limits for the flyover flight condition of $C_L = 0.4$. It was essential that the low speed design should not compromise unduly the high-speed performance. BAE and DERA investigated hinged leading-edge devices and Alenia slotted leading-edge devices. The same design criteria were used for the high and low speed designs.

Alenia completed 2D and 3D analyses[13] of leading-edge slats. Encouraging results from the 2-D work led to 3D analysis of a slatted wing on the ESCT datum wing without a fuselage. Thin-layer Navier-Stokes calculations indicated that slat deflection delayed flow separation in the outboard region and hence a large L/D increase was achieved at high incidence.

Deflection angles for a hinged leading-edge flap were defined by BAe, using a manual design method based on extensive low-speed wind tunnel results. RANSMB Navier-Stokes calculations were done by BAE and Euler calculations by DERA (on coarse and fine grids). Because of the markedly different solutions obtained it was considered inappropriate to use numerical shape optimisation methods. Figure 8 summarises the work on the hinged devices:

the manual design method and the Navier-Stokes computations indicated that the target value of L/D could be achieved at $C_L = 0.4$. It was concluded[12] that further evaluation of the CFD methods is required for this class of flow. Leading-edge flap deflection angles for the 5 flap segments were chosen for the wind tunnel model using the BAE manual design method, and the geometry was supplied to NLR for model design and manufacture.

Experimental Evaluation

The aim of the wind-tunnel model testing was to verify the wing design shapes at the supersonic, transonic and low-speed design points, and to understand the wing flow development to assess the validity of the design approach used. A sting-mounted model (1/80 scale of ESCT) consisting of a fuselage with a cylindrical afterbody and three wings was designed, manufactured and tested by NLR[14] (figure 9). Three wing configurations were tested:
(1) the supersonic twist / camber / thickness design
(2) the supersonic design with LE and TE flaps deflected for transonic cruise
(3) the supersonic design with LE flaps deflected for low-speed fly over.

Figure 10 shows the geometry of the upper surface of the transonic wing configuration. Overall forces and surface pressures were measured. In addition some flow visualisation was completed to aid the understanding of flow development with incidence on the wing. Because the model was to be tested at high dynamic pressures the manufactured shapes of the transonic and supersonic wings were modified so that in the wind tunnel, under load, the design geometry was approximately recovered. Calculations to estimate the wing deformation indicated a wing twist of about 1.2° for the supersonic and transonic wing geometries at maximum dynamic pressure. DERA adapted the wing twist of the designed geometry accordingly, provided the modified shape to BAE for final smoothing, who then supplied the geometry to NLR for manufacture. Figure 11 compares the predicted wing deformation at transonic conditions with that measured in the wind tunnel. The total pressure for the supersonic test was chosen to produce a similar deflection to that obtained at transonic conditions. A selection of the measurements of surface pressures is shown in figures 12, 13 and 14, and drag polars in figures 15 and 16. These results are discussed in detail in the following sections.

Flow analysis and comparison with experiment

The three wing configurations for the supersonic design point, for the transonic design point (with graduated deflections of leading- and trailing-edge flaps), and for low speed (with different graduated deflections of leading-edge flaps) were analysed[15] by NLR, CIRA, and DLR using their CFD methods. This provided an independent check on the design computations by Alenia, ONERA, DERA, Saab and BAE.

Analysis methods

Two methods, linear and Euler, were used by CIRA[16] to predict the supersonic performance of both the upswept-tail full-scale aircraft and wind tunnel model including sting configurations. The linear method includes both far-field and near-field approaches to the calculation of zero-lift wave drag. Skin friction drag was estimated using a flat-plate reference skin friction for a given Mach number, Reynolds number and adiabatic wall temperature, assuming a fully turbulent boundary layer. The multiblock grids used in the Euler analysis by CIRA had a C-O topology, and over 1 million cells in the fine grid. A rounded wing tip was added to close the flat tip of the supplied geometry (The wind tunnel model had a square flat tip).

DLR used a Navier-Stokes method[17] with two turbulence models: the simple algebraic Baldwin-Lomax model with the Degani-Schiff modification, and the more complex two-equation k-ω model. For supersonic cases, a single block C-O grid was used, with a near conical far field. Three grid densities were used with the finest grid having 2¼ million cells. For transonic calculations DLR used the NLR transonic grid described below. DLR generated the subsonic multi-block grid. The medium density grid contained over 700 thousand cells, and consisted of 6 blocks, incorporating a large C-grid around the deflected wing leading edge.

NLR used a Navier-Stokes flow solver[18] with two artificial dissipation models; the robust scalar dissipation model employing non-isotropic artificial dissipation by means of cell aspect-ratio scaling, and the more refined matrix model. The latter was less adequate for the grids adapted for high Reynolds number computations hence all final computations employ the scalar model. The algebraic Baldwin-Lomax turbulence model was used with the Degani-Schiff modification as for DLR. The multi-block grid had 12 blocks in a C-H topology. Four levels of grid density were investigated, the finest containing around 3¼ million cells. All Navier Stokes solutions were fully turbulent. NLR performed an initial computation with transition matching the strips applied to the wind tunnel model. Laminar separation was present on the outer wing, in agreement with flow visualisation in the wind tunnel, which also indicated a separation on the outer wing. Using a fully turbulent boundary layer improved the convergence behaviour of the flow solver with only a minor effect on the computed drag.

ONERA performed a computation at the supersonic design case using a viscous-coupled Euler method. In this method a three-dimensional laminar-turbulent boundary layer code is coupled to a structured multi-block Euler (and Navier-Stokes) solver. A simple turbulence model (eddy viscosity and a mixing length formulation) was used within this study. The effects of the boundary layer were simulated via transpiration velocities. The coupling procedure was only applied on the wing, with a slip condition being applied to the fuselage. For this work the CIRA Euler grid was used. Approximately 300 Euler iterations were done between each boundary layer coupling procedure with convergence being reached after 5-6 cycles.

Analysis Results

Surface pressures at supersonic conditions

Predictions by the Navier-Stokes Baldwin-Lomax and Euler solvers were in reasonable agreement. The Euler solution has a higher leading edge peak than the viscous solutions, with slightly lower pressures continuing over the entire chord. The Navier-Stokes and viscous coupled Euler results agree well. A comparison of the predictions of the two Navier-Stokes turbulence models and measurements is shown in Fig 12. Good agreement with experiment was obtained. Fig 12 shows very little change in the Navier-Stokes solution by moving from the simple Baldwin-Lomax turbulence model to the more complex k-ω model. This suggests that the flow is not close to separation, so the simple turbulence models prove adequate.

Surface pressures at transonic conditions

A comparison of DLR and NLR fine-grid transonic pressure distributions is given in Fig 13, which also includes measured tunnel data and lines of local Mach number (M_{local} = 1.3 for the front half of the chord and 1.1 over the remaining chord). These Mach lines, generated using the approximate Lock theory[19], represent the flow attachment constraint as applied during the transonic design. The computational results agree very well, the largest differences being over the leading-edge flap surface and around the rear shock. The comparison with the wind-tunnel model data is excellent, including the pressure peak at the flap knuckle, although the rear shock is located too far aft. Both the predicted and measured pressures largely conform to the local Mach number limit imposed during design. The pressure 'shoulder' at mid-span touches the limit, whilst the flap knuckle pressure peak is seen to violate the limit at the tip. Both the Navier-Stokes solutions and model test flow visualisation showed a region of separated flow near the tip, with the remainder of the wing flow being attached. Thus the approach used to design for attached flow at transonic speeds using inviscid Euler methods has been validated. The likelihood of separated flow at the tip was anticipated at the design stage[12] but was accepted as a practical wing design would incorporate a swept tip.

Significant grid density effects were found in the DLR computations; the rear shock was smeared with the medium grid, and the leading edge flap knuckle pressure peak was reduced. As a result the small separations noted above only appear in the fine grid solutions. The two DLR turbulence models show close agreement for the fine-grid pressure computations, again presumably because the flow is largely attached.

Surface pressures at subsonic conditions

Pressures are plotted in Fig 14, for the design condition. The comparisons with model measurements are encouraging. The mixed attached / separated flow at the design point is well predicted. The basic vortex structures match those obtained in the wind-tunnel model flow visualisation well, though some of the secondary structures do not appear in the computed results. It should be noted that the flow structure alters rapidly around the design point, changing from an attached to a highly separated character, and is thus very sensitive to small changes in incidence.

Overall forces at supersonic conditions

The computed and measured drag polars are plotted in Fig. 15. The CIRA Linear method predicts a higher drag and higher lift-curve slope than the Euler solver. The viscous-coupled Euler solution of ONERA gives an additional 3-4 counts of drag compared to the CIRA results (inviscid Euler + skin friction), largely because of the friction drag component. Overall, including the two DLR Navier-Stokes results, the predicted drags at the design point are within 8 counts and the design lift is reached at similar angles of incidence (within 0.1°). Both CIRA and DLR performed grid refinement studies. Generally the fine grid results appear to give a good approximation to coefficient values for a grid of vanishing cell size.

The Euler drag prediction by CIRA is the closest to the model measurements, with the Navier-Stokes result being some 10 counts higher for both turbulence models. The reasons for the poor Navier-Stokes predictions of C_{D0} are not understood at present and are the subject of further studies, however the Navier-Stokes method gives the closest match to the lift-dependant drag component and the best prediction of the lift curve. Studies of the effect of grid density sensitivity on the predicted aerodynamic coefficients by CIRA and DLR indicated that the fine grid results again offer a good approximation to a grid of vanishing cell size.

Overall forces at transonic conditions

The computed and measured drag polars are plotted in Fig. 16. Two of the predictions are for different implementations of the Baldwin-Lomax turbulence model (by NLR and DLR) and the third uses the DLR k-ω turbulence model. The comparison of computed transonic drag shows a difference of over 10 counts for the solutions of NLR and DLR. This is almost entirely due to friction drag differences, pressure drags being within one count of each other. Lift and pitching moments show a little scatter between the codes despite the good agreement of the wing pressure distributions noted above. Comparing computations and wind tunnel measurements of the drag polar shows that the DLR Baldwin-Lomax solution over predicts the drag by about 7 counts, the equivalent NLR solution is under by 3½ counts, while the DLR k-ω prediction is within 1¼ counts. This suggests that use of the more complex turbulence model may be justified. Lift-dependent drag computed by the NLR and DLR methods follow the measured polar well.

There were some significant differences between the measured and computed values for pitching moment. Several potential causes were examined to explain these differences. A grid refinement exercise by NLR showed a strong effect on predicted pitching moment, with lift moving outboard with increasing grid density. Noting the differences between predicted and measured wing twist on the wind-tunnel model, an approximate re-twist of the CFD geometry was carried out by DLR to model more accurately the model shape in the transonic tunnel. Model manufacture and aeroelastic twist deformations resulted in the mid-region of the model wing having a greater nose down twist than the CFD geometry, by about 0.3°. Approximate scaling of the pressure levels reduced C_L from 0.19438 to 0.18955, and the C_m from –0.057717 to –0.05. This appeared to be the possible cause for the C_m discrepancy. Thus the wing C_m would appear to be sensitive to small changes in twist.

Overall forces at low-speed conditions

DLR performed medium grid Navier-Stokes computations for the low speed, high lift configuration, largely using the Baldwin-Lomax turbulence model. Comparison with a single k-ω computation was good, with only minor differences in the friction drag estimates.

Reynolds Number and Configuration Effects

Results for the overall forces on the aircraft configuration at flight Reynolds number are tabulated in Fig 17. For supersonic conditions, tunnel-to-flight aerodynamic effects were examined using CIRA's Linear and Euler methods. The effect of the configuration change (tunnel-model to aircraft) suggests that modelling the former is sufficient to understand what may be expected in terms of pressure distributions on the aircraft. According to the Euler solution the inviscid drag increases by about 5 counts when changing from the tunnel sting geometry to an upswept tail; 3 counts come from the fuselage and the remainder from the wing. Lift for the flight configuration is reduced slightly due to the smaller contribution from the fuselage tail, and there is a corresponding change in pitching moment. There is good agreement between Euler and Navier-Stokes predictions at supersonic conditions, suggesting that scale effects are likely to be small.

At transonic conditions the effect of the increase in Reynolds number was computed by NLR for the model geometry, with the sting present. The rear shock increases in strength and moves aft slightly (because of the reduced de-cambering of the wing due to the thinner boundary layers), and the pressure drag is predicted to fall by 1 to 3 drag counts. Lift is increased as Reynolds number is increased, due to the more aft shock location.

The NLR results show that the upswept tail geometry creates an overspeed resulting in a significantly enlarged shock across the entire span, giving rise to an extra 4 counts of pressure drag. The change in wetted area and pressure distribution result in a further 2 counts of friction drag. There is an increase in lift curve slope with the upswept tail. The effect of the configuration change may not have been so apparent if the analysis had been done at tunnel scale due to the thicker boundary layers. It is possible that only at the flight Reynolds number is the full impact of the configuration change revealed. Thus it is important that the two effects be understood together.

Overall Assessment

Performance of CFD methods for analysis

Supersonic conditions

Euler methods are more than adequate for predictions of pressures. Navier-Stokes also perform well, with the algebraic Baldwin-Lomax turbulence model giving similar results to the 2-equation k-ω model for the attached-flow design case. Part of the difference in prediction of drag could be due to changes in model geometry under load.

Transonic conditions

The Navier-Stokes methods give excellent predictions for the transonic pressure distributions, with the shock a little too far aft. Indicated areas of flow separation agree well with those noted in the wind tunnel flow visualisation. The solution degrades significantly with a medium size grid. Little advantage is gained in using the more complex 2-equation k-ω turbulence model. The drag predictions appear to be better than for supersonic conditions when compared to tunnel measurements, though the comparison between the methods is not so good.

Low-speed conditions

Analysis at the low-speed high lift condition proved challenging. Grid generation was problematic. The final results were extremely encouraging, with the DLR N-S (Baldwin-Lomax) medium grid solutions modelling the changing flow structure with incidence relatively well. The low alpha attached flow condition was well predicted, as was the design condition with mixed attached and separated flow. At the higher incidence where the flow is completely separated the predictions were not so good. Both drag and lift were fairly well predicted up to the design alpha. Predicted lift and pitching moment were a little low at the highly separated conditions, possibly because of an under-estimation of the vortex lift.

Performance of CFD methods for design

All the Euler-based design methods used coarse grids for design and fine grids for analysis to keep execution time within reasonable limits. This might have been expected to have an effect on the ability to predict the flow correctly, but in fact the effect on pressure drag and pitching moment is small in the supersonic case. In the transonic case the DERA coarse grid design shows a 5 counts deviation in C_{Dp} from the NLR N-S results and the DERA fine grid Euler calculation. This is within the variation of the results from the different fine grid analyses.

Dual point optimisation proved too compromising for reasonable aerodynamic performance levels. Leading and trailing edge flaps show considerable potential as a means of achieving acceptable transonic cruise performance without penalising the supersonic performance. The analysis at flight Reynolds number indicates that the supersonic and transonic L/D design targets of 10 and 15 respectively were both achieved with attached flow. The Inverse method was not directly used in the main design exercise. Instead the capabilities were demonstrated in a test case using the datum configuration as starting point and the DERA supersonic design as target. Good results were achieved. The DERA geometry could be quite accurately reproduced over most of the wing, the root section showing the largest deviation, mainly due to the use of a greatly simplified fuselage geometry. Constraints could not be incorporated, but were checked as part of the post-processing.

Detailed geometric constraints have been successfully applied in the design process, resulting in a practical wing envelope. Care is required in defining aerodynamic constraints to be applied in an inviscid based design method, to meet a real-life constraint. Thus the aircraft incidence requirement was not met in the final design. Care is also required in applying wing surface-pressure constraints to limit shock induced separation within an inviscid design process using limited grid density. The final analysis does indicate that the desired effect was achieved.

Low speed design

The Navier-Stokes predictions at the design point showed significant areas of separated flow even with the leading-edge deflected and this gave doubts about the validity of using Euler solutions at this design condition. In addition poor convergence due to the highly complex nature of the flow, and the need to use a coarse grid for the Euler code within the CFD optimisation method, prevented a useful outcome from the automatic optimisation design approach. There was no computational means of determining the level of suction liable to cause separation and the mechanism for drag reduction evident in the CFD results was not fully understood. In this particular study, optimisation techniques did not show a sufficient level of maturity for application to the low speed high lift design problem. A manual approach has, however, been relatively successful.

General conclusions on the design process

Linear methods offer an efficient means of improving supersonic performance in the early stages of design. The use of CFD optimisation coupled to an Euler code has proved to be an effective design process, giving slight improvement over the linear method at supersonic speeds, but offering significant advances at transonic conditions, where linear methods are not applicable. Much experience has been gained in the use of such optimisation techniques.

For detailed studies and transonic design exercises non-linear methods are required. Optimisation techniques can be successfully applied with both geometric and aerodynamic constraints. The design is easier to manage if done in stages, with thickness, twist and camber design being separated. Aerodynamic constraints should include pitching moment, as well as incidence and surface pressure limits. When applying aerodynamic constraints using inviscid methods, careful grid dependency and flow resolution studies are required for a fully successful design. Leading and trailing edge flap deflections can be included as geometric variables for transonic optimisation.

The need for further evaluation of the CFD methods for the highly complex flow at low speed coupled with a full assessment of the design sensitivities means that it is not possible to propose a design method at this stage. A manual design using a matrix of CFD predictions may be the best approach unless the execution times of acceptability accurate CFD methods for this type of flow can become fast enough for automatic optimisation.

Reynolds number effects on the surface pressures appear to be relatively small at supersonic speeds. At transonic conditions the effect is greater, with the wing rear shock strengthening slightly and moving aft, however this is relatively minor compared with the effect of the upswept tail. Changing from the constant section sting to the actual aircraft geometry significantly increases shock strength across the entire span. Realistic aircraft shape and flight Reynolds number calculations are important in order to gain a full picture of the final design. It is recommended that, as in the EUROSUP project, design should be done on the upswept tail geometry, thus requiring extra CFD computations for comparison with wind tunnel data.

Concluding remarks

The results achieved by the EUROSUP project indicate that considerable progress has been made in the aerodynamic technology to support a second-generation supersonic civil transport. The state of the art aerodynamic analysis and design tools employed were validated by experiment, and enabled significant improvements to be made in the L/D performance at transonic and supersonic cruise conditions. The situation for the aerodynamic design of the SCT class of configuration is far less satisfactory at low speed - the need for further computational and experimental research in this area has been clearly identified.

Acknowledgements

The research work described in this paper is 50% funded by the Commission of the European Union, under contract BRPR-CT95-0082. The remaining 50% is funded by the industry partners and National governments. The UK DERA research is 50% funded by the Department of Trade and Industry CARAD programme. The author would like to acknowledge the contributions to this paper by many members of the EUROSUP consortium, including T Surply and D Prat (Aerospatiale), M Averardo and P Vitagliano (Alenia), S Rolston, A Mann and K Nicholls (BAE Systems Airbus), A Vicini (CIRA), U Herrmann (DLR), A Elsenaar and J van Muijden (NLR), D Destarac, P Lemee and R Grenon (ONERA), E Totland (Saab), and J Doherty, C Newbold and T Evans (DERA).

References

1 Rolston S.C., Doherty J.J., Evans T.P., Grenon R., Averado M.A. *"Constrained aerodynamic optimisation of a supersonic transport wing: a European Collaborative Study"* AIAA-98-2516, 16th Applied Aerodynamics Conference, 1998

2 Lovell D.A. *"Aerodynamic Research to support a second generation supersonic transport aircraft - the EUROSUP project"* ECCOMAS 98, Athens, Sept 1998
3 Lovell D.A., *"Reduction of wave and lift-dependent drag for supersonic transport aircraft"* ICAS 98, Melbourne, Sept 1998
4 Destarac D, Lemee P, *"EUROSUP Task 2 Final Report: CFD Method Evaluation"* EUROSUP/ONERA/T003/1 (ONERA RTS 11/2733AY), July 1997
5 Averardo M.A., *"Drag sensitivity to shape change."* EUROSUP/ALN/T-001/1, July 1996.
6 Rolston S.C, *"Criteria for Aerodynamic Design - EUROSUP Tasks 3 and 4"* EUROSUP/BAe/T001/1 May 1996
7 Evans T.P., Doherty J.J., *"The aerodynamic design of the EUROSUP configuration"* DERA/AS/ASD/CR97620/1 December 1997
8 Averado M.A., Vitagliano P.L. *"Supersonic wing design through linear methods and transonic analysis through an Euler flow solver for a supersonic civil aircraft (ESCT) wing-body configuration"* EUROSUP/ALN/T-002/1 September 1997
9 Grenon R., *"EUROSUP Task 3 : ESCT high speed design"* ONERA Technical Report RTS 15/3733 DAAP/Y April 1998
10 Shaw J.A., Peace A.J., May N., Pocock M., *"Verification of the CFD simulation system SAUNA for complex aircraft configurations"* AIAA paper 94-0393 1994
11 Mann A, *"Design of a Supersonic Transport Wing at three Mach numbers using Optimisation Techniques"*, EUROSUP/BAe/T-003/1, Feb 1999
12 Nicholls K, *"EUROSUP Task 4 Overall Report : Low Speed Design"*, EUROSUP/BAe/T-005/1, Feb 1999
13 Averardo M A, Vitagliano P L, *"Aerodynamic study of a leading-edge slotted device for supersonic civil transport aircraft through 3-D Navier-Stokes Solver"*, EUROSUP/ALN/T-003/1, April 1998
14 Elsenaar A, *"Wind Tunnel Test of the EUROSUP SCT Configuration"*, EUROSUP/NLR/T017/1,
15 Mann A, *"CFD analyses of EUROSUP supersonic, transonic and low-speed configurations."* , EUROSUP/BAe/T-003/2, Feb 1999
16 Vicini A, "*Euler and Linear Analysis of the EUROSUP Supersonic Configuration"*, EUROSUP/CIRA/T-003/1', Nov 1998
17 Herrmann U, *"Analysis of the Optimised EUROSUP Supersonic Transport Configuration (ESCT) at Three Design Mach Numbers using a Navier-Stokes Solver"*, EUROSUP/DLR/T-002/1, Feb 1999
18 Van Muijden J,"*CFD Analysis of the EUROSUP Transonic Wing/Body Configuration Using the NLR ENFLOW Navier-Stokes Method"*, EUROSUP/NLR/T-021/1, Feb 1999
19 Lock R C, "*An Equivalence Law Relating Three and Two Dimensional Pressure Distributions"*, Aeronautical Research Council R&M No 3346, May 1962

Partner / Activity	DERA	AS	Alenia	BAe	CIRA	DASA	DLR	NLR	ONERA	Saab
Programme co-ord	✓									
CFD evaluation			✓		✓		✓	✓	✓	
Low-speed design	✓		✓	✓						
High-speed design	✓		✓	✓					✓	✓
Low-speed analysis							✓	✓		
High-speed analysis					✓		✓	✓		
Model D, M & test								✓		
W/T data analysis		✓				✓			✓	
Overall assessment	✓	✓	✓	✓	✓	✓	✓	✓	✓	✓

Fig 1 EUROSUP partners and research activities

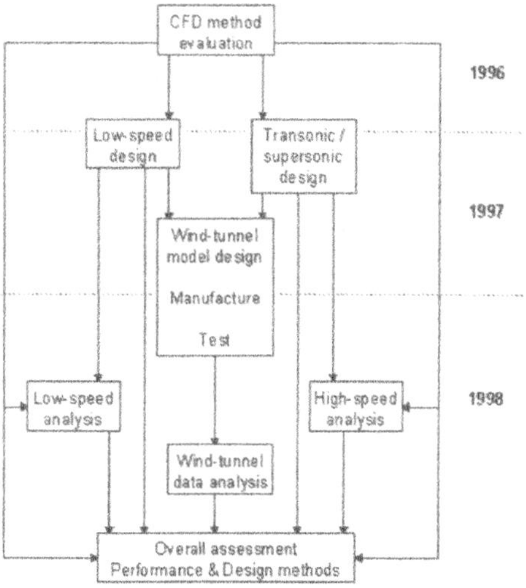

Fig 2 Research task interdependencies

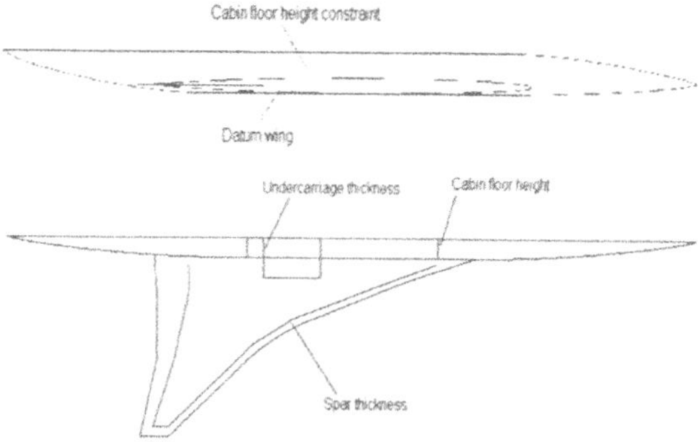

Fig 4 Geometric constraints applied in design

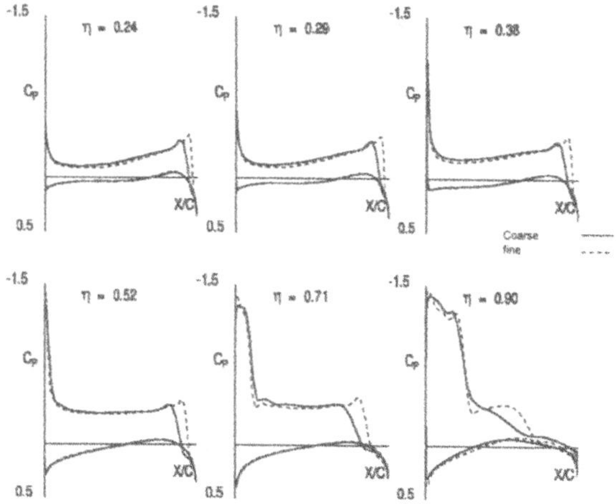

Fig 5 Cp comparison for fine and coarse grids. M=0.95, α=4°

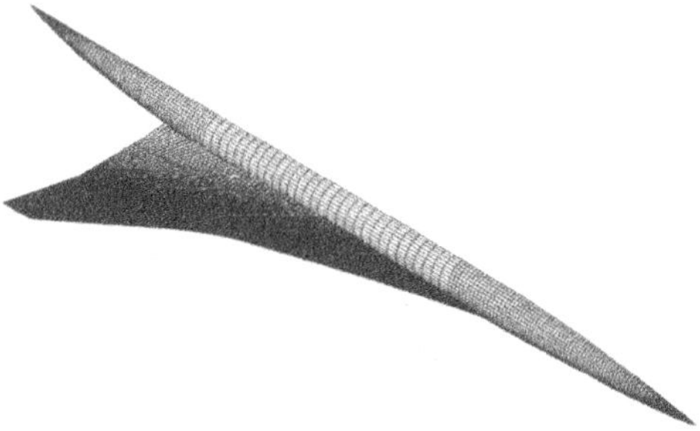

Fig 3 Geometry of the ESCT configuration

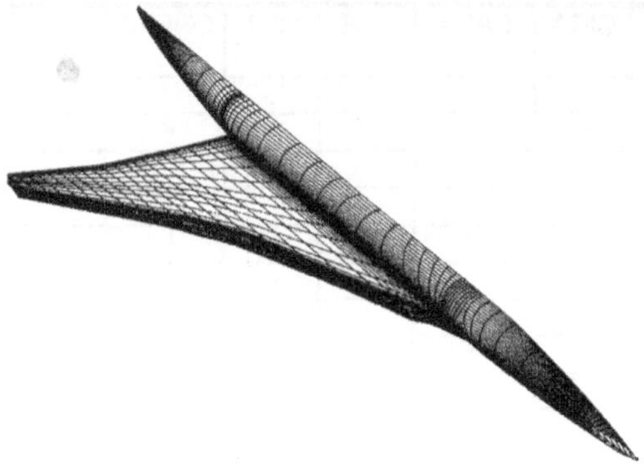

Fig 6 Coarse grid for wing design with deflected leading-edge flaps

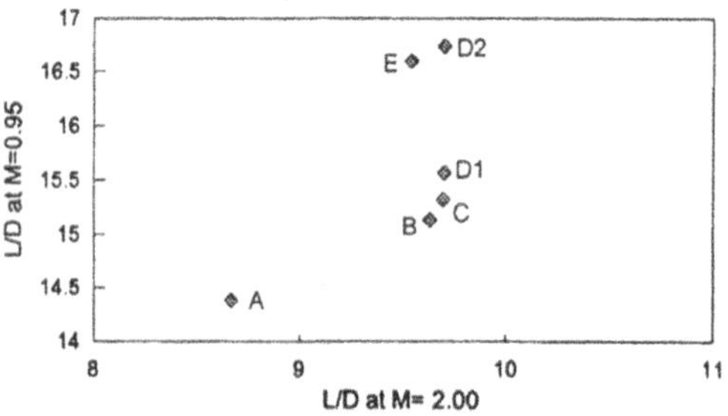

A - Datum B - Supersonic design by linear theory (Alenia)

C - Initial supersonic design by Euler optimisation method

D1 - Final supersonic design by Euler optimisation method

D2 - Final supersonic design with optimised LE deflection for transonic design point; Euler optimisation method

E - Dual -point supersonic / transonic design by Euler optimisation method

Fig 7 Summary of results of high-speed design

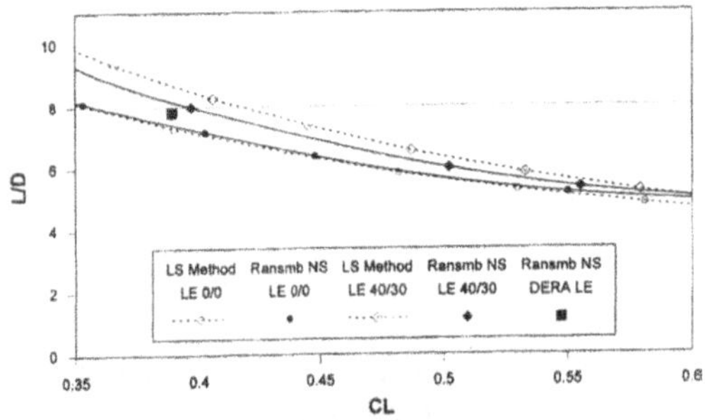

LS method uses a model experimental data base. Ransmb NS is Reynolds-averaged Navier Stokes multi-block method

Fig 8 Low speed L/D estimates

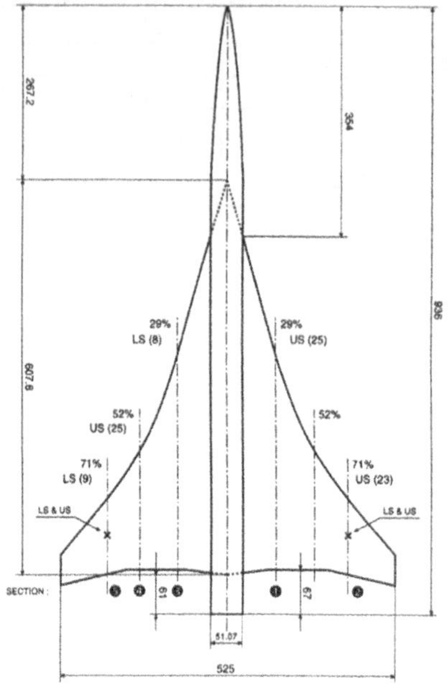

Fig 9 Planform of wind-tunnel model

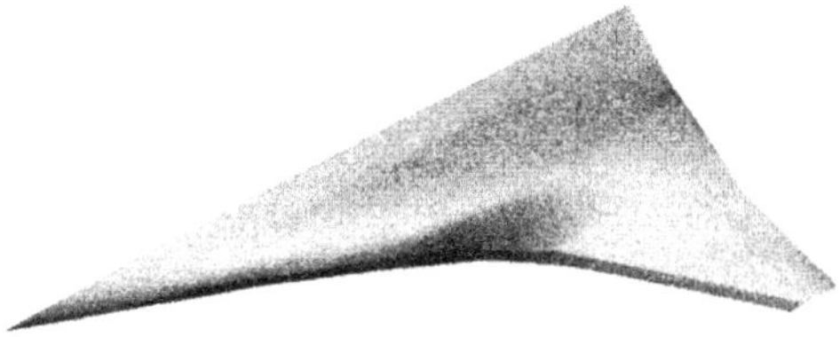

Fig 10 Wing upper surface, transonic configuration

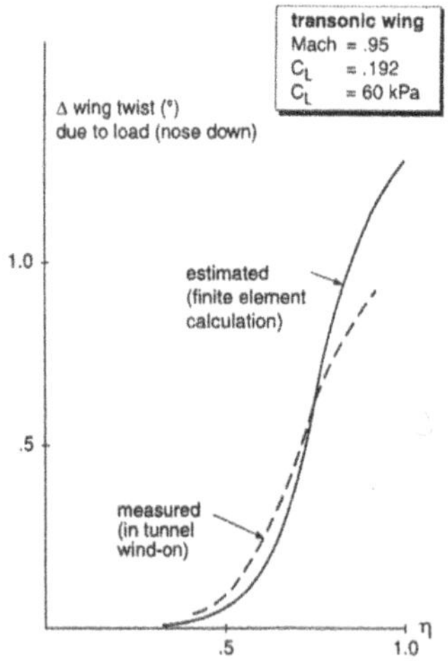

Fig 11 Wing deformation at transonic conditions

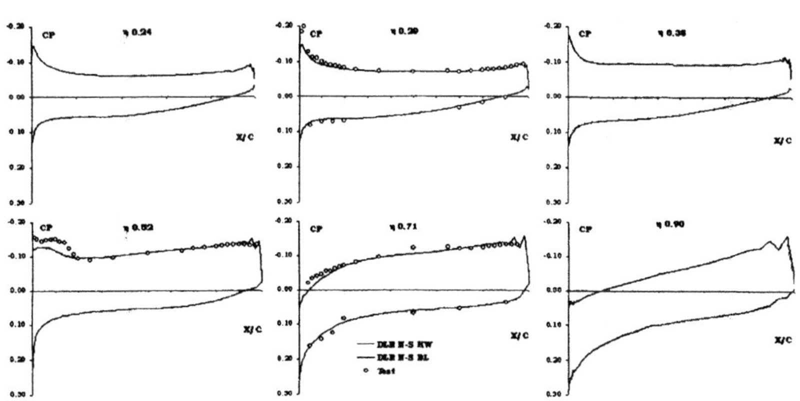

Fig 12 Cp from model and CFD
M=2.0 design point

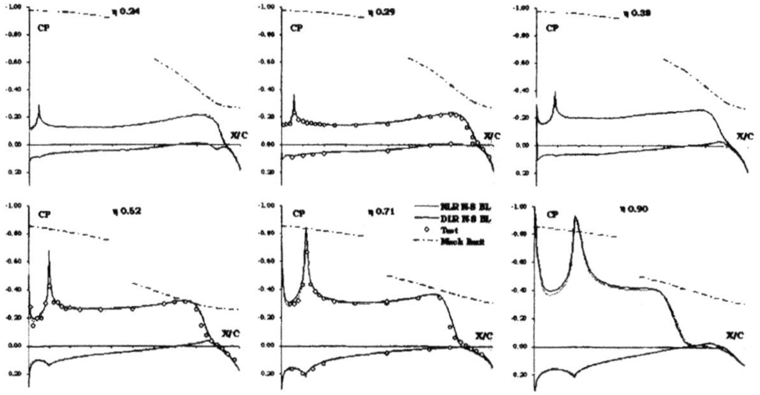

Fig 13 Cp from model and CFD
M=0.95 design point

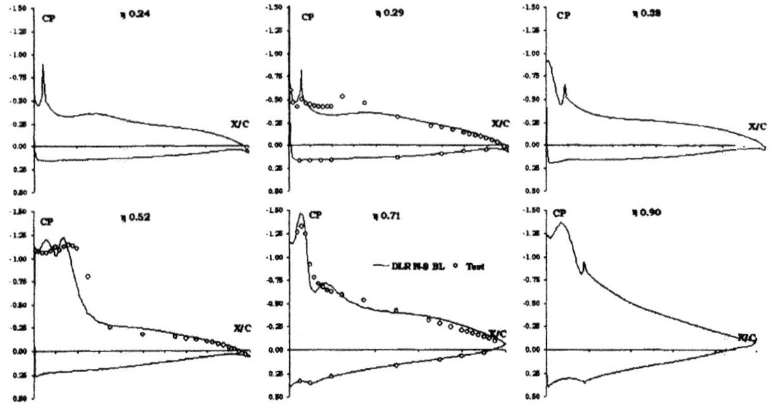

Fig 14 Cp from model and CFD.
M=0.25, α=10°

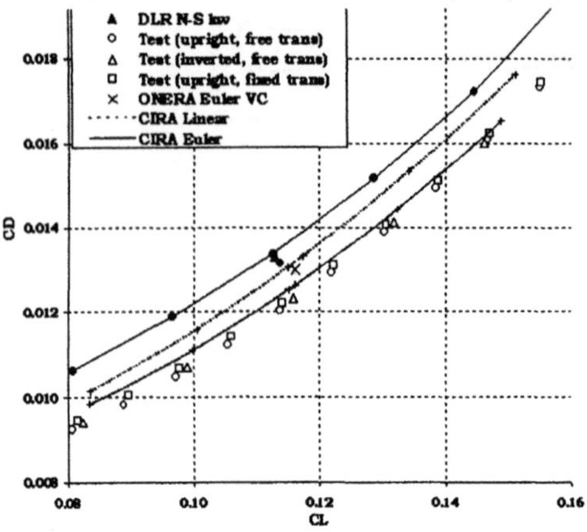

Fig 15 Drag polar at M=2.0, wind-tunnel conditons

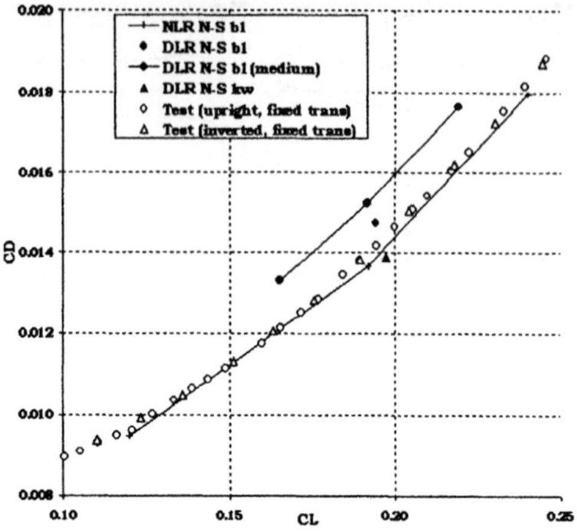

Fig 16 Drag polar at M=0.95, wind-tunnel conditions

Partner	Data	Mach	Alpha	C_L	C_D pressure	C_D friction	C_D total	C_M	L/D
	Target	2.00	4.0	0.115	~	~	0.0115	-0.0298	10.0
CIRA	Linear	2.00	4.0200	0.11500	0.00869	0.00348*	0.01217	-0.03293	9.45
CIRA	Euler	2.00	4.0100	0.11505	0.00798	0.00348*	0.01146	-0.03201	10.04
	Target	0.95	4.0	0.192	~	~	0.0128	-0.0553	15.0
NLR	N-S bl	0.95	3.8515	0.19200	0.00755	0.00446	0.01201	-0.06335	15.99

Fig 17 Predicted aerodynamic coefficients at cruise design points for aircraft

Attachment Line Transition in Supersonic Flow

D. ARNAL * and J. RENEAUX **

* ONERA/DMAE, BP 4025, 2 av. Edouard Belin, 31055 Toulouse Cedex 4, France
** ONERA/DAAP, BP 72, 29 av. de la Division Leclerc, 91322 Châtillon Cedex, France

Summary

According to a criterion proposed by Poll, leading edge contamination by a turbulent boundary layer occurs when a characteristic Reynolds number $\bar{R}^*$ exceeds a critical threshold. Two series of experiments have been performed in order to assess the validity of this criterion. The first one was carried out on a swept cylinder at Mach 3. In the second series of experiments, a swept wing was designed with a leading edge providing values of $\bar{R}^*$ close to the flight values of the inner wing of a supersonic transport aircraft at Mach 2. Both experiments indicated that the leading edge flow became turbulent for values of $\bar{R}^*$ in rough agreement with Poll's criterion. Additional experiments were performed in order to investigate the effects of roughness elements placed on and off the attachment line.

1 Introduction

Reduction of skin friction drag represents a very important challenge for subsonic and supersonic transport aircraft. Among the different techniques which could be applied to reach this objective, laminar flow technology seems to offer the greatest potential. For subsonic transport aircraft, many investigations conducted through national or international programs in Europe and in the U.S. led to encouraging results, in wind tunnel as well as in free flight conditions. Extended regions of laminar flow were observed on swept wings by applying suction around the leading edge and by optimizing the pressure distribution further downstream. In principle, similar techniques could be used for supersonic transport aircraft. In all cases, however, it is first necessary to ensure that the flow along the so-called *attachment line* is laminar.

In the case of swept wings, the attachment line is a particular streamline which divides the flow into one branch following the upper surface and another branch following the lower surface, see figure 1. Along this line, transition can be the result of the amplification of "natural" waves excited by the available disturbance environment; this phenomenon will not be discussed in this paper. Other possibilities for the appearance of turbulent flow along the attachment line are *leading edge contamination* and *boundary layer tripping* by large roughness elements.

Leading edge contamination is likely to occur when a swept body is attached to a solid surface (fuselage, wind tunnel wall). Under certain circumstances, the turbulent boundary layer developing along the surface can propagate along the attachment line, thus making the boundary layer on the wing fully turbulent. Some years ago, Poll proposed an empirical criterion in order to predict the appearance of this phenomenon. Quite recently, two series of experiments were carried out in the R1Ch and S5Ch wind tunnels at ONERA for the purpose of validation of this criterion in supersonic flow. The results are described in paragraph 2. Additional data were also obtained concerning the problem of *boundary layer tripping* by roughness elements placed on and off the attachment line; the most important findings are presented in paragraph 3.

2 Leading edge contamination

2.1 Leading edge contamination criterion

For the sake of simplicity, let us consider the case of a swept cylinder of constant diameter D. In the coordinate system (X, Z, y) linked to the cylinder, Z coincides with the attachment line, X is normal to Z and the y axis is normal to the wall, see figure 1. U_e and W_e are the free stream velocity components X and Z. Close to the attachment line, U_e depends linearly on X: $U_e = kX$. If the infinite span assumption is used, W_e is constant. The Reynolds number $\bar{R}$ and the reference length scale η are defined as:

$$\bar{R} = \frac{W_e\,\eta}{\nu_e} \qquad \text{and} \qquad \eta = (\nu_e/k)^{1/2}. \tag{1}$$

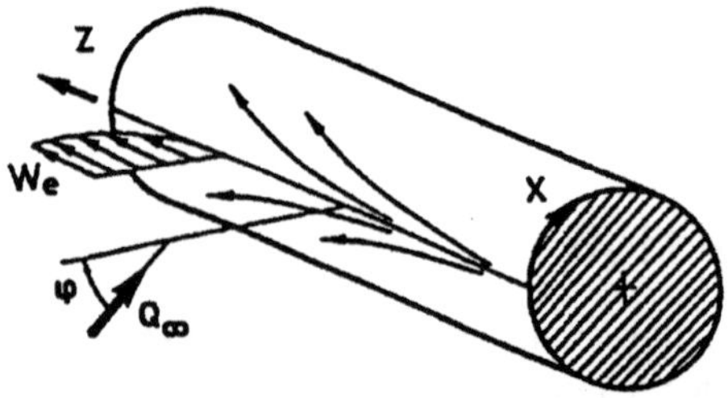

Figure 1: Experimental set-up in the R1Ch wind tunnel

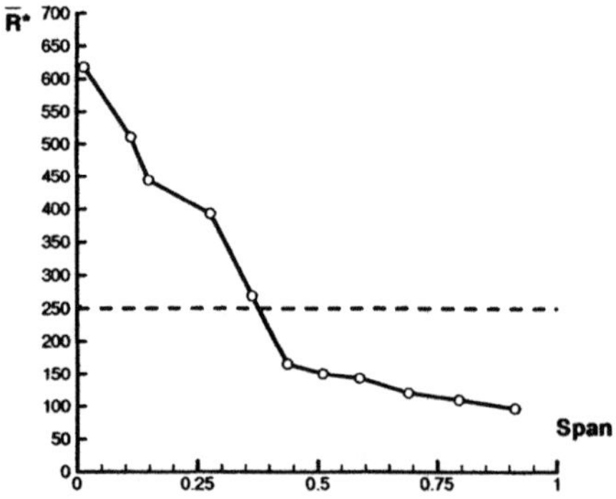

Figure 2: Spanwise variation of $\bar{R}^*$

For low speed flows, η is close to the boundary layer displacement thickness. For high speed flows, Poll [11] introduced a modified length scale η^* and a modified Reynolds number $\bar{R}^*$ which have the same definition as η and $\bar{R}$, except that ν_e is replaced by ν^*. The latter quantity is the kinematic viscosity computed at a reference temperature T^* which may be estimated from the following empirical relationship:

$$T^* = T_e\,[1 + 0.1\,(T_w/T_e - 1) + 0.6\,(T_{aw}/T_e - 1)]\,. \tag{2}$$

T_w and T_{aw} denote the wall temperature and the adiabatic wall temperature, respectively. T_e is the static temperature at the boundary layer edge.

The problem of leading edge contamination has been widely investigated for low speed flows (see [9], [7], [10], [1] for instance), and a simple criterion, based on the value of $\bar{R}$, was developed by Pfenninger (1965, [9]) and carefully validated by Poll [10]: if $\bar{R}$ is lower than 250, the bursts of turbulence convected along the wall are damped and vanish as they travel along the attachment line. However, for $\bar{R} > 250$, these bursts are self-sustaining. They grow, overlap and the leading edge becomes fully turbulent.

Several series of experiments were devoted to the study of leading edge contamination at high speeds (see [3], [4], [6], among others). In most of these studies, end plates were used as sources of gross disturbances. Poll [11] analyzed available experimental data for

$0 < M_e < 6$ and showed that leading edge contamination occurs for $\bar{R}^* = 245 \pm 35$. One of the objective of this study was to check the validity of the latter criterion in supersonic flow for two different models.

2.2 Values of $\bar{R}^*$ in free flight conditions

In flight conditions, leading edge contamination is expected to occur in the inner part of the wing downstream on the round leading edge while the sharp supersonic leading edge in the outer part is not affected by this phenomenon. A specific study has been launched in order to determine the $\bar{R}^*$ values on the attachment line of the AS1 configuration of a supersonic transport aircraft supplied by Aerospatiale Matra-Airbus. The mesh which has been defined using the ICEM system has 28 blocks and 1 700 000 grid points with C-type topology on the wing and some grid refinements around the attachment line region. Euler computations have been carried out with the code described in [14] allowing the $\bar{R}^*$ Reynolds number to be computed.
Figure 2 shows the computed variation of $\bar{R}^*$ along the span for the cruise conditions: $M_\infty = 2$ and a 60 000 ft altitude. The $\bar{R}^*$ values reach 600 in root region and decrease up to values around 100 near the tip. These large variations justified the need for fundamental experiments in order to have a precise idea of the leading edge contamination threshold.

2.3 Experiments in the R1Ch wind tunnel

2.3.1 Experimental set-up

The first series of experiments was performed in the supersonic R1Ch wind tunnel of the ONERA Chalais-Meudon Center. It is an intermittent, blowdown wind tunnel with a circular nozzle, the exit diameter of which is 31 cm. The free-stream Mach number M_∞ is 3. The stagnation temperature T_i was between 340 and 370 K, and the stagnation pressure P_i was varied between 0.8 and 3 bar.
A cylinder of diameter $D = 4$ cm equipped with an hemispherical nose was tested at variable sweep angles φ. The model was placed in the nozzle jet, as shown in figure 3. It was made of steel and equipped with a row of thermocouples and with a row of static pressure taps distributed along two lines parallel to the spanwise direction Z. The azimuthal distance $\Delta\theta = \Delta X/D$ between the two rows is 25°.
During the run, the cylinder rotates around its axis, so that it is possible to measure the wall heat flux and the pressure distributions in the X and Z directions. The boundary layer state is determined from the value of the wall heat flux coefficient.

2.3.2 Leading edge contamination by end plates

For these experiments, end plates of length $l = 0.2$ m were fixed to the model at the junction between its spherical and its cylindrical parts (figure 3). These end plates were placed at zero angle of attack with respect to the incoming flow direction.
Two values of φ were used for these experiments: $\varphi = 20°$ and 30°. As the corresponding values of the normal Mach number $M_n = M_\infty \cos\varphi$ are 2.8 and 2.6, a bow shock is formed upstream of the cylinder. Assuming that this shock is parallel to the attachment line, the static pressure P_e and the static temperature T_e at the boundary layer edge can be deduced from the two-dimensional oblique shock relations.

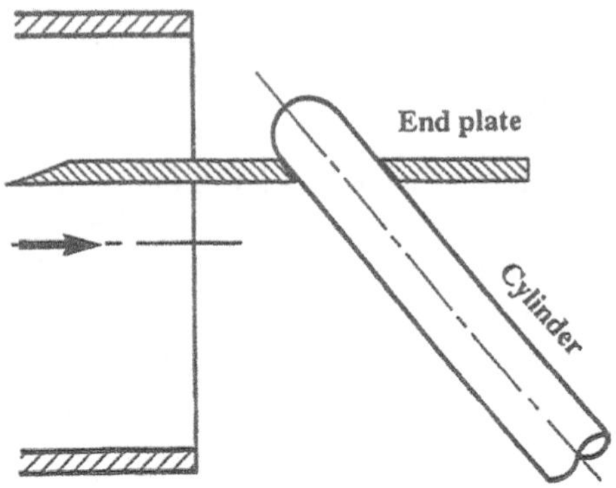

Figure 3: Experimental set-up in the R1Ch wind tunnel

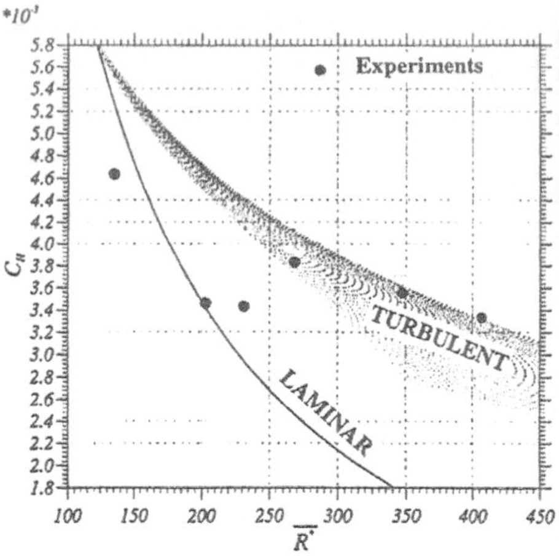

Figure 4: Example of experimental results (R1Ch experiments)

Pressure measurements along the attachment line indicate that the Kp distribution is uniform for $Z/D > 1$, where $Z = 0$ corresponds to the junction between the end plate and the cylinder. When the asymptotic conditions are reached, the measured pressure P_m is slightly different from the theoretical pressure P_e because the "effective sweep angle" φ_{eff} is not exactly equal to the geometrical sweep angle φ. φ_{eff} was computed by looking at the best fit with the measured Kp values. In all cases, φ_{eff} was lower than φ, with a difference between 3° and 5°.
Due to the rather large values of the normal Mach number, the velocity gradient k was deduced by using the modified newtonian law. As soon as k is known, it is possible to compute the attachment line Reynolds numbers $\bar{R}$ and $\bar{R}^*$.
The free stream Mach numbers M_e at the edge of the attachment line boundary layer are around 0.45 and 0.85 for $\varphi = 20°$ and 30°, respectively. On the other side, the ratio T_w/T_{aw} is close to 0.9 ($T_w \simeq 290$ K). It follows that the compressibility effects are weak and that $\bar{R}$ and $\bar{R}^*$ are close together: the difference is of the order of 1% for $\varphi = 20°$ and 5% for $\varphi = 30°$. The length scale η^* is close to the displacement thickness δ_1.
The time history of the wall temperature measured by the thermocouples (at the beginning of the run) made it possible to compute the wall heat flux Φ_w. For the sake of comparison with theoretical laws, the Stanton number C_H was used:

$$C_H = \frac{\Phi_w}{\rho_e U_e C_p (T_w - T_{aw})} \tag{3}$$

where ρ_e and U_e are the density and the free stream velocity along the attachment line. Close to the plate-cylinder junction, C_H was very large, then it decreased up to a nearly constant value which was reached at $Z/D \simeq 2$. This value was used to determine the state of the boundary layer by comparison with the empirical relationships proposed by Poll [10] for laminar and turbulent boundary layers. For laminar flows, C_H depends on $\bar{R}$ only; for turbulent flows, it depends on $\bar{R}$, T_e and T_{aw}.
Because the wall was nearly adiabatic, the measured wall heat flux was rather low, but it was large enough to allow a clear distinction between laminar and turbulent flows. The

experimental values of C_H are plotted as function of $\bar{R}^*$ in figure 4. It is important to keep in mind that each point corresponds to a different run, with different values of φ, P_i and T_i. In other words, M_e, P_e, T_{aw} and T_e are not the same for the different points. This explains why the theoretical values of C_H for turbulent flows do not correspond to a single line, but to a shaded area. The important result is that the measured Stanton number begins to deviate from the laminar law for $\bar{R}^* \simeq 200$ and that the fully turbulent regime is reached for $\bar{R}^* \simeq 260$. These results are not in contradiction with Poll's criterion.

2.4 Experiments in the S5Ch wind tunnel

2.4.1 Experimental set-up

In this second series of experiments, the wing was mounted on the floor of the supersonic S5Ch wind tunnel which has a closed circuit and a 0.3m × 0.15m test section. This facility is of the continuous type. Test Mach number can vary from 0 to 3.15 with a stagnation pressure from 180 mbar to 850 mbar and a stagnation temperature of 310 K. The model is a constant chord swept wing generated from a symmetrical airfoil. The chosen sweep angle ($\varphi = 74°$) is the leading edge sweep angle of the inner wing of the AS1 supersonic transport aircraft. Three dimensional Euler computations were carried out to determine the pressure distributions and to choose the leading edge radius. A radius of 14 mm was selected in order to provide values of $\bar{R}^*$ lower than 150 at low stagnation pressure and close to flight values existing on the inner wing of a supersonic transport aircraft at high stagnation pressure.

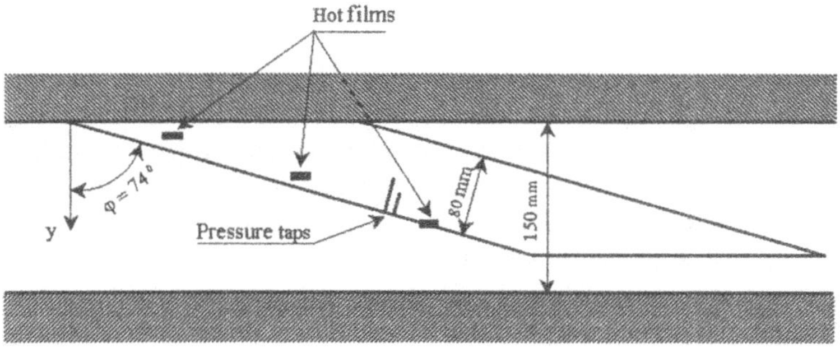

Figure 5: Experimental set-up in the S5Ch wind tunnel

Figure 5 presents the experimental set-up. The measurements are performed when the thermal equilibrium at the wall is reached ($T_w/T_{aw} = 1$), and the boundary layer state is determined through hot film signals. Three hot films were glued along the model span. The hot film 1 is close to the wing root. The second hot film is located in the mid span region slightly downstream of the attachment line while the hot film 3 is installed farther from the root on the attachment line. Two rows of pressure taps are aligned perpendicular to the leading edge. These pressure taps are regularly spaced in curvilinear coordinates in order to allow the effective sweep angle φ_{eff} and the leading edge $\bar{R}^*$ value to be computed. In these experiments, it was found that φ_{eff} was very close to the geometrical sweep angle.
The boundary layer of the wind tunnel wall was triggered in order to have a turbulent

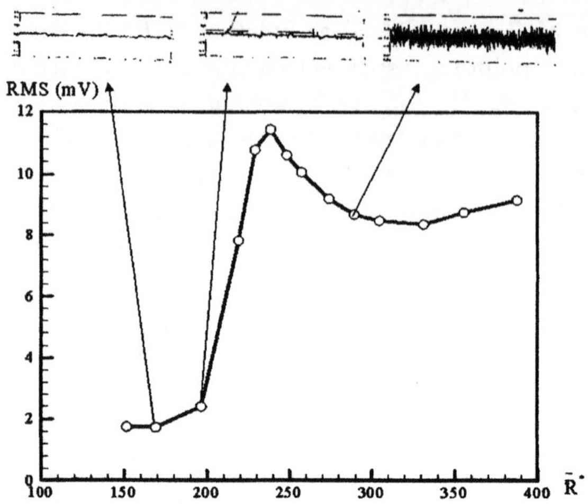

Figure 6: Example of experimental results (S5Ch wind tunnel)

boundary layer at the model root for the complete range of investigated stagnation pressures. A probe has been used during the test to check the state of this boundary layer having a thickness of about 5 mm.

2.4.2 Leading edge contamination by the wind tunnel wall

As an example, the RMS values computed from the signals of hot film 3 are plotted in figure 6 versus the $\bar{R}^*$ values. The free stream Mach number M_∞ is equal to 1.97, and the Mach number M_e parallel to the leading edge is 1.83. The obtained results show a change of the slope of the curve, corresponding to the appearance of the first turbulent spots indicating the onset of contamination. Leading edge contamination appears for $\bar{R}^*$ values slightly greater than 200. The instantaneous signals delivered by hot film 3 are also plotted in figure 6, showing that the number of spots increases until the leading edge boundary layer becomes fully turbulent. The same critical value of $\bar{R}^*$ has been obtained for a free stream Mach number M_∞ of 2.55.

It is interesting to notice that the results obtained in this experiment are close to the ones measured on the cylinder described in section 2.3, although the test conditions are completely different:

- The S5Ch experiment has been carried out with a subsonic leading edge ($M_n = 0.55$ and 0.70, no bow shock).
- The free stream Mach numbers M_∞ and the Mach numbers M_e parallel to the leading edge are different; in the R1Ch experiments, the attachment line flow is subsonic ($M_e =$ 0.45 and 0.85) while it is supersonic ($M_e = 1.83$ and 2.34) for the S5Ch experiments.
- A 74° sweep angle was considered in the S5Ch experiments, while the two values of φ tested in the R1Ch wind tunnel are 20° and 30°.

This demonstrates that the $\bar{R}^*$ Reynolds number is the right parameter which governs the leading edge contamination phenomenon in supersonic flow.

3 Roughness elements on and off the attachment line

These experiments were carried out in the R1Ch wind tunnel ($M_\infty = 3$) with the cylindrical model previously described in paragraph 2.3.1. The end plate was removed and isolated roughness elements were fixed on a ring which remains at rest during the rotation of the model. These roughness elements consist of small steel cylinders (height = diameter = d), their axis being normal to the wall. Their spanwise location is $Z = 1.5\,D$, and their size is such that d/η^* varies between 3 and 7 (let us recall that $\eta* \simeq \delta_1$).
The roughness elements were first placed on the attachment line, i.e. $\theta_R = 0°$. It was observed that a turbulent wedge started to develop immediately downstream of the roughness element as soon as $\bar{R}^*$ exceeds 240-250. This means that turbulent contamination by "large" roughness elements occurs at the same Reynolds number as turbulent contamination by a turbulent boundary layer, a result which is in agreement with previous investigations, see [2], [4], [6], [13] for instance.
Additional experiments were performed by placing the roughness elements at non-zero values of θ_R. The main result is that, for given values of $\bar{R}^*$ and d, the efficiency of the roughness element rapidly decreases as soon as the protuberance is slightly displaced off the attachment line. This implies that boundary layer tripping at non-zero values of θ_R (values of θ_R equal to 5°, 10° and 15° have been investigated) requires to increase $\bar{R}^*$ up to values which are much larger than 250. In other words, for a constant value of $\bar{R}^*$, the minimum roughness height which is necessary to trigger transition increases with θ_R. As previously stated by Morrisette [8], the attachment line is the location where a laminar boundary layer is the most sensitive to roughness elements.

4 Conclusion

The two series of experiments reported in this paper demonstrated that the onset of leading edge contamination occurs at $\bar{R}^* \simeq 200$ and that a fully turbulent regime is reached for $\bar{R}^* \simeq 250$. Therefore the critical value proposed by Poll seems to be close to the limit corresponding to the completion of leading edge contamination. Let us notice that a similar threshold was found during experiments performed with the R1Ch model in another wind tunnel at Mach 10 [2]. This shows that the criterion can be applied with confidence for a large range of aerodynamic conditions.
Since the attachment line contamination is likely to occur on the inner wing of the supersonic transport aircraft wing, large research effort is required to study this phenomenon and to develop anti-contamination devices. In subsonic flow, Gaster bumps and leading edge suction have proved their efficiency for delaying leading edge contamination, see for instance [12]. On the other side, passive relaminarisation devices applicable to compressible flows have been developed at NASA [5]. Such devices need to be tested in a systematic manner at supersonic speeds.
The experiments reported in this paper also demonstrated the sensitivity of the attachment line flow to roughness elements, a result which was previously established for hypersonic flows, see [2].

Acknowledgments

These studies have been supported by the "Service des Programmes Aéronautiques" (SPAé).

References

[1] D. Arnal and J.C. Juillen. Etude de la transition et de la contamination de bord d'attaque sur ailes en flèche. AGARD CP 438, 1988.

[2] D. Arnal, F. Vignau, and F. Laburthe. Recent supersonic transition studies with emphasis on the swept cylinder case. Conf. on Boundary Layer Transition and Control, Cambridge, April 1991.

[3] D.M. Bushnell and J.K. Huffman. Investigation of heat transfer to leading edge of a 76° swept fin with and without chordwise slots and correlations of swept-leading-edge transition data for Mach 2 to 8. Technical Report TM-X-1475, NASA, 1967.

[4] J.L. Da Costa. Contribution à l'étude de la transition de bord d'attaque par contamination en écoulement hypersonique. Master's thesis, Poitiers University, France, 1990.

[5] T.R. Creel. Effects of sweep angle and passive relaminarisation devices on a supersonic swept-cylinder boundary layer. AIAA 91-0066, January 1991.

[6] L. Gaillard. Etude de la transition de bord d'attaque sur un cylindre en flèche en écoulement hypersonique. Master's thesis, Poitiers University, France, 1993.

[7] M. Gaster. On the flow along leading edges. *The Aeron. Quarterly*, XVIII, Part2, May 1967.

[8] E.L. Morrisette. Roughness induced transition criteria for space shuttle-type vehicles. *J. of Aircraft*, 13(2), February 1976.

[9] W. Pfenninger. Flow phenomena at the leading edge of swept wings. AGARDograph 97, Part 4, 1965.

[10] D.I.A. Poll. Some aspects of the flow near a swept attachment line with particular reference to boundary layer transition. Technical Report 7805./K, Cranfield, College of Aeronautics, August 1978.

[11] D.I.A. Poll. Boundary layer transition on the windward face of space shuttle during reentry. AIAA Paper 85-0899, 1985.

[12] J. Reneaux, J. Preist, J.C. Juillen, and D. Arnal. Control of attachment line contamination. 2nd European Forum on Laminar Flow Technology, Bordeaux, June 1996.

[13] A.S. Skuratov and A.V. Fedorov. The laminar-turbulent transition past roughness at the attachment line of a swept cylinder in supersonic flow. *Izv. AN SSSR, Mekh. Zhidk. i Gasa (translated in english)*, (6):28–35, 1991.

[14] A.M. Vuillot, V. Couaillier, and N. Liamis. Turbomachinery Euler and Navier-Stokes calculations with a multidomain cell-centered approach. AIAA Paper 93-2576, 1993.

Laminar Design for Supersonic Civil Transport

Achmed Traoré
Institute of Design Aerodynamics, DLR, German Aerospace Center
D-38108, Lilienthalplatz 7, Braunschweig, Germany

Pascal Lemée
ONERA, Office National d'Etudes et de Recherches Aérospatiales
29 avenue de la Division Leclerc 92322 Chatillon, France

SUMMARY

The paper presents two design methodologies for supersonic wing sections with hybrid laminar flow control, one for the inner wing and one for the outer wing. For the inner wing with subsonic leading edge an inviscid method with a conical flow approximation is used to determine the pressure distribution for the boundary layer computation. The Tollmien-Schlichting and crossflow instabilities are computed with a database method. Suction distributions are optimized in order to reduce friction drag at minimum suction flow rate. For the outer wing the approach is based on coupled Euler/boundary layer flow simulation and linear stability analysis for transition prediction. The investigations show that combinations of simple pressure distribution shapes can be used to optimize airfoils for maximum extent of laminar flow and hence minimum friction drag.

NOMENCLATURE

α	=	angle of attack
c_f	=	skin friction coefficient
c_l	=	lift coefficient
c_p	=	pressure coefficient
c_q	=	suction coefficient; $c_q = (V_s\, \rho)_w$
Ma	=	Mach number
Re	=	Reynolds number based on freestream condition
Φ_{LE}	=	Leading edge angle
N_{tr}	=	Transition N-factor (envelope method)

INTRODUCTION

Mission requirements of second generation supersonic civil transport (SCT) demand friction drag reduction which probably cannot be achieved with fully turbulent configurations. The technological potential of laminar control for SCT is up to now not entirely evaluated. Problems concerning suction rates or passive/active boundary layer control are at now subject of investigations. This is the context for the development of tools for, and the validation of, a laminar design methodology for SCT.

The aim of a laminar flow design is to move the region of the laminar/turbulent transition as far downstream as possible. In supersonic flow laminar control is feasible using a combination of profile shaping and active laminar flow control. Appropriate measures could be air suction or wall cooling. On the inner wing of SCT high leading edge sweep angles in combination with blunt leading edges cause an enhancement of crossflow in the boundary layer in the leading edge region. Crossflow disturbances are amplified in this area and can provoke early transition. Therefore suction has to be performed in an area from the leading edge to the front spar. Whereas on the outer wing of SCT the leading edge is very thin. Hence it will be technically difficult to integrate suction panels in this area to control crossflow instability. To overcome this difficulties two possible solutions exists for the laminar design. First a sharp nose can be chosen as with the BCA-SST [1] concept. On such wedge shaped leading edges no crossflow disturbances are amplified in the nose region. But at low speed the flow separates at the leading edge. This generates additional vortex drag and may change the aircraft handling qualities. Another design option is to retain a blunt nose profile and to control the crossflow instabilities with an adequate leading edge shape. In both options Tollmien-Schlichting (TS) waves have to be damped with suction.

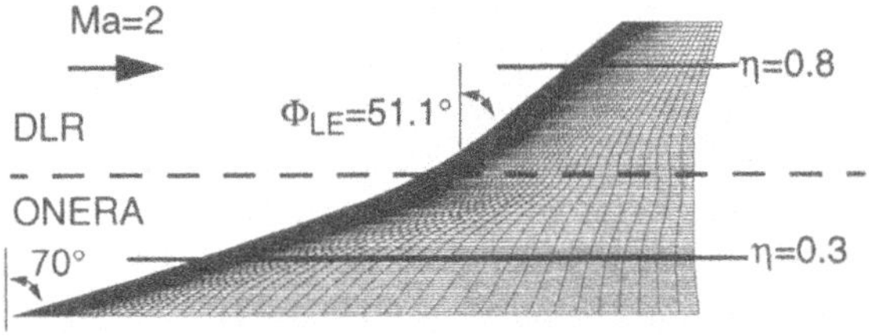

Fig. 1: Wing of the AS-1 configuration.

Since 1997 DLR and ONERA have been cooperating in the supersonic laminar design with the goal to design and assess a three dimensional laminar wing. The investigations are carried out using the Aerospatiale Matra-Airbus AS-1 configuration for Ma=2 as a baseline wing shape (see **Fig. 1**). DLR investigates the outer wing with blunt supersonic leading edge. Similar investigations are performed by ONERA for the inner wing.

1.PRELIMINARY ASSESSMENT

In the first step of this research the total drag reduction on an existing geometry with regard to the application of LFC is evaluated, and the suction flow rate and suction power needed to achieve a significant run of laminar flow on the wing surface are assessed. The survey on the attachment line contamination reveals that turbulent flow at the wing root is propagated along the attachment line of the inner wing surface. In the following, we assume that this phenomenon can be prevented by using passive devices. The assessment procedure of transition onset was based on the following computation sequences: An inviscid method (Euler) determines the pressure distribution and its output is input to the three-dimensional boundary layer code. The Tollmien-Schlichting and cross-flow instabilities are computed with the streamwise database method [2] and the «C1» criterion [3] respectively. This approach is verified in certain cases, with additional computations, based on the solution of the linear stability equations. Suction distributions are optimized in order to reduce on maximum friction drag for a minimum flow rate. The suction area is defined from the leading edge to the front wing spar on the upper and lower wing surface.

A correct computation of the pressure distribution near the attachment line (**Fig. 2**) is important in this kind of study in order to properly initialize the computation of the boundary layer because, without suction, natural transition occurs near the attachment line. To this end, special attention was paid on the mesh generation near the leading edge as far as grid density and topology are concerned. Wall suction stabilizes the boundary layer and delays the transition onset (**Fig. 3**).

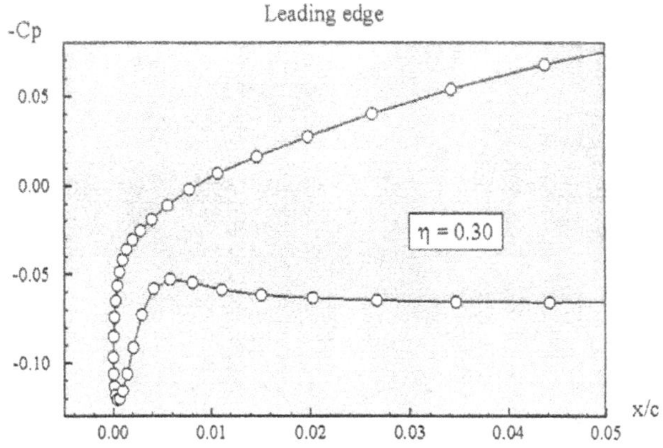

Fig. 2: Computed pressure distribution at the leading edge

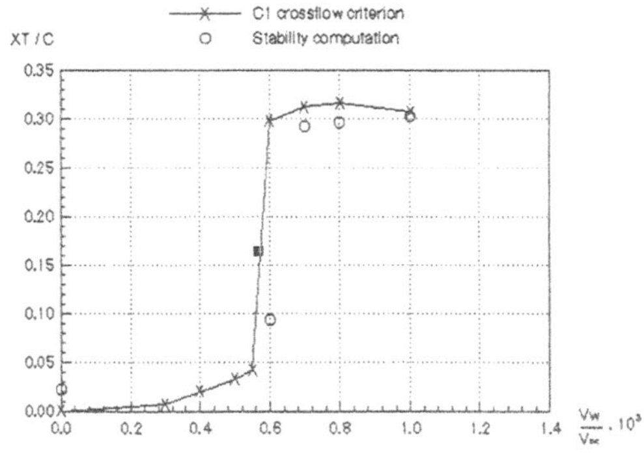

Fig. 3: Evolution of the transition location vs. suction rate at η=0.3

The evolution of the total amplification factor for uniform suctions from 0% to 20% of the local chord on an inner upper wing surface section is presented in **Fig. 4**.

Without suction, cross-flow instabilities causes the transition near the leading edge. The most unstable frequency and direction at transition point are respectively about 7000 Hz and 75°. Suction decreases significantly the total amplification factor, and major instabilities get more and more in the cross-flow direction. Above a certain suction velocity, the transition position remains unchanged however the local laminar skin friction coefficients increase thereby reducing the gains. Appropriate suction laws delay significantly the transition and assure considerable benefits.

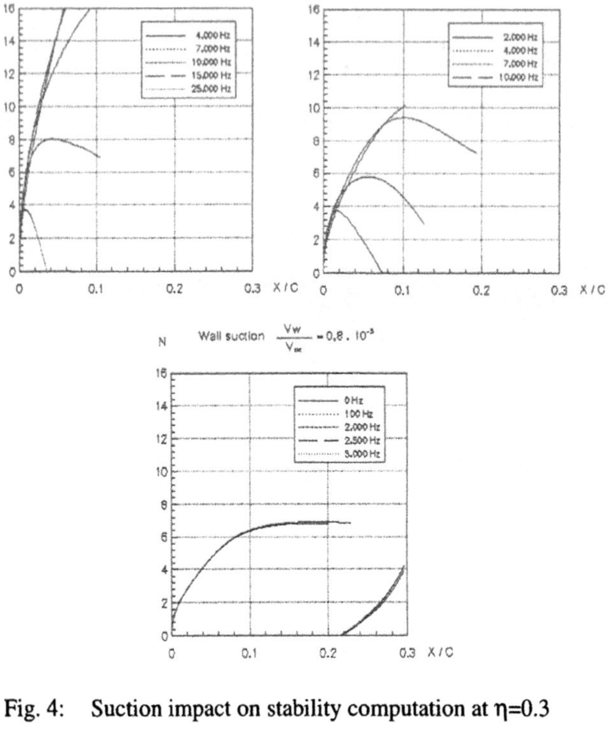

Fig. 4: Suction impact on stability computation at η=0.3

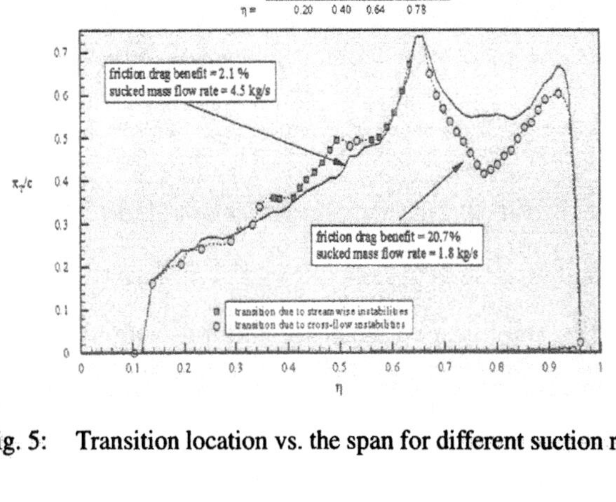

Fig. 5: Transition location vs. the span for different suction rates.

Fig. 5 illustrates the importance of optimizing the suction distribution. Uniform suction is compared with variable suction velocity in span and streamwise directions. For the same suction area and finally quite the same transition position, benefits on the friction drag have been improved by 20% and the suction flow rate has decreased by 60% for the optimized suction distribution. Transition location now ranges from 25% of the chord at the root up to 60% of the chord at the tip and the viscous drag reduction reaches 20.7% on the upper wing surface. The total aircraft drag reduction has been estimated as 4.1%, and the total suction mass flow rate is 6.26 kg/s assuming that suction flow is ejected in adaptive condition. Taking into account the suction power absorbed by the suction pump the final benefit was evaluated as 2.7% of the propulsion power.

This preliminary study has shown the importance of suction distribution which depends on the pressure distribution. An adequate flow field would allow to further delay the boundary layer transition onset with a lower suction flow rate. Benefits will be multiple. First, the friction coefficient is reduced by the extension of the laminar flow and a lower suction flow rate, moreover the power absorbed by the suction pump decreases. That is the reason why the next step is to determine the adequate pressure distribution for LFC application and define the wing shape that produces these pressure distributions. The research work to determine favorable pressure distributions for inner wing section with subsonic leading edges are performed by ONERA whereas DLR concentrates on the outer wing with supersonic leading edge.

2.INNER WING DESIGN

Design Method

The computation procedure for inner wing design is based on the following sequences (see **Fig. 6**):

- first, a pressure distribution is defined. Some constraints were fixed to avoid purely academic aerodynamic solutions. The acceleration at the leading edge is kept similar to the reference case in order to keep the nose radius reasonably large.
- the velocity distribution is deduced from pressure distribution using conical-flow approximation as originally proposed by Kaups and Cebeci. This hypothesis was confirmed with the reference case.
- then, boundary layer computation is performed in considering a suction area from the leading edge to the front spar, it makes about 8% of the local chord. Transition location is predicted with criteria [3] .

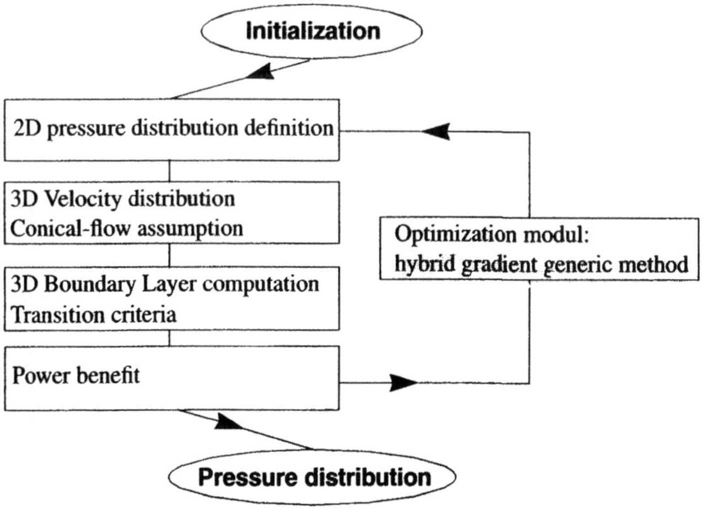

Fig. 6: Flowchart of the optimization process

Design parameters

Pressure distribution as obtained from the inviscid calculation on the previous work SCT wing body presented above is used as reference along this study. The first step carried out a parametric study in order to define sensitivities of the major design variable before optimization would be attempted. A steep initial acceleration is a means to avoid cross-flow development in the leading edge. Nevertheless, if the acceleration is too high, then a strong suction rate is necessary to limit cross-flow instabilities with penalties in the global result. Moreover, Tollmien-Schlichting disturbances grow rapidly with a deceleration (as observed for the reference pressure distribution). That is to say, it seems beneficial to break down the pressure evolution to the transition location into 2 accelerations (see **Fig. 7**):

- a first one, steep and short to control the cross-flow instabilities and;
- a second one, more moderate to limit the streamwise disturbances.
-

Several suction distribution have been tested, like a high level on the first acceleration and a lower on the moderate one. But, it appears that the suction intensity to control cross-flow development in the first part is similar to the distribution level to control TS instabilities in the second part.

On the lower wing a similar strategy is applied. But, streamwise disturbances increase when Mach number decrease, by compressibility effect. Therefore, it is difficult to succeed in reaching the same benefit on the lower wing surface as on the upper. However, to take advantage of the compressibility effect on the streamwise disturbances, it seems interesting to decrease the pressure level on the lower wing. Note that this approach is limited by the constraints on lift and pitching moment coefficient.

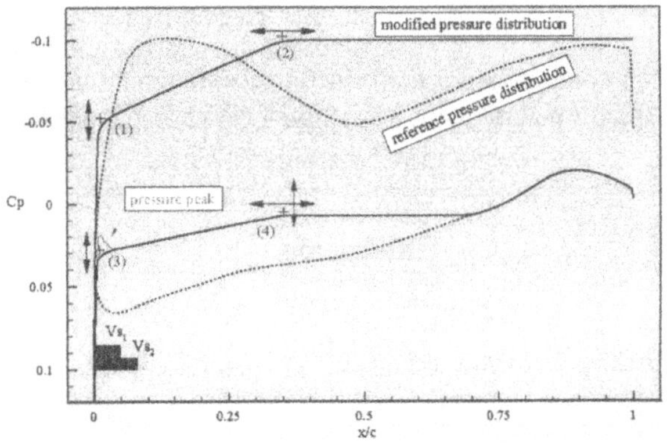

Fig. 7: **Optimized pressure distribution and design parameters**

Based on suggestion by Pfenninger, pressure peaks were also tested near the leading edge. The pressure recovery downstream of the peak pressure generates a cross-flow in the opposite sense to that generated in the initial flow expansion and hence significantly lowers the cross-flow into the wing box area.

Optimization

Fig. 8 presents the laminar flow control implementation results on the reference and optimized pressure distribution. For both cases, suction is applied over the first 8% of the chord, and lift and pressure drag coefficients are similar.

Suction from the leading edge to the front spar (8.1% of the local chord on section y/b=30%)

■ **Reference pressure distribution (y/b=30%)**, CD_p=0.0058, CL=0.1, CM=-0.019

	$V_{suction}$	$x_{transition}/c$	Friction drag benefit	Power benefit
upper wing	0.625 m/s	0.24	7.31 %	5.04 %
lower wing	0.340 m/s	0.20	7.79 %	6.15 %

■ **Optimised pressure distribution**, CD_p=0.0059, CL=0.095, CM=-0.025

	$V_{suction}$	$x_{transition}/c$	Friction drag benefit	Power benefit
upper wing	0.405 m/s	0.34	15.59 %	14.08 %
lower wing	0.181 m/s	0.28	12.76 %	11.95 %

Fig. 8: Friction drag and power benefits for the inner wing

The new pressure distributions enables to delay the transition onset by about 10% of the local chord with 40% less suction mass flow rate. Optimized pressure distributions on the inner wing therefore double the benefit on friction drag compared to the reference inner wing. And, in taking account suction penalties, the discrepancy grows to reach roughly a ratio of 3 for the upper wing and a ratio of 2 for the lower wing.

3.OUTER WING DESIGN

Design method

The design methodology is based on a systematic study of different simple shaped pressure distributions with the goal to find the beneficial characteristics for laminarity. A combination of the most favorable distributions will yield the optimized laminar profile.

Starting from the reference airfoil new target pressure distributions are graphically defined and corresponding profile geometries are generated with an inverse design tool. After performing a laminar boundary layer computation and a stability computation the transition location can be determined with an assumed transition N-factor. Then a new boundary layer computation with known transition location provides the friction drag coefficient C_f.

According to **Fig. 9** the inverse design process starts with a graphic prescription of a target pressure distribution. Then an optimization loop starts. Note that the object function for the optimizer is calculated by a summation of local c_p-differences between actual and target distribution.

For the present application the FLOWer code solves the Euler equations to provide the pressure distribution at each optimization step. This computation is performed on a grid of 122x20 cells. When the inverse design process stops a final Euler computation is achieved on a grid of 224x40 cells. This inviscid field solution is then used as the outer boundary condition for the boundary layer computation.

The tool PROFIX generates profiles by catenating geometric support points with Bezier curves. Each support point is defined by 3 parameters: the locations x, y and the local curve slope. To define a blunt nose a ramp function is used with a prescribed infinite slope at x/c=0. This tool uses the basic curve function routines introduced by H. Sobieczky [6] . **Fig. 10** sketches an example how the curve functions are used to define the airfoil.

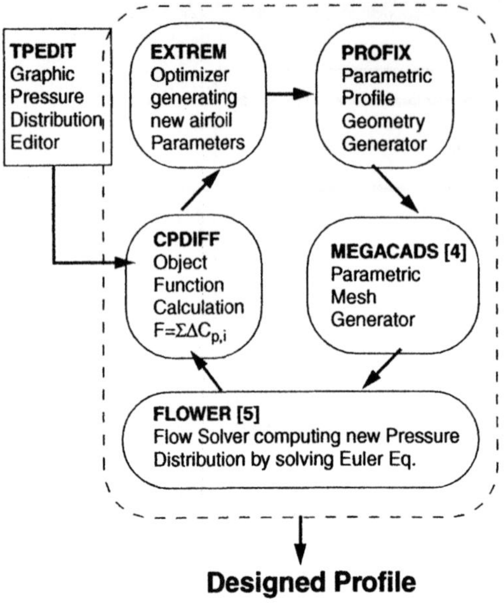

Fig. 9: Flowchart of the inverse design process.

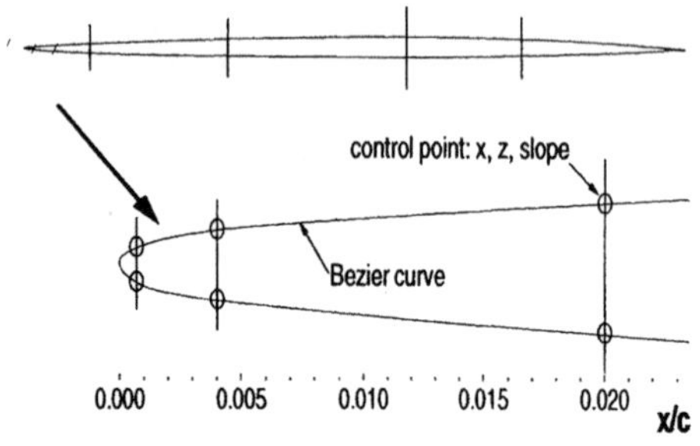

Fig. 10: Sketch of profile parametrization.

The number of support points can be changed by user input according to local profile shaping requirements. For the present application 14 support points are used which are controlled by 32 geometric parameters.

The location of the support points are chosen in a way to allow an accurate reproduction of the target pressure distribution. During the inverse design the angle of attack is a free parameter. For reason of structural constraints the profile thickness of the reference airfoil has to be retained. During the inverse design this is achieved by local adaption of the target pressure distribution.

The code SOBOL [7] used here solves the second-order boundary-layer equations. Solution of this set of equations can be regarded as classical boundary-layer solutions corrected so that wall curvature, viscous interaction and wall-normal inviscid flow gradients are taken into account. The solution of the system of partial differential equation is performed with a space marching finite-difference method. The simulation of suction is implemented in the code. For the present design study a uniform suction panel with a suction power corresponding to a c_q=0.33 10^{-3} over a length of $\Delta x/c$=0.06 is selected. This suction power amount emerges from preliminary suction investigations.

The code COAST [8] analyses the instability of compressible boundary layers with the semi-empirical e^N-method [9] . COAST solves the linear stability equations of compressible, parallel, three-dimensional flow along curved surfaces. The envelope N-factor integration method is implemented and used for the present application. Curvature effects are not considered. The transition N-factor for free flight conditions is assumed to be N_{tr}=6.

Design parameters

Starting from the reference profile denoted by *case0*, three different series of generic pressure distributions were generated. The variation of the pressure distribution is restricted to the first 25% of chord. The first series denoted by *ramp* (**Fig. 11**) consists of a variation of the slopes of a linear pressure drop without suction peak. The second series denoted by *lev* shows a parallel pressure level shift of the linear pressure drop (**Fig. 12** top).

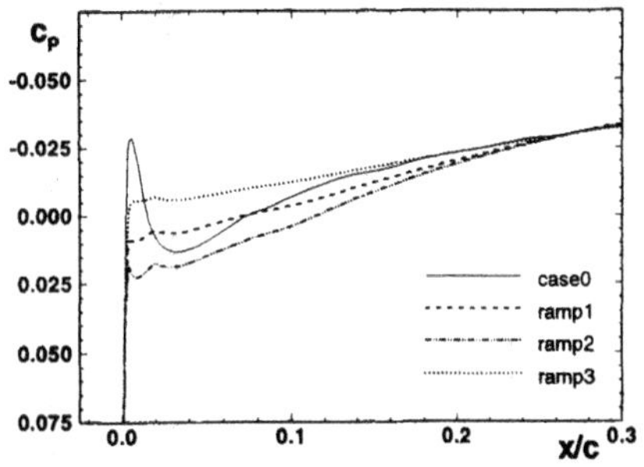

Fig. 11: Generic *ramp* pressure distributions.

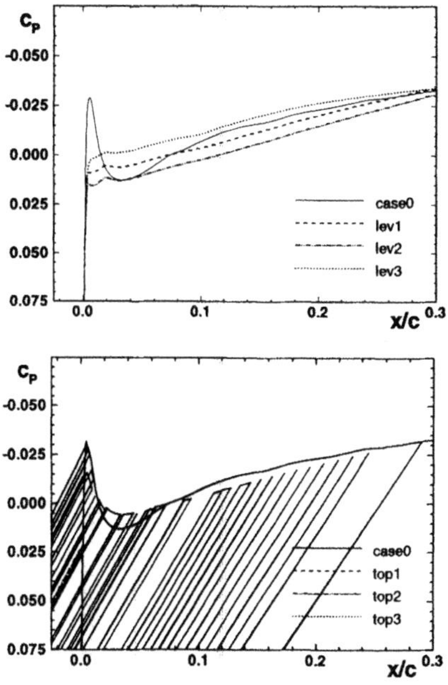

Fig. 12: Generic *lev* and *top* pressure distributions.

The last series denoted by *top* combines three different suction peaks with a *ramp* distribution (**Fig. 12** bottom)

First the effect of the suction panel location on wave damping/amplification behavior was studied. This was done by moving the suction panel with the same suction power up and downstream. The optimized suction location was found in the area where TS waves just begin to grow. It was found necessary, for the suction fitness evaluation to perform an optimization of the suction panel location for each case separately.

The second basic investigation concerns the variation of the suction panel length. This is done by a variation of the local suction coefficient keeping the total suction power constant and the suction panel location unchanged at 5% chord. Three suction panel lengths are investigated on the *ramp1*-distribution whose natural stability is shown in **Fig. 13**. Note that the lack of the suction peak causes a distinct dominance of crossflow waves.

As given in **Fig. 14** increasing the suction length causes a dilution of the suction efficiency and so yields no better damping behavior on TS-waves. This instance meets the technical constraint which demand short suction panels.

Design results

All stability results shown here are optimized regarding the suction panel location. The most favorable *ramp* distribution is *ramp3* **Fig. 15** (top). Note that the lower slope of the pressure drop yields a longer laminar extent.

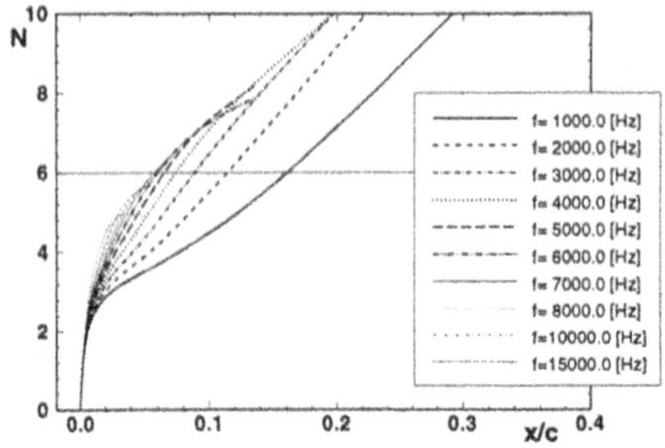

Fig. 13: Stability of the *ramp1*-distribution (no suction).

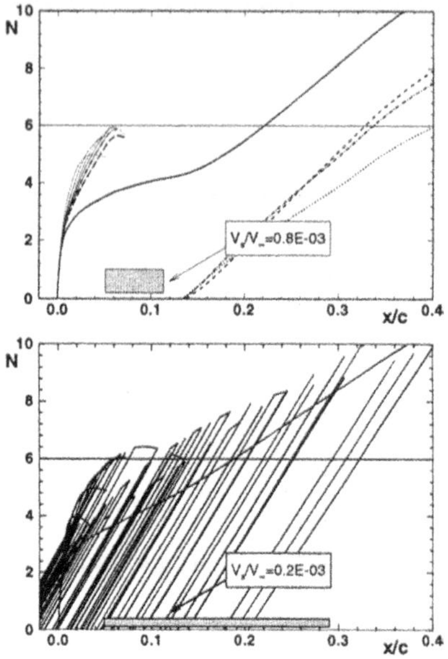

Fig. 14: Variation of the suction panel length

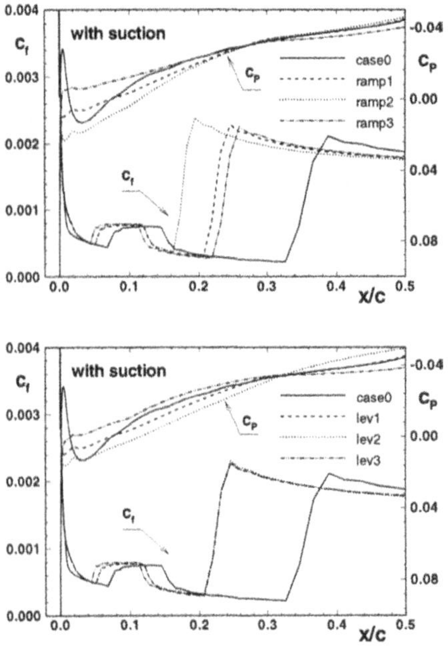

Fig. 15: Transition on *ramp* and *lev* distributions.

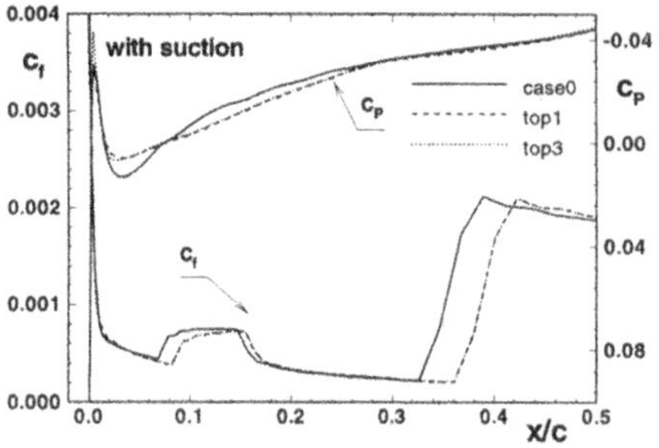

Fig. 16: Transition on the *top* distributions.

The transition location of the three *lev* distributions do not differ from each other (see **Fig. 15** bottom). This shows that the pressure level cannot be used as an effective design attribute. On this account the *lev* distributions are not considered for the further design.

The *top1* and *top3* distribution have the same transition location at 36% chord (see **Fig. 16**) and so they represent the most favorable designs due to the property that suction peak damps crossflow disturbances.

Optimized design

Combining the characteristic shape of *top1* and *ramp3* a new pressure distribution is generated. With an optimized suction location the new designed profile has a laminar flow region of 46% chord, see **Fig. 17**.

Fig. 18 illustrates the impact of the new pressure distribution on the stability. The higher suction peak has a damping effect on crossflow waves. The shorter pressure rise extent after the peak damps the TS-waves with a parallel shift of the N-curves downstream maintaining the effect of suction.

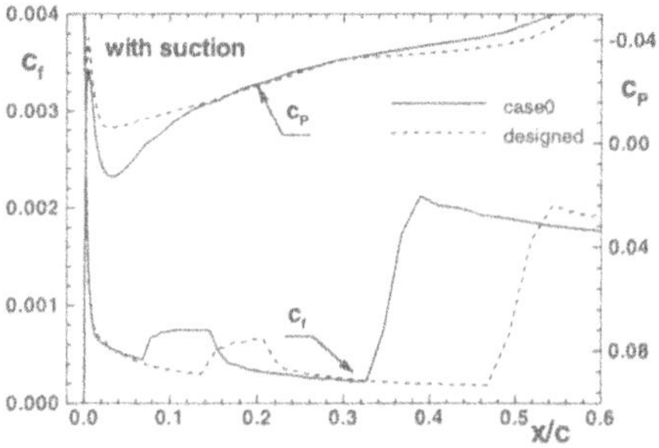

Fig. 17: Designed pressure distribution.

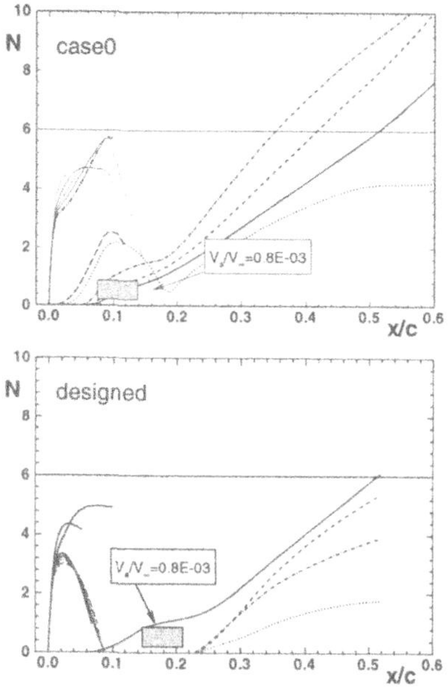

Fig. 18: Stability of the designed profile.

A comparison of both airfoil geometries in **Fig. 19** shows that the new designed profile has about the same curvature at the leading edge and the same thickness as the reference airfoil *case0*.

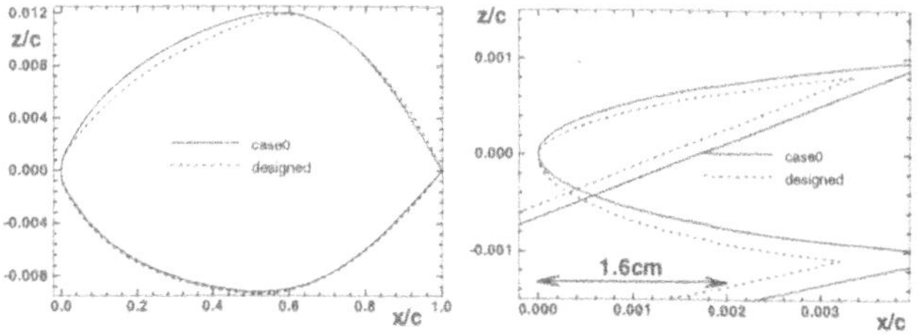

Fig. 19: Geometry of the designed profile.

Table 1 shows that with the new distribution the total drag can be reduced by about 5.9% compared with the reference distribution *case0*. In the third column the total airfoil drag of upper and lower surface was conservatively estimated by using twice the skin friction drag of the upper surface.

Table 1: Comparison of total drag

case	c_{dp}	c_{df}	$c_{dp}+2c_{df}$
case0 (no suction)	0.006828	0.001651	0.01013
case0 (suction)	0.006828	0.001280	0.009388
designed (suction)	0.006780	0.001039	0.008821

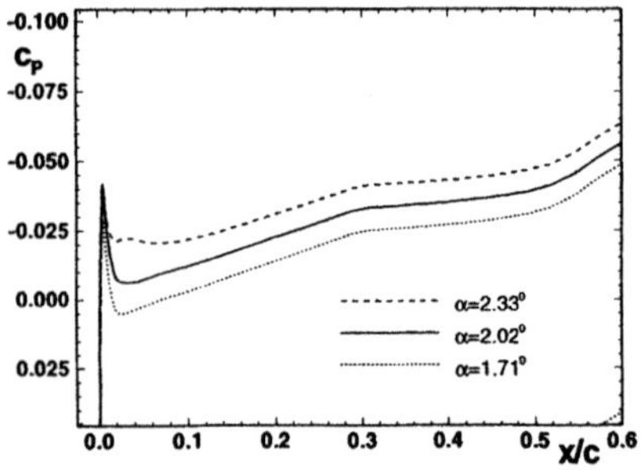

Fig. 20: Variation of the pressure distribution with the angle of attack.

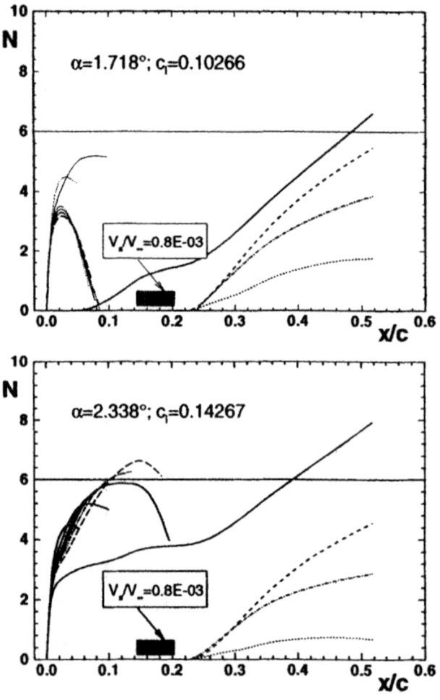

Fig. 21: Laminar pocket evaluation

The spanwise range of the validity of laminar characteristics of the designed profile is also evaluated. This is done by increasing the Reynolds number about 1.5 times and decreasing 0.75 times with reference to the design Reynolds number of Re=4.8 $10^6[m^{-1}]$. This simulates a typical variation of the profile chord length. The transition location is not strongly affected by the given Reynolds number range. This indicates that the designed pressure distribution could be used over the whole outer wing of the AS-1.

For the evaluation of the laminar pocket of the designed profile the angle of attack is changed in two directions by $\Delta\alpha$=0.3°. This causes an alteration of the lift coefficient of Δc_l=20%. The corresponding pressure distributions gives **Fig. 20**. The design angle of attack is α=2.02°

The stability results in **Fig. 21** show that the lower angle causes a negligible decrease of the laminar length of $\Delta x/c=0.015$. Increasing the angle of attack has a stronger impact. Here we have an amplification of crossflow waves due to the reduced suction peak and this causes a significant loss of laminar length shifting the transition location to about x/c=0.1.

4.CONCLUSION

The present study has shown that suction laws needed for extended laminar flow over the inner wing depend on the pressure distribution. A favorable pressure distribution was designed to further delay the boundary layer transition onset with a lower suction rate. The results show that the benefit concern both friction coefficient reduction and lower suction flow rate which decreases the suction pump power.

For the outer wing a successful design of laminar flow airfoils with a blunt supersonic leading edge was described. Simple shaped pressure distributions were generated and their stability analyzed. A combination of suction peak at leading edge to control crossflow and favorable pressure gradient behind the leading edge suction peak to delay TS-instability were used to design an optimized pressure distribution for hybrid laminar flow control. The study shows that crossflow instability can be controlled only through profile shaping on a 51° swept wing with blunt leading edge at Ma=2. The location of a given suction panel was optimized to increase the efficiency of suction in order to enhance the extent of laminar flow. Compared to the turbulent reference design the new design has less wave drag and less skin friction drag, both for natural laminar flow and hybrid laminar flow control.

REFERENCES

[1] R.D. Wagner, m. C. Fischer, F.S. Collier, W. Pfenniger: *Supersonic Laminar Flow Control on Commercial Transports,* ICAS-90-3.6.3, 1990

[2] D.I.A. Poll: *Boundary layer transition on the windward face of space shuttle during reentry,* - AIAA Paper 85-0899, 1985

[3] T.R. Creel: *Effects of sweep angle and passive relaminarization devices on a supersonic swept-cylinder boundary layer* - AIAA Paper 91-0066, 1991

[4] Brodersen, O.; Hepperle, M.; Ronzheimer, A.; Rossow, C.C.; Schöning, B., *The Parametric Grid Generation System MegaCads.* Proc. of the 5th Inter. Conference on Numerical Grid Generation in Computational field Simulations, 1996, Mississippi, Ed.: Soni, B.K., Thompson, J.F., Häuser,J., Eisemann, P., pp. 353-362, 1996

[5] Koll, N.; Rossow, C.C.; Becker, K.; Thiele, F., *Megaflow A Numerical Flow Simulation System*, Proceedings of the 21st ICAS Congress Melbourne, 1998

[6] Sobieczky, H., Chourdhry, S., Eggers, T., *Parametrized supersonic transport configuration,* 7th European Aerospace Conference EAC, 1994, Toulouse, France, Paper 3.14

[7] Monnoyer, F., *Second Order BOundary Layers SOBOL Mk 2.7 Handbook*, UVHC-LMF-NT-001, 1994

[8] Schrauf, G.: *COAST2 - A Compressible Stability Code, Users Guide and Tutorial.* Report No. EF1-1973, 1993

[9] Arnal, D., *Boundary-Layer Transition: Prediction based on Linear Theory, Special Course on Progress in Transition Modeling*, AGARD-R-793, Brussels, Belgium, (1993)

The stability results in Fig. 47 show that the lower angle causes a negligible decrease of the laminar region to x/L≈0.[illegible]. Increasing the angle of attack has a stronger impact. Here we have an amplification of crossflow waves due to the [illegible] suction peak and this causes a significant loss of laminar length shifting the transition location to about x/c≈0[illegible].

CONCLUSIONS

[illegible]

[illegible]

[9] [illegible] Conference on [illegible] Transition Modelling, [illegible], Brussels, Belgium, [illegible]

Session 6.1
Transition Prediction

Linear Stability Theory Applied to Natural and Hybrid Laminar Flow Experiments

G. Schrauf*, K.-H. Horstmann†
* DaimlerChrysler Aerospace Airbus, 28183 Bremen, Germany
† DLR Institut für Entwurfsaerodynamik, Lilienthalplatz 7, 38108 Braunschweig, Germany

Summary

The e^N-method is applied to three large-scale experiments performed in the ONERA S1MA wind tunnel and two flight experiments with the ATTAS VFW614 and a Fokker F100 aircraft. N-factor correlations using incompressible as well compressible linear stability theory obtained for natural and hybrid laminar flow surfaces are discussed.

1 Introduction

The e^N-method, based on linear local stability theory, is the standard tool for the design of natural or hybrid-laminar-flow surfaces for transport aircraft. It is based on the computation of growth rates of wave-like disturbances in the boundary layer using linear stability theory. Transition is assumed to take place where the most unstable disturbances are amplified by a factor e^N, with N determined by correlations with experiments.

The e^N-method was first developed for two-dimensional, incompressible flows, and later extended to three-dimensional, compressible flows. In two dimensions, the disturbances can be characterized by their frequencies allowing for an unambiguous calculation of their amplification rates. The extension to three-dimensional boundary layers introduces an additional degree of freedom so that, in addition to the frequency, a second quantity must be specified to identify an instability wave for the N-factor calculation. Possible choices are

(A) the wave propagation direction,

(B) the wave length,

(C) the spanwise wave number,

(D) or to choose the wave with maximal local amplification.

For each of the four possibilities there are four different procedures or "strategies" to calculate N-factors. Details can be found in [1].

Boundary layer transition can be caused by three different mechanisms on a typical swept wing of a transport aircraft. A not yet fully understood mechanism can cause transition directly at the attachment line. Leaving the attachment line, the boundary layer exhibits a strong, inflectional cross-flow velocity profile causing the cross-flow instability. Further downstream, the boundary layer becomes unstable with respect to Tollmien-Schlichting waves. Both, cross-flow and Tollmien-Schlichting instability, can be captured by the e^N-method.

In this paper, we consider the results of five large-scale laminar flow tests: three wind tunnel experiments in the ONERA S1MA wind tunnel, flight tests with the ATTAS VFW614 NLF glove [2, 3] and with the Fokker F100 NLF glove.

2 Evaluation with Linear Stability Theory

There are two different transition prediction methods based on correlated N-factors: a method based on only one N-factor and methods that use two N-factors, one to capture the cross-flow amplification and a second one to model the Tollmien-Schlichting amplification. Each N-factor is computed by one of the strategies listed above.

2.1 Envelope Method

The envelope method, choice (D), is a one N-factor method. The flight tests with the Fokker F100 NLF glove have been extensively evaluated with this method [5, 6]. The outcome was that the envelope method can be applied for natural laminar flow and yields the smallest scatter of all N-factor methods if compressible stability theory including surface curvature effects is used. For hybrid laminar flow the cross-flow amplification depends strongly on the boundary layer suction, much less than the Tollmien-Schlichting amplification occuring behind the suction panel. Due to the reduction of cross-flow amplification by suction, transition is generally caused by Tollmien-Schlichting waves. As long as cross-flow transition is prevented, the transition location is not much influenced by moderate changes in the suction rates. Because both, cross-flow and Tollmien-Schlichting amplification rates are integrated into one N-factor, its value at the transition location depends on the suction rate and is, thus, not "universal" [7].

2.2 Two N-Factor Methods

It was shown in [5], that for Tollmien-Schlichting amplification on a transonic wing with high aspect ratio it is sufficient to consider the 0°-direction even though this direction is not the direction of largest TS-amplification for transonic flow. This is due to the fact that the local TS-amplification rates vary only slowly with propagation direction. Therefore, we choose strategy (A) with 0°-direction for Tollmien-Schlichting waves. An example of the application of this strategy for the calculation of N_{TS}-factors with incompressible as well as compressible theory is given in Figure 1(a).

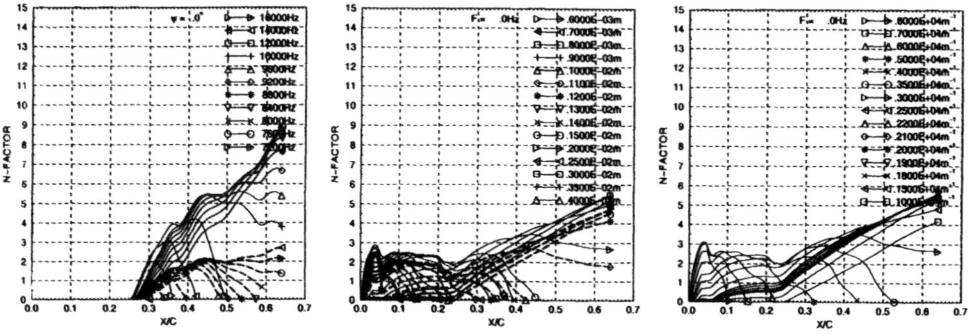

(a) N_{TS}-factors (b) N_{CF}-factors (c) N_β-factors

Figure 1 N-factors obtained with incompressible (solid lines) and compressible (dashed lines) theory ($M_\infty = 0.82$)

In a low-turbulence environment, stationary cross-flow waves, or cross-flow vortices, dominate the transition process [8, 9]. Furthermore, it is shown in [10] that traveling cross-flow waves are more efficiently damped by suction than stationary cross-flow vortices. Both observations justify our approach of considering only the stationary cross-flow disturbances for the design of laminar surfaces.

The influence of compressibility on the N-factors is larger for Tollmien-Schlichting waves than for cross-flow vortices as can be seen from Figure 1 and also in Figure 2(b). The same behavior has been shown for the Fokker F100 and the ATTAS flight tests.

Flow visualizations show that cross-flow vortices have approximately constant wave length so that (B) is the appropriate strategy. For wings with high aspect ratio we can also apply strategy (C) as it is the mathematically correct strategy for infinitely long swept wings. Both strategies, give equivalent results for typical wings of transport aircraft as was shown in [4, Figure 15] for the ATTAS glove and in [6, Figure 14] for the Fokker F100 glove. For the ELFIN II performance wing, sample calculations with strategy (B) and (C) are given in Figures 1(b) and 1(c). The equivalence of correlated (N_{TS},N_{CF}) and (N_{TS},N_β) pairs is shown in Figure 2(a) with N_{CF} given by strategy (B) and N_β given by strategy (C).

3 Comparison of Laminar Flow Experiments

In Figure 3 we compare the correlated (N_{CF},N_{TS})-factor pairs of the S1MA tests with the ELFIN II HLF performance wing and the ELFIN I HLF experiment. We observe that the N-factor pairs are in agreement, even though the pressure distributions were of different types.

A small section of the outer part of the ELFIN II model had a polished surface without suction holes in order to simultaneously produce NLF results for identical flow conditions. The few results available for the NLF section are also shown in Figure 3. They coincide well with the

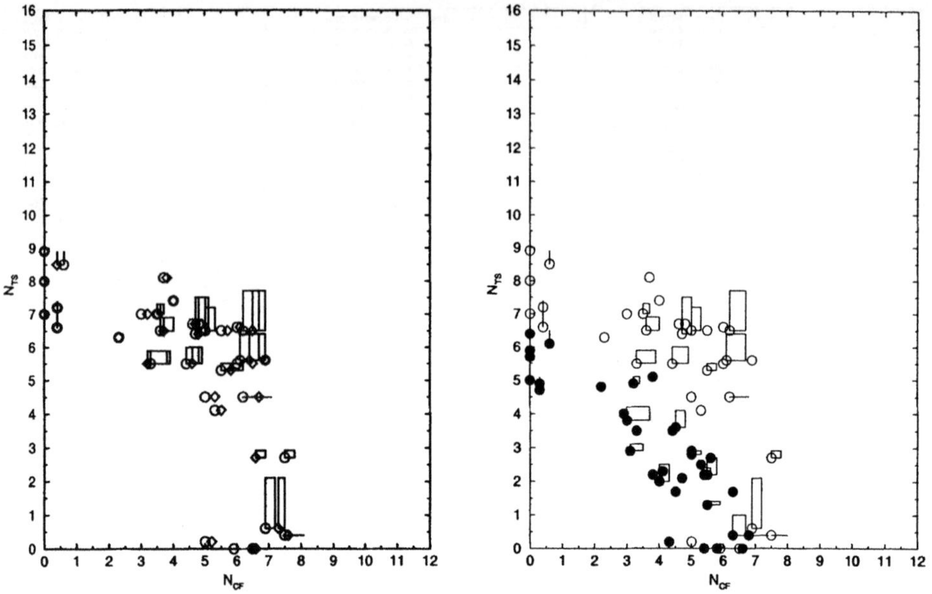

(a) (N_{CF}/N_{TS}) pairs ○ and (N_β/N_{TS}) pairs ◇ obtained with incompressible stability theory

(b) (N_{CF}/N_{TS}) pairs obtained with incompressible ○ as well as compressible stability theory •

Figure 2 Correlated N-factor pairs for the ELFIN II HLF test

ones obtained for the part with suction. This contradicts the findings of [11] in which was stated that the correlated N_{CF}-factors for the HLF cases should be smaller than the ones for the NLF-cases due to higher initial amplitude caused by the greater roughness of the suction panel.

Furthermore, we include in Figure 3 the results of the TLF S1MA wind tunnel experiment with the ATTAS NLF glove. This is the earliest S1MA experiment presented in this paper. The correlated N_{CF}-factors agree well with those of the two other experiments, whereas the N_{TS}-factors are much larger than in both of the previous wind tunnel experiments.

In Figure 4 we compare the the ELFIN II HLF wind tunnel results with the two NLF flight tests performed with the ATTAS VFW614 and the Fokker F100 aircraft. The F100 results do not include the so-called "pathological" cases [5, 6] for which transition occurs behind the N-factor maximum.

The correlated N_{CF}-factors obtained with both incompressible as well as compressible stability theory show the same trend: they are smallest for the ELFIN II HLF experiment, larger for the Fokker F100, and largest for the ATTAS tests. It was expected the ELFIN II experiment would render the smallest correlated N_{CF}-factors because the wind tunnel flow contains more vortical disturbances than the atmosphere and a perforated suction surface is rougher than a highly polished NLF-glove. The difference between the Fokker F100 and the ATTAS was surprising because the surface quality of both gloves was comparable and the disturbances in the on-coming free-stream conditions were supposedly similar.

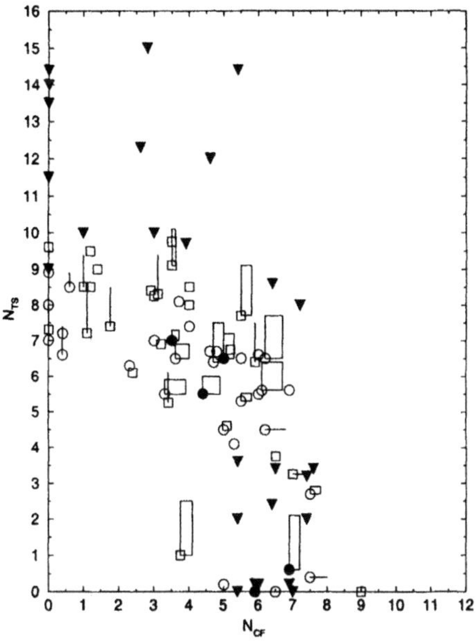

Figure 3 Correlated (N_{CF},N_{TS})-factor pairs of the ELFIN II S1MA test on the HLF panel ∘, as well as on the NLF panel •, the ELFIN I HLF test □, and the TLF NLF test ▼ obtained with incompressible theory

As shown in Figure 4(a), the correlated N_{TS}-factors of the Fokker F100 tests coincide well with those of the ATTAS flight tests using incompressible stability theory, whereas with compressible theory, the Fokker F100 N_{TS}-factors agree better with the ELFIN II experiment.

4 Conclusions

The correlated (N_{CF},N_{TS})-factor pairs of three experiments in the S1MA wind tunnel and two flight tests have been presented. Using incompressible stability theory sufficiently consistent correlations that can be used for design purposes were obtained. There are, however, open questions:

1. Why do we get the same correlated N_{CF}-factors for the HLF and the NLF panel of the ELFIN II wind tunnel experiment?
2. Why are the correlated N_{CF}-factors different for the ATTAS and the Fokker F100 flight tests?
3. Why are the correlated N_{TS}-factors for the TLF S1MA experiment larger than for the ELFIN I/II S1MA experiments?
4. Why do the correlated N_{TS}-factors for the ATTAS and the Fokker F100 flight tests agree for incompressible stability theory and disagree if compressible stability theory is used?

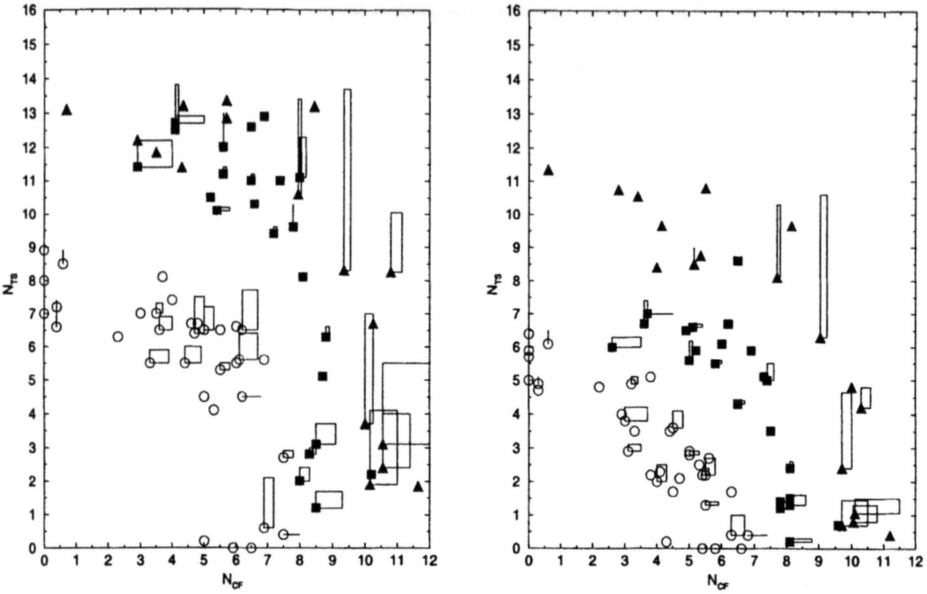

(a) Incompressible theory (b) Compressible theory

Figure 4 Correlated (N_{CF}/N_{TS})-factor pairs of the ELFIN II HLF test ○, the Fokker 100 ■, and the ATTAS ▲ flight tests

5 Acknowledgments

This work has been supported by the German TLF program as well as the ELFIN, ELFIN II, and HYLDA projects of the Industrial and Materials Technologies Program of the European Commission.

References

[1] *Arnal, D.* Boundary Layer Transition: Predictions based on Linear Theory. In: Special Course on Progress in Transition Modeling, AGARD Report No. 793, 1994, pp. 2-1 – 2-63.

[2] *Redeker, G., Hortsmann, K. H., Köster, H., Thiede, P., Szodruch, J.* Design of Natural Laminar Flow Glove on a Transport Aircraft, AIAA-90-3043-CP (1990), 375 – 384.

[3] *Hortsmann, K. H., Redeker, G., Quast, A., Dreßler, U., Bieler, H.* Flight Tests with a Natural Laminar Flow Glove on a Transport Aircraft. AIAA-90-3044-CP (1990), 385 – 392.

[4] *Schrauf, G.* Transition prediction using different linear stability analysis strategies. AIAA-Paper 94-1848, 12th AIAA Applied Aerodynamics Conference, Colorado Springs, CO, USA, June 1994.

[5] *Schrauf, G.; Perraud, J.; Vitiello, D.; Lam, F.; Stock, H.W. and Abbas, A.* Transition Prediction with Linear Stability Theory - Lessons Learned from the ELFIN F100 Flight Demonstrator. Proceedings of 2nd European Forum on Laminar Flow Technology, AAAF Rep., 1996, 8-58 – 8-71.

[6] *Schrauf, G.; Perraud, J.; Vitiello, D. and Lam, F.* Comparison of boundary layer transition prediction using flight test data. AIAA J. Aircraft **53**, 891–897.

[7] *Schrauf, G.* Linear stability theory applied to wind tunnel and flight experiments. In: Computational Fluid Dynamics '98, Invited Lectures Minisymposia and Special Technological Sessions of the Fourth European Computational fluid Dynamics Conference, 7-11 September 1998, Athens, Greece, **2**, 126–131, John Wiley & Sons Ltd., 1998.

[8] *Bippes, H.* Instability Features Appearing on Swept Wing Configurations. In: Laminar-Turbulent Transition. IUTAM Symposium Toulouse/France, September 11-15, 1989, Springer-Verlag, Berlin/-Heidelberg/New York, 1990.

[9] *Deyhle and H., Bippes, H.* Disturbance growth in an unstable three-dimensional boundary layer and its dependence on environmental conditions. JFM 316 (1996), 73–113.

[10] *Abegg, C., Bippes, H. and Bertolotti, F. P.* On the Application of Suction for the Stabilization of Crossflow Instability over Perforated Walls. In: W. Nitsche, H.J. Heinemann, R. Hilbig, New Results in Numerical and Experimental Fluid Mechanics II, Contributions to the 11th AG STAB/DGLR Symposium Berlin, Germany 1998, 1–7. Friedr. Vieweg & Sohn Verlagsgesellschaft mbH, Braunschweig/Wiesbaden, 1999.

[11] *Crouch, J.D. and Ng, L.L.* Variable N-Factor Method for Transition Prediction in Three-Dimensional Boundary Layers. AIAA Journal 38, 2000, 211–216.

Modern Transition Prediction Techniques Based on Adjoint Methods

Ardeshir Hanifi
The Aeronautical Research Institute of Sweden (FFA)
Box 11021, SE-161 11 Bromma, Sweden

Summary

This paper gives an overview of the recent activities of the transition group at FFA where new tools for prediction and control of transition based on adjoint methods have been developed. Some applications of these methods for identifying the optimal disturbances and computing the receptivity properties in boundary layers are discussed. Moreover, a procedure for design of optimal suction distribution for laminar flow control is given. This can be seen as the first step towards incorporating transition prediction in automatic aerodynamic shape optimization. The results presented here demonstrate the efficiency of adjoint methods in providing relevant physical information about gradients of disturbance amplitude with respect to variations of initial and boundary conditions, mean flow and external forcing.

1 Introduction

The laminar-turbulence transition in boundary layer flows is often thought to be caused by breakdown of small growing disturbances inside the boundary layer. The growth of such disturbances is highly dependent on a number of parameters like the geometry, surface finish and external forcing. Therefore, an efficient and reliable transition prediction method should be able to account for effects of such parameters.

A robust and powerful tool to operate on the flow in order to achieve a desired objective (*i.e.* to minimize the downstream amplitude of an emerging instability mode) relays on the knowledge of the sensitivities of flow perturbations to boundary and external forcing. Recently, new theories based on the adjoint methods have been developed which attack this issue in an efficient way.

The adjoint methods have widely been used in optimization problems. Their strength are in providing the gradient of the objective function with relatively low computational cost. In context of flow instability, the adjoint methods have been used in works by *e.g.* Hill [7], Luchini & Bottaro [11], Andersson, Berggren & Henningson [3], Luchini [10], Airiau [1] and Pralits *et al.* [12]. In these works, which include the local and nonlocal instability theory for both compressible and incompressible flows, the response of instability waves to initial condition, surface roughness and external forcing have been addressed.

To familiar you with the adjoint technique, let us consider the following problem. We wish to minimize the objective function

$$J = \frac{1}{2}\int_0^\infty u(x_1,y)^2\, dy,$$

where the state variable u satisfies the state equation

$$u_x = gu + hu_y + f; \quad u(x_0,y) = u_0, \quad u(x,0) = u_w.$$

The derivation of the adjoint equation is done in the following steps: deriving equation for δu (the first variation of u w.r.t. input/boundary condition and forcing f), multiplying that with the co-state variable p, integrating it over the whole domain $(x_0,x_1) \times (0,\infty)$ and removing derivatives of δu by means of partial integration. Then, the terms valid in the whole domain give the adjoint (or co-state) equation

$$-p_x = gp - \frac{\partial}{\partial y}\Big(hp\Big).$$

Using initial and boundary conditions $p(x_1,y) = u(x_1,y)$ and $\lim_{y\to\infty} p(x,y) = 0$, the reminding terms yield

$$\int_0^\infty \delta u_1 u_1\, dy = \int_0^\infty \delta u_0 p(x_0,y)\, dy - \int_{x_0}^{x_1} \delta u_w h(x,0)p(x,0)\, dx + \int_{x_0}^{x_1}\int_0^\infty \delta f p\, dy\, dx.$$

Since the first term in the equation above is equal to δJ, the expressions for gradients of J with respect to initial condition, wall boundary condition and forcing are identified to be

$$\nabla_{u_0} J = p(x_0,y), \quad \nabla_{u_w} J = -h(x,0)p(x,0), \quad \nabla_f J = p(x,y).$$

These gradients are easily obtained by first integrating the state equation (from x_0 to x_1) and then backward integration of the adjoint equation (from x_1 to x_0). In this way, a single loop of calculations provides the information about the gradients with respect to all parameters at the same time.

In the present paper, we will present results obtained by adjoint tools developed by the transition group at FFA. The receptivity analysis have been performed in collaboration with the University of Paul Sabatier and IMFT in Toulouse, France. The studies are performed in the framework of the nonlocal stability theory using the parabolic stability equations (PSE), see *e.g.* [4] and [15].

2 Transition in environments with high free-stream turbulence

In environments with low free-stream turbulence, the laminar-turbulence transition is thought to be caused by breakdown of Tollmien-Schlichting waves or cross flow vortices. Therefore, in traditional transition prediction methods, the transition location is correlated to the streamwise position where the amplification of these disturbances reach a critical value, *e.g.* e^9. However, in presence of high free-stream turbulence the transition often occurs upstream of the critical Reynolds number where all eigenmodes are stable. Experimental works, among others Klebanoff *et al.* [9], Kendall [8] and Alfredsson & Matsubara [2], have revealed the existence of alternating longitudinal high and low velocity streaks inside the boundary layer which according to the linear stability theory are stable. Figure 1 shows a flow visualization by Alfredsson & Matsubara [2], where these streaky structures, their secondary instability and breakdown to turbulence spots

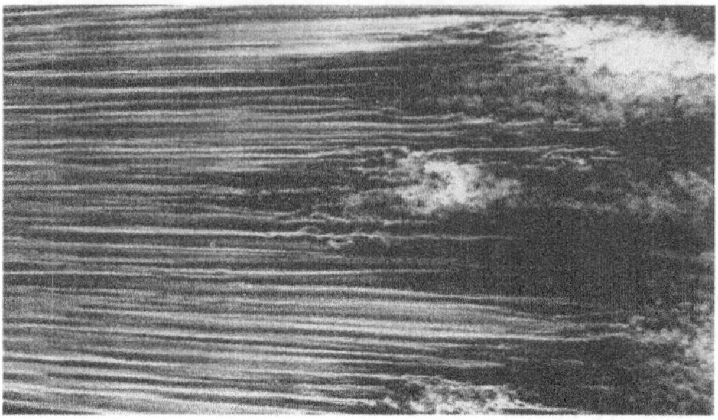

Figure 1 Flow visualization of streaks in a flat plate boundary layer with free-stream turbulence of 2.2% and $U_\infty = 6.0\ m/s$. Streamwise extent of photograph is $220 - 700\ mm$. From Alfredsson & Matsubara [2].

can be seen. The transient growth of streaky structures in parallel shear flows have also been studied theoretically by *e.g.* Butler & Farrell [5], Reddy & Henningson [14]. Recently, Andersson *et al.* [3] calculated the initial profiles for stationary disturbances which grow most in a flat plate boundary layer. In this work, the gradient of disturbance energy with respect to the initial disturbance profiles were given by the solution of the adjoint stability equations. The optimal initial conditions, those which give the largest disturbances at a point further downstream (say x_1), were obtained by first integrating the parabolic stability equations downstream from x_0 to x_1. Then, the adjoint equations, with initial condition given by state variables at x_1, were integrated backward in streamwise direction. The adjoint quantities at x_0 provide the gradient with respect to the initial conditions (see the example given in the introduction) which was used to update the initial disturbance profiles. This loop was repeated until converged solution was find.

The most growing initial disturbances found by Andersson *et al.* were streamwise vortices which at downstream developed streaky structures. These structures were similar to those found in the experiments with high free-stream turbulence. Figure 2a shows a comparison of the calculated optimal disturbances to the experimental data from Westin *et al.* [17].

Based on these findings, Anderson *et al.* suggested a simple relation between the transition Reynolds number and the energy of optimal disturbances, E,

$$Re_{tr} \sim E_{tr}.$$

Assuming the amplitude of these disturbances to be proportional to the free-stream turbulence level, Tu, they suggested the following relation

$$Re_{tr} = K \times Tu^{-2}.$$

Figure 2b shows a comparison between the experimental data and the theoretical results calculated on basis of the above relation. There even the predicted transition Reynolds numbers given by van Driest & Blumer's [16] empirical formula are plotted.

3 Receptivity analysis

Here we use the adjoint PSE to compute the receptivity properties in a growing flat plate boundary layer. The results presented here are concentrated on the effects of the surface inhomo-

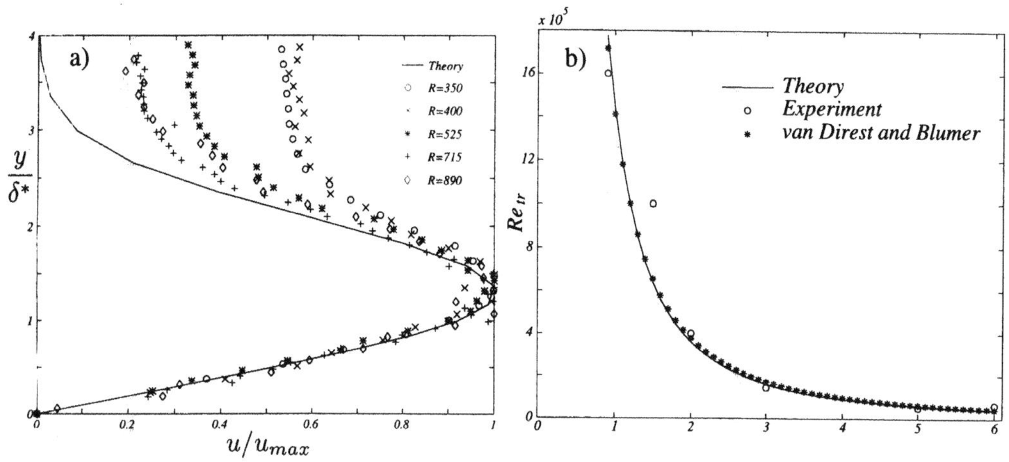

Figure 2 a) Comparison between streamwise component of optimal profiles and experimental data. b) Correlation of predicted transition point with experimentally detected one. From Andersson *at al.* [3].

geneities, however also the problem of the acoustic receptivity can be studied in a similar way, see *e.g.* [7] and [1].

Small roughness elements (or periodic suction/blowing) can be modeled by non-zero disturbance velocities at the wall. Therefore, one can estimate the sensitivity of disturbance growth to small surface imperfections by calculating the gradient of disturbance amplitude to velocity disturbances at the wall. As we demonstrated for the example problem in the introduction, this is easily and efficiently done by using adjoint of the stability equations.

Figure 3a shows the sensitivity functions for an oblique wave in a flat plate boundary layer as a function of the source location. For the reference, the location of the neutral points are marked by $+$ sign. The changes in the amplitude of the disturbances are found by multiplying the value of sensitivity function by the value of the streamwise and normal velocity components, w and u, at the wall. As can be seen there, the normal component of wall disturbances have a larger effect on the growth of the disturbances. Furthermore, the efficiency of wall disturbances to excite disturbances inside the boundary layer decay as the Mach number increases. An important observation is that the disturbances are most sensitive to excitation close to the lower branch of the neutral stability curve. This is in agreement with previous studies, *e.g.* Crouch [6].

In figure 3b the contour plot of the sensitivity function to streamwise momentum forcing is given. The response to the normal momentum forcing (not plotted here) was found to be one order of magnitude lower. The maximum of response is located close to the critical layer (thick line) in vicinity of the lower branch of the neutral stability curve which also has been observed in other investigations, *e.g.* Hill [7].

4 Laminar flow control

An efficient way of maintaining laminar boundary layer on the airfoils is to apply flow suction at the wall. This approach also has been examined with success within the European projects on Laminar Flow Technology. Since the suction devices also require some power input, in order to maximize the total energy gain, optimization of suction rate is necessary. Here, we present an

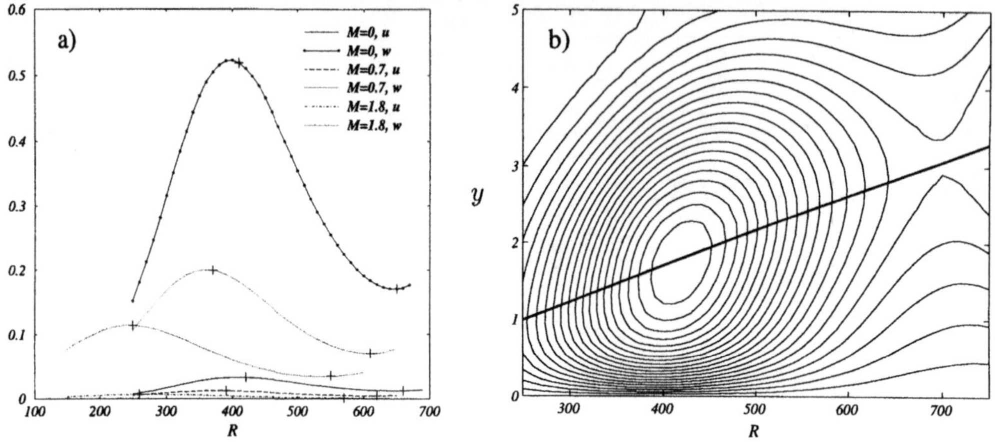

Figure 3 Sensitivity coefficient for the maximum of streamwise disturbance velocity at $R = 750$ to a) wall disturbances u and w for different Mach numbers b) momentum forcing in Blasius boundary layer. The thick line shows the position of the critical layer. $\beta = 0.05$, $F = 10^{-4}$. From Pralits *et al.* [12].

approach based on the adjoint method to find the optimal distribution of suction such that the disturbance energy is minimized. Let us measure the size of a disturbance by its kinetic energy defined as

$$E = \frac{1}{2} \int_0^\infty (|u|^2 + |v|^2 + |w|^2) dy,$$

where y is in the wall normal direction. The optimal distribution of the suction rate is obtained by minimizing the objective function

$$J = \frac{1}{2} \int_{x_0}^{x_1} \int_0^\infty (|u|^2 + |v|^2 + |w|^2) dy \, dx + \frac{\epsilon}{2} \int_{x_0}^{x_1} W_{\mathrm{w}}^2 dx,$$

with respect to the normal velocity at the wall, W_{w}. Here x is in the streamwise direction. The second term in the above expression is to ensure that the suction rate does not grow unbounded. Note that here we search for optimal mean normal velocity at the wall rather than the disturbance normal velocity. This means that also the boundary-layer equations should be considered in the optimization procedure. If the nonlocal stability and boundary layer equations in symbolic form are written as

$$\mathcal{L}_{\mathrm{PSE}} \, \Phi = 0, \quad \int_0^\infty \Phi \bar{\Phi}_x \, dx = 0 \quad \text{and} \quad \mathcal{L}_{\mathrm{BL}}(\mathbf{U}) \, \mathbf{U} = 0,$$

then the adjoint system can be written as

$$\mathcal{L}^*_{\mathrm{PSE}} \, \Psi = g(\Phi, r), \quad \frac{\partial}{\partial x} \int_0^\infty h(\Phi, \Psi, \alpha) \, dx = 0 \quad \text{and} \quad \mathcal{L}^*_{\mathrm{BL}}(\mathbf{U}) \, \mathbf{U}^* = \mathbf{f}(\Phi, \Psi).$$

Here, Φ and α are the disturbance quantities, Ψ and r co-state variables and superscript * refers to adjoint quantities. The solution of the adjoint boundary layer equations, when appropriate initial and boundary conditions are used, gives the desired gradient of the objective function with respect to the mean flow quantities. Then, a gradient-based method can be used to find the optimal distribution of normal velocity at the wall. In figure 4 a schematic of design procedure

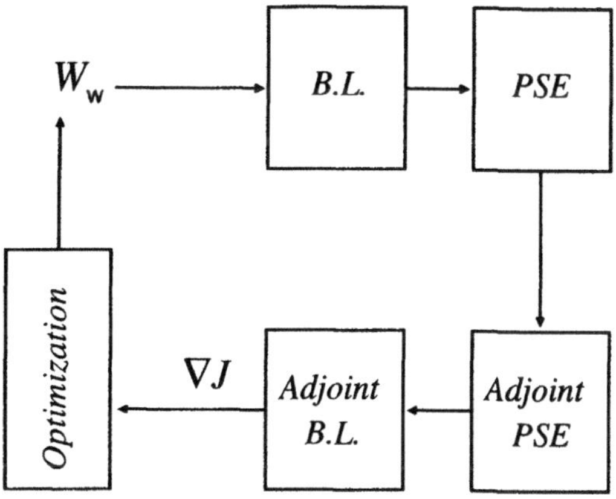

Figure 4 Schematic of the optimization procedure for suction distribution.

for optimal suction distribution is given. In figure 5a the derivative of the objective function, as a function of the streamwise position, calculated using the adjoint technique and finite differences, are compared. As can be seen there the agreement is excellent. Note that each bullet corresponds to two runs of PSE and boundary layer codes. While the solid line is obtained by only a single run of PSE and boundary layer codes followed by a run of adjoint boundary layer and adjoint PSE codes. In figure 5b the N-factors for cases of zero and optimal suction are given. There, also the optimal suction profile is given. The amplification of the disturbance has been eliminated totally.

In aeronautical applications, due to manufacturing difficulties, the suction distribution is given by a number of chambers with a given constant pressure. The procedure described above can easily be modified to produce an optimal distribution of such chambers. This theory can also be developed further to optimize geometry in order to stabilize the flow.

5 Conclusions

A brief overview of the recent activities in the field of transition at FFA has been given here with emphasis on the application of adjoint methods for transition prediction, receptivity and flow control.

The transition prediction methods for high free-stream turbulence environments based on the transient growth of optimal disturbances, suggested by Andersson *et al.* [3], has shown a good correlation between the experimental data and predicted transition Reynolds number.

Surface roughness elements and periodic suction/blowing can be modeled by non-zero boundary conditions for disturbance velocities. Then, the receptivity of instability waves to surface inhomogeneities are found from the solution to the adjoint of nonlocal stability equations.

An efficient procedure for optimal design of suction distribution for disturbance control has been presented. The procedure is based on the solution of nonlocal stability equations, their adjoint and adjoint of boundary layer equations. This is the first step towards incorporating transition prediction in automatic aerodynamic shape optimization.

The applications discussed in this paper reveal the efficiency of adjoint based tools in obtaining the gradients of disturbance amplitude with respect to variation of initial conditions, mean flow, and forcing.

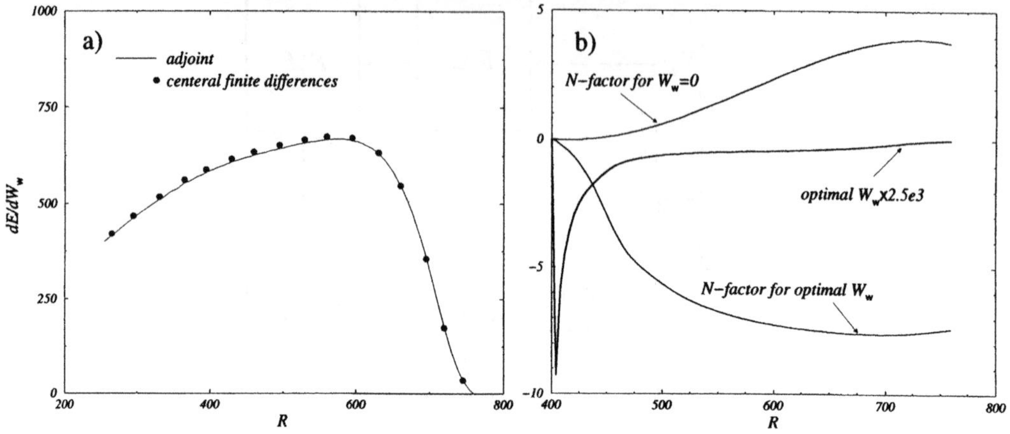

Figure 5 a) Comparison of calculated derivatives of disturbance energy using finite differences and adjoint method, b) N-factor with and without suction and the optimal suction distribution for an oblique TS wave in a Blasius boundary layer. From Pralits *et al.* [13].

References

[1] C. Airiau. Non-parallel acoustic receptivity of a Blasius boundary layer using an adjoint approach. To appear in *Flow Turbulence and Combustion.*

[2] P. H. Alfredsson and M. Matsubara. Free-stream turbulence, streaky structures and transition in boundary layer flows. AIAA Paper 2000-2534, 2000.

[3] P. Andersson, M. Berggren, and D.S. Henningson. Optimal disturbances and bypass transition in boundary layers. *Phys. Fluids*, 11:134–150, 1999.

[4] F. P. Bertolotti, T. Herbert, and P. R. Spalart. Linear and nonlinear stability of the Blasius boundary layer. *J. Fluid Mech.*, 242:441–474, 1992.

[5] K. M. Butler and B. F. Farrell. Three-dimensional optimal perturbations in viscous shear flow. *Phys. Fluids A*, 4:1637–1650, 1992.

[6] J. D. Crouch. Receptivity of three-dimensional boundary layers. AIAA paper 93-0076, 1993.

[7] D. C. Hill. Adjoint systems and their role in the receptivity problem for boundary layers. *J. Fluid Mech.*, 292:183–204, 1995.

[8] J. M. Kendall. Experimental study of disturbances produced in a pre-transitional laminar boundary layer by weak freestream turbulence. AIAA Paper 85-1695, 1985.

[9] P. S. Klebanoff, K. D. Tidstrom, and L. M. Sargent. The three-dimensional nature of boundary layer instability. *J. Fluid Mech.*, 12:1–34, 1962.

[10] P. Luchini. Reynolds-number-independent instability of the boundary layer over a flat surface: optimal pertubations. *J. Fluid Mech.*, 404:289–309, 2000.

[11] P. Luchini and A. Bottaro. Görtler vortices: A backward in time approach to the receptivity problem. *J. Fluid Mech.*, 363:1–23, 1998.

[12] J. O. Pralits, C. Airiau, A. Hanifi, and D. S. Henningson. Sensitivity analysis using adjoint parabolized stability equations for compressible flows. To appear in *Flow Turbulence and Combustion.*

[13] J. O. Pralits, A. Hanifi, and D. S. Henningson. Adjoint-based suction optimization for 3D boundary layer flows. *In progress.*

[14] S. C. Reddy and D. S. Henningson. Energy growth in viscous channel flows. *J. Fluid Mech.*, 252:209–238, 1993.

[15] M. Simen. Local and nonlocal stability theory of spatially varying flows. In M.Y. Hussaini, A. Kumar, and C.L. Streett, editors, *Instability, Transition and Turbulence*, pages 181–201. Springer-Verlag, 1992.

[16] E. R. van Driest and Blumer C. B. Boundary layer transition: Free-stream turbulence and pressure gradient effects. *AIAA J.*, 1:1303, 1963.

[17] K. J. A. Westin, A. V. Boiko, B. G. B. Klingmann, V. V. Kozlov, and P. H. Alfredsson. Experiments in a boundary layer subject to free-stream turbulence. Part I: Boundary layer structure and receptivity. *J. Fluid Mech.*, 281:193–218, 1994.

Receptivity Processes and Transition Scenarios for Swept-Wing Flows with HLF Technology

Janke E., Bertolotti F.P., Hein S., Koch W., Stolte A., Theofilis V., Dallmann U.

German Aerospace Center (DLR), Institute of Fluid Mechanics,
Bunsenstr. 10, D-37073 Göttingen, Germany

Summary

At DLR-Göttingen, theoretical models and numerical methods are being developed in joint efforts with experiments (see CEAS/Dragnet-paper by Abegg et al.). These models and methods cover the transition process from instabilities at the attachment line via the receptivity phase and the regions of linear and nonlinear disturbance growth up to the late nonlinear stages with an instability to secondary disturbances. Considerable progress towards substantiating our objective of a physically reliable transition prediction is reported. The results underline the importance of physical processes near the leading edge on the laminar-turbulent transition downstream, as well as the need for precise knowledge about environmental disturbances from measurements as an input for receptivity analysis. With the initial disturbances available, the downstream disturbance evolution into the final stages of transition can be predicted reliably using the NOLOT/PSE-code, a joint development of DLR-Göttingen and FFA, Sweden. The paper summarizes results achieved in the EC-funded program EUROTRANS [7], in GARTEUR-AG 27 and within the German DFG-program "Transition".

Introduction

Increasing the aerodynamic efficiency in flows of industrial interest is closely linked to improved methods for simulation, prediction and control of laminar-turbulent transition. Particularly in three-dimensional flows, where the empirical e^N-method (solely based on linear disturbance growth rates) can neither incorporate receptivity dependent transition scenarios nor the essential effects of nonlinear disturbance saturation, it is necessary to develop models that account for both effects in order to provide physically consistent transition prediction tools. Hence, the motivation behind the work presented is to identify physical mechanisms relevant for the transition process by simulating complex flows. In turn, this approach allows for making simplifying assumptions that can be incorporated in effective transition prediction tools.

Provided no early turbulent breakdown occurs due to bypass or attachment line transition, the laminar-turbulent transition process in three-dimensional boundary layer flows dominated by an instability to crossflow disturbances can be described by five main phases. First, in the *receptivity phase*, disturbances are generated. Their initial amplitudes depend on environmental conditions, e.g. surface roughness, suction velocities, freestream turbulence or acoustic noise. Many of the initially present disturbances decay, however, and only a few are amplified in the downstream flow. Second, the phase of exponential amplification is characterized by an independent growth of these few unstable disturbances and is, hence, called *phase of linear growth.* Once the individual disturbances have grown large enough they start to interact and initiate the third *phase of nonlinear growth and saturation.* The main feature of this phase is the development of co-rotating vortices that modulate the otherwise spanwise uniform mean flow until the disturbances saturate and reach an equilibrium state. The location of the saturation onset, as well as the saturation amplitude level are in general functions of the initial disturbance amplitude, frequency and phase which are determined by receptivity mechanisms. In the saturation region, the mean flow profiles are being distorted and the well known co-rotating crossflow vortices are fully developed. From here, **two different routes** leading to the fifth *phase of the break-down to turbulence* can been distinguished. Whereas the strong growth of small-scale structures can directly trigger the turbulent break-down in the first case, a second route via a fourth *phase of an instability to secondary, three-dimensional and high-frequency disturbances* can lead to an explosive growth of these secondary disturbances, thus initiating a rapid break-down to turbulence as well. The different stages are depicted in Figure 1 and will be adressed in the following sections.

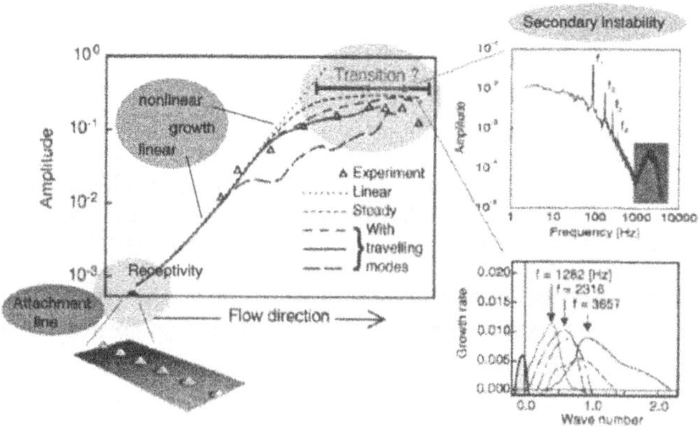

Figure 1 The transition process in crossflow instability dominated boundary layers: comparison of experimental and numerical results from DLR's model flow for a swept wing [2]. High-frequency disturbances from experiment and computation are plotted on the right (e.g. [11]).

Disturbances at and near the attachment line

The goal of maintaining laminar flow over a significant chordwise distance of swept geometries can be doomed by triggering transition already at its attachment line (AL, see Figure 2a). Hence, the main objective of studying instability mechanisms at the AL is the establishment of a global critical sweep Reynolds number $\bar{Re}_{crit}$ below which all disturbances are damped and transition can be avoided by configuring the AL accordingly.

In small disturbance environments, e.g. low freestream turbulence levels and polished swept cylinder surfaces, theoretical studies based on a reduced 2D-model agree well with experimental results [16] ($\bar{Re}_{crit,exp}$=570; $\bar{Re}_{crit,theory}$=583.1). This agreement can be seen in Figure 2b which shows measured and computed growth rates of amplified disturbances versus R_θ, the Reynolds number based on the momentum thickness. Moreover, based on solving the AL-instability as 2D-eigenvalue problem (2D-EVP, see [12] [17]) recent progress lead to the discovery of so-called *polynomial modes*. This term reflects the dependence of different 3D-disturbances on powers of the chordwise coordinate, revealed by analytically modelling the 2D-EVP results [17]. The good agreement of experimental and theoretical results may be explained by the fact that even though 3D (polynomial) disturbances are unstable at the AL, they are less amplified than 2D-disturbances. However, with increasing chordwise distance from the AL, the polynomial disturbances grow faster than the least damped 2D-disturbance at the AL and, hence, are expected to dominate the disturbance evolution in the chordwise direction [12][17]. Despite the progress, two questions of importance to the transition process at and near the AL are still open. The first regards the problem of how any AL-instabilities connect to the CF-vortices observed downstream, a question that is motivated by the high sensitivity of CF-disturbances to stimulation near the leading edge. The second question addresses the existence of a global theoretical $\bar{Re}_{crit,theory}$ near the experimentally observed $\bar{Re}_{crit,exp}$=250 in the case of large disturbance environments.

a) b)

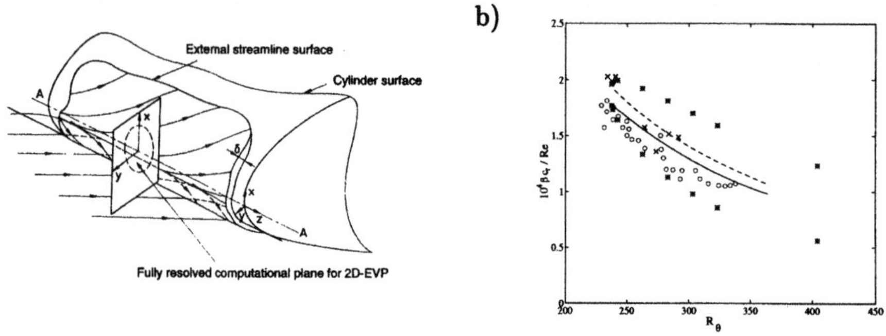

Figure 2 **a)** Sketch of AL-flow [16] and **b)** computed linear (dashed) and nonlinear (solid) maximum amplification rates vs. R_θ=0.404·$\bar{Re}$ in comparison with experimental results by Pfenninger&Bacon (1969) 'o' and Poll (1979) '×' [16]. The '∗' denote the EVP-neutral loop.

Regarding the first question, the derivation of a new class of solutions to the AL-instability problem which exists in addition to the formerly found 2D and polynomial modes [17], as well as their link to the CF-modes [4] represents the first step towards understanding the connection between AL and CF-instabilities. Considering swept Hiemenz flow, the model for flows near the AL, this connection was recently established in terms of eigenvalues and growth rates [4]. With respect to the second question, there are different opinions about the relevance of subcritical instability of 2D-finite-amplitude equilibrium states, i.e. the existence of a global critical sweep Reynolds number lower than predicted by linear stability theory due to the presence of nonlinear 2D-finite-amplitude equilibrium states. Whereas such subcritical instability was documented, for example, by directly computing 2D-equilibrium solutions to the Navier-Stokes equations written as a nonlinear eigenvalue problem and thereby indeed decreasing the critical sweep Reynolds number to $\bar{Re}_{subcrit}$=511 [1], earlier results from Direct Numerical Simulations (DNS) [15] and recent results ([16] and ref.'s therein) did not report these 2D equilibrium states. However, the results on the role of 3D-secondary instabilities [1] and comments regarding three-dimensionality [16] indicate that the latter must be considered in future efforts to explain the discrepancy between experimental and theoretical values for $\bar{Re}_{subcrit}$.

Receptivity processes

The generation of disturbances within the boundary layer due to forcing from environmental disturbances is generally referred to as *receptivity*. The work outlined in this section is directed towards closing the modelling gap between the environmental disturbances to be provided by measurements and necessary inflow conditions (spectral modes and amplitudes) for computing the further disturbance evolution with PSE methods.

The receptivity of incompressible 2D boundary layers has enjoyed strong interest in the past and considerable progress regarding the effects of acoustic and vortical freestream disturbances, as well as their interaction with surface roughness has been achieved at DLR and elsewhere (see [3] and ref.'s therein). As an example, Figure 3a depicts the effect of vortical structures in the freestream, say originating from a micro-wing, on the growing 2D boundary layer along a flat plate. It is seen that the low-level freestream turbulence generates disturbances within the boundary layer that closely resemble experimental results denoted by the symbols. On the contrary, much less effort has been devoted to 3D boundary layers. Henceforth, this section briefly describes developed methods to compute the generation of CF-vortices by surface roughness or steady wall suction.

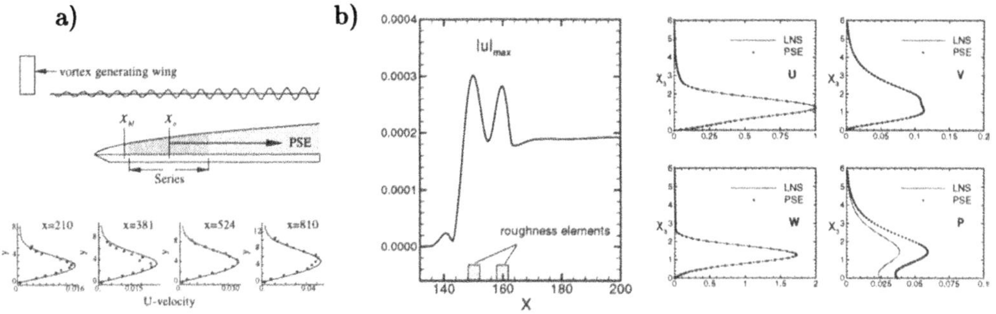

Figure 3 Receptivity mechanisms to **a)** freestream turbulence originating from a micro-wing and generated Klebanoff-mode disturbances (Blasius flow)[3], and **b)** to cylindrical surface roughness and generated crossflow disturbances (swept Hiemenz flow).

Provided there is no transition at the AL, the adjacent region of a strong pressure decrease makes the flow highly unstable to CF-disturbances. These are stimulated by spanwise inhomogenities in surface finish or suction distribution. The wide agreement in the literature (see review in [6]) about the dominant role of stationary versus traveling CF-vortices on the transition process under free-flight conditions supports the limitation of the work cited here to investigations of steady wall-forcing. The most recent work on the effects of steady suction is provided by very accurate, but expensive DNS [14]. There, the physical effects of small, moderate and even over-suction on the flow near and downstream of the suction holes is documented in detail. Taking a numerically more feasible approach, an *equivalent forcing method* considering only selected disturbances with influence on the transition downstream was developed at DLR in the small disturbance limit [2]. This linear method is currently being extended in order to account for nonlinear effects by solving the full Navier-Stokes equations for a given basic flow. The significant decrease in grid resolution needed to represent the *equivalent* in comparison with the *original* forcing amounts to substantial savings in terms of required memory and CPU-time. However, these advantages are counteracted by a deficiency of the method to resolve the complete physics *near* the forcing location. Hence, only if the interest is on the far-field response to small and moderate localized forcing, the method of equivalent forcing provides an efficient and accurate tool. For the investigation of the near-field response and over-suction, in contrast, DNS is the most accurate alternative. Figure 3b depicts disturbance amplitudes due to the presence of two spanwise rows of roughness elements versus chord and the resulting profiles of the disturbance components far downstream of the roughness elements at x=200 in comparison with PSE-solutions. From the good agreement of the far-field receptivity results from a Linearized Navier-Stokes computation (LNS) with the PSE solutions, it is seen that coupling the two methods is straightforward: The receptivity code delivers disturbance amplitudes and profiles that can be directly used as inflow conditions for PSE-computations.

Transition scenarios for swept transonic wings

In this section, the importance of correct initial conditions will be discussed for Tollmien-Schlichting (TS) instability dominated transition scenarios. These are characterized by the nonlinear interaction of either TS-TS, or TS-CF instabilities resulting in the generation of higher harmonics (small-scale structures). The generation and strong growth of these small-scale structures eventually triggering transition is a nonlinear effect and, hence, cannot be captured using linear approaches. The dependence of the disturbance

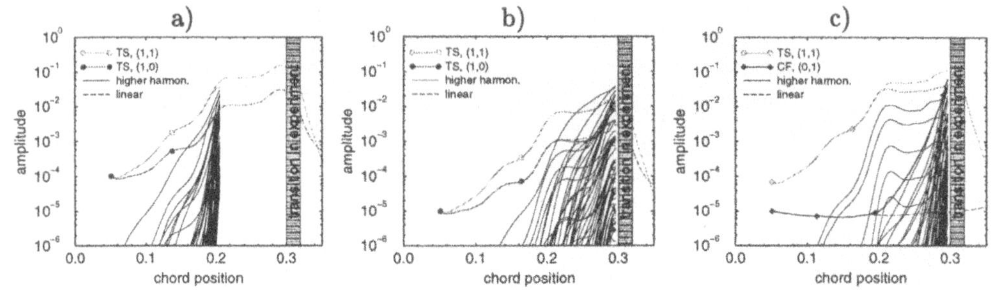

Figure 4 Comparison of measured and predicted transition location in dependence of chosen transition scenario and initial amplitudes; nonlinear NOLOT/PSE results for the Fokker-100 flight test obtained within the European BRITE-EURAM project EUROTRANS.

evolution downstream on the initial conditions is indicated in Figure 4. Arbitrarily varying the initial disturbance amplitudes of two TS-waves at 5% chord from higher values in Figure 4a to smaller values in Figure 4b, the transition location detected in the Fokker-100 flight test can be matched by the steep amplitude increase of the higher harmonics. A similar result is obtained when initializing a CF-disturbance and a TS-wave, as shown in Figure 4c. Also depicted in Figures 4a,b,c are the results of corresponding linear PSE-computations (dashed). The almost constant disturbance amplitudes of both disturbances beyond 20% chord (the CF-disturbances in particular, Figure 4c) underline the limitations of the classical linear e^N-method which would predict a constant N-factor in this case, missing the essential physics of this transition scenario. A more complete discussion of the transition scenarios mentioned here is given in [8]. An indication for the

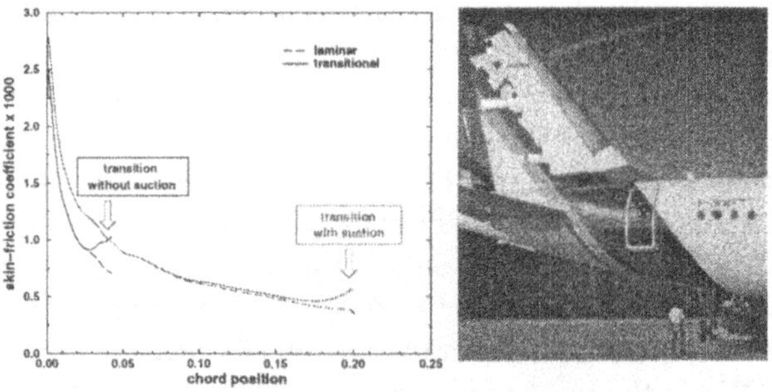

Figure 5 Transition study for the A320 fin showing the effect of suction on the transition location; nonlinear NOLOT/PSE results. Right: The modified fin with porous leading edge.

immediate breakdown to turbulence other than the fast growth of small-scale structures is the departure of the skin-friction coefficient from its laminar value due to nonlinear disturbance interactions. As an example for the cooperation with Daimler-Chrysler Aerospace in Bremen (see [5]), Figure 5 shows the stabilizing effect of suction through porous panels for the Airbus A320 fin. For fixed initial conditions, the downstream shift of the skin-friction rise with suction indicates a significant delay of transition.

High-frequency secondary instability

The last section addresses the second known route to transition via the explosive growth of high-frequency secondary disturbances in the region of nonlinear disturbance saturation. In contrast to the TS-dominated scenarios of the previous section, this mechanism is typical only for CF-instability dominated transition and has not yet been modeled with PSE-methods. Nevertheless, recent numerical results based on linear Floquet theory [13] [11] [9] and experimental observations at DLR and Arizona State University (ASU) [19] consistently document its onset and phase of linear growth. Two main families of high-frequency disturbances can be distinguished. The *Mode I*-family is associated with regions of high spanwise shear, whereas the *Mode II*-family originates from highly inflectional regions of the instantaneous modified meanflow profiles in the direction of the CF-vortex [6]. These regions are depicted in Figure 6a.

a) b) c)

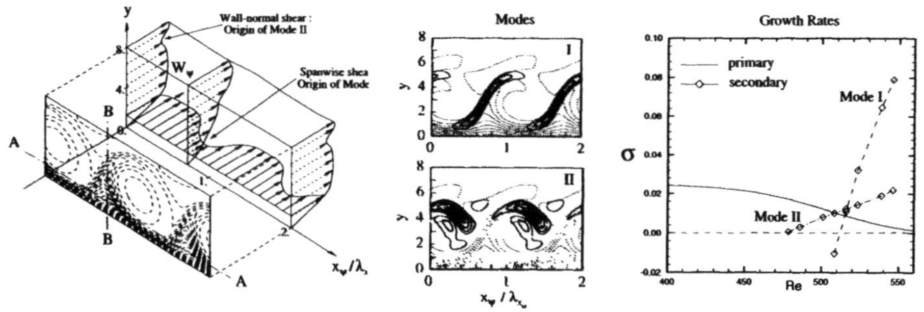

Figure 6 **a)** Contours of and cuts through the velocity profiles along the stationary crossflow vortex; regions of origin for secondary instability modes [11], **b)** Iso-contours of velocity component along the stationary crossflow vortex (dotted) and superimposed secondary eigenfunctions (solid) from a simulation of the DLR transition experiment [11], as well as **c)** growth rates for primary and secondary high-frequency disturbances (swept Hiemenz flow [9]).

The structure of the *Mode I* and *Mode II*-eigenfunctions is shown in Figure 6b, where they are superimposed on the CF-vortex in the plane perpendicular to its axis, corresponding to plane A–B in Figure 6a. The secondary instability structures are located away from the wall, and by comparing Figures 6a and 6b, their maxima can be traced to the aforementioned regions of spanwise and wall-normal shear. Both observations underline the inviscid character of this instability. Further, Figure 6c shows temporal growth rates of Mode I and Mode II secondary instabilities, also obtained using Floquet theory, in comparison with spatial growth rates of the most amplified primary CF-disturbance for swept Hiemenz flow (PSE-results). In qualitative agreement with the experiments (frequency range, steep growth, cf. [19], CEAS/Dragnet-paper by Abegg et al.), the Mode I disturbance is the most amplified disturbance with growth rates one order-of-magnitude higher than the primary growth rates in the region of nonlinear disturbance saturation. The Floquet theory applied at different chordwise locations captures the initially linear growth of the secondary disturbances. Whereas the theoretical framework for the secondary instability other than DNS is presently limited to linear theory, recent experimental results at DLR (see [19] and CEAS/Dragnet-paper by Abegg et al.) document a nonlinear saturation of secondary disturbances that is followed by an immediate turbulent breakdown. Despite the good agreement of theory and experiment with respect to the frequency range, structure and chordwise location of this instability in the linear range, its nonlinear character poses significant problems for modelling this transition scenario accurately. Open questions regard initial amplitudes of the secondary disturbances and a nonlinear mathematical formulation.

Motivated by the wake-like structure of the mean velocity profiles that are modified due to the presence of a stationary CF-vortex (see Figure 6a: origin of Mode I), yet another open question in this field addresses the absolute/convective nature of the high-frequency instability. The existence of a self-sustained absolute instability mechanism would qualitatively change present approaches to transition control due to its independence of upstream conditions after initiation. However, despite the agreement on the existence of an algebraic instability that enhances temporal disturbance growth and underlines the rapid growth characteristics of the high-frequency disturbances [9], an absolute instability has not been confirmed yet in the flows investigated. In fact, new DNS-results [18] for the DLR swept flat plate experiment indicate that the high-frequency secondary instability is of convective nature. Nevertheless, the search for an absolute instability is still in progress theoretically in Göttingen [10] and experimentally at ASU.

Conclusions

Continued joint efforts of experiments and theory are necessary to accomplish non-empirical transition prediction in 3D boundary layers. The paper briefly reviews selected theoretical and numerical topics and highlights methods developed at DLR-Göttingen which aim at identifying transition scenarios in boundary layers of industrial interest.

The presented solution of the receptivity problem to steady wall forcing reveals the urgent need for precise quantitative information regarding environmental disturbances. These have to be provided by measurements in wind-tunnels, as well as under free-flight conditions. With the correct initial disturbances at hand, coupling the developed receptivity tools with the NOLOT/PSE-code has the potential of reliably computing the disturbance evolution into the late nonlinear stages. There, in order to correctly predict the location of the final breakdown to turbulence, different theoretical and numerical tools have to be applied according to the relevant transition scenario. For transition with Tollmien-Schlichting (TS) instabilities involved, one scenario has been identified that predicts the transition location in satisfactory agreement with experimental data (Figure 4). However, the receptivity problem of 3D boundary layers to unsteady disturbances in the freestream is yet unsolved. For CF-instability dominated transition, in contrast, current experimental and numerical work is still directed at identifying high-frequency secondary disturbances relevant for the initiation of the final breakdown. Using Floquet theory, the detection of a Mode I-disturbance with sufficiently high frequency and growth rate might serve as a strong indication for an immediate breakdown to turbulence.

In conclusion, the methods discussed represent a significant step forward in understanding and modeling the laminar-turbulent transition in 3D boundary layers. Essential building blocks missing are the input of correct environmental disturbances and the solution of the receptivity problem for freestream disturbances. With these tools and data available, design-accompanying transition studies for swept-wing boundary layers under free-flight conditions can be envisioned.

Acknowledgements

The authors wish to acknowledge the generous support by several national research programs on transition, namely DFG-contracts, by "Alexander-von-Humboldt"-fellowships for international collaborations, as well as support from industry and European projects.

References

[1] Balakumar P., Trivedi P.A.: *Finite amplitude stability of attachment-line boundary layers*, Physics of Fluids, vol. 10 (9), pp. 2228-2237, 1998.

[2] Bertolotti F.P.: *On the birth and evolution of disturbances in three-dimensional boundary layers*, Proceedings: IUTAM Symposium on Nonlinear Instability and Transition in 3D Boundary Layers, Kluwer Academic Publishers, pp. 247-256, 1996.

[3] Bertolotti F.P.: *Response of the Blasius boundary layer to freestream vorticity*, Physics of Fluids, vol. 9 (8), pp. 2286–2299, 1997.

[4] Bertolotti F.P.: *On the connection between crossflow vortices and attachment-line instabilities*, IUTAM Symposium on Laminar-Turbulent Transition, Sedona, AZ, U.S.A., Sept. 1999, to appear with Springer.

[5] Bieler H., Alewelt R., Hein S.: *Industrielle Randbedingungen für den aerodynamischen Entwurf von Grenzschichtabsaugesystemen*, DGLR-JT-043, pp. 365–372, 1998.

[6] Bippes H.: *Basic experiments on transition in three-dimensional boundary layers dominated by crossflow instability*, Progress in Aerospace Sciences, vol. 35, pp. 363-412, 1999.

[7] "EUROTRANS, European Program on Transition Prediction, Final Report", Contract BRPR-CT-95-0069, submitted 1999.

[8] Hein S., Hanifi A., Casalis G.: *Nonlinear transition prediction*, European Congress on Comp. Meth. in Appl. Sciences and Eng., ECCOMAS 2000, Barcelona, Sept. 2000.

[9] Janke E., Balakumar P.: *On the secondary instability of three-dimensional boundary layers*, Theoretical and Computational Fluid Dynamics, vol. 14 (3), pp. 167–194, 2000.

[10] Koch W.: *Absolute/convective instability analysis of secondary crossflow vortices in a three-dimensional boundary layer*, IUTAM Symposium on Laminar-Turbulent Transition, Sedona, AZ, U.S.A., Sept. 1999, to appear with Springer.

[11] Koch W., Bertolotti F.P., Stolte A., Hein S.: *Nonlinear equilibrium solutions in a 3D boundary layer and their secondary instability*, JFM vol. 406, pp. 131–174, 2000.

[12] Lin R.S., Malik M.R.: *On the stability of attachment-line boundary layers. Part 1: The incompressible Hiemenz flow*, JFM, vol. 311, pp. 239–255, 1996.

[13] Malik M.R., Li F., Choudhari M.M., Chang C.-L.: *Secondary instability of crossflow vortices and swept wing boundary layer transition*, JFM, vol. 399, pp. 85–115, 1999.

[14] Messing R., Kloker M., Wagner S.: *DNS of suction through discrete holes in a 3D boundary layer*, Notes on Numerical Fluid Mechanics, vol. 72, Vieweg Press, pp. 323–330, 1999.

[15] Spalart P.R.: *Direct numerical study of leading-edge contamination*, AGARD CP-483, pp. 5.1–5.13, 1988.

[16] Theofilis V.: *On linear and nonlinear instability of incompressible swept attachment line boundary layer*, JFM, vol. 355, pp. 193–227, 1998.

[17] Theofilis V.: *On the verification and extension of the Görtler-Hämmerlin assumption in 3D incompressible swept attachment line boundary layer flow*, DLR-IB-223-97 A44, 1997.

[18] Wassermann P., Kloker M.: *DNS of laminar-turbulent transition in a 3D aerodynamics boundary layer flow*. In E. Krause, W. Jäger (Eds.): High Performance Computing in Science and Engineering '00. Springer-Verlag, 2000.

[19] White E.B., Saric W.S.: *Application of variable leading-edge roughness for transition control on swept wings*, AIAA-Paper 2000-0283, 2000.

Influence of Acoustic Excitation on 3D Boundary Layer Instabilities

Sandrine Aubrun*, Alain Seraudie, Daniel Biron and Daniel Arnal

ONERA Centre de Toulouse,
2 av Edouard Belin, BP 4025, 31055 Toulouse cedex 4, France.

Summary

The effect of sound on laminar-turbulent transition process is studied for 3D boundary layers governed by streamwise or crossflow instabilities. We quantify the upstream shift of the transition location versus frequency range of acoustic excitation and versus sound level in both cases. Then, the comparison between linear stability theory results and experiments enables us to show that the excitation modifies the initial amplitude of instabilities but not their amplification factor. Finally, we propose a relationship between the loss of N factor and sound level, which is valid for streamwise and crossflow cases.

1 Introduction

The laminar-turbulent transition of a boundary layer is dependent on parameters such as velocity and pressure fluctuations of the external flow (turbulence intensity and acoustic level). In flight conditions, velocity fluctuations have low level but pressure fluctuations are not negligible and could be responsible of an early transition from laminar to turbulent flow. In order to predict the transition on swept wings, it is interesting to study the effect of sound on transition location and to try to quantify this effect on the N factor which is used to predict transition location. If the investigations dealing with the effect of sound on instabilities and on transition are common for 2D boundary layers [1, 2, 3, 4], the few works devoted to effect of sound on 3D boundary layers seem to conclude that sound has no effect on crossflow instabilities [5, 6]. In order to complete these results, we present an experimental study with a swept ONERA D airfoil placed at several sweep angles and angles of attack. It allows us to generate two types of 3D boundary layer transition on swept wing. The first one is characterized by a transition due to travelling streamwise instabilities, located in a positive gradient pressure area. It is close to 2D configuration with Tollmien-Schlichting (TS) instabilities. The second one is characterized by crossflow (CF) instabilities, in a negative pressure gradient. These crossflow instabilities include travelling waves and stationary longitudinal vortices. In this case, we want to check if the acoustic excitation, which acts only on travelling waves, sufficiently modifies the proportion between travelling and stationary instabilities to move the transition location in the upstream direction. The excitation is generated using a compressor driver. Information at the wall (hot films) allows us to determine the transition location versus sound level and frequency range of the acoustic excitation. Then, we compare the amplification of unstable waves deduced from experimental results to computations based on linear stability theory. This calibration enables us to confirm the linearity of the receptivity process. Finally, we propose a law of loss of N factor versus the excitation level.

2 Experimental set-up

The model is a symmetric swept wing of ONERA D type, its normal chord c is 0.35 m and its span is 2 m. The sweep angle is $\varphi = 60°$ and the tested angles of attack are $\alpha = -4°$ and $-8°$. The model is fixed in the middle of the test section (0.35*0.6 m^2) of an open low speed wind tunnel at ONERA Toulouse. The circuit is carefully designed to reduce the contribution of freestream disturbances : the turbulence level $Tu = u'/Q_\infty$ is 0.05%, and the acoustic noise $p'/1/2\,\rho Q_\infty{}^2$ is 0.3% (with $p' = 113\,\text{dB}$). The reference velocity Q_∞ is between 50 and 80 m/s (associated chord Reynolds number Re_c between 1.2 10^6 and 1.9 10^6).

A compressor driver JBL 2490H, coupled with a specific horn (figure 1) generates the acoustic excitation as close as possible to the wing. The horn is designed in order to produce acoustic plane waves parallel to the leading edge. It has been checked that they remain plane below 2500 Hz at least on 30% of chord. The natural frequency response due to the loudspeaker coupled with the wind tunnel is not flat and so, some frequencies are favored. We build the generator signal (with the spectral analyzer Bruel & Kjaer 3550) to correct the frequential distribution of sound and make it flat in the area near the leading edge (location of the highest receptivity). This corrected signal is function of the angle of attack, of the sweep angle of the wing and of the flow velocity. The frequency range lies between 300 and 3000 Hz and the global sound pressure levels can reach 130 dB.

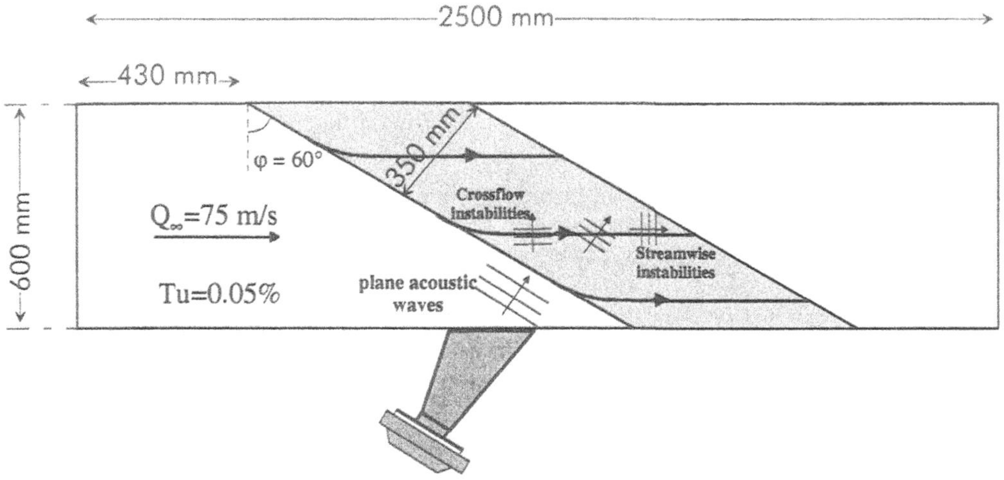

Figure 1: Swept wing in wind tunnel and compressor driver with specific acoustic horn.

To study the evolution of the transition location, 12 DANTEC hot film probes are glued on the upper side of the model between 15 % and 80 % chord. The chordwise evolution of the root-mean-square level (of the output voltage of hot films) indicates the state of the boundary layer and the transition location. The rms level, particularly low in laminar flow, grows up, reaches a maximum in the intermittency region and comes down to the turbulent rms level. The latter stays higher than the laminar one. The transition criterion is based on the intersection between the line with the highest slope of velocity fluctuations and the laminar level. Nevertheless, some turbulent spots can appear upstream of this position.

The spectral analysis of hot film signals allows us to identify unstable frequency ranges and the effect of sound on spectral content of the boundary layer.

In addition, some velocity measurements are performed with a single hot wire probe of DANTEC type. Detailed measurements obtained with hot wire anemometry can be found in [7].

3 Linear stability theory

We perform the stability calculations with the linear stability code CASTET. We use parallel flow approximation and we choose two methods to compute the integrated amplification factor N :

$$N = \max_f \left[\ln\left(A(x)/A_0\right)\right] \tag{1}$$

where A_0 is the amplitude of the initial disturbance on the neutral curve.
These two methods are :

- the *envelope* method
- the *fixed* β^* method where β^* is a wave number in the spanwise direction.

Knowing the experimental transition location, the envelope method provides the frequency of the most unstable wave at transition. The *fixed* β^* method provides the couple (frequency - β^*) for the most unstable wave at transition. This wave is considered as responsible of transition. So, we can also deduce from stability computations the theoretical amplification N factor and the direction of the wave number ψ at transition. The envelope method gives a frequency range of the most unstable instabilities in good agreement with experiments. On the other hand, the associated N factor is not really suitable to prediction of transition. In many cases, a given frequency is amplified first through the effect of crossflow instability ; downstream of the point of minimum pressure, it is then amplified through the effect of streamwise instability and the N factor represents the cumulative contributions of crossflow and streamwise instabilities. This is all the more not representative of reality since previous works (GARTEUR AG27 group [8]) have found that it probably does not exist any non-linear interactions between these both cases of instabilities. The *fixed* β^* method gives lower frequency ranges of unstable waves but the transition N factors, N_T, have a clearer physical meaning. Consequently, all our results based on theoretical N factor will be performed with the *fixed* β^* method.

4 Studied configurations

In order to compare the effect of sound on transition due to streamwise or crossflow instabilities, we choose two configurations :

- The streamwise case is obtained with a sweep angle $\varphi = 60°$, an angle of attack $\alpha = -4°$ and a reference velocity $Q_\infty = 75 m/s$ ($Re_c = 1.8\ 10^6$). The free transition location is close to $x_t/c = 0.7$. The velocity distribution on the upper side of the model (fig. 2) shows that the velocity gradient is negative at transition. The linear stability theory (with envelope strategy) indicates that the direction of the wave vector of the most amplified mode at transition is $\psi = -10°$. The experimental frequency range of unstable waves deduced from hot wire measurement at transition in boundary layer ($x/c = 0.7$ and $y = 0.4mm$) is between 1200 and 3000 Hz, which is in agreement with theory (fig. 3.a).

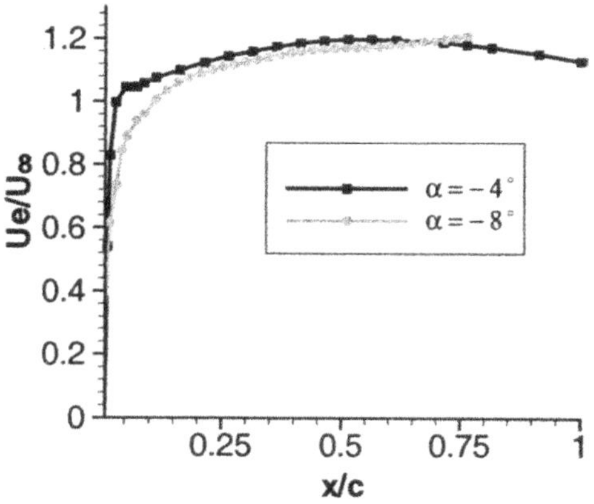

Figure 2: Velocity distributions on the model upper side.

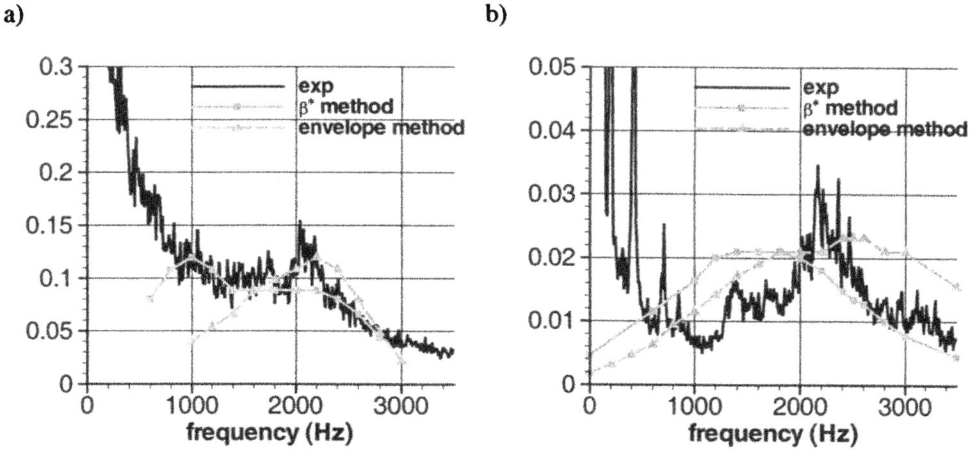

Figure 3: Comparison between experimental and theoretical spectral distribution of rms level at transition a) for streamwise case and b) crossflow case.

• The crossflow case is obtained with a sweep angle $\varphi = 60°$, an angle of attack $\alpha = -8°$ and a reference velocity $Q_\infty = 75m/s$. The transition location is at $x_t/c = 0.25$. The velocity gradient is positive at transition. The linear stability theory indicates that $\psi = +65°$ at this point. The experimental frequency range of unstable travelling waves measured at transition ($x/c = 0.25$ and $y = 0.3mm$) is between 1200 and 3500 Hz, which is in agreement with theory (fig. 3.b). In addition, the presence of stationary instabilities characterized by longitudinal vortices has been proved in [9].

5 Effect of sound on transition location

In both configurations, we excite the flow with a $\Delta f = 200$ Hz wide excitation in the range 300-3000 Hz with a constant sound level 120 dB. Using the transition criterion described in § 2, we determine the transition location x_t, it is plotted versus frequency range of acoustic excitation for both configurations in figure 4.a. The most efficient frequency range is 2100-2300 Hz for streamwise case and 2300-2500 Hz for crossflow one. These are centered on the middle of the natural frequency range of unstable waves and are in agreement with linear stability theory. In the configuration generating streamwise instabilities (see § 4), we measure an upstream displacement of the transition location close to 25% of the length chord. In the crossflow case, the same sound level produces a transition motion of only 5%.

Then, we study the effect of the sound level on transition location for the most efficient excitation : figure 4.b presents transition location versus sound level in the range 2100-2300 Hz for the streamwise case and 2300-2500 Hz for the crossflow one.

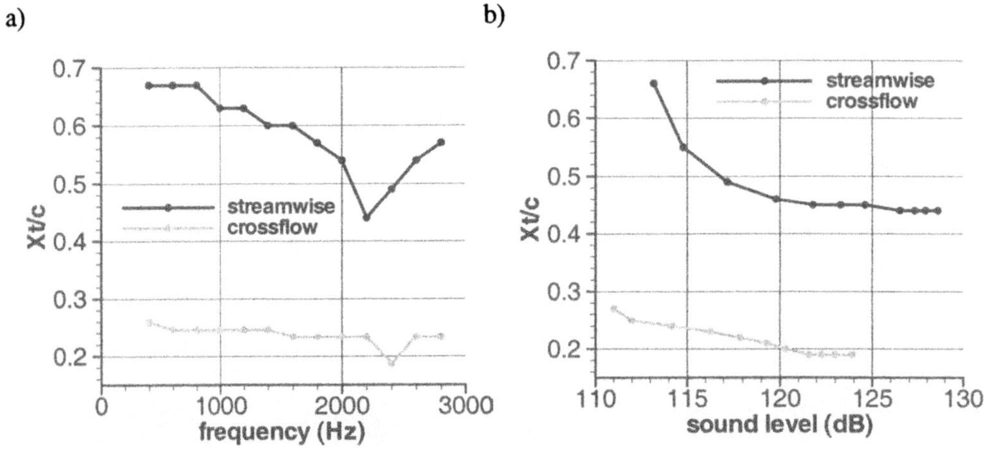

Figure 4 : Transition location evolution a) versus frequency range of acoustic excitation and b) versus sound level.

It is important to notice that, even if a saturation of the sound effect on transition location is visible in both cases, hot films rms levels continue to move slowly versus sound level. As the spacing between two successive hot films is 5% chord, the measured transition movement seems to be more discontinuous as it is in reality, and the details of the transition shift between two sensors are ignored.

Nevertheless, it is interesting to remark that our results about crossflow case are different from previous studies [4, 5]. Two reasons are possible :

- In these studies, the transition is only driven by stationary instabilities and the acoustic excitation of travelling instabilities can not go beyond them,
- the acoustic source is not optimized (distance between source and wing, sound level or direction propagation of waves...)

In a next future, we plan to investigate the effect of the source location and of the direction of propagation to try to explain this difference.

6 Comparison with linear stability theory

Our goal is to compare the experimental amplification rate (like N factor) in the streamwise direction to the theoretical one. The experimental N factor is deduced from hot film signal spectra. Figure 5.a presents the evolution of these experimental N factors that are compared to the theoretical N factor computed with the fixed β^* strategy.

$A(x)$ is the rms level of spectra in the excited frequency range, which includes the acoustic and instability contributions. The initial amplitudes A_0 are chosen so that all N factors are close to each other and to the theoretical one in the first sections. On the other hand, in figure 5.b, we see that the absolute initial amplitude is linearly proportional to sound level expressed in a linear scale. It demonstrates that the receptivity is a linear process. The same result has been obtained by Shapiro in a 2D boundary layer [4].

a) b)

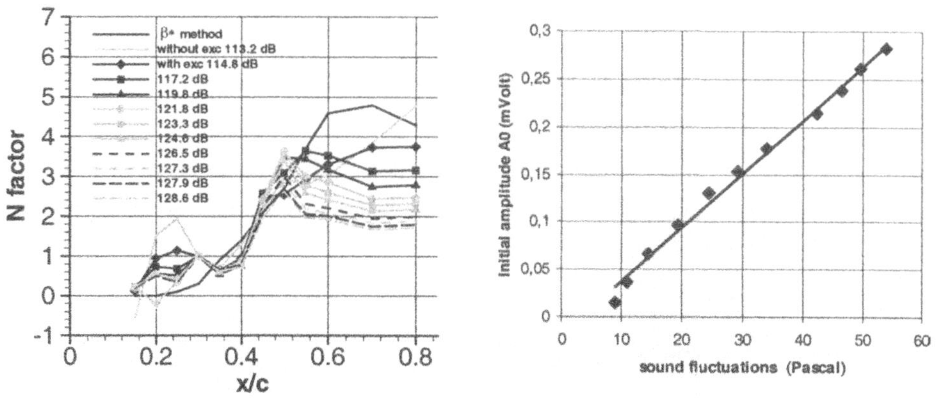

Figure 5 : a) N factor evolution versus sound level for streamwise case and b) associated experimental initial amplitudes A_0.

Moreover, figure 5.a proves that the experimental slopes of the N factors depend neither on the excitation presence nor on its level. The sound acts as an amplifier of initial amplitude but does not modify the amplification process. The curves are different from each other's at last sections because transition is reached at distinct locations. Nevertheless, there always exists a part of curves which is close to the theoretical one. In streamwise case, experiments and theory diverge for $x/c \geq 0.5$. We can assume that it is due to non-linear phenomena. Same investigations performed on crossflow case seem to bring the same conclusion but some additional measurements are needed.

7 N_T loss versus the excitation level

In practice, the interest of this kind of study is to define whether we could predict the new transition location (or its associated N_T factor) in presence of an efficient excitation. From the present results, we can define a relationship that links the loss of N factor at transition versus the efficient excitation level. We present in figure 6 the difference ΔN_T between N factor at transition, N_T, with and without excitation versus the excitation level, for both cases. The N_T values are deduced from the fixed β^* method.

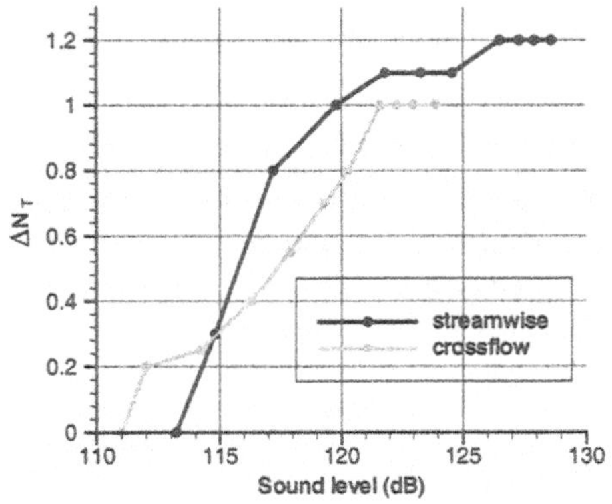

Figure 6 : ΔN_T evolution versus sound level.

It seems that the same level of efficient excitation leads to the same loss of N_T factor, even if the upstream shift of transition location is higher for streamwise case than for crossflow one (25 % instead of 5%). This is explained by the fact that :

$$\frac{\partial N}{\partial x}_{streamwise} \leq \frac{\partial N}{\partial x}_{crossflow}$$

8 Conclusion

We have studied the effect of sound on transition in 3D boundary layers on swept wing. We have chosen a case in which the transition is due to travelling streamwise instabilities and the second one when it is due to travelling and stationary crossflow instabilities. In both cases, we showed that the transition location shifts upstream when the acoustic excitation is performed in frequency ranges of natural unstable waves. For the most efficient frequency ranges, with an excitation around 10 dB higher than free stream sound level, we measured an upstream motion around 25% of length chord for streamwise case and only 5% for crossflow one. The study of the evolution of transition location relative to sound level informs us about a possible saturation.

Then, we could quantify the amplifications in the streamwise direction of instabilities versus sound intensity and compare them with linear stability theory. It appears that the amplification factor does not depend on sound level, but the initial amplitude of instability is different. Furthermore, this initial amplitude is linearly proportional to sound level expressed in a linear scale. It means that the receptivity mechanism is linear. The amplification factor is also in good agreement with theory as far as non-linear effects are negligible.

Finally, we can propose a relationship between the loss of amplification N factor at transition location and the sound level applicable for our both streamwise and crossflow configurations. This result allows us finally to take into account the presence of measured added sound level to correct the N factor for the transition location prediction. This result is not a general criterion but only a correlation for present data.

References

[1] A.S.W. Thomas and S.G. Lekoudis. Sound and Tollmien-Schlichting waves in a Blasius boundary layer. *Research notes of Phys. Fluids* **21**(11), November 1978.

[2] W.S. Saric, E.B. White and H.L. Reed. Boundary layer receptivity to freestream disturbances and its role in transition. *AIAA* 99-3788.

[3] W.S. Saric, E.B. White. Influence of high amplitude noise on boundary layer transition to turbulence. *AIAA* 98-2645.

[4] P. Leehey and P. Shapiro. Leading edge effect in laminar boundary layer excitation by sound. IUTAM Symposium on Laminar-Turbulent Transition, Stuttgart 1979, Springer Verlag (1980).

[5] R.H. Radeztsky, M.S. Reibert, W.S. Saric and S. Takagi. Effect of micro-sized roughness on transition in swept-wing flows. *AIAA* 93-0076

[6] H. Deyle and H. Bippes. Disturbance growth in an unstable three-dimensional boundary layer and its dependence on environmental conditions. *J. Fluid Mech.* (1996), vol. 316, pp. 73-113.

[7] S. Aubrun, A. Seraudie, D. Biron, D. Arnal. Effect of sound on laminar-turbulent transition on swept wings. 22nd International Congress of Aeronautical Sciences. 27 Aug – 1 Sept 2000. Harrogate, United Kingdom.

[8] D. Arnal and G. Casalis. Calculs de couche limite et de stabilité sur les cas-tests du groupe GARTEUR AG27 (deuxième année). ONERA report. R.F. 2/5100.22 DMAE. Novembre 1999.

[9] D. Arnal. Boundary layer transition : predictions based on linear theory. Presented at an *AGARD-VKI Special Course on 'Progress in transition modelling'*. March-April 1993.

Acknowledgments

This work has been supported by the "Service des Programmes Aéronautiques" (SPAé).

Session 6.2
HLF Experimental Techniques

Study of Wind Tunnel Simulation Methodology for HLFC Wings

by
P. W. C. Wong, M. Maina
Aircraft Research Association Ltd., Bedford, MK41 7PF, UK.

Summary

This paper describes a theoretical investigation into the development of a wind tunnel testing methodology for HLFC (Hybrid Laminar Flow Control) wings. An important aspect of testing HLFC wings is the ability to simulate the extent of laminar flow that would occur in flight. Investigation has shown that the use of surface cooling may be a feasible technique for stabilising the boundary layer in order to delay transition in the tunnel tests. With the assumption that the laminar extent at flight scale can be maintained at wind tunnel scale using cooling, the issue of simulating the turbulent boundary layer still needs to be addressed. A viable technique for simulating the viscous flow development at flight scale is the application of surface suction to the model in order to thin the turbulent boundary layer. The application of this technique combined with surface cooling for a number of aerofoils has achieved a close simulation of the flight scale results in terms of pressure distributions and aerodynamic forces.

Notation

c	Aerofoil chord.
C_D	Drag coefficient.
C_{DV}	Viscous drag coefficient.
C_{DW}	Wave drag coefficient.
C_L	Lift coefficient.
C_p	Pressure coefficient.
M_n	Mach number normal to sweep.
n	N-factor in stability analysis (amplitude ratio).
Re	Reynolds number.
T_∞	Ambient temperature.
T_w	Wall temperature.
V_s	Suction velocity, non-dimensionalised by freestream velocity.
x/c	Chordwise ordinate, non-dimensionalised by chord.
x_t	Transition location, non-dimensionalised by chord.
α	Angle of incidence.
δ^*	Boundary layer displacement thickness.
Λ	Angle of sweep.

1. Introduction

There has been considerable research in recent years into control techniques for delaying transition from laminar to turbulent flow in order to reduce drag for the improvement of the aerodynamic efficiency of aircraft. One such technique that has been widely considered for application to civil transport aircraft is hybrid laminar flow control (HLFC), which involves the combination of surface suction in the leading edge region and wing geometry shaping. Over the years, much theoretical and experimental research has been directed at the technologies required for HLFC design, including flight test programmes for demonstrating the viability of such a technique for full scale aircraft application, such as the A320 HLFC fin research programme [1]. Apart from theoretical design methods and flight tests, wind tunnel tests are also important in the process of the design and development of HLFC aircraft. Wind tunnel testing of aircraft designed for fully turbulent flow and techniques for extrapolating the results to flight scale are well established. A methodology for testing such aircraft in conventional transonic wind tunnels and the extrapolation of the results to full scale was proposed by AGARD WG 09 in 1988 [2]. However, for aircraft that employ HLFC, the simulation problem is more complex and not well understood. Information on methods of testing HLFC aircraft available in the open literature is somewhat limited. This paper describes a theoretical study, carried out at the Aircraft Research Association (ARA), to address a number of issues involved in the wind tunnel testing of HLFC wings and some of the problems related to scale effects. An aim of the research study is to develop a testing methodology for HLFC wings using conventional tunnels such as the 2.7m × 2.4m tunnel at ARA and the 2.4m × 2.4m pressurised tunnel at DERA Bedford. Due to the complexity of the problem, the analysis considered here is for infinite swept wings and equivalent two-dimensional flow conditions, with the aim of developing the basic principles involved in the simulation. The investigations have been carried out using the transonic aerofoil code BVGK [3] modified for HLFC applications [4] in conjunction with the linear stability analysis code CoDS [5] for transition prediction.

2. Background

An important aspect of testing HLFC wings is the ability to simulate the extent of laminar flow and the transition location that would occur in flight. The transition location may be predicted theoretically by using linear stability analysis and the 'e^n' method. Freestream flow quality in terms of noise and turbulence affects transition location. For flight, the N-factor for transition may be in the range of 15 to 20. In contrast, for wind tunnels, where freestream flow quality is worse than in flight conditions, this value is likely to be in the region of 6 to 10. Despite the more favourable conditions for maintaining laminar flow at the lower Reynolds numbers associated with tunnel scale, the extent of laminar flow expected at flight conditions may not be achievable in wind tunnels with a much lower transition N-factor. Apart from the issue of freestream flow quality, Ref 6 has shown that premature transition can also occur due to disturbances created by particles in the flow impacting on the model surface. It was found that a single roughness element on the aerofoil surface could cause a significant spanwise variation in drag.

The AGARD methodology [2] recommends two possible approaches that can be followed in a model test for extrapolating data to full scale. A 'Reynolds number sweep' approach, where the tests are carried out over a range of Reynolds numbers with transition fixed at a position close to where it would occur in flight; and a 'transition sweep' approach, where the boundary layer in the tunnel test is manipulated by aft transition fixing to produce viscous flow behaviour closer to that at flight Reynolds number. The main aim of these tests is to establish the trends against Reynolds number of quantities such as shock strength, shock position, drag and trailing edge pressure at a given Mach number and lift coefficient. These trends are extrapolated to flight scale Reynolds number with guidance from CFD calculations. Details of the simulation techniques involved are given in Refs 2 and 7, therefore will not be repeated here. Experience with the application of the methodology is described in Ref 8.

For HLFC wings, it is clear that with the first approach there may be transition movement due to Reynolds number variation and consequently the flight laminar extent may not be achievable at tunnel scale. With the second approach there will be a limit to simulation of the viscous flow behaviour in view of the fact that the actual transition position at full scale is already well aft. These simulation issues are illustrated by the results shown in Fig 1, which shows the variation in transition position with tunnel Reynolds number for two aerofoils, M2303 and Lock. Throughout this paper, these two aerofoils have been chosen as examples to illustrate the issues involved. The aerofoils exhibit favourable and adverse 'rooftop' pressure gradients at their respective design conditions as shown in Figs 2 and 3. For the purposes of the exercise described here, these aerofoils have been assumed to be the two-dimensional sections equivalent to an infinite yawed wing with 35° leading edge sweep. For these sections, transition onset at flight Reynolds number of 35×10^6 occurs at about 30% chord, where a suction velocity of –0.0007 from 0 to 5% chord and –0.0002 from 5% to 15% chord has been applied for suppressing the various modes of instability.

With the Reynolds number sweep approach (applicable to the DERA tunnel), it can be seen that if the tunnel Reynolds number is increased above 4×10^6 for the M2303 aerofoil, the extent of laminar flow which would occur at flight scale will no longer be achievable. The N-factor for transition at tunnel conditions is assumed to be 8, which is consistent with the value correlated for the tunnel at DERA Bedford [9]. For the Lock aerofoil, severe turbulent separation occurred at Reynolds numbers below 10×10^6, causing flow break down. Figs 2 and 3 indicate the limited margin available for aft transition fixing, as the transition trip is required to be sufficiently forward of the shock location, typically 10% to 15% chord, in order to avoid any local interaction between the flow over the trip and the shock. With the improvement in wing design standard and consideration of the design of larger aircraft with higher chord Reynolds numbers, the results for the Lock aerofoil illustrate the problems in testing at lower Reynolds numbers for aircraft designed for full-scale Reynolds number.

3. Simulation Methodology

3.1 Laminar flow extent

It is clear from the above discussion that, in general, the transition location in flight will not be reproduced in the wind tunnel without the use of some method of suppressing the instabilities in order to delay transition. Surface cooling is well known to have a stabilising effect on boundary layers. A theoretical study of the use of the cooling technique as an alternative to suction for wind tunnel simulation has been carried out, although the practicality of using this technique needs to be assessed. Calculations have been performed for a number of aerofoils covering a range of pressure distributions that may be relevant to HLFC design. For these sections, at wind tunnel scale, transition is mainly due to crossflow instability close to the leading edge. It was found that if cooling is applied in the initial 10% chord region to suppress these instability modes, transition could be delayed to the position at which it occurs at flight scale. Calculations have been carried out to determine the level of cooling required for tunnel Reynolds numbers of 3×10^6 to 12×10^6, thereby covering the range applicable to the ARA and DERA wind tunnels. The results are shown in Fig 4 for the M2303 and Lock aerofoils, where $n = 8$ for crossflow transition has been assumed, as correlated for the DERA tunnel [9]. The results indicate that the cooling rates required for delaying transition to 30% chord as occurs at flight scale increases rapidly with tunnel Reynolds number. The required cooling

rates for the M2303 aerofoil are much higher than for the Lock aerofoil. This is due to the fact that the type of pressure distribution shown in Fig 2 leads to highly unstable crossflow disturbances.

3.2 Turbulent boundary layer

With the assumption that the laminar extent at flight scale can be achieved at wind tunnel scale using surface cooling, the issues of simulating the turbulent boundary layer still need to be addressed. As can be seen in Figs 2 and 3, for the same transition location at flight and tunnel scales, there are differences in the pressure distributions, shock positions and drag values between the two Reynolds numbers, as would be expected. As already noted, the ability to manipulate the turbulent boundary layer by aft transition fixing may be limited for HLFC wings, since transition is already well aft, therefore there may be difficulties in extrapolating data to full scale particularly for cases with turbulent separation at tunnel scale. A technique for reducing the turbulent boundary layer thickness in tunnel scale to that of flight scale is the application of suction. The idea of using boundary layer suction to simulate higher Reynolds numbers was proposed in Ref 10. It was suggested that if the current theoretical models of the turbulent boundary layer structure are reliable, it would be possible to determine an appropriate distribution of suction over the entire wing surface of the model to enable the boundary layer behaviour at a higher Reynolds number to be simulated precisely over virtually the whole wing. From an engineering standpoint, this would not be practical. However, for the purposes of simulation, consideration of this technique could be limited to the three most important regions of boundary layer development: the rear upper surface of the wing, the rear lower surface, and in the region of interaction with the shock wave on the upper surface.

For the types of pressure distribution and shock positions shown in Figs 2 and 3 for the M2303 and Lock aerofoils, an appropriate location for the application of suction for the reduction of the turbulent boundary layer thickness would be in the region between 65% and 85% chord. The application of suction in the turbulent flow region may be used in combination with the AGARD simulation methodology. The key premise assumed here is that the correct simulation of the boundary displacement thickness at the trailing edge will result in a close simulation of the pressure distribution and hence overall aerodynamic performance. With the Reynolds number sweep approach, it would be preferable to keep the level of suction constant as the tunnel Reynolds is increased. Using the M2303 aerofoil as an example, for a case with a suction velocity of -0.004 applied to the upper surface, the effect of increasing Reynolds number on the displacement thickness distribution is shown in Figs 5 and 6 for the upper and lower surfaces, respectively. It can be seen that at a tunnel Reynolds number of 12×10^6, the trailing edge displacement thickness at tunnel scale has been reduced to that of flight scale. Fig 7 shows that the corresponding pressure distribution and drag values now nearly match the flight scale results. The transition position on the lower surface at tunnel scale for this case is taken at 30% chord instead of 1% chord as in flight, in order to reduce the scale effects on the lower surface. It could be argued that suction could also be applied to the lower surface instead of aft transition fixing, but for simplicity and illustrative purposes, only the upper surface with suction in the turbulent region has been considered in the current investigation. The above simulation technique has also been applied successfully to the Lock aerofoil as shown in Fig 8. For this aerofoil, the flight scale results have been achieved at 12×10^6 tunnel Reynolds number with a much lower suction velocity (V_s = -0.0015). The investigation has shown that the above simulation method can be applied for a range of angles of incidence. This is shown

in the results for the $C_L \sim \alpha$ and $C_D \sim C_L$ curves presented in Figs 9 and 10, respectively, for the M2303 aerofoil. It should be noted that in the results shown, the same suction velocity is applicable across the incidence range, which is of practical benefit for testing.

For tunnels with constant Reynolds number, such as the ARA tunnel, where the transition sweep approach would normally be adopted, it may not be possible to fix transition further aft than the flight scale location, since transition is already close to the shock. To simulate the turbulent boundary layer at flight scale, the suction quantities need to be increased. For a tunnel Reynolds number of 3.5×10^6, which may be representative of the ARA tunnel, a suction velocity of -0.01 is required to match the flight scale results for the M2303 aerofoil. This is illustrated in the results presented in Figs 11 and 12, where the $C_L \sim \alpha$ and $C_D \sim C_L$ curves at flight scale have also been simulated successfully.

4. Concluding remarks

The investigation reported here has shown the potential of combining surface cooling in the leading edge region with surface suction in the turbulent region for testing HLFC wings. The results have demonstrated the successful use of this technique in conjunction with the AGARD simulation methodology. The practicality of using cooling and suction needs to be assessed. Further investigation is required to extend the technique to predict buffet onset. The issues involving the simulation of the transition zone should also be studied in the future.

Acknowledgement

The work reported in this paper was carried out under contract to BAE SYSTEMS.

References

[1] Henke R, Capbern P, Davies A J, Hinsinger R and Santana J L. The "A320 HLF Fin – Programme": Objectives and Challenges. Proc. of *2nd European Forum on Laminar Flow Technology*, Bordeaux, France, June 1996.

[2] Laster M L (ed). Boundary layer simulation and control in wind tunnels. AGARD AR 224, April 1988.

[3] Ashill P R, Wood R F and Weeks D J. An improved, semi-inverse version of the viscous Garabedian and Korn method (VGK). RAE TR 87002, 1987.

[4] Wong P W C and Maina M. Modification to the semi-inverse Garabedian and Korn program (BVGK) for application to laminar flow aerofoils. ARA TM 410, 1994.

[5] Atkin C J. Unpublished work at BAe Regional Aircraft Ltd.

[6] Elsenaar A. The windtunnel as a tool for laminar flow research. Proc. of *17th ICAS Congress*, Beijing, China, September 1990.

[7] Haines A B. Scale effects on aircraft and weapon aerodynamics. AGARD-AG-323, July 1994.

[8] Haines A B. Experience in the use of a viscous simulation methodology for tests in transonic tunnels. AIAA 90-1414, June 1990.

[9] Ashill P R, Betts C J and Gaudet I M. A wind tunnel study of transition flows on a swept wing at high subsonic speed. Proc. of *2nd European Forum on Laminar Flow Technology*, Bordeaux, France, June 1996.

[10] Green J E. Second Goldstein Lecture – Modern developments in fluid dynamics. *Aeronautical Journal*, March 1992.

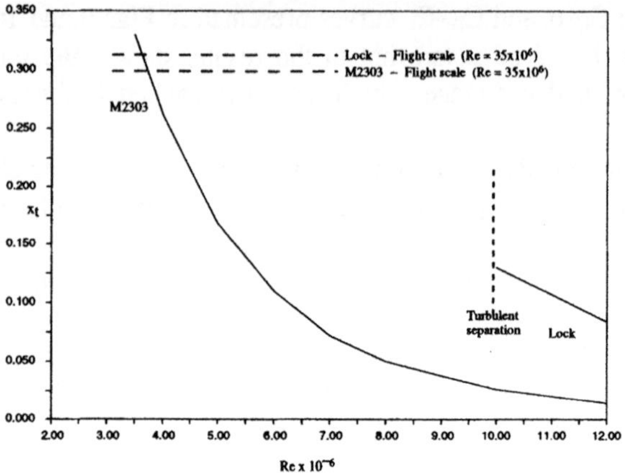

Fig 1. Transition locations in wind tunnel
$n = 8, \ \Lambda = 35^{o}$

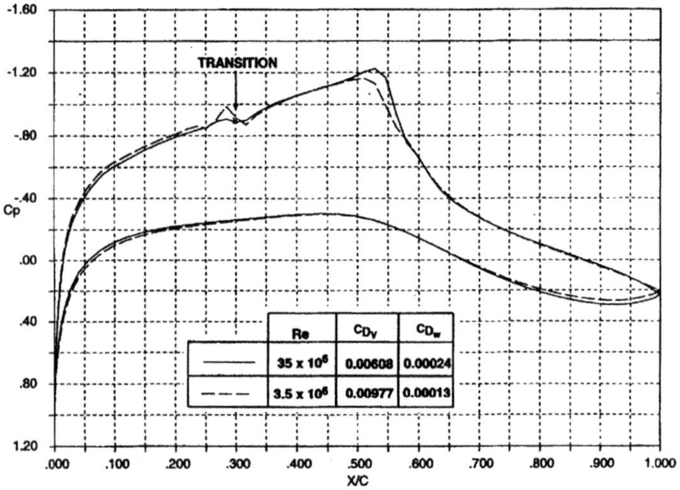

Fig 2. Comparison of pressure distributions
M2303 aerofoil, $M_n = 0.68$, $C_L = 0.5$

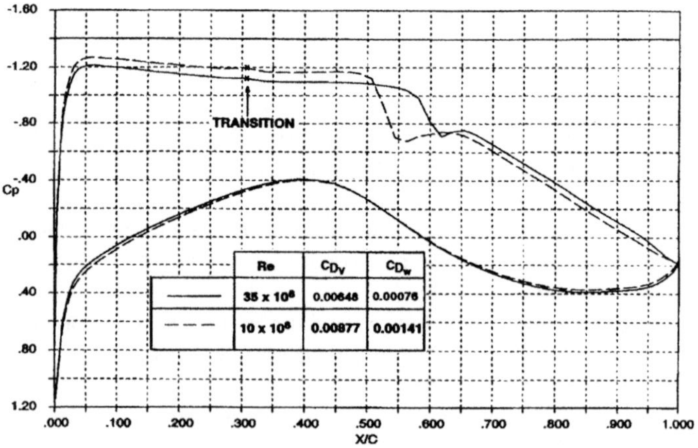

Fig 3. Comparison of pressure distributions
Lock aerofoil, $M_n = 0.72$, $C_L = 0.84$

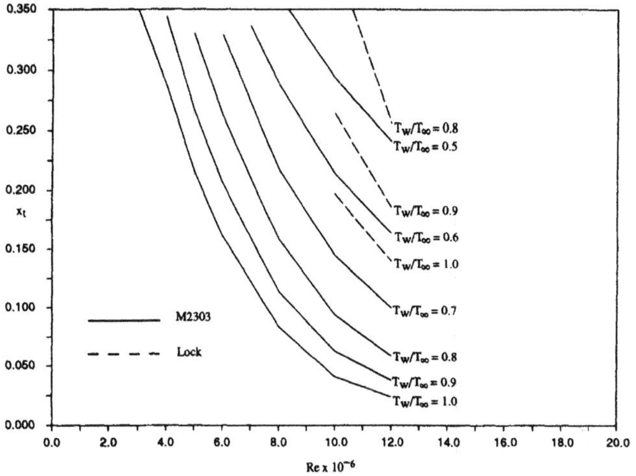

Fig 4. Effect of surface cooling on transition
$n = 8, \quad \Lambda = 35^o$

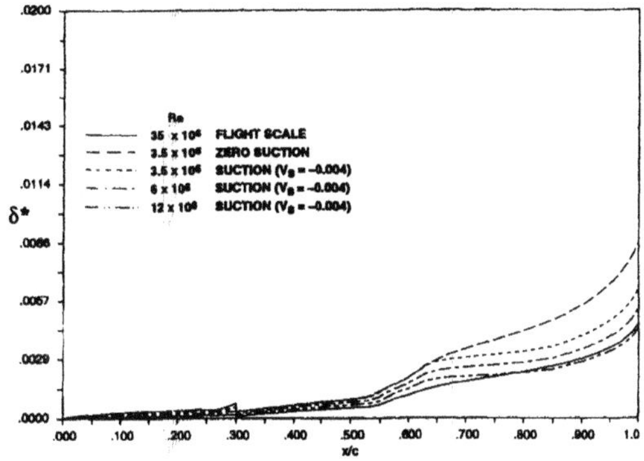

Fig 5. Upper surface displacement thickness, Reynolds number sweep
M2303 aerofoil, $M_n = 0.68$, $C_L = 0.5$

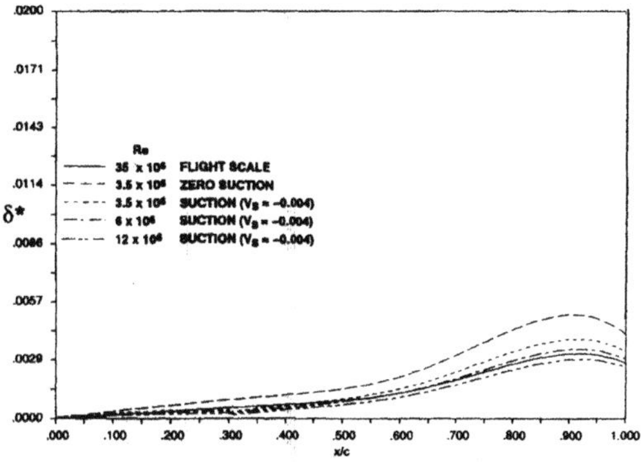

Fig 6. Lower surface displacement thickness, Reynolds number sweep
M2303 aerofoil, $M_n = 0.68$, $C_L = 0.5$

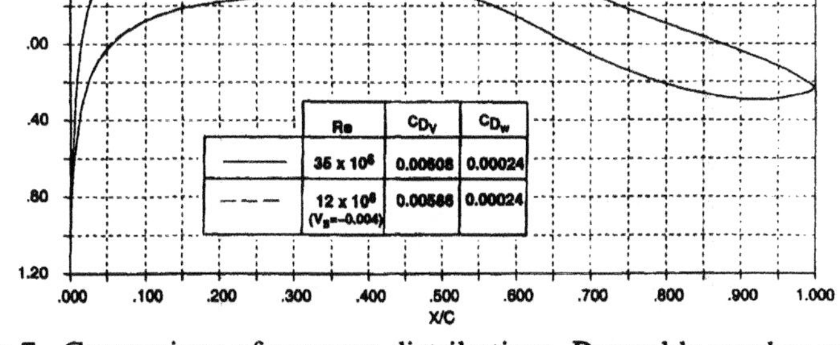

Fig 7. Comparison of pressure distributions, Reynolds number sweep
M2303 aerofoil, $M_n = 0.68$, $C_L = 0.5$

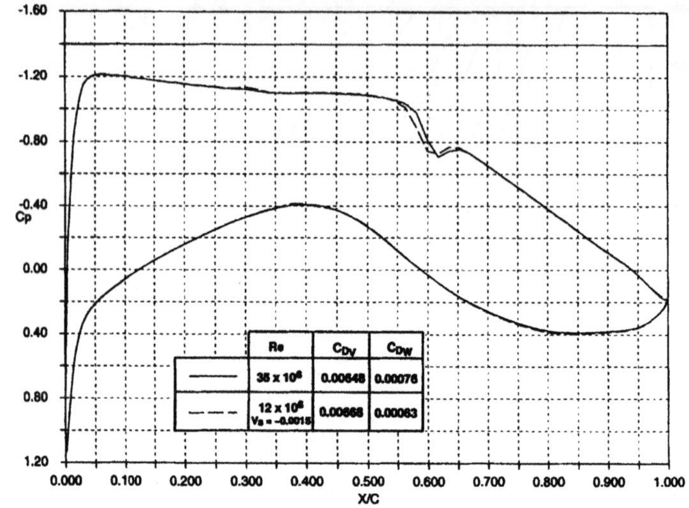

Fig 8. Comparison of pressure distributions
Lock aerofoil, $M_n = 0.72$, $C_L = 0.84$

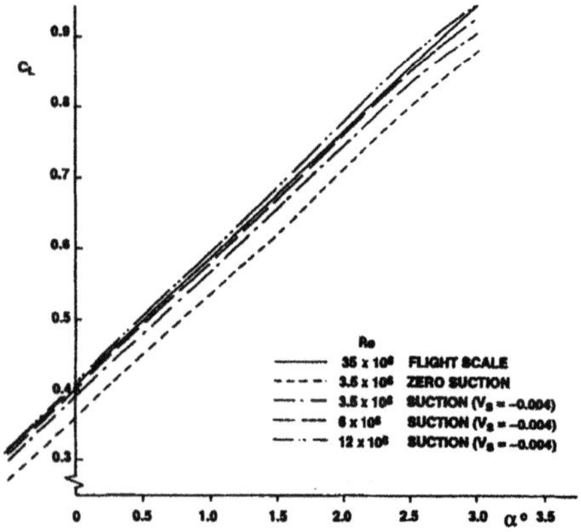

Fig 9. Variation of C_L with α,
Reynolds number sweep
M2303 aerofoil, $M_n = 0.68$

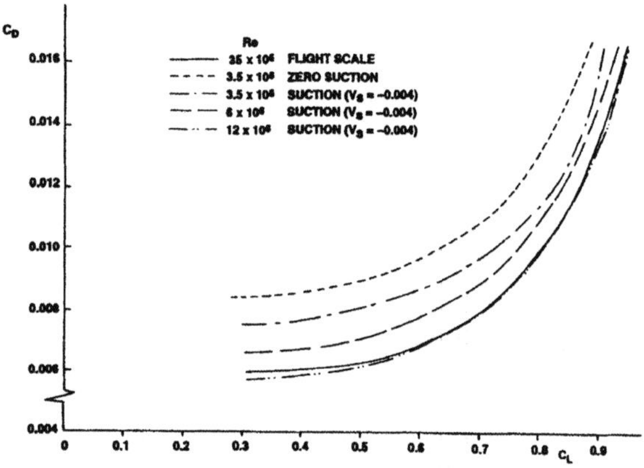

Fig 10. Variation of C_D with C_L,
Reynolds number sweep
M2303 aerofoil, $M_n = 0.68$

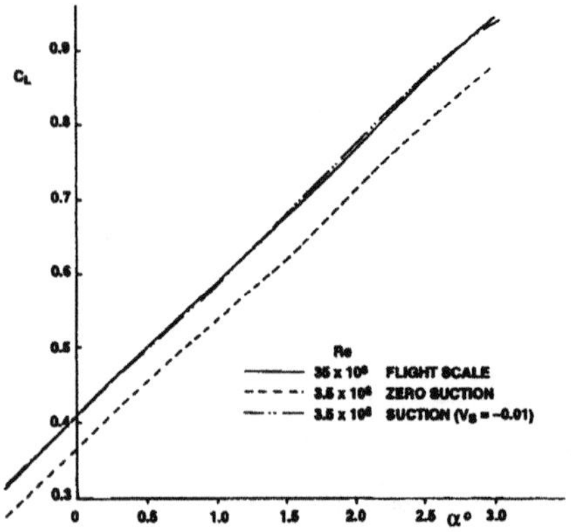

Fig 11. Variation of C_L with α,
transition sweep
M2303 aerofoil, $M_n = 0.68$

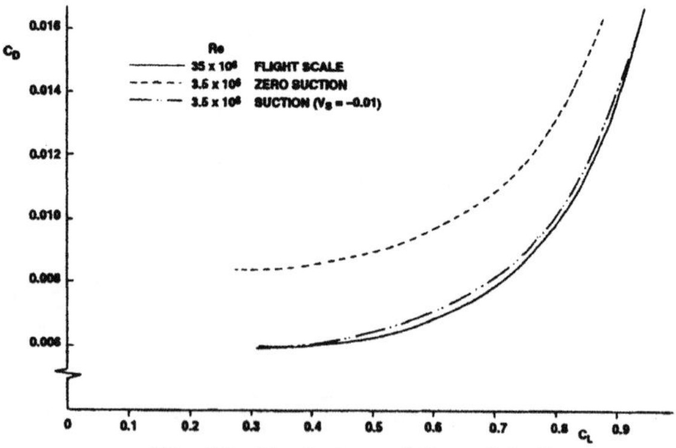

Fig 12. Variation of C_D with C_L,
transition sweep
M2303 aerofoil, $M_n = 0.68$

Transitional Flow Physics and Flow Control for Swept Wings: Experiments on Boundary-Layer Receptivity, Instability Excitation and HLF-Technology

C. Abegg, H. Bippes(†), A. Boiko, V. Krishnan, T. Lerche, A. Pöthke, Y. Wu, U. Dallmann
German Aerospace Center (DLR), Institute of Fluid Mechanics
Bunsenstr. 10, D-37073 Göttingen, Germany

Dedicated to the memory of Hans Bippes, our friend and colleague as well as long-time *spiritus rector* for experimental laminar-turbulent transition research at DLR in Göttingen.

Summary

Basic experiments on transitional flow physics for swept wings are performed at DLR-Göttingen supporting the theoretical-numerical work (see the corresponding CEAS/DragNet-contribution by Janke et al.). One common aim is to predict and to control boundary-layer transition as reliable and simple as possible. The DLR swept flat plate transition experiment provides the setup with a crossflow dominated three-dimensional boundary-layer flow, to investigate the stabilization with the aid of surface suction through holes and slots. An effective and resistant attenuation of the disturbances initiating transition can be reached. One objective of this study is to quantify the effectiveness and robustness, i.e. the flow control sensitivity on suction device imperfections and variations. Another objective is the prediction of transition for the design of Hybrid Laminar Flow (HLF) transport-aircraft wings. This attempt requires experimental informations for physical modeling of the origin of boundary-layer instability. Therefore, the impact of vortical freestream disturbances on the receptivity of three-dimensional boundary layers is evaluated. A major goal is to identify the mechanism(s) that cause the final breakdown of a swept wing's three-dimensional boundary layer into a fully turbulent flow regime. With a controlled excitation of primary and secondary traveling disturbances, the appearance and growth of high-frequency secondary instability are investigated upstream of breakdown. Two high-frequency secondary modes are identified in agreement with theoretical predictions. Finally experiments using a wing with a thickness ratio of 15%, the damping effect of the convex surface curvature on disturbance growth is demonstrated and transition scenarios different from the DLR swept flat plate transition experiment become obvious.

1 Introduction

For the identification of the most effective and as robust as possible control and for a more reliable but, nevertheless, as simple as possible prediction of transition in flows across swept wings, a deeper understanding of the main physical mechanisms of the receptivity to any disturbances (freestream turbulence, wall roughness, suction holes, etc.) and of the final transition breakdown into turbulence of three-dimensional boundary layers is essential. The presented experimental results concentrate on transition dominated by crossflow instability as appearing in the accelerated region of swept-wing flows. As experimental setups we choose for the generation of a generic three-dimensional boundary-layer flow the DLR swept flat plate transition experiment with displacement body [1] and a swept wing with large thickness ratio serves in addition for the investigation of the influence of wall curvature.

A successful way for boundary-layer stabilization is surface suction, one of the key technologies in the European Hybrid Laminar Flow (HLF) efforts. The DLR swept flat plate transition experiment provides a generic flow configuration which allows very detailed suction experiments with regard to the optimization of suction panel geometries. Every theory that might predict a

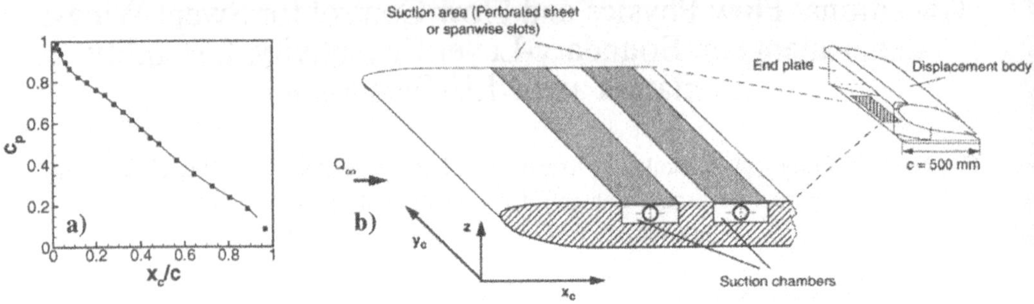

Figure 1 (a) Pressure distribution and (b) experimental setup of the DLR swept flat plate transition experiment. Here the flat plate is equipped with two suction chambers covered by slotted or perforated suction surfaces.

wing flow's transition to turbulence requires experimental results on the receptivity of a three-dimensional boundary layer to vortical freestream disturbances in order to model the disturbed flows initial conditions and on the evolution of high-frequency secondary instability in order to identify and predict the final stages of transition to turbulence. Despite of the successes achieved during the last two decades by utilization of the generic DLR swept flat plate experiment for a better understanding and controlling transitional flows, we must nevertheless consider additional wing curvature effects on laminar-turbulent transition. Hence, additional "real" swept-wing instability and transition experiments (up to transonic free-flight experiments) are important to validate any result achieved via generic flow configurations.

2 Boundary-layer stabilization with the aid of suction

Surface suction is a powerful method to stabilize 3d boundary layers. Experiments with the swept flat plate transition experiment have been carried out at DLR-Göttingen ([3], [4]), to elucidate not only the physical mechanisms of surface suction, but also to optimize the suction surface geometry (holes or slots) with respect to a resistant and effective transition control.
The flat plate is equipped with two suction chambers between 16% and 35% chord (see Fig.1b). In this experiment, where attachment-line instability is avoided by a sufficiently small nose radius, suction chambers more upstream — especially in the region of the neutral point of flow instability — increase the disturbance content significantly. For constant porosity of 1% and suction rate of $c_q = 0.1\%$, three different suction panels are used: spanwise slots with a chordwise width of 100 μm, perforated sheets with hole diameters of 120 μm and diameters of 80 μm. The tests are performed in the 1MG, the 1m low speed wind tunnel of the DLR-Göttingen with an open test section (Tu = 0.15%).
Suction holes or slots in the wing's surface produce perturbations in the flow even in the absence of suction (like roughness elements). Without suction the spatial growth of the traveling disturbances (Fig. 2b) for the perforated surface with 80 μm holes and the slots reach the same saturation amplitude of 10% at 75% chord, while the disturbances observed at the polished surface reach the same saturation amplitude just at 85% chord. The effect of the different surface roughnesses of the models becomes more evident in the growth of the stationary disturbances (Fig. 2a). The amplitudes measured on the models with the perforated surface and with the slots start to decrease at 70% and 75% chord, respectively, due to the appearance of turbulent spots which (most likely) grow at the cost of stationary flow energy. On the polished surface no turbulent spots are observed. Hence, suction device dependent flow disturbances lead to earlier transition if the suction system should fail. However, surface suction can overcome this penalty. The full efficiency of surface suction for the damping of stationary crossflow disturbances can be seen in Fig. 2c. The traveling disturbances (Fig. 2d) are stronger damped than the stationary disturbances. To optimize the efficiency of the suction surface, three-dimensional disturbances

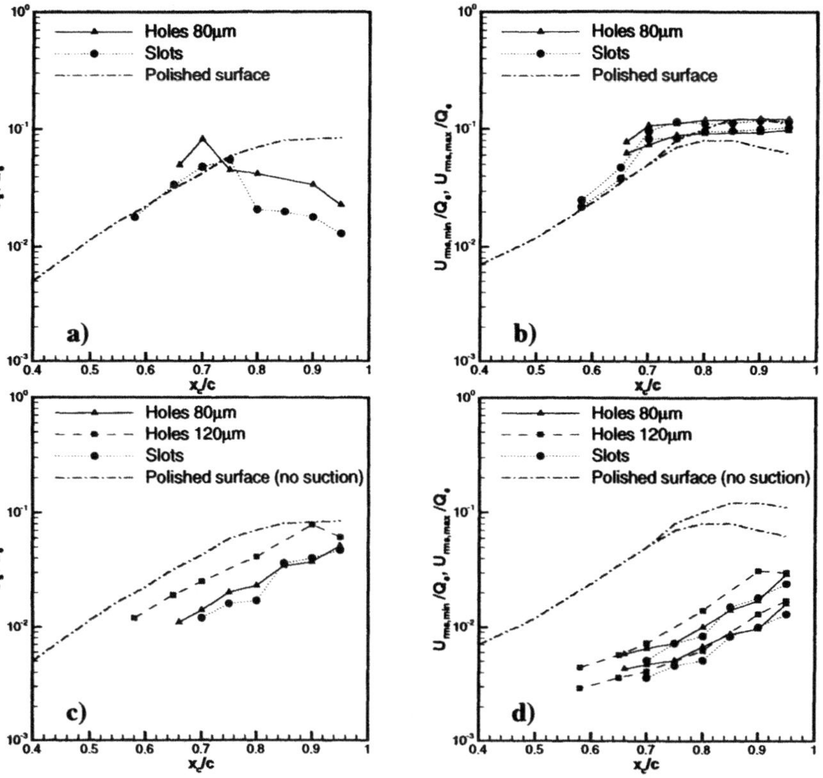

Figure 2 Spatial growth of the stationary (U_s/Q_e) and minimal/maximal traveling ($U_{rms,min}/Q_e$, $U_{rms,max}/Q_e$) disturbances amplitudes for different suction surface geometries: (a) stationary and (b) traveling disturbances without suction ($Q_\infty = 19m/s$,$c_q = 0.0\%$). (c) stationary and (d) traveling disturbances with suction ($Q_\infty = 19m/s$,$c_q = 0.1\%$). Additionally, the reference case of the polished surface [5] without suction is shown.

in the suction velocity distribution which would support the excitation of stationary crossflow vortices have to be minimized. The spanwise slots, which produce a two-dimensional suction velocity distribution show the strongest attenuation of the disturbances.

3 Boundary-layer receptivity to freestream vortices

The initial conditions for every disturbance development is given by the environmental disturbances and the boundary-layer receptivity and these strongly influence the disturbance growth in a three-dimensional boundary layer. Hence, laminar-turbulent transition prediction and control have to pay attention to the different types of disturbances present in wind tunnels and in free-flight, respectively. An increase in freestream turbulence decreases the saturation amplitudes of the stationary modes and finally leads to the domination of the traveling modes [5]. However, such a specification is not sufficient to characterize the differences of transition in wind tunnels and under free-flight conditions. For comparisons with theoretical approaches it is crucial to characterize and measure all of the details of the external, environment dependent turbulence field in any experiment.

To reach a deeper understanding of the impact of vortical freestream disturbances (possibly present in wind tunnels but absent in free-flight) on the behavior of a crossflow dominated swept wing boundary layer, experiments with the DLR swept flat plate transition experiment have been carried out following a suggestion of J. Kendall [6] (see also the CEAS/DragNet-paper by Janke et al.). A vortex originates at the tip of a micro wing located above the plate and propagates into the boundary layer. The vortex strength is controlled by varying the wing's angle

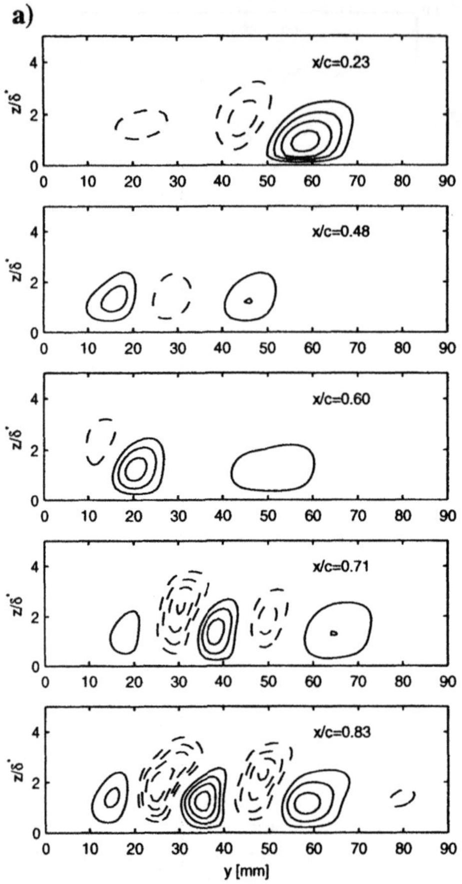

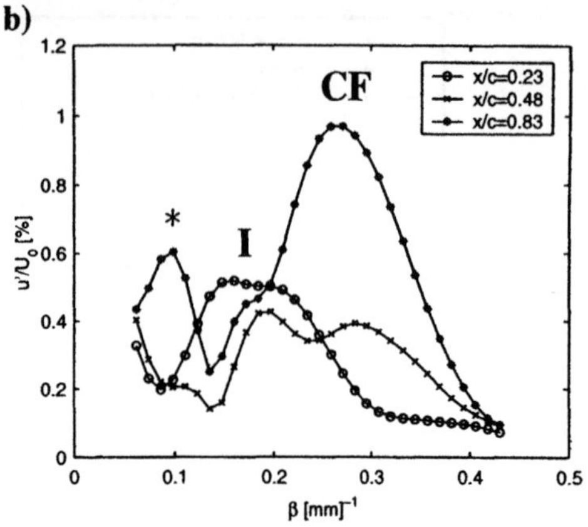

Figure 3 (a) Isolines of the streamwise velocity deformation caused by a single axial vortex penetrating into the three-dimensional boundary layer of the DLR swept flat plate transition experiment for different chordwise locations. Solid lines are positive values, dashed lines negative. During the propagation downstream a multiplication of the vortex can be observed. (b) Spanwise wavenumber spectra of the boundary-layer distortion at different chordwise locations. At the most downstream location $x_c/c = 0.83$ the crossflow wavelength (CF) dominates and wavelengths with smaller wavenumbers ($*$) appear. At all positions the initial wavelength (I) is present.

of attack and the flow velocity [7].

The freestream vortex penetrates into the boundary layer and generates an initial stationary vortical boundary-layer structure. Due to the presence of crossflow instability in the three-dimensional boundary layer, the spanwise flow structure develops from the initial response at $x_c/c = 0.23$ to a multi-vortex structure at $x_c/c = 0.83$ (Fig. 3a). In spectral space this process is characterized by an amplification of the disturbances with wave numbers characteristic for crossflow instability (CF in Fig. 3b). Meanwhile, the continuous downstream forcing of the boundary layer by the freestream vortex (distributive receptivity) leads to the presence of the initial wavelength (I in Fig. 3b) over the whole model chord, similar to observations in experiments on a non-swept plate [7]. Additionally, a generation of smaller wave numbers occurs ($*$ in Fig. 3b). Its nature is not well understood, but its appearance is probably a reflection of the whole wave-packet scale in the spectral space or a result of the disturbance nonlinear activity.

4 Convex surface curvature effect on swept wing boundary-layer instability and transition

Instability analyses predict a significant damping effect of convex curvature, i.e. the spatial growth of the stationary disturbances is much smaller, if a convex surface curvature is taken into account. Thus, a reliable quantitative transition prediction can only have success, if such an effect is taken into account.

To investigate the effect of convex curvature in the presence of dominating crossflow instability, experiments with a swept airfoil with 15% thickness ratio are carried out. The used profile

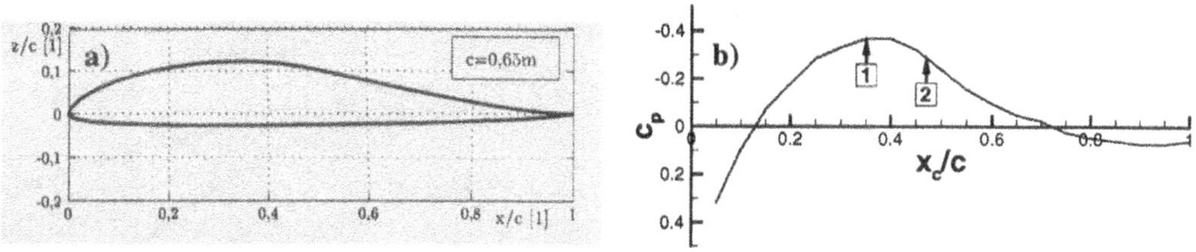

Figure 4 (a) Profile of the swept wing with convex curvature HQ26 and (b) pressure distribution of the upper side ($Q_\infty = 28m/s$). At location [1] crossflow instability dominated transition scenarios observed, while at [2] TS instability plays a significant role.

HQ26 and the appropriate pressure distribution of the upper wing side are shown in Fig. 4a. The airfoil is mounted in the closed test section of the NWB — the DNW low speed wind tunnel located at the DLR-Braunschweig (Tu $= 0.08\%$) — with a sweep angle of 45° and an angle of attack of $-3°$. Up to a location of 35% chord a negative pressure gradient at the upper side of the profile can be observed (Fig. 4b). This accelerated range of the boundary layer is mainly unstable to crossflow instability. Thus, stationary crossflow vortices develop together with unsteady obliquely propagating waves. Downstream of 35% chord, the pressure gradient is positive. Here Tollmien-Schlichting (TS) wave instability together with crossflow instability appear in a more complex transition scenario.

Depending on freestream velocity (Reynolds number) the transition location can be shifted towards different regimes of the wing's pressure field, i.e. towards boundary-layer flow regimes subject to different primary instabilities. In the crossflow dominated regime upstream of 35% chord a gradual widening of the disturbance spectrum is found, after the stationary and traveling disturbances have reached saturation. Laminar-turbulent transition occurs with a continuous broadening of the spectrum as shown in Fig. 5. The different flow behavior, compared with the DLR swept flat plate transition experiments where secondary high-frequency disturbances appear in a crossflow dominated transition process, might be caused by the small stationary mode's saturation amplitude due to the wing's convex surface. In the regime downstream of 35% chord TS and crossflow instabilities are amplified simultaneously, high-frequency secondary disturbances are observed in a similar way as in the DLR swept flat plate transition experiment and lead to a sudden breakdown of the transitional flow (Fig. 6). The transition location relative to the mean-flow profiles, however, is different compared to the flat plate experiments. On the wing the high-frequency disturbances occur at locations, where the mean crossflow velocity component almost vanishes. The streamwise U_s-profiles show no inflection point at these locations. In order to identify the origin of secondary instabilities of three-dimensional boundary layers further experimental flow analyses are required following the theoretical work (see the corresponding CEAS/DragNet-contribution by Janke et al., Fig. 6).

5 Experiments on high-frequency secondary instability

After the primarily amplified disturbances have reached saturation an occasional appearance of high-frequency disturbances can be observed with frequencies one order of magnitude larger than those of the fundamental waves. The wave characteristics of these secondary instabilities, the conditions for their appearance and their growth as well as their role in the final breakdown of the laminar flow are deduced from data of the DLR swept flat plate transition experiments. These experiments are done under controlled condition, i.e. steady crossflow vortices and oblique traveling waves are initiated [8]. For this purpose artificial disturbances for stationary and primary traveling modes are introduced at $x_c/c = 0.12$ and additionally for a controlled excitation of secondary instability modes high-frequency disturbances are introduced at $x_c/c = 0.75$.

For freestream velocities of 17 m/s the frequency of the excited primary traveling mode is 82 Hz

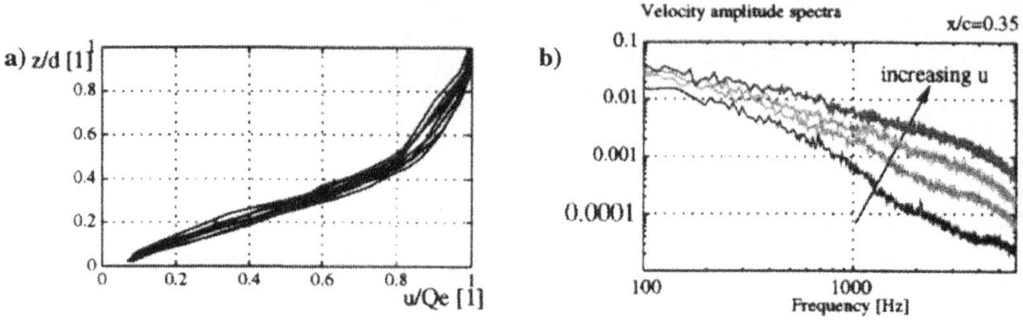

Figure 5 Pure crossflow dominated transition at $x_c/c = 0.35$ (location 1 in Fig. 4b) of the model with convex curvature. (a) The mean flow profiles show weak distortion. (b) A gradual widening of the spectrum and no occurrence of high-frequency secondary instability are observed.

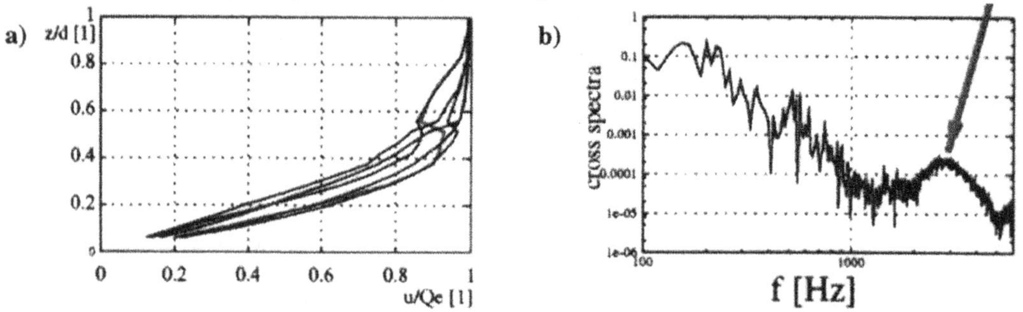

Figure 6 Mixed TS/crossflow dominated transition at $x_c/c = 0.47$ (location 2 in Fig. 4b) of the model with convex curvature. (a) The mean flow profiles show strong distortion. (b) Occurrence of high-frequency secondary instability are observed (marked with the arrow).

and two high-frequency secondary modes at 2050 Hz and 820 Hz are identified. The 2050 Hz mode can be associated with high shear of the streamwise velocity profiles in wall-normal direction near the edge of the boundary layer (Fig. 7b). The second high-frequency mode at 820 Hz is found in the lower part of the boundary layer and can be correlated with the shear of the wall-normal velocity profiles near the wall (Fig. 7b). There are indications taken from other experiments [10] that high-frequency instability correlated with the shear of spanwise velocity profiles can also dominate the laminar-turbulent breakdown. In our experiments where the spanwise position of the high-frequency excitation is fixed relative to the stationary vortex, no such correlation of the high-frequency disturbances has been identified yet.

Fig. 7a shows the chordwise amplitude growth for the stationary, primary traveling and the two secondary high-frequency modes. Introducing stationary and primary traveling artificial disturbances as in case (0,1)+(1,1) the amplitudes of the stationary vortices and the primary traveling (82 Hz) modes reach saturation at 75% chord. Beyond this location, the secondary high-frequency modes which are introduced at 75% chord show a strong amplitude growth. In agreement with theoretical predictions [9] the high-frequency modes grow much stronger than primary traveling modes.

6 Conclusions

Basic experiments on transitional flow physics and flow control for swept wings have been presented. It has been shown that surface suction can attenuate crossflow disturbances significantly even if the suction is applied downstream of the leading edge region, provided attachment-line instability and swept-leading edge flow transition can be avoided. The traveling disturbances are stronger damped than the stationary disturbances. To optimize the efficiency of the suction surface, three-dimensional disturbances in the suction velocity distribution have to be minimized.

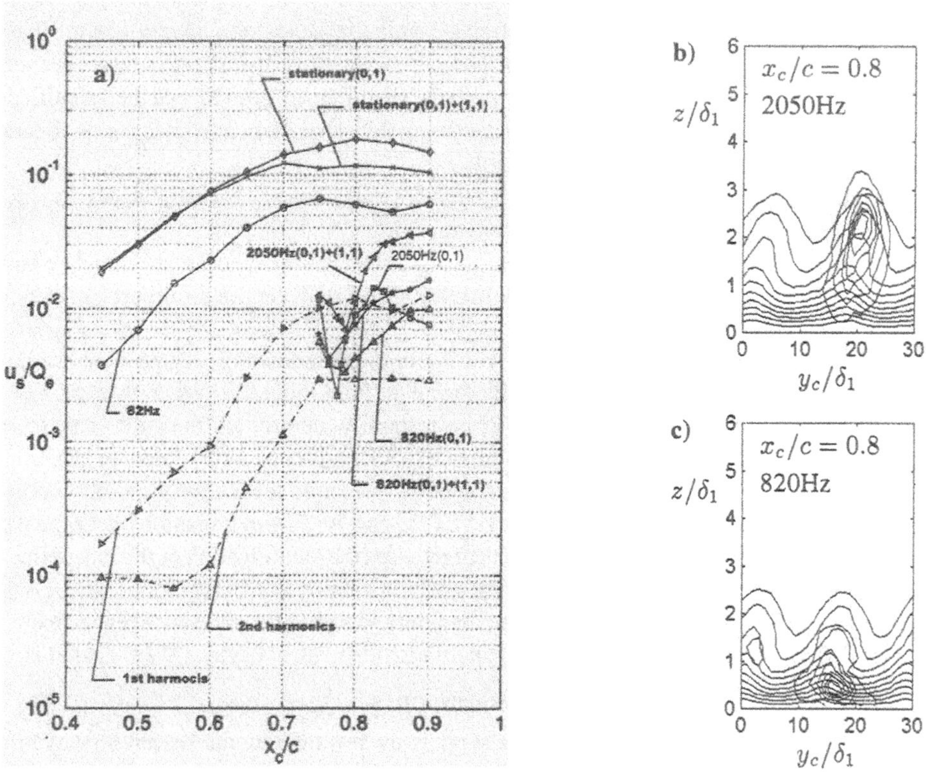

Figure 7 (a) Spatial downstream growth for the case (0,1) where only a stationary mode was initiated and for the case (0,1)+(1,1) where additionally a traveling mode was initiated. The amplitudes of the stationary mode, the primary traveling (82 Hz), the 1st and 2nd harmonics of the primary mode and the two identified secondary modes (2050 Hz and 820 Hz) are shown. (b) and (c) show the isotachs of the streamwise velocity component at $x_c/c = 0.8$ and the location of the appearance of secondary high-frequent instabilities for the 2050 Hz and the 820 Hz mode.

The receptivity of a three-dimensional boundary layer to quasi-stationary vortical freestream disturbances has been studied. It became obvious, that due to the crossflow instability a continuous selection of corresponding unstable wave numbers occurs in streamwise direction. This leads to a generation of vortical structures inside the boundary layer, which are characteristic for crossflow instability.

The transition experiments with a swept wing, which were focused on the disturbance damping effect of convex surface curvature, provide evidence for two different transition scenarios. For transition to take place in the crossflow dominated instability regime a gradual widening of the disturbance spectrum could be observed while transition to turbulence takes place in the mixed TS/crossflow instability regime with a rapid growth of selected secondary high-frequency disturbances. The different behavior compared with the DLR swept flat plate transition experiment, where secondary disturbances appear in a crossflow dominated transition process, is likely due to the small magnitude of the stationary mode's saturation amplitudes enforced by the convex wing surface.

By controlled excitation of primary low and secondary high-frequency disturbances in the DLR swept flat plate transition experiment, details of the amplifications of two different secondary modes could be identified. Both seem to be correlated with regions of high shear of the wall-normal streamwise velocity profiles, one near the edge of the boundary layer and the second near the wall. Both high-frequency secondary modes are predicted by theoretical investigations.

For swept wings crossflow instability is quite often a dominating transition phenomenon. However, the final breakdown to turbulence appears where the crossflow disturbances have reached nonlinear amplitude saturation and no further primary disturbance growth can be identified. Hence, transition prediction for crossflow dominated wing flows based on any linear or nonlinear disturbance growth (e^n-methods) cannot be reliable.

Three experimental efforts for further focused research appear necessary for reliable and as simple as possible flow predictions as well as effective transitional flow control:

1. Progress in physical-mathematical modeling of the disturbance flow spectrum to initialize any instability analyses either with the nonlinear NOLOT/PSE or DNS requires support by further generic receptivity experiments [5] [7].

2. Identification of the mechanisms responsible for the final breakdown of a three-dimensional transitional flow around a swept wing into a fully turbulent flow should leed to a physical transition criterion. Our DLR experiments [1] [2] [8] on instability controlled transition match the corresponding theoretical efforts (see corresponding CEAS/DragNet-paper by Janke et al.).

3. Hybrid Laminar Flow technology has demonstrated its principle feasibility via suction control under free-flight conditions (EU/EC-project HYLTEC and A320-Fin transition experiment). However, any economy consideration of flow control requires the identification of the most effective means (and modes) of (suction) control and/or novel instability controls beyond the present HLF-technology. Such objectives can be substantiated only via generic wind tunnel experiments with the goal to find a robust and simple swept wing transition control device [1] [3] [10] [11].

Acknowledgements

The authors wish to acknowledge the generous support by several national research programs on transition, namely DFG-contracts, "Alexander-von-Humboldt"-fellowships and support via DLR/CAE (China) and DLR/NLR (India) programs for international collaborations, as well as support from European projects.

References

[1] Bippes H. (1999) Basic experiments on transition in 3D-boundary layers dominated by crossflow instability. Progress in Aerospace Sciences 35: 363–412.

[2] Pöthke A. (1999) Grundlegende Untersuchungen zum laminar-turbulenten Umschlag an der konvex gekrümmten Oberseite eines schiebenden Tragflügels. Diss. thesis Physics Dep., Univ. Göttingen.

[3] Bippes H., Wiegel M., Bertolotti F. (1998) Experiments on the control of crossflow instability with the aid of suction through perforated walls. Proc. IUTAM Sym. FLOWCON, Göttingen, Sept. 7-11, 1998.

[4] Abegg C., Bippes H., Janke E. (1999) Stabilization of boundary-layer flows subject to crossflow instability with the aid of suction. Proc. IUTAM Sym. Lam.-Turb. Transition, Sedona, Sept. 13-17, 1999.

[5] Deyhle H., Bippes H. (1996) Disturbance growth in an unstable three-dimensional boundary layer and its dependence on environmental conditions. J. Fluid Mech. 316: 73-113.

[6] Kendall J.M. (1990) Boundary-layer receptivity to freestream turbulence. AIAA 21st Fluid Dyn. Conf., Seattle, June 18-20, 1990: AIAA 90-1504.

[7] Boiko A.V. (1999) Flat plate boundary-layer response to controlled freestream axial vortices, In Proc. 11th Int. Couette-Taylor Workshop, Bremen, July 20-23, 1999: 111-112.

[8] Lerche T. (1997) Experimentelle Untersuchung nichtlinearer Strukturbildung im Transitionsprozeß einer instabilen dreidimensionalen Grenzschicht. Fortschr.-Ber. VDI Reihe 7 Nr. 310. Düsseldorf.

[9] Koch W., Bertolotti F.P., Stolte A., Hein S. (2000) Nonlinear equilibrium solutions in a three-dimensional boundary layer and their secondary instability. J. Fluid Mech. 406: 131-174.

[10] White E.B., Saric W.S. (2000) Application of variable leading-edge roughness for transition control on swept wings. AIAA 38th Aerosp. Sci. Meeting, Reno, Jan. 10-13, 2000: AIAA 2000-0283.

[11] Abegg C. (2000) Experimental investigation on the effectiveness, robustness and sensitivity of surface suction panels with respect to HLF control. To appear as DLR-Report.

New Developments in Surface Flow Sensor Technology within the Framework of AEROMEMS

O. Burkhardt+, U. Dinata+, C. Warsop*, W. Nitsche+

+ Technical University of Berlin (TUB), Institute of Aeronautics and Astronautics (ILR), Department of Aerodynamics, Marchstr. 14, 10587 Berlin, Germany

* Advanced Technology Centres – Sowerby, BAE SYSTEMS, PO BOX 5, Filton, Bristol, UK

SUMMARY

A number of flush mounted surface flow sensor devices (MEMS surface hot-film, surface hot-wire, PVDF sensor and surface fence) have been developed and investigated for applications of wall shear stress measurements in a transonic flow. In order to maximise their robustness, the MEMS hot-film and PVDF sensor were coated with abrasion resistant coatings. Testing of the sensors was carried out in boundary layer flows up to M_∞=0.9 in a high speed wind tunnel of the ILR. The sensor characteristics including sensitivity, dynamic response and cut-off frequency were measured. The test results indicate application ranges and advantages of the improved sensor devices. Of the sensor types studied, MEMS hot-films seem to be the best solution for the flow conditions investigated and, with some development, could be used to realise an active boundary layer control system in the medium term.

INTRODUCTION

Recent developments in micro-fabrication technology have led to the possibility for materials incorporating arrays of Micro-Electro-Mechanical Systems (MEMS) to be manufactured. These materials may contain sensors, electro-mechanical actuators and electronic circuitry (with size of a few microns to a few millimetres). One potential application for MEMS technology is in the measurement and manipulation of fluid flows with an interactive distributed control. It has been known that under certain conditions, the flow over a surface can be significantly affected by extremely small geometric features on the surface itself. For example, theoretical studies have shown that with the appropriate technology, turbulent skin friction drag could be reduced by 50-80%. However, the MEMS technology required to achieve such an objective on a commercially viable aircraft is unlikely to be available for many decades. Therefore, from an industrial perspective it is felt that it is only possible to consider the limited, less ambitious application of 'MEMS' to control other aspects of flow around an aircraft.

One realistically realisable application of MEMS technology is the flow separation control by active enhancement of mixing in the lower regions of a turbulent boundary layer. 'MEMS' devices could be placed in localised regions on a wing where the boundary layer is receptive to small scale disturbances. Sensors would detect the condition of the boundary layer and provide information for the control of the 'MEMS' actuators. This would enable the overall performance of the wing to be better optimised throughout the flight envelope, for example by reducing buffet and cruise drag and increasing maximum lift. A project funded by the European Commission, AEROMEMS [1], has been set up to examine the viability of applying MEMS for boundary layer flow control on aircraft and it is within this framework that the work presented in this paper was undertaken.

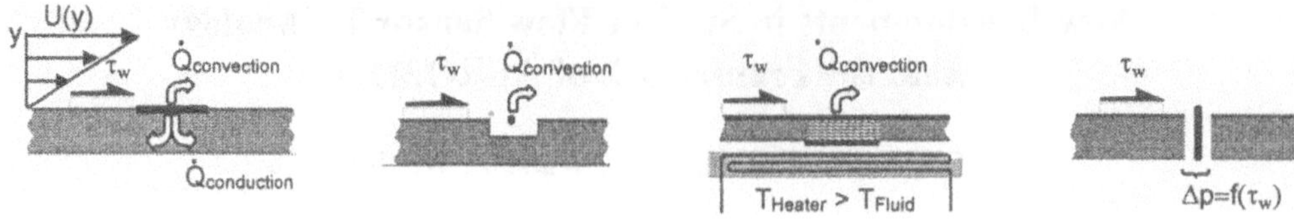

Figure 1: Wall shear stress measurement techniques;
(a) Surface hot-film, (b) Surface hot-wire, (c) PVDF sensor, (d) Surface fence

Such flow control technologies require small, robust non-intrusive surface shear stress sensors, which are able to operate in harsh environments with a minimal maintenance. While conventional laboratory type sensors such as hot-films are generally considered too fragile for general engineering application, micro-fabrication has opened up possibilities for improving general robustness and susceptibility to abrasion.

Sensor demands were focused on robustness, local and global wall shear stress detection and high frequency turbulence measurements using small, coated sensors with a highly dynamic response in an array arrangement. In order to ascertain the suitability of various wall shear stress measurement technologies, four types of sensor techniques have been improved and tested up to M_∞=0.9 to investigate the sensor sensitivity and dynamic response characteristics.

WALL SHEAR STRESS SENSOR DEVICES

The wall shear stress sensor devices developed within the framework of AEROMEMS include:
- coated and non-coated surface MEMS hot-film sensor developed by BAE SYSTEMS,
- surface hot-wire with a sensor welded over a very small cavity,
- coated and non-coated PVDF (Polyvinylidene fluoride) sensor,
- surface fence incorporating MEMS different pressure sensor.

Figure 1(a) shows the measurement principle of a surface hot-film based on convective heat transfer from a heated sensing element into the flow. The element can be heated in constant temperature mode by the closed loop control of a Wheatstone bridge, in which the sensor is one of the bridge resistors. The sensor convection heat loss rate is a function of wall shear stress, τ_w, on the sensor. Figure 1(b) illustrates a surface hot-wire bonded over a cavity that is almost analogues to the surface hot-film but with negligible substrate effects [7]. Figure 1(c) represents the measurement principle of a heated PVDF sensor. The sensor is able to measure only wall shear stress fluctuation, which induces the heat flux fluctuation of the thin sensor material. Temperature changes of the PVDF material cause a pyro-effect, an effect of electrical charge of the material [9]. As shown in Figure 1(d), the surface fence principle is based on a pressure difference in front of and behind a fence placed within the boundary layer flow that can be correlated to τ_w using the similarity law of the viscous sublayer [3, 10].

The instantaneous wall shear stress contains mean and fluctuating parts as written by equation $\tau_w = \bar{\tau}_w + \tau'_w$. In the case of fluctuation measurements, a sensor dynamic response is important especially in high Reynolds number flows. Investigations of the dynamic response take place via an analysis from measuring data as a function of the time or frequency using a Fast Fourier Transformation and could characterise the influence of flow stream velocities on a signal-to-noise ratio (SNR) of sensor systems.

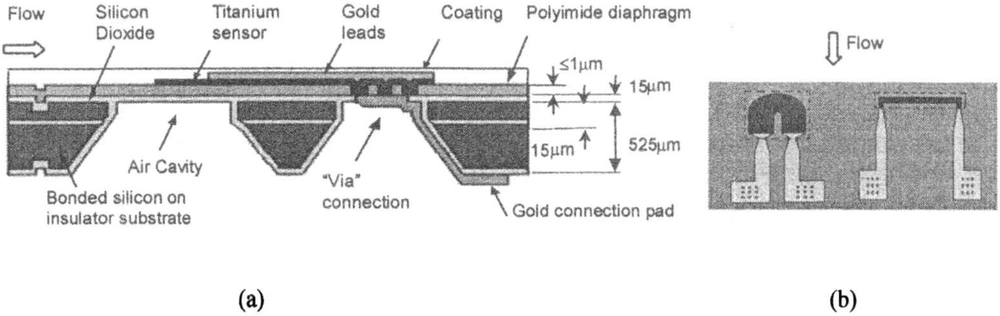

Figure 2: (a) MEMS hot-films of BAE SYSTEMS; (b) Improved designs (horse-shoe, left; slit type, right)

MEMS Surface Hot-Films

A schematic layout of a MEMS surface hot-film sensor developed by BAE SYSTEMS is shown in Figure 2(a). The sensors are fabricated on a silicon wafer with an air cavity below the sensor element to reduce heat conduction from the sensor through the substrate. After the sensors fabrication, the entire top surface of the wafer is coated with a thin layer (<1 μm) of silicon dioxide to provide resistance to surface abrasion. The coating also helps to eliminate the very small roughness effect of the sensor and its gold contacts. The sensor elements themselves are made of Titanium and have nominal cold resistances of 14-25Ω.

A 0.5μm-thick Titanium resistor is deposited on a 15μm-thick Polyimide diaphragm on a 500μm-thick, 1-0-0-orientated silicon substrate. Electrical leads and contacts were of gold. Typically hot-film resistors were placed over the centre of a 500×500×15μm-thick Polyimide diaphragm. Electrical connections were taken from the front face of the silicon wafer to the rear side using a technique developed by BAE SYSTEMS. Subsequent sensor designs have been fabricated where the sensor area has been reduced substantially to fit on a 100×100μm diaphragm. In the first design iteration the resistor and cavity were not optimised in any way and after initial testing, it was noticed that although the sensitivity was reasonable (250mV/Pa) the frequency response was extremely poor (initial roll-off at 20Hz).

To fully understand the problem, a thermal model was developed in association with researchers at Queen Mary & Westfield College, London [5] which proved that the poor frequency response was caused by a number of factors associated with the thermal characteristics of the diaphragm and connecting leads. In the next design iteration, a number of improvements were made to the sensor design in order to increase sensitivity and frequency response. The improved designs had either a horseshoe or slit shape, where the sensor area relative to the diaphragm area was maximised, see Figure 2(b). The area of the gold tracks covering the diaphragm was also minimised and their area was locally reduced to minimise conduction along the leads. Overall these modifications improved overall sensitivity to 750mV/Pa and the frequency response to several kilohertz.

Figure 3: Silicon shadow mask and wafer, developed at BAE SYSTEMS

There was a need for a thin insulating oxide layer under the electrical pads on the rear of the wafer because the resistance of the sensors (about 20Ω) was comparable to the resistance of the parent silicon between the two sets of pads. A dry oxide etch process using a contoured shadow mask was developed which only removed oxide covering the "via" connections. The deposition and patterning of the gold contacts on the rear of the wafer was also achieved through the use of a separate shadow mask, where the deposition and patterning process could be achieved in a single operation. The shadow mask technique (Figure 3) developed at BAE SYSTEMS involved the creation of a separate silicon wafer having a surface topology, which is the exact inverse of that of the sensor wafer. Windows were incorporated into the shadow mask corresponding to the areas on the device wafer, where etching or deposition was required.

Surface Hot-wire

The developed surface hot-wire (SHW) sensor is bonded over a tiny spanwise cavity (Figure 4) to extremely reduce the substrate effect in order to increase SNR. The sensor was welded on a printed circuit board with a Copper layer over the cavity made with an etching process. The used wire was of Tungsten (α_{20}=0.0036K^{-1}) with cold resistance of 7.5Ω.

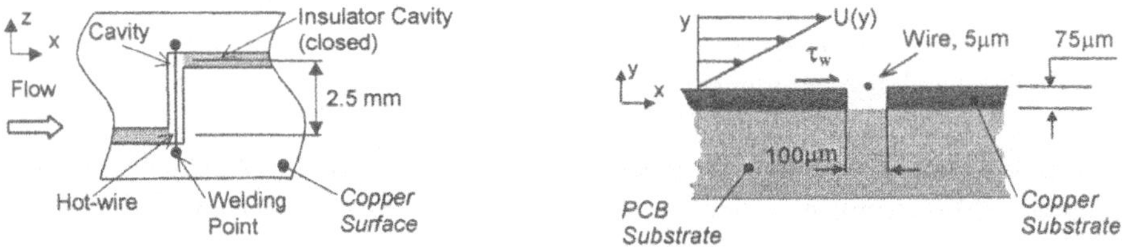

(a) (b)

Fig. 4: Surface hot-film (SHW) developed by TUB; (a) top view, (b) side view [7]

Coated PVDF Sensor

The used PVDF sensor is made of a ferroelectrical polymer film, which has both piezo and pyroelectrical properties. Pyroelectricity is an effect caused by temperature changes of the material resulting in a change of electrical surface charge. The material has a crystalline structure, which results of a drawing process in an electrical poling field to achieve an alignment of crystalline dipoles. The mode delivers a moving of charge that depends on the film temperature gradient due to the heater placed on its lower surface. The sensor allows no static measurements due to charge loss caused by leakage currents in the amplifier input [9].

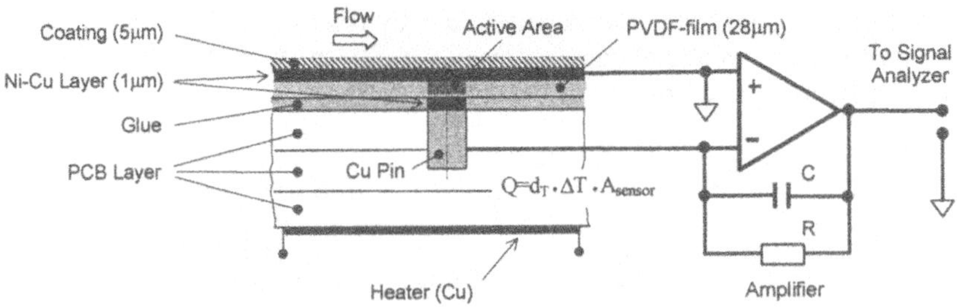

Figure 5: Sketch of a coated PVDF sensor heated for pyro-effect

The tested PVDF sensor used a 28μm-thick polymer film (with a pyro-coefficient, d_T, of -30.10^{-6}C/m²K) sandwiched between two 1μm-thick Ni-Cu layers (Figure 5). In the present study, the upper layer was coated using a 5μm-thick TESA™ film for protection from abrasive erosion. The diameter of the sensor active area, A_{sensor}, remaining from an etching process on the lower Ni-Cu layer was 3 mm. This sensitive area was connected with a copper contact pin to a charge amplifier, while the upper coated layer is used as a ground. For the pyro-effect, the sensor was heated to achieve a temperature difference, ΔT, of approximately 20°C between the sensor and flow.

Surface Fence using an Integrated, Piezoresistive Pressure Sensor

For a conventional surface fence, taps on both sides of the fence are connected with tubes to a differential pressure transducer [3, 10]. However, these tubes may cause an additional damping effect resulting in a low sensitivity in sensor dynamic signal. Such a disadvantage can be minimised by placing a small sensor very close to the taps as shown in Figure 6. For this purpose, a small, piezoresistive pressure sensor based on MEMS is used that allows measurements of mean and fluctuating parts of wall pressure and shear stresses.

The used transducer was a commercially available piezoresistive sensor (from Honeywell™, 26pc type). The sensing element consists of four almost identical piezoresistors buried in a silicon diaphragm surface placed in a sensor die [Figure 7(a)]. The piezoresistors are connected in a Wheatstone bridge to an amplifier and signal conditioner [Figure 7(b)]. Referring to [6], a pressure applied to the sensor causes the diaphragm to flex, resulting in stress or strain in the diaphragm and also in the buried piezoresistors. Due to the pressure, all four resistors will change by approximately the same value ΔR. Two resistances will increase and the other two decrease depending on their orientation with respect to the crystalline direction of the silicon material. The signal voltage generated by the Wheatstone bridge is proportional to the pressure. The sensor has a specification including a sensitivity of 16.7mV/psi under a bridge voltage of 10V, a pressure range of ±6.9kPa and a response time of t=1ms. The sensing element has a temperature compensation for a span of 0-50°C.

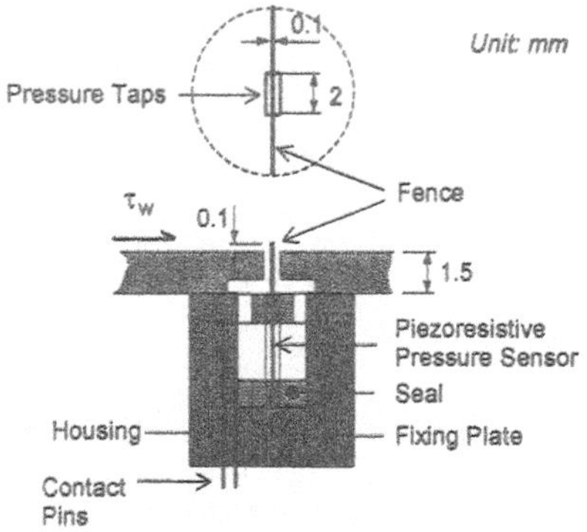

Figure 6: Sketch of the surface fence with an integrated, piezoresistive pressure sensor

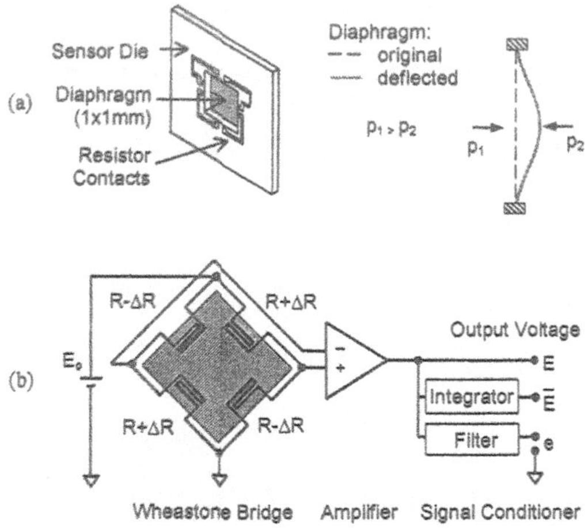

Figure 7: (a) Pressure diaphragm in a sensor die and its deflection; (b) Piezoresistor Wheatstone bridge

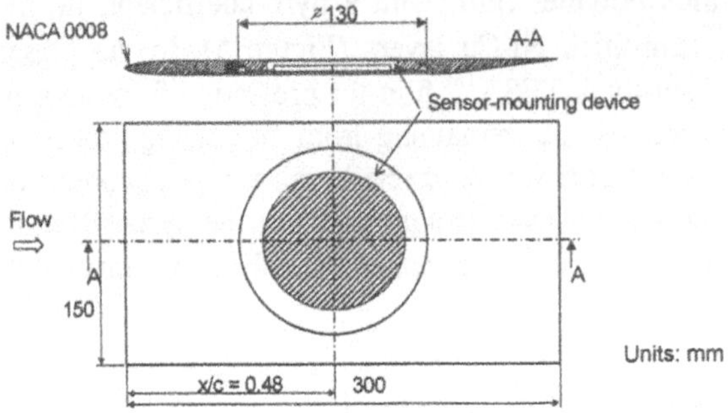

Figure 8: Flat plate with a sensor mounting device

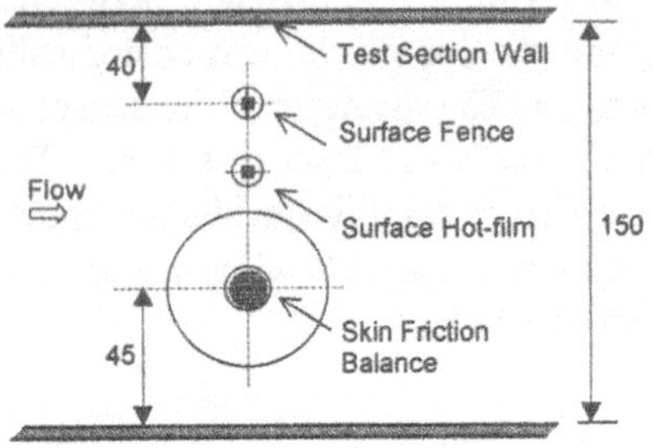

Figure 9: Arrangement for surface fence test

MEASUREMENT SET-UP

The MEMS surface hot-film, surface hot-wire and PVDF sensor were investigated to know the dynamic response characteristics and compared to a TAO-Systems™ surface hot-film (made of Nickel with cold resistance of 6Ω and α_{20}=0.0029K^{-1}). To operate the hot-film and hot-wire sensors, a self-built constant temperature anemometer (CTA) were employed. This anemometer was one of the channels of a small, low cost multi-channel anemometer designed by TUB within the AEROMEMS project for measuring a spatial evolution of a flow phenomena. All sensors were placed on a flat plate with a removable sensor wafer mounting (Figure 8) and tested in boundary layer flows up to M_∞=0.9 (Re=1.8×10^6 at x/c=0.4) in a transonic wind tunnel. Three Preston tubes (CPM3) with different diameters (d_1=0.3, d_2=0.65, d_3=0.9mm) were used as reference. The steady state wall shear stress was up to 135N/m² with a calculated friction velocity u_τ of below 13m/s.

The piezoresistive surface fence was tested for static and dynamic response characteristics and compared to the TAO surface hot-film. The calibration was conducted with a skin friction balance (from Electrotechnologies Selem™, SM-251 type), which had a floating head diameter of 10mm and a sensitivity 1mV/mgf with resolution of less than 1mgf and an accuracy of below 0.5%. These devices were mounted in a flat wall section of the transonic wind-tunnel and placed spanwise to enable simultaneous measurement (Figure 9). The test was carried out in turbulent boundary layer flows up to M_∞=0.9 (Re=14.5×10^6 at x=0.8m). The steady state wall shear stress was up to 80N/m² with u_τ<10m/s.

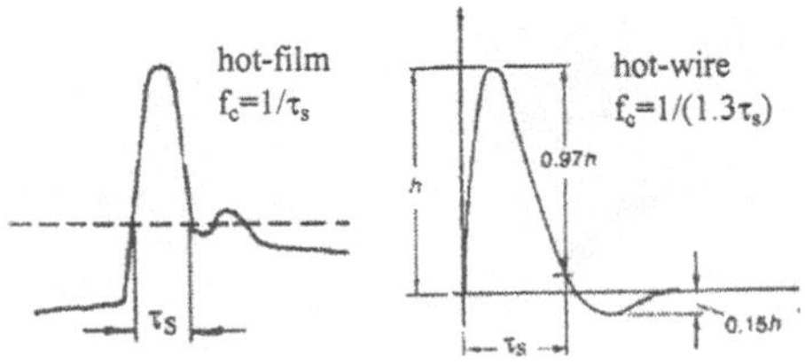

Figure 10: Cut-off frequency estimates using an optimised square-wave test [2]

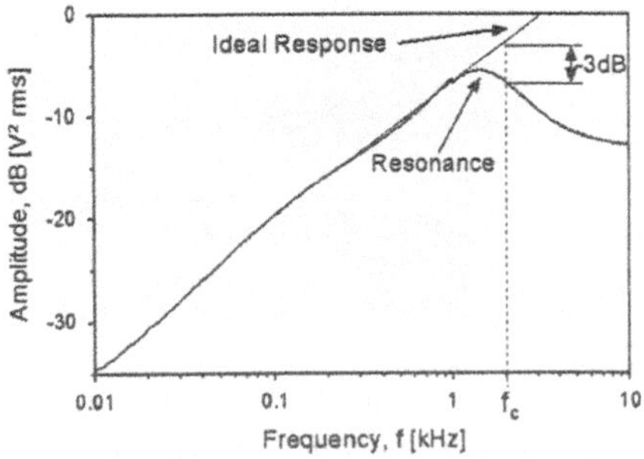

Figure 11: Frequency response and cut-off frequency estimate using a sine-wave test [4]

Dynamic Response and Cut-off Frequency

A square-wave voltage-perturbation test is used as the preferred means by which the system response could be determined in term of cut-off frequency, f_c. The anemometer output is fed into an oscilloscope for monitoring the response subjected to the test signal input. An adjustment could be made to the anemometer until a response as shown in Figure 10 is obtained. Here, τ_s is the optimum time interval for the pulse (with 15% undershoot respecting to the maximum, h) to decay to 3% of the maximum value, as promoted by Freymuth [4]. To obtain f_c, a sine-wave test can be also used, which is analogous to flow fluctuations. This excitation should be carried out with a sweeping signal of sine-waves after the square-wave test and then, f_c is determined at a position where the amplitude has decreased by 3dB from an ideal response (see Figure 11). Additionally, a resonance frequency is indicated at maximum amplitude.

RESULTS AND DISCUSSIONS

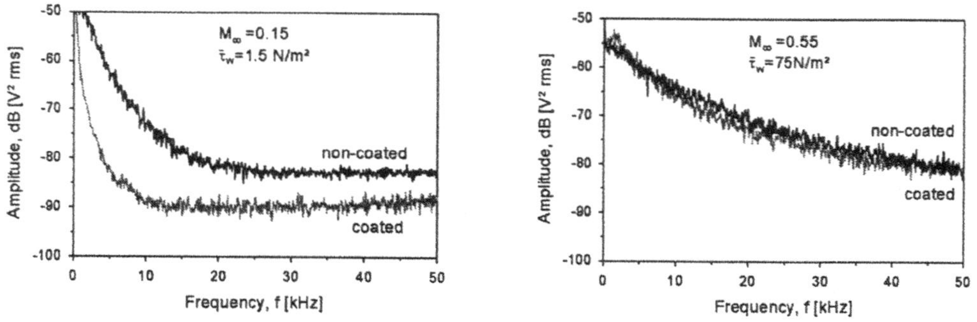

Figure 12: Typical power spectra of the coated and non-coated MEMS hot-film

Coated MEMS Surface Hot-Films

Figure 12 shows two typical turbulent power spectra for the coated and non-coated MEMS hot-film at M_∞=0.15 (left, laminar) and 0.55 (right, turbulent). The results suggest that the coating layer has a clear influence on the dynamic response of the sensor at the lower speed, but almost no effect at the higher speed. It is interesting that at the lower speed, the coated sensor responds with lower signal amplitudes than the non-coated one. This suggests a damping effect at the low speed caused by heat capacity of the coating layer.

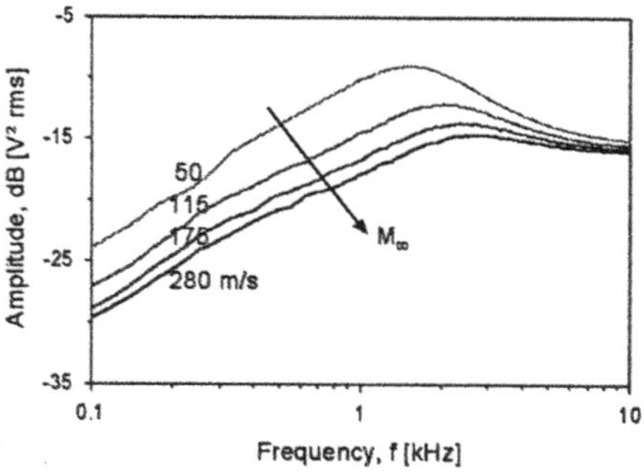

Figure 13: Typical frequency response of a coated MEMS hot-film (horseshoe, R_0=14Ω, O.H.R.=1.2)

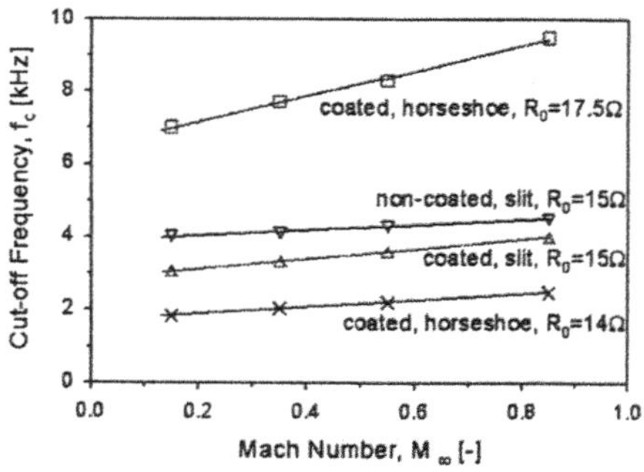

Figure 14: Typical cut-off frequency of coated and non-coated MEMS hot-films (O.H.R=1.2)

Figure 13 illustrates typical results for the frequency response of the coated MEMS hot-film using the CTA with a ratio of sensor hot-to-cold resistance (overheat ratio) of 1.2 at various flow velocities. The sensor was excited using a 5mV peak-to-peak sine-wave signal applied with a linear sweep from 0.1 to 100kHz through a 10kΩ resistor to the Wheatstone bridge. At low frequencies the response amplitudes are low and then raise with a slope of 14dB/decade up to maximum points. As the frequency is further increased, the amplitudes decrease and reach a saturation. Additionally, the sensor amplitudes attenuate as the free-stream flow velocity increases, probably due to the CTA characteristics itself.

Figure 14 shows the Mach number, M_∞, influence on the cut-off frequency of the MEMS-HF (hot-film) with different cold resistances at an overheat ratio, O.H.R., of 1.2. The trend of all sensors indicates a linear increase of f_c as M_∞ gets higher. The diagram also shows an increase f_c for an increasing cold sensor resistance, but it is interesting that cut-off frequencies of the slit-type sensor attenuate due to the coating. Therefore, the silicon dioxide coating has an effect to decrease the sensor cut-off frequency.

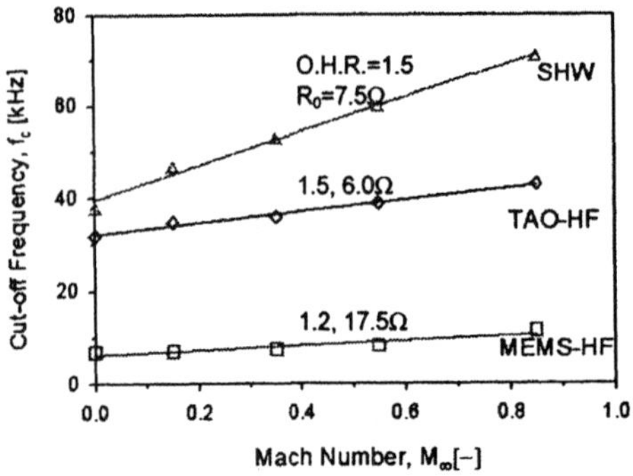

Figure 15: Typical cut-off frequencies

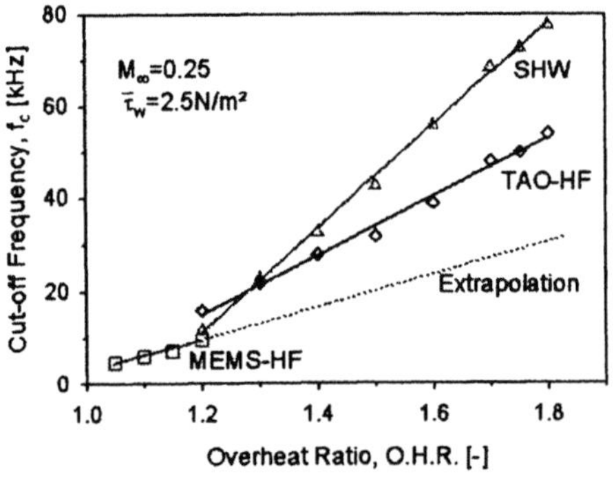

Figure 16: Influence of O.H.R. on f_c

Surface Hot-Films and Surface Hot-Wire

Figure 15 shows a Mach number influence on f_c of the coated MEMS hot-film, surface hot-wire and non-coated TAO hot-film operated with the CTA. It can be seen that f_c gets higher as M_∞ increases that is caused by increasing sensor heat conduction losses into the flow. Figure 16 suggests that an increase in O.H.R., which translates directly into the increasing sensor operating temperatures and heat transfer losses into the flow, tends to increase f_c. For this test, the employed O.H.R. of TAO-HF and SHW sensor varied from 1.2-1.8. However, due to the higher cold resistance of the MEMS-HF, their O.H.R. with the self-built CTA could only be varied from 1.05-1.2. The achieved cut-off frequencies for the MEMS-HF (2-10kHz) are clearly lower than for the two other sensors (TAO-HF: 16-52kHz, SHW: 12-77kHz). As comparison, the f_c's of SHW are dominant due mainly to the negligible substrate effect.

At O.H.R.'s above 1.2, the MEMS sensors with high cold resistances (18-25Ω) were burnt out due to instabilities in the CTA bridge resulting in very high currents from its feedback transistor to the sensors. With appropriate redesign of the anemometry circuitry and/or modification of the MEMS sensors to reduce their cold resistance to say 10Ω, the MEMS-HF could be operated at higher overheat ratios. Examination of Figure 16 indicates that it may be possible to linearly extrapolate the cut-off frequency with overheat ratio. It looks likely that they could then be expected to achieve f_c in excess of 10kHz for the MEMS sensor.

Coated PVDF Sensors

Figure 17 shows two power spectra at different M_∞ measured by the coated and non-coated PVDF sensor for comparison. These results suggest that the coating layer has almost no effect at both low and high speeds. However, the PVDF sensors can only be used for dynamic measurements. Further investigations, within the frame of the AEROMEMS, showed that these sensors were very sensitive to electric and acoustic noises.

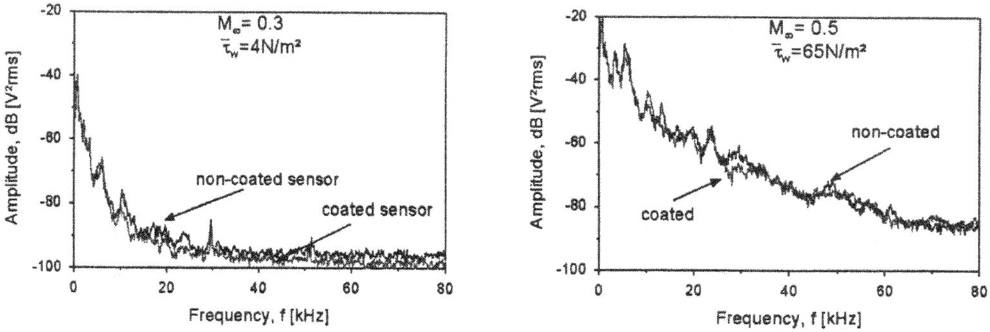

Figure 17: Typical power spectra measured with the coated and non-coated PVDF sensors

Piezoresistive Surface Fence

A sensitivity obtained as a derivative of a static response function between the sensor output voltage and wall shear stress is necessary in order to know a sensor's characteristics when influenced by various wall shear stresses. A typical, normalised sensitivity curve for the tested piezoresistive surface fence in comparison to a surface hot-film is shown in Figure 18. While the sensitivity of the hot-film attenuates with increasing τ_w (due to a decreasing heat loss rate), that of the piezoresistive surface fence increases until a maximum is reached at about 67N/m^2.

Furthermore, Figure 19 shows power spectra of the piezoresistive surface fence up to 20kHz for M_∞<0.9. The spectrum diagram indicates a clear influence of the flow velocity on the piezoresistive sensor signal. An increase in the Mach number, indicating that the wall shear stress fluctuation gets higher, results in an increase of the power spectrum amplitude level. As shown in the diagram, the signal levels of wall shear stress fluctuations for free stream velocities are well above the electronic noise at M_∞=0 that shows a high SNR of the whole sensor system.

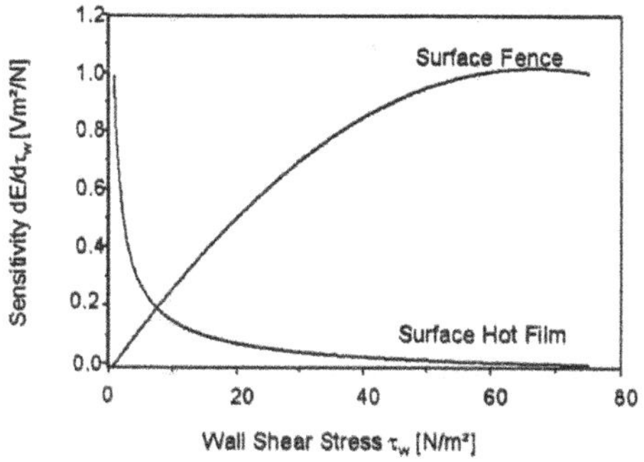

Figure 18: Normalised sensitivity of the piezoresistive surface fence and the surface hot-film

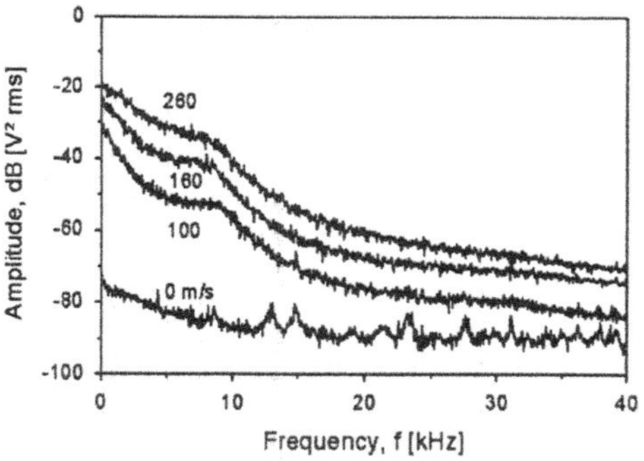

Figure 19: Typical turbulent power spectra using the piezoresistive surface fence

CONCLUSIONS

Four improved surface devices for static and dynamic wall shear stress measurements have been presented. MEMS hot-films and a PVDF sensor were coated to protect against surface erosion. A surface hot-wire was welded over a cavity to reduce heat losses into its substrate. A surface fence was enhanced with a piezoresistive pressure sensor. All devices were developed in order to better satisfy the requirements for application as part of a MEMS-based active flow control system on an aircraft (mechanical robustness, local area resolution, dynamic response). The sensor performances have been characterised up to high speed flow conditions.

The test results for silicon dioxide-coated MEMS hot-films, which are small enough for local and global flow effect measurements in arrays, indicate a clear coating influence on the dynamic response due to an added damping effect of the coating heat capacity, especially at low speed (M_∞=0.15). At higher speeds ($M_\infty \geq 0.55$), the effect would be very small and the overall SNR is good. Furthermore, despite the low overheat ratios, which could be utilised, the measured power spectra have shown that the hot-films are good enough for potential applications, e.g., for flow control applications on an aircraft. The currently designed sensor and, perhaps more importantly, the optimisation of the constant temperature anemometry to the sensor only allowed operations at low overheat ratios (below 1.2). Measured operating cut-off frequencies were therefore only in the region of 2-10kHz compared to the surface hot-wire. Results obtained in this study suggest that a reduced sensor cold resistance and a more optimised constant temperature anemometry circuit could allow the sensors to be operated at higher overheat ratios and achieve cut-off frequencies in excess of 10-30kHz.

The surface hot-wire indicated higher cut-off frequencies (12-77kHz) due to high O.H.R. The overall SNR was very good caused by a negligible heat loss into the substrate, but such a sensor was difficult to be coated. Moreover, results of the PVDF sensor showed no significant coating influence on its performance at either low or high speed. However, this sensor can only be used for dynamic measurements. Further investigations within the AEROMEMS project have shown that this sensor was very sensitive to electric and acoustic influences.

The piezoresistive surface fence is potentially useful for static and dynamic wall shear stress measurements, especially due to its increasing high sensitivity. Unfortunately, the presented design is too large to realise local array measurements and could potentially cause significant disturbances to the flow being measured due to its intrusiveness. The measured performance of the different sensor types suggests guidelines for their application, where hot-films would be useful for very localised measurements of low wall shear stress levels, while a piezoresistive surface fence is suitable for measurements up to higher speeds.

Summarising the advantages and disadvantages of the tested sensors, the coated MEMS hot-films by BAE SYSTEMS seem to be a potential solution for industrial aerospace applications based on their ability to measure local and global static and dynamic wall shear stress especially in abrasive environments. However, frequency response could be improved by reducing the sensor cold resistance and/or by better optimisation of the anemometry circuitry. Additional performance could also result from a reassessment of the sensor design in order to optimise the thermal matching of the sensor, diaphragm, connecting leads and substrate materials.

REFERENCES

[1] AEROMEMS Homepage, *Investigation of the Viability of MEMS Technology for Boundary Layer Control on Aircraft Aerospace Technology, Industrial Manufacture, Materials Technology*, Update Date: 4 Apr. 1998. http://www.tra3.com/dragnet/aeromems.html

[2] Brunn, H. H., *Hot-wire Anemometry-Principles and Signal Analysis*, Oxford University Press, 1995

[3] Fernholz, H. H., Janke, G., Schober, M., Wagner, P.M., Warnack, D., New Developments and Applications of Skin-friction Measuring Techniques, *Measurement, Science and Technology* 7, 1996

[4] Freymuth, P., Frequency Response and Electronic Testing for Constant-Temperature Hot-Wire Anemometers, *J. Phys. E: Sci. Instrum. 10*, 705–10, 1977

[5] Gorman, J., Hussain, F., Gaster, M., The Development of a Micromachined Hot-Film Sensor, *Pro ceedings of 13th Australasian Fluid Mechanics Conference*, Monash University, Melbourne, Dec. 98

[6] Honeywell Catalog 15, *Micro Switch, Sensing Control, Environmental Condition Sensors*, Honeywell Inc., Freeport, Illinois 61032, September 1998

[7] Sturzebecher, D., Nitsche, W., Visualization of the Spatial-Temporal Instability Wave Development in a Laminar Boundary Layer by Means of a Heated PVDF Sensor Array, *Notes on Numerical Fluid Mechanics*, Vol. 60, Vieweg Verlag Braunschweig (Germany), 1997, pp. 335-342

[8] Sturzebecher, D., Anders, S., Nitsche, W., The Surface Hot-Wire as a Means of Measuring Mean and Fluctuating Wall Shear Stress, Submitted for publication

[9] Tichý, J., Gautschi, G., *Piezoelektrische Meßtechnik, Physikalische Grundlagen, Kraft-, Druck- und Beschleuigungsaufnehmer*, Verstärker, Springer Verlag, New York, 1980

[10] Weiser, N., Nitsche, W., Renken, F., Wall Shear Stress Determination by Means of Obstacle-wires, *Proceeding of 8th Symposium on Turbulent Shear Flows*, Technical University of Munich, Germany, 1991, pp. 4-5-1 – 4-5-6

Session 7
Realisation of Adaptive Wing Concepts

Aspects of Shock Control and Adaptive Wing Technology

E. Stanewsky and H. Rosemann

DLR, Institute of Fluid Mechanics,
Bunsenstrasse 10, 37073 Göttingen, Germany

Abstract

The development of the boundary layer and shock boundary layer interaction essentially determine the performance boundaries of transonic transport aircraft. Employing shock and boundary layer control has therefore a large potential for improving flight performance in terms of cruise drag and with regard to the drag-rise and buffet boundaries. Various means of shock and boundary layer control are described and aerodynamic performance improvements due to control discussed. Shock conditions and critical stages in the development of the boundary layer on a wing are, however, strongly dependent on freestream conditions which change considerably during a typical aircraft flight mission. In order to optimize performance at all points of the flight envelope, the effective wing geometry and other means of control must therefore adjust to these changing conditions. Adaptive wing aspects are, consequently, being addressed.

1. Introduction

The development of the boundary layer and the interaction of the wing upper-surface shock with the boundary layer essentially establish the flight performance of transonic transport aircraft at cruise as well as at high-speed off-design conditions. This is demonstrated in Figure 1: as, for instance, the Mach number for a given lift coefficient is increased — and similar considerations hold for increasing lift at a constant Mach number — shock waves develop, resulting in an increase in drag mainly due to the occurrence of wave drag. Subsequently, viscous drag also strongly increases essentially caused by a thickening of the boundary layer due to the shock and the sustained rear adverse pressure gradients on the airfoil or wing upper surface. At a later stage shock-induced flow separation develops which leads, finally, to the buffet process [1]. Employing shock and boundary layer control has, consequently, a large potential for improving flight performance in terms of cruise drag, hence speed and/or fuel consumption, and with respect to the drag-rise and buffet boundaries.

Generally, aircraft follow specific flight profiles consisting of take-off, climb, the mission proper, descend and landing. The mission proper is again characterized by changing freestream conditions which result for a transport aircraft, for instance, from fuel-burn and the corresponding weight reduction during cruise [2], but also due to off-design cruise conditions which cannot be avoided. The various requirements are, of course, accounted for by multiple but fixed design points which compromise, however, the overall aircraft performance. Employing adaptive wing technology — and flow control in the above sense — where the effective wing geometry can be adjusted to the changing flow and load requirements, allows, on the other hand, not only to fully explore the aerodynamic flow potential at each point of the flight envelope — resulting in aerodynamic performance gains during cruise and off-design conditions — but also to improve structural design.

In the following chapters, characteristic examples of shock and boundary layer control will be discussed and adaptive-wing schemes and associated benefits presented.

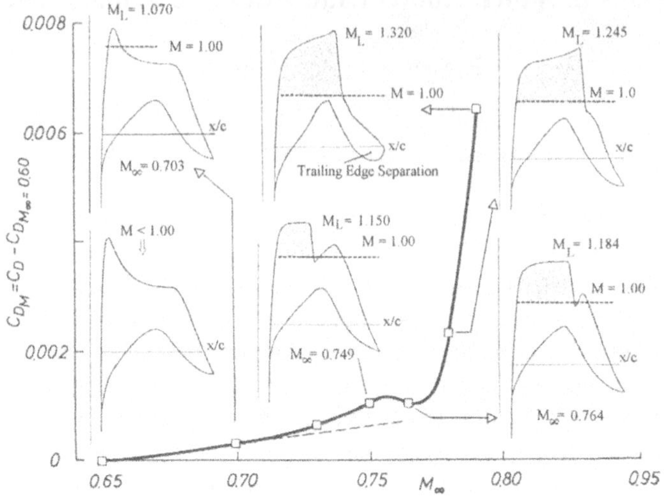

Figure 1 Transonic drag development, airfoil CAST 10/DOA2
$C_L = 0.50$, $Re_c = 30\text{x}10^6$ [1]

2. Shock boundary layer interaction control

One may distinguish two ways of controlling the shock and the shock-associated boundary layer development: one may either try to influence the shock strength, e.g., weakening the shock by spreading the shock associated pressure rise over a certain streamwise distance, or energize the boundary layer making it more resistant to the adverse pressure gradients prevailing on an airfoil or wing. Some of the corresponding control devices, to be considered and referred to in the pursuing discussion, are sketched in Figure 2.

a. Geometric devices b. Pneumatic devices

Figure 2 Examples of shock and boundary layer control devices

2.1 Direct boundary layer control

Here, we shall consider as examples of direct shock-related boundary layer control the effect of discrete suction on total airfoil drag [3] and the effect of the latter and of vortex generators [4], respectively, on the boundary layer development in the shock region.

Experiments on the effect of discrete shock-upstream suction on drag have been carried out by DERA utilizing the laminar-type airfoil DRA-2303 [5]. Characteristic results are depicted in Figure 3: at the freestream Mach number of $M_\infty = 0.68$, considered here as a typical example, drag is noticeably reduced over the entire lift range investigated with the maximum reduction amounting to 7.5%. This decrease in total drag is essentially due to a decrease in viscous drag as can be derived from the development of the boundary layer displacement and momentum thicknesses which are considerably reduced by suction, Figure 4 [6]. It should be noted that a worthwhile drag reduction remains even when accounting for 'pump' drag [5]. It was also found that slot suction may have a favorable effect on the buffet boundary [3].

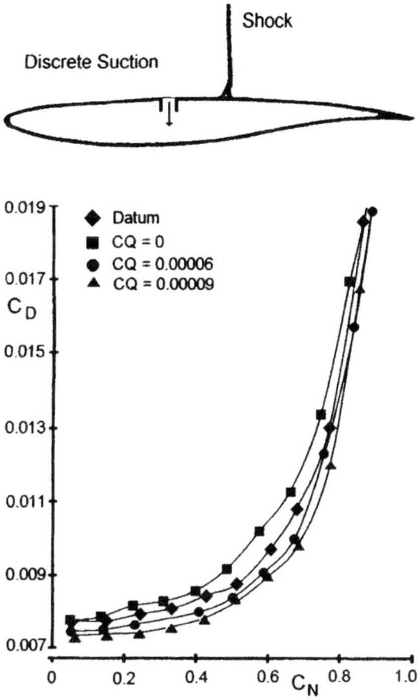

Figure 3 Effect of discrete suction on drag for the airfoil DRA-2303, $M_\infty = 0.68$, $Re = 19 \times 10^6$ [5]

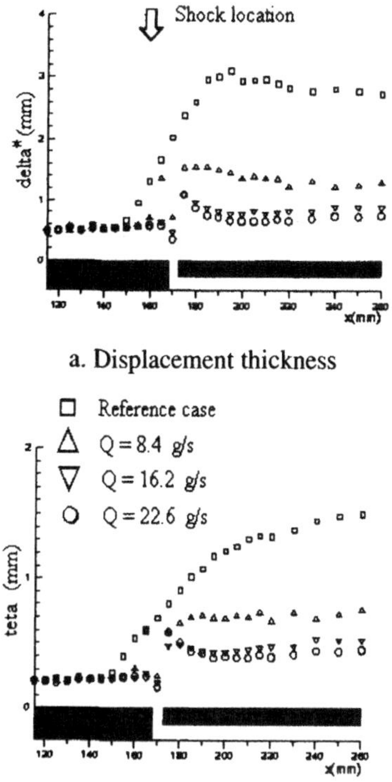

a. Displacement thickness

b Momentum thickness

Figure 4 Effect of discrete suction on the boundary layer development [6]

The effect of vortex generators upstream of the shock to energize the boundary layer is similar to shock-upstream suction. This is demonstrated in Figure 5 [4] which shows that shock-upstream vortex generators result in a reduction in displacement thickness downstream of the interaction, Figure 5c, although the shock strength is increased, Figure 5b, quite similar to the development depicted in Figure 4. Figure 5 also indicates that passive control by cavity ventilation — which is a means of reducing shock strength as will be discussed below — generally increases the displacement thickness downstream of the interaction region, although shock strength is reduced.

2.2 Control of shock strength by cavity ventilation and contour bumps

In extensive studies of shock control by passive and active (part-suction) cavity ventilation via a single cavity/perforated plate arrangement, and by hybrid control where the passive cavity in the shock region is followed by discrete suction (Figure 2), it was found that, although shock strength and correspondingly wave drag was reduced (see, e.g., Figure 5b [4]), total drag always increased [7, 3]. The reason for the drag increase is the generally stronger initial increase in the displacement thickness due to control, Figure 5c, and the amplification of this increase by the sustained rear adverse pressure gradients prevailing on airfoils and wings leading to a dominating increase in viscous drag. Control by ventilation seems, however, well suited for the purpose of stabilizing the shock and avoiding shock-induced separation, a feature required, for instance, in engine intakes [5].

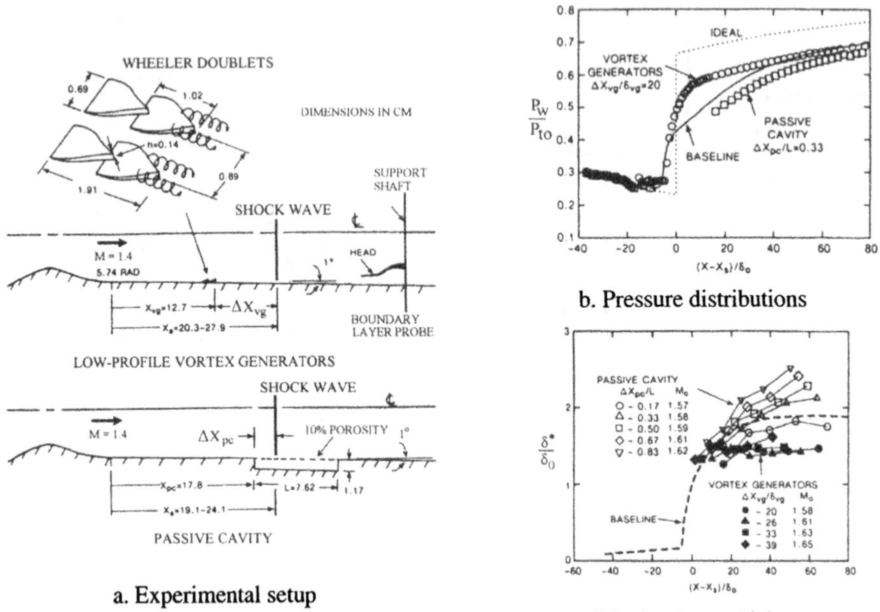

a. Experimental setup

b. Pressure distributions

c. Displacement thickness

Figure 5 Effect of VGs and passive cavity ventilation on shock boundary layer interaction [4]

Bumps — and bumps combined with shock-upstream discrete suction — for shock control have been experimentally investigated by DLR in conjunction with the A340-type 'ADIF' airfoil and an equivalent infinitely-swept sheared wing [3, 8, 9]. For the airfoil and the sheared wing, bumps of various geometries were designed in numerical optimization studies. The modular airfoil model investigated is shown in Figure 6a to indicate the various bumps and the bump and suction-slot locations studied. The equivalent sheared-wing model test setup, also showing the contoured end walls required to ensure infinitely swept-wing conditions, is depicted in Figure 6b. Here, we shall only consider sheared-wing results.

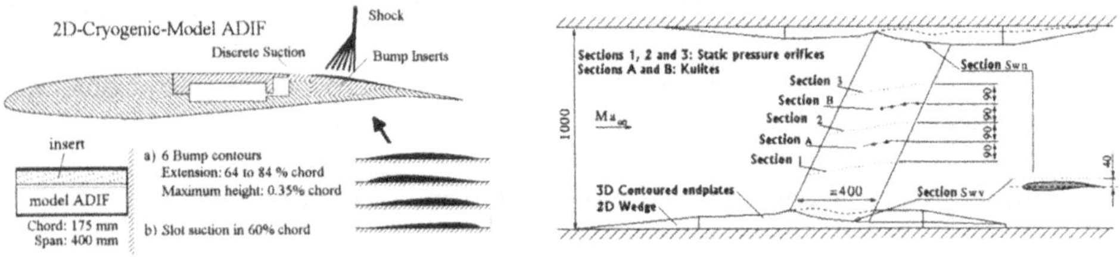

a. ADIF airfoil model [8]

b. Sheared-wing model test setup [9]

Figure 6 ADIF airfoil model and test setup for the equivalent sheared wing model

The conditions selected for demonstrating the bump effectiveness are a freestream Mach number of M_∞ = 0.852, a bump height of 0.1573% chord and a suction-slot location of $(x/c)_{Suction}$ = 0.55. Figure 7 shows that at these conditions relatively large drag reductions are achieved at the higher lift coefficients while a drag increase occurs at lift coefficients of $C_L <$ 0.50. This behavior is characteristic for the effectiveness of contour bumps which depends strongly on the bump location with respect to the shock and on the shock strength (also see Figure 9). At the design condition of the bump — and the contoured end plates — ($C_L \approx 0.56$), the reduction in drag due to the bump amounts to 8% with the relatively low reduction in drag mainly being due to the thick boundary layer approaching and already spreading the shock. Applying suction upstream of the bump (shock) reduces drag up to 22% indicating the rather large potential of such an arrangement. Corresponding surface and wake pressure distributions in Figure 8 reveal that the bump reduces wave drag due to the spreading of the shock as well as viscous drag; adding suction does not increase wave drag — thanks to the now more effective bump — but reduces viscous drag as is indicated by the better pressure recovery over the rear of the wing but also by the narrowing of the wake.

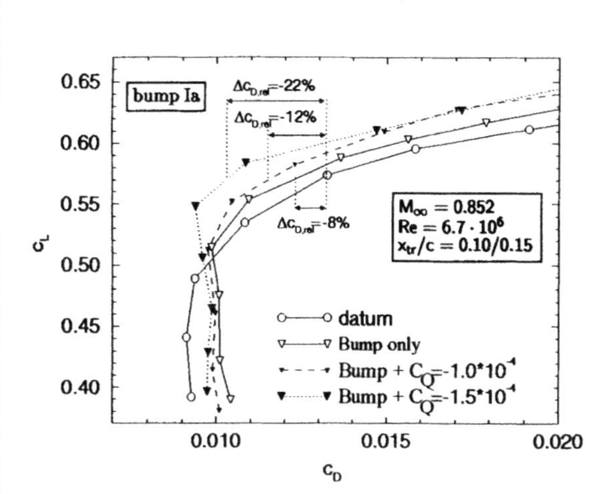

Figure 7 Drag polars for the ADIF swept wing with control by bump and bump plus upstream suction [9]

3. Wing adaptation

Wing-adaptation concepts range from local contour modifications to variable-camber leading- and trailing- edge flaps to the variation of the complete airfoil or wing contour. The objective is, of course, to adjust the wing geometry to changing flight conditions to gain optimum performance, e.g., minimum drag at varying but prescribed cruise conditions. We will consider here a variable-height bump, a variable-camber trailing-edge flap, and the combination contour-bump/trailing-edge flap. The work has been performed within the German adaptive-wing project ADIF [10] and the EUROSHOCK II project [3].

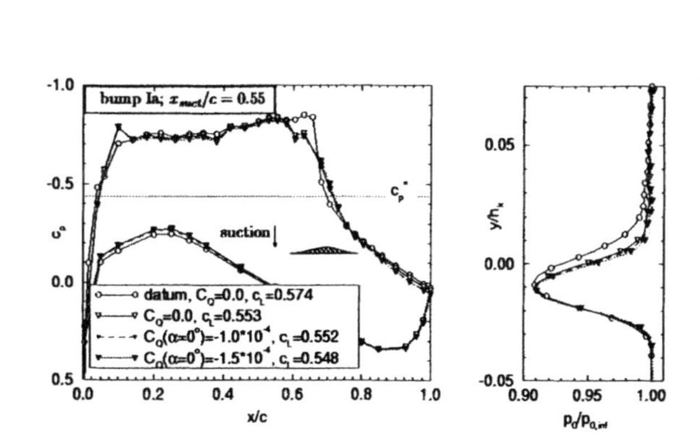

Figure 8 Pressure distributions corresponding to conditions of Figure 7 at $C_L \approx 0.56$ [9]

3.1 Bump adaptation and mission benefits

A contour bump was applied by DASA-Airbus to a hybrid-laminar-flow wing of a long-range A340-type aircraft to study the benefits arising from such an application [2]. The bump was optimized for a near-cruise Mach number of $M_\infty = 0.84$ and the wing-section pressure distributions depicted in Figure 9a: the pressure distributions are (typically) characterized by highly accelerating flow and relatively strong shock waves on the upper wing surface but limited shock movements with changing lift coefficient. The corresponding optimized drag polars are shown in Figure 9b indicating that the bump height has to be adapted to achieve minimum drag, changing in height from 0.5% chord to 0.2 % chord as lift is reduced, required, for instance, due to fuel-burn during a long-range mission. Transferring these results — and similar results for the designated cruise Mach number of $M_\infty = 0.82$ — to the complete aircraft establishes the flight polars and the drag balance for the long-range mission, Figure 10.

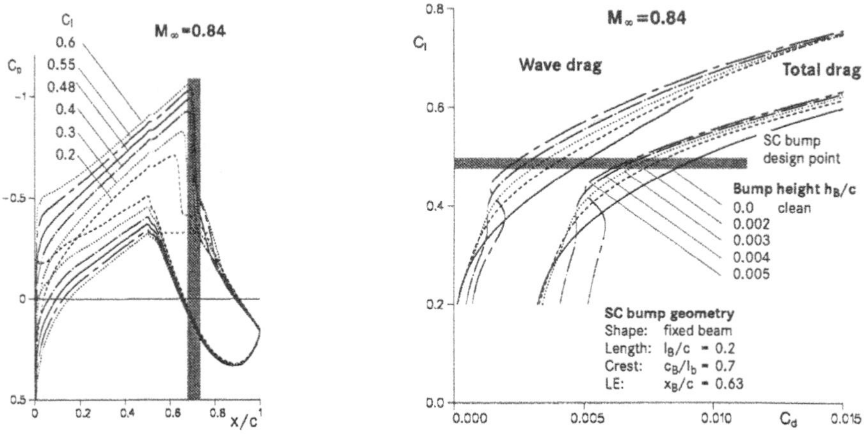

a. Wing-section pressure distribution

b. Drag polars with and w/o bump control

Figure 9 Pressure distributions and bump-optimized polars for the HLF wing section at flight conditions, wing section PHLF1, $Re_c = 35x10^6$, $(x/c)_{trans} = 0.50$, $M_\infty = 0.84$ [2]

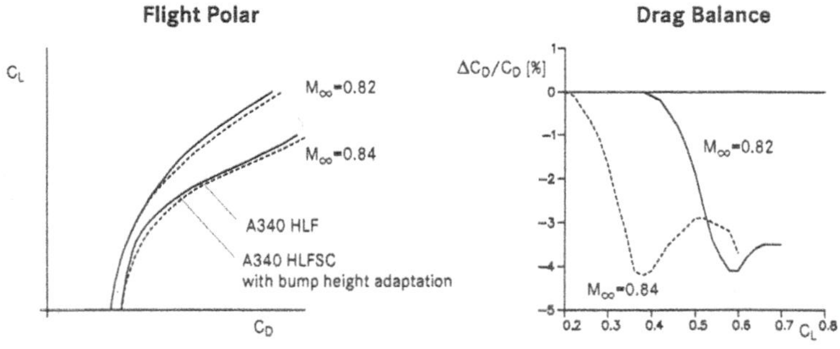

Figure 10 Predicted flight polar and drag balance for the A340 HLF-wing aircraft with bump control [2]

Based on these flight polars and an estimated increase in aircraft weight of 0.25 tons for the adaptive-bump installation, a standard North Atlantic flight mission with a range of 3500 NM was investigated assuming 600 trips per year at the cruise Mach number $M_\infty = 0.82$ and 620 trips per year at the off-design Mach number $M_\infty = 0.84$. At these conditions, a reduction in fuel consumption per year due to adaptive shock control of about 353 tons (1.23%) at $M_\infty = 0.82$ and of about 792 tons (2.11%) at $M_\infty = 0.84$ was obtained from flight-mission computations.

3.2 Adaptation of bump/flap arrangements

The effectiveness of an adaptive bump in reducing drag and the corresponding mission benefits have been demonstrated above. A variable-camber trailing-edge flap may be employed to generate or reduce lift when weight or altitude conditions change during a transatlantic flight, or to optimize L/D for a given lift coefficient. It may also be used to raise the buffet boundary to establish the required margin with respect to cruise conditions [10].

Due to the potential of the contour bump and the flap, the optimization of a bump in conjunction with a variable-camber flap to achieve minimum drag was considered by DLR in co-operation with DASA-Airbus [11]. The optimization was carried out for the airfoil 'VC-opt', designed by DASA-Airbus to be used with variable-camber trailing-edge flaps. The airfoil geometry, indicating typical bump locations and flap deflections, is sketched in Figure 11. Drag-minimized polars for separate bump and flap optimizations, respectively, and for the combined flap/bump optimization are compared in Figure 12a: the bump — as already indicated above — results in a considerable drag reduction at higher lift coefficients where stronger shocks occur, but the gains naturally deteriorate as lift decreases and shocks weaken. Variable camber, on the other hand, results in noticeable drag reductions at low to medium lift coefficients while at higher lift drag with and w/o optimization is essentially the same. The latter is due to the flap not being able to achieve such high lift without strong shock waves. When combining contour bump and flap in the optimization process, considerable gains are attainable over the entire lift range investigated with maximum drag reductions amounting to 24%.

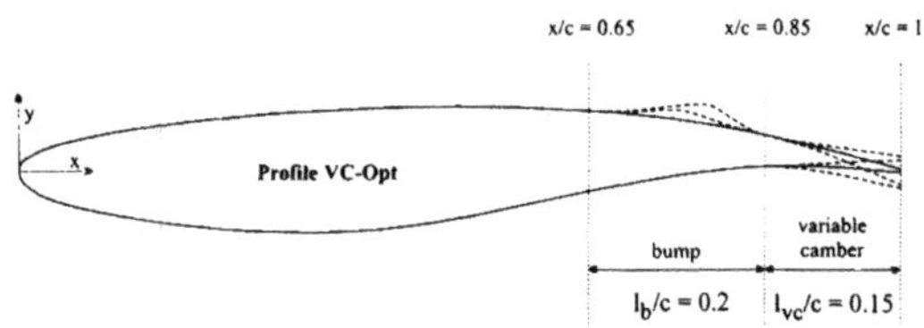

Figure 11 Airfoil 'VC-opt' with bump and variable camber contour modifications [11]

Figure 12b shows the corresponding effect on L/D indicating a considerable increase in the lift-to-drag ratio and a shift of the maximum L/D to higher lift coefficients. Also shown is the noticeable increase in the useful cruise range assuming a minimum required (airfoil) L/D.

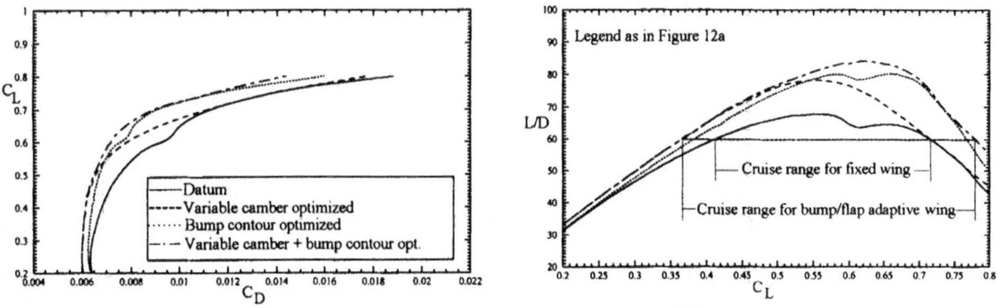

a Drag polars b. Lift to drag ratio L/D

Figure 12 Effect of local contour optimizations on drag and L/D, airfoil 'VC-opt', $M_\infty = 0.793$ [11]

4. Summary and conclusion

The development of the boundary layer and shock boundary layer interaction essentially determine the high-speed performance boundaries of transonic transport aircraft in terms of cruise drag and the drag-rise and buffet boundaries. It has been shown that shock control, either by weakening the shock or by energizing the boundary layer, employing contour bumps, discrete suction or vortex generators, respectively, reduces drag and delays the detrimental development of the boundary layer. Especially the contour bump was found to be most effective with drag reductions of up to 22% for the sheared wing in the presence of a thin shock-upstream boundary layer. Since the occurrence and strength of shock waves and critical stages in the development of the boundary layer are dependent on freestream conditions, which may change considerably during long-range cruise, an adaptive wing is required for optimum aircraft performance. Adaptive elements shown to be effective in improving wing (or aircraft) performance were adaptive contour bumps, trailing-edge flaps and a combination of the two. An adaptive bump (in height) resulted, for instance, for an A340-type aircraft equipped with a hybrid-laminar-flow wing, assuming typical long-range-missions, in fuel savings of up to 2.11%, while the combination bump/flap resulted for a variable-camber-optimized airfoil in drag reductions of up to 24% with a corresponding increase in maximum L/D and a considerable extension of the useful cruise range.

5. References

[1] Stanewsky, E. and Krogmann, P. "Transonic Drag Rise and Drag Reduction by Active / Passive Boundary Layer Control", AGARD Report No. 723, Lecture Series Aircraft Drag Prediction and Reduction, July 1985

[2] Thiede, P., Dargel, G., Assessment of Shock and Boundary Layer Control Concepts for Hybrid Laminar Flow (HFL) Wing Design, in: EUROSHOCK II Final Technical Report, TR BRPR-95-76/1, 1999 and DASA-Airbus DA-Report No. EF-069/99, 1999

[3] Stanewsky, E., Délery, J., de Matteis, P.P., Fulker, J., Doe, R., EUROSHOCK II – Drag Reduction by Shock and Boundary Layer Control, EUROSHOCK II Final Technical Report, TR BRPR-95-76/1, 1999, and Synthesis Report TR BRPR-95-76/2,1999 (to be published in Notes on Numerical Fluid Mechanics)

[4] McCormick, D.C., Shock/Boundary-Layer Interaction Control with Vortex Generators and Passive Cavity, AIAA Journal, Vol. 31, No. 1, pp. 91 – 96, January 1993

[5] Fulker, J., Simmons, M. J., An investigation of active, suction, shock and boundary layer control, in: EUROSHOCK II Final Technical Report, TR BRPR-95-76/1, 1999, and DERA Report MSS4X/CR980817/1.0, 1999

[6] Bur, R., Benay, R., Corbel, B., Soares-Margadinho, R., Soulevant, D., Study of control devices applied to a transonic shock wave/boundary layer interaction, in: EUROSHOCK II Final Technical Report, TR BRPR-95-76/1, 1999, and ONERA Technical Report RT 126/7078 DAFE/Y, 1999

[7] Stanewsky, E., Délery, J., Fulker, J., Geissler, W. (Edit.), EUROSHOCK – Drag Reduction by Passive Shock Control, Notes on Numerical Fluid Mechanics, Volume 56, Friedr. Vieweg & Sohn Verlagsgesellschaft mbH, Braunschweig/Wiesbaden, 1997

[8] Knauer, A., Performance improvement of transonic airfoils through contour modifications in the shock region, DLR-Research Report 98-03, 1998 (also Ph.D.-Thesis University Hanover), and EUROSHOCK II Final Technical Report, TR BRPR-95-76/1, 1999

[9] Birkemeyer, J., Drag minimization on a transonic wing by ventilation and adaptive contour bumps, DLR-Research Report 1999-28, 1999 (also Ph.D.-Thesis University Hanover), and EUROSHOCK II Final Technical Report, TR BRPR-95-76/1, 1999

[10] Mertens, J., Aerodynamic objectives of the adaptive wing ADIF, in: DGLR-Yearbook 1998, 1998

[11] Richter, K., Rosemann, H., Numerical investigation of transonic airfoil drag reduction by the combined application of a contour bump and variable rear camber, DLR Internal Report IB 223-1999 A 31, December 1999

DRAG REDUCTION ON GURNEY FLAPS AND DIVERGENT TRAILING EDGES

D. W. Bechert, R. Meyer and W. Hage
DLR, Department of Turbulence Research,
Mueller-Breslau-Str. 8, 10623 Berlin, Germany

Summary

Miniflaps at the trailing edges of airfoils (e.g., Gurney flaps or divergent trailing edges) change the Kutta condition and thus produce higher lift. Unfortunately, however, the drag is also increased due to the flow separation downstream of this particular type of trailing edge. Therefore, the trade-off between beneficial and detrimental effects is considered in this paper. Various aspects of the flow on airfoils with Gurney flaps are addressed:

(1) *Transonic flow.* Wind tunnel experiments have been carried out with a CAST 10-2/DOA2 airfoil with Gurney flap and at high subsonic flow speed. The lift to drag ratio is improved and the test wing behaves like one having a 20% larger surface area. In addition, buffeting becomes less critical.

(2) *Selection of flap size.* Detailed wind tunnel studies have been carried out with a low-drag glider wing at low subsonic velocities. The Gurney flap height was varied in six steps so that the most relevant parameter regime was covered. For the lift increase and for the device drag, simple empirical laws were obtained. Subsequently, for practical applications, a procedure for the selection of a suitable flap height providing a beneficial effect has been devised.

(3) *Drag reduction by wake stabilization.* In the separation regime downstream of a Gurney flap, an absolute instability, i.e., a Karman vortex street occurs, even if the incident boundary layers are turbulent. Therefore, one approach towards drag reduction is to stabilize the wake and hence eliminate the Karman vortex street. This can be achieved with a variety of trailing edge modifications, e.g., with slits or holes in the Gurney flap. A particular structure which exhibits spade-like protrusions at the trailing edge produces also good results. Actually, the latter structure has been adopted from the trailing edge of dragonfly wings. Eliminating the Karman vortex street with these various trailing edge modifications reduces the Gurney flap device drag by about 22 – 30% without a perceivable change of the enhanced lift. Obviously, vibration and noise radiation are also reduced together with the suppression of the Karman vortex street.

1. Introduction

Miniflaps such as Gurney flaps are small extensions at the trailing edge of an airfoil which point downwards at an angle of about 90°. The size is typically between 0.5% and 2% of the airfoil chord length. These very small flaps enhance the lift of an airfoil considerably, typically by 6-25%. These devices have been first used in the seventies on racing cars by Dan Gurney. Since that time, they are attached to the rear wings on those vehicles, producing enhanced negative lift. That causes better road contact and hence permits increased speed on road bends. After Gurney's invention for racing cars, Liebeck [1] recognized the potential of Gurney flaps for aircraft wings. This led to further investigations at McDonnell Douglas on modified trailing edges, also referred to as divergent trailing edges (DTE). Henne [2] has given a detailed survey on this work and on previous related ideas. The commercial transport airliner MD11 is the first aircraft equipped with a diverging trailing edge which permits an improved performance at transonic cruise conditions.

The shape of the trailing edge is of paramount importance for the lift. In addition, it is worth mentioning (see also Henne [2]) that the static pressure on the lower side upstream of the Gurney flap or the diverging trailing edge is significantly higher than the value which is found at the end of the upper side of the airfoil. Consequently, traditional closure principles such as the condition of the continuity of static pressure at the trailing edge have to be abandoned here. Recent detailed investigations on the flow on diverging trailing edges have shed some light on the local flow structure [3-5]. At present, further technological applications of miniflaps on aircraft for high lift conditions [6] and for transonic cruise [2] are likely. Enhanced performance of wind turbines has also been demonstrated [7]. Numerous other applications are conceivable such as, e.g., in turbomachines.

2. Transonic flow

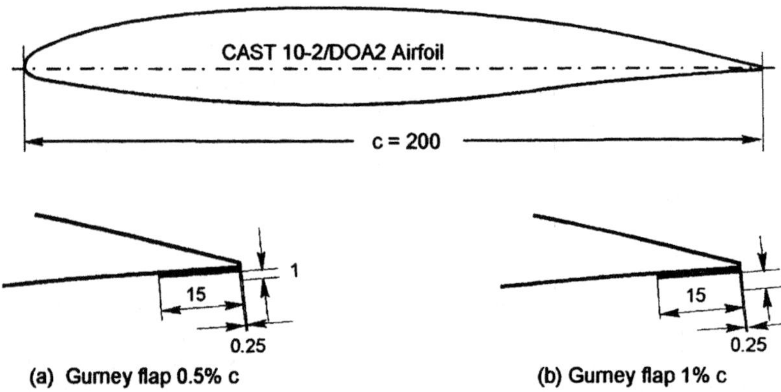

Fig. 1. Test wing with CAST 10-2/DOA2 airfoil with either (a) a flap height of 0.5% chord or (b) 1% chord. Dimensions in mm. Adhesive film and Gurney flaps each 0.25 mm thick.

Investigations on the effect of Gurney flaps on the transonic airfoil CAST 10-2/DOA2 have been carried out in the DLR transonic wind tunnel in Göttingen. This airfoil was designed for a Mach number of 0.765 and an angle of attack of 2°. The geometrical configuration of the airfoil and of the attached Gurney flaps can be seen in Fig. 1. The wind tunnel data shown here have been collected by pressure distribution (lift) and by wake rake (drag) measurements. The experiment was essentially two-dimensional with the (unswept) wing spanning the whole width of the quadratic 1 x 1 m^2 test section of the wind tunnel. The wind tunnel wall was perforated in order to minimize wall effects and to permit comparisons with a previous comprehensive data set [8] which had been collected with the same wing in the same wind tunnel configuration.

In Fig. 2, it can be seen that the tiny Gurney flaps cause large effects on both lift and lift to drag ratio. It is interesting to see that even with a Gurney flap height of 0.5% of the chord length the lift increase persists whereas the parasitic drag caused by the separation downstream of the flap is reduced according to the reduction of the device height form 1% to 0.5% chord. Thus, one obvious way to reduce the parasitic drag of Gurney flaps consists of the selection of an optimal, and, actually very small, height of the device.
In Fig. 3, we show pressure distributions without and with Gurney flap. The lift coefficient is at about $C_L \approx 0.9$ where differences between the two cases are most obvious. Due to the presence of the Gurney flap, the angle of attack at which this lift coefficient is obtained, is lower. As a consequence thereof, the local supersonic Mach numbers on the forward part of the upper side of the airfoil are lower. This ensues a weaker shock at the downstream end of that supersonic flow regime. Thus, the drag is significantly reduced.

One may summarize the properties of the altered wing as being comparable *to one having an about 20% increased wing area.* In addition, this altered wing has an *improved lift to drag ratio.* The beneficial effects in this particular case are mostly due to compressibility. With increasing supersonic Mach numbers on the upper side of the airfoil, it becomes increasingly difficult to keep shock strengths and ensuing separation losses at a low level. Thus, enhancing the loading of the lower surface with a Gurney flap (or a diverging trailing edge) helps to reduce those losses. However, this requires a subtle design of the airfoil and its trailing edge.

In our experiments with various modifications of the test wing, vibrations (buffeting) of the wing always started to occur at about the same angle of attack for a given Mach number. However, for a given angle of attack, the lift *with* Gurney flap is significantly higher that *without.* This means that with a Gurney flap (or DTE) in place, the buffeting limit is shifted away from the cruise conditions of a commercial aircraft flying at high subsonic speeds.

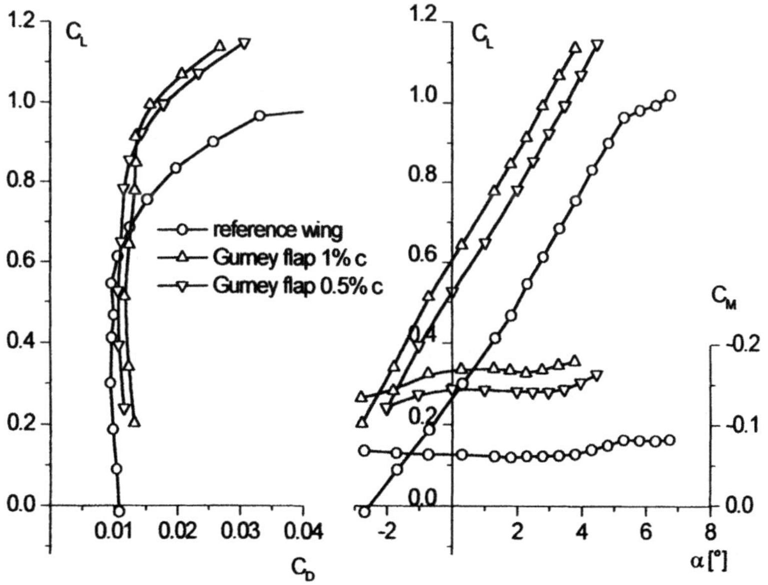

Fig. 2. Drag polars, lift coefficient C_L and moment coefficient C_M plotted versus the angle of attack α. Airfoil CAST 10-2/DOA2, Mach number $M = 0.73$, Reynolds number $Re = 2.7 \times 10^6$. Data from [9].

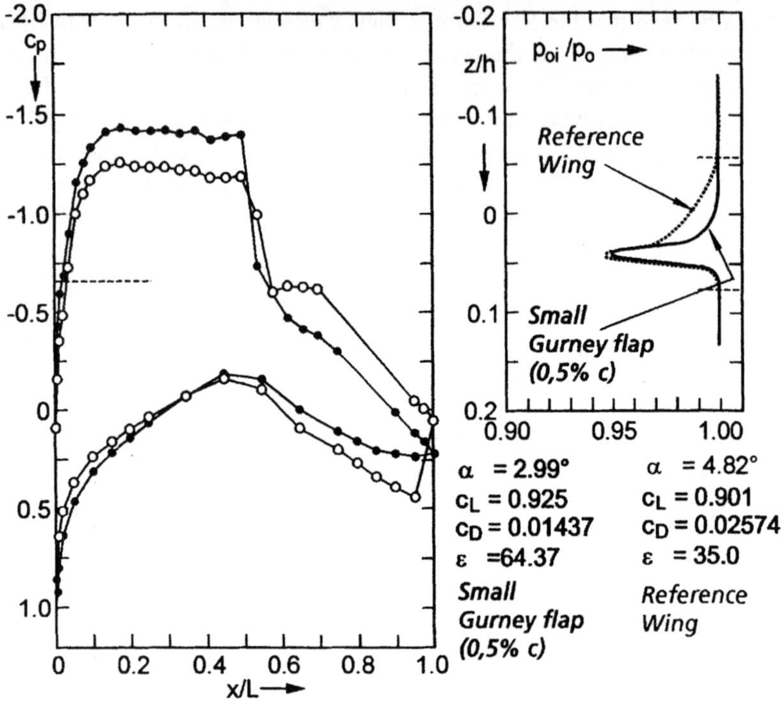

Fig. 3. Pressure distribution and wake profile of reference wing (solid symbols) and wing with small Gurney flap (0.5%)(open symbols). Airfoil, Mach and Reynolds numbers as in Fig. 2. Data from [9]. ε is the lift to drag ratio $\varepsilon = C_L/C_D$

3. Selection of flap size

3.1. Basic considerations and test arrangement

The aim of the experiments described below is to find ways to minimize the parasitic drag of Gurney flaps. This does not necessarily require experiments at transonic speeds. Therefore, these investigations have been carried out in a slow-speed wind tunnel ($M \approx 0.1$) with low operating costs and easy access, i.e., the wind tunnel at the Technical University of Berlin. For the tests, a wing with a laminar airfoil was selected, because the low drag of these airfoils enables us to detect conveniently even incremental changes of the parasitic drag of various types of Gurney flaps. The HQ17 airfoil being used had been designed by Horstmann and Quast from DLR for application on glider aircraft. The airfoil co-ordinates are given in a previous paper [10]. The lift and drag data in this second test series were collected by direct force measurements with a 6-component strain gauge wind tunnel balance with high resolution. Some of the Gurney flap devices being tested have a three-dimensional shape. Thus, it was assumed that here, direct force measurements should produce more reliable data than a wake rake would.

The test wing is mounted between side plates which extend between floor and ceiling of the test section (1400 x 2000 mm^2) of the wind tunnel. The advantage of the side plates is that the boundary layer on their surface is thinner than the one which develops on the wind tunnel walls. On the other hand, one has to deal with variable bypass flow rates between side plates and wind tunnel walls due to variable blockage effects caused by the test wing operating at variable angles of attack. By measuring the bypass flow rates we took into account this influence on the velocity incident on the wing. In addition to that, the usual wind tunnel corrections [11] have been applied.

Nevertheless, there are additional side plate effects which contaminate in particular *the drag force measurements.* Even with the side plates extending over the whole height of the test section, we cannot exclude that there is still a residual contribution of induced drag. In addition, the laminar flow regimes on the wing are locally contaminated by turbulent "wedges" caused by the turbulent flow on the side plates. Further, there is a small leakage flow in the gap between wing and side plates. However, due to the narrow width of that gap (0.2 mm) that influence is considered as not very important.
On the other hand, flow vizualization shows that local flow separation occurs in the corners between wing and side plates. These separation regimes could be eliminated by vortex generators on the wing. That reduced the total drag of the wing typically by about 10%. Obviously, under these circumstances, our wing cannot exhibit the low drag values which had been obtained by previous wake rake measurements in Delft [12] and Stuttgart [13]. A

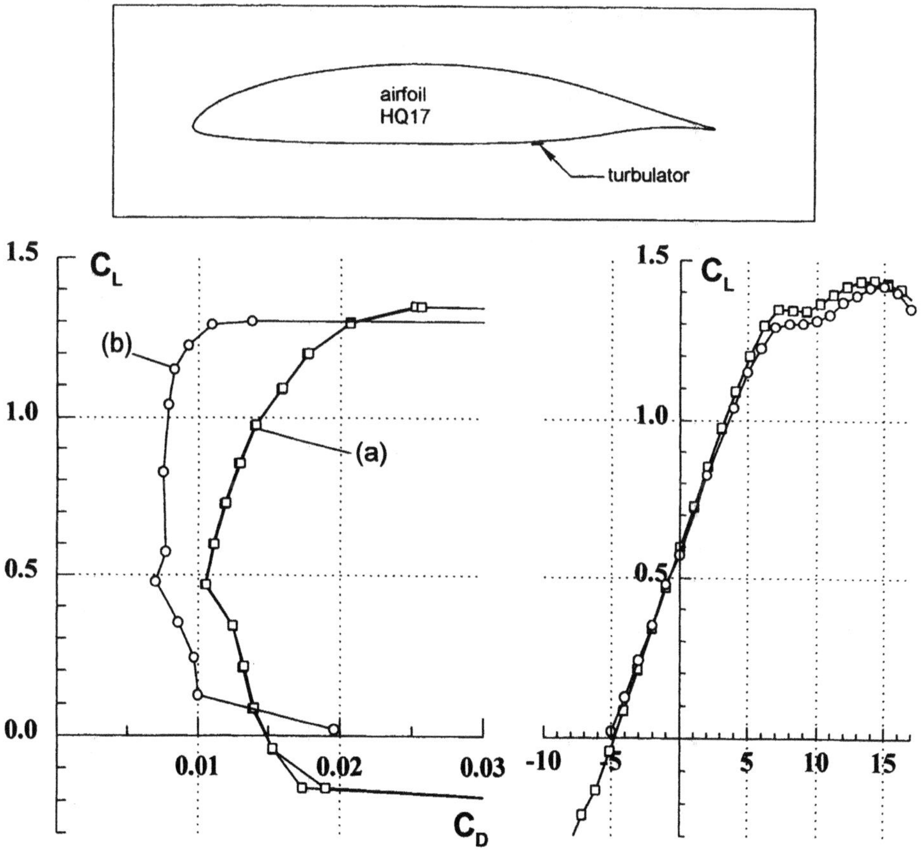

Fig. 4. Comparison of present direct force measurements (a) with previous data [12,13], based on pressure distribution and wake-rake measurements (b). Airfoil: HQ17, Re = 1 x 10^6.

comparison of these data is shown in Fig. 4.

In our measurements, we used a turbulator on the lower side of the airfoil at 65% chord. It consisted of a plastic zigzag strip with 0.75 mm thickness. Location and thickness were determined experimentally based on flow visualisation with paint on the lower airfoil surface. This particular parameter choice eliminated the laminar separation bubble there for the Reynolds numbers at which we collected data, i.e., for 0.5 and 1 x 10^6. In addition, a small drag reduction was observed. The use of a turbulator seemed important because the boundary layer incident on the Gurney flap should be well-defined. We also checked that the transition on the lower side of the airfoil did not perceivably move once a Gurney flap was installed.

The aim of the present investigation, however, is not to "streamline" data so that all differences vanish. We want to investigate trailing edge modifications on a wing with low drag. For this purpose, the curve (a) in Fig. 4 suits us well as a reference drag polar.

3.2 Experiments with two-dimensional trailing edge geometries

In order to obtain more information on a suitable size of a Gurney flap, we have carried out wind tunnel tests with different heights of those devices. Data for Re = 1.0 x 10^6 can be seen in Fig. 5. We have chosen the particular trailing edge configuration of Fig. 5 in order to produce a somewhat tapered shape in front of the vertical part of the flap.

We have also collected additional data with the Gurney flap attached to the trailing edge like in Fig. 1 (not shown in Fig. 5). It turned out that the relevant flap height Δh is in all cases the *flap height which is added* to the finite thickness of the trailing edge of the reference wing. Usually, we will normalise this *effective flap height* Δh with the chord length c of the airfoil, i.e., we will deal mostly with the expression $\Delta h/c$. Further, we will define the *device drag coefficient* ΔC_D as the difference between the minimum drag coefficient of the airfoil *with* Gurney flap and of that *without,* i.e., that of the reference wing. It turns out to be useful to normalise ΔC_D with the normalised flap height $\Delta h/c$. The dependence of drag increase and lift increase on $\Delta h/c$ can be seen in Fig. 6. In order to highlight the increase of lift due to the Gurney flap, we normalise the maximum lift *with* Gurney flap C_{Lmax} by the maximum lift *without* Gurney flap C_{L0max}. These normalised data can be also seen in Fig. 6. For comparison, we have also plotted the displacement thickness of the boundary layer δ on the lower side of the reference wing just at the trailing edge and for the angle of attack of lowest drag (α = -1°).

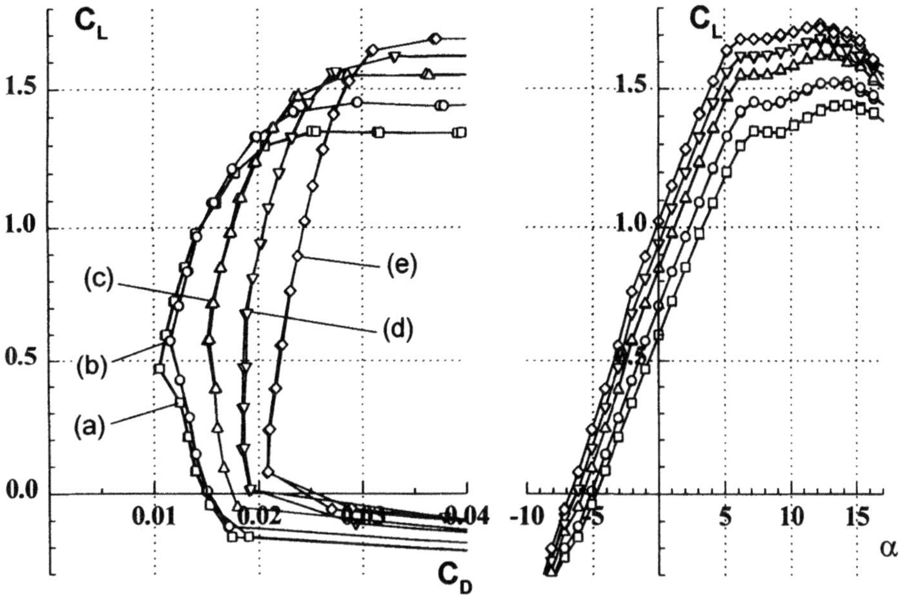

	Reference wing / Gurney flap h [%]	Δh [%]	$C_{L\,max}$	$C_{D\,min}$	$\varepsilon_{max} = \left[\frac{C_L}{C_D}\right]_{max}$
(a) —□—	Reference wing		1.438	0.0105	69.6
(b) —○—	0.5	0.18	1.524	0.0115	70.0
(c) —Δ—	1	0.67	1.627	0.0151	63.6
(d) —∇—	1.5	1.17	1.686	0.0183	58.3
(e) —◇—	2	1.66	1.740	0.0208	53.4

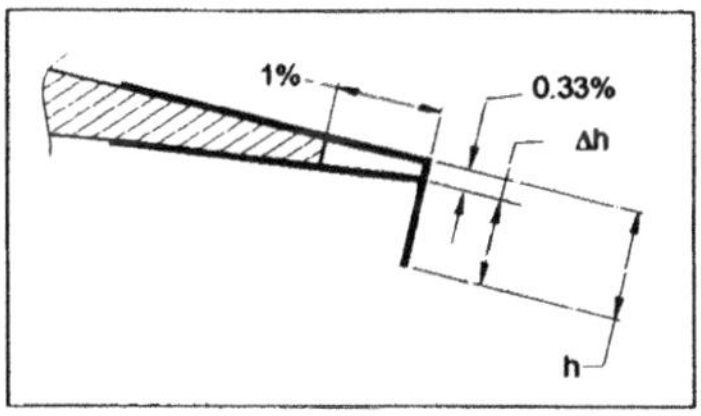

Fig. 5. Drag polars of the reference wing and of the same wing with various Gurney flap heights. Airfoil: HQ17, Re = 10^6

If the device drag ΔC_D is normalised with the normalised Gurney flap height Δh/c, one obtains an almost constant value of about 0.6. There are, however, slight differences between the data collected at the two Reynolds numbers. Much more surprising is the fact that the increase in lift does not show any difference between the two Reynolds numbers.

Of course, data collected at just two Reynolds numbers can be hardly generalised. Obviously, for significantly thicker boundary layers there should exist at least some deviation. Probably,

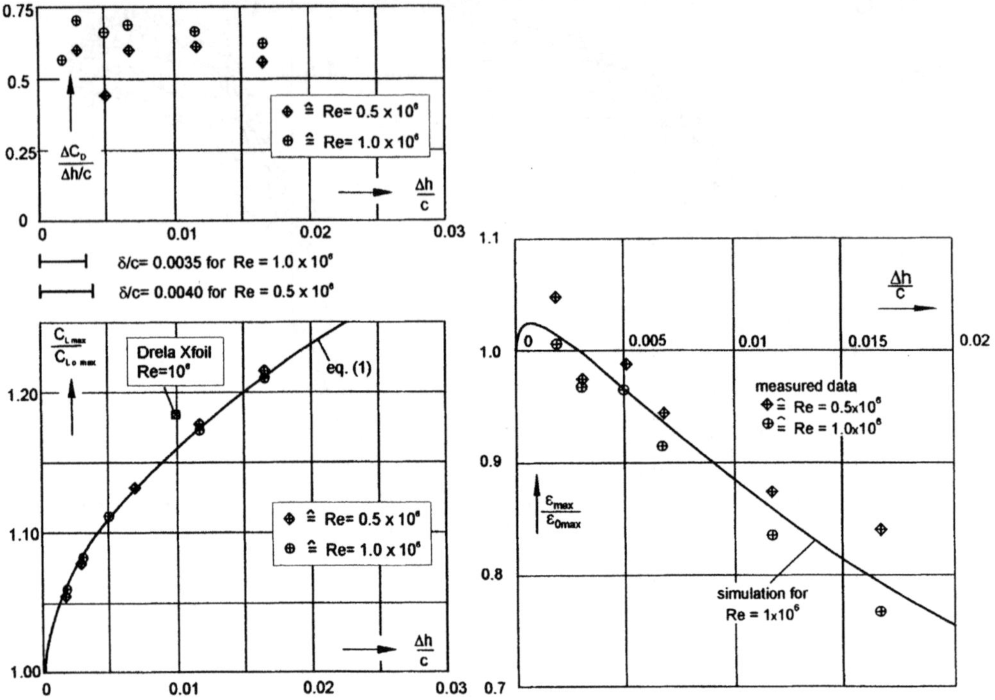

Fig. 6. The dependence of device drag and lift increase on Gurney flap height (left); Lift to drag ratio of various Gurney flap heights (right).

with a thicker boundary layer, both the device drag and the lift increase are likely to be reduced. This is what we see if we replace the Gurney flap by a quarter circle and calculate theoretical data with the Drela code [14]. Possibly, in this particular application, this code may be already stretched beyond its validity range. In addition, a turbulator was not present in our calculation with this code. We know from previous experiments that, with a laminar separation bubble being present on the lower side of the airfoil, one obtains results differing from the ones being shown here.

Nevertheless, playing with the Drela code [14] confirms another hunch of the authors. For a potential flow solution where the flow can creep around the upper sharp edge of the Gurney flap, one obtains about *double the lift increase* than for the case with viscosity and boundary layer development taken into account. Roughly speaking, this means that given a sufficiently thin boundary layer on the lower side of the airfoil, this lower side behaves similar to the potential case in its ability to create a downwash downstream of the trailing edge. On the other hand, the flow from the upper side does not care and separates from the upper edge of the Gurney flap. Of course, this cannot be exactly correct, but it provides a feeling for what happens. And, of course, we are aware of the complexity of the real flow, as measured and calculated by Sauvage [5].

Returning to an interpretation of our own measured data: One may wonder whether or not it is possible to improve the lift to drag ratio ε of an airfoil at low subsonic flow speeds with the help of a Gurney flap. In Fig. 6 we show data of ε_{max}, i.e., the best lift to drag ratio being achieved for various Gurney flap heights, normalised with the best lift to drag ratio ε_{0max} of the reference wing. It appears as if at very low Gurney flap heights there may be a very small regime where an improvement could be possible. However, looking at the data in Fig. 6, one may argue that this could be a spurious effect as well. Our AIAA-paper [15] sheds more light on this issue.

3.4. Other shapes of mini-flaps

Obviously, there is a small separation regime which starts just upstream of the Gurney flap on the lower side of the airfoil. One may wonder what the device drag reduction is if this corner regime is filled with model clay so that a quarter of a circle is formed. As Fig. 7 shows, there is an improvement, but it is marginal. We observe a very slight decrease of the drag and an equally slight increase in lift.

On the other hand, a diverging trailing edge behaves rather differently, see Fig. 7. The drag is considerably decreased but the lift is also decreased, if the diverging trailing edge is compared with a Gurney flap of identical height ($h = 1\%$). Actually a smaller Gurney flap with $h = 0.82\%$ (or slightly less) produces very similar results. Thus, a diverging trailing edge works indeed very similar to a Gurney flap, but a smaller Gurney flap produces already the same effect as a bigger diverging trailing edge. On the other hand, *small is not always beautiful*: The greater height of the diverging trailing edge provides a greater mechanical stiffness and stability of the wing, which can be translated into a lower weight of the wing, which is certainly desirable.

In our paper we will, however, stick to the simple Gurney flap shape. All devices for drag reduction which we develop in the following sections will be, nevertheless, also valid for diverging trailing edges.

4. Drag reduction by wake stabilisation.

4.1. Concept, test arrangement and proof of absolute instability

In the separated flow downstream of a Gurney flap we expect that an absolute instability [16-18] can be observed. One well-known example of an absolute instability is the Karman vortex street which occurs in the wake of a cylinder. Such instabilities occur not only in laminar

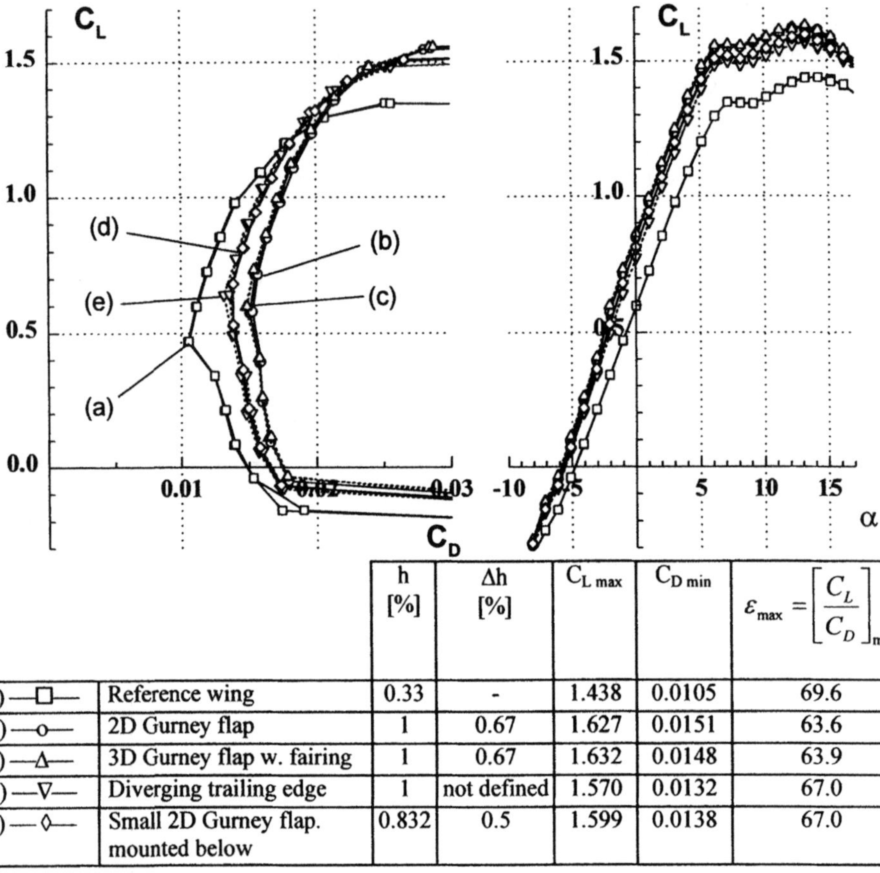

		h [%]	Δh [%]	$C_{L\,max}$	$C_{D\,min}$	$\varepsilon_{max} = \left[\frac{C_L}{C_D}\right]_{max}$
(a) —□—	Reference wing	0.33	-	1.438	0.0105	69.6
(b) —o—	2D Gurney flap	1	0.67	1.627	0.0151	63.6
(c) —Δ—	3D Gurney flap w. fairing	1	0.67	1.632	0.0148	63.9
(d) —∇—	Diverging trailing edge	1	not defined	1.570	0.0132	67.0
(e) —◊—	Small 2D Gurney flap. mounted below	0.832	0.5	1.599	0.0138	67.0

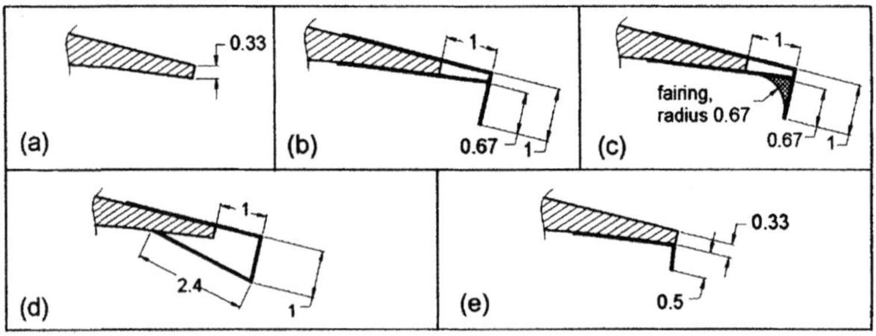

Fig. 7. Comparison of a diverging trailing edge with a Gurney flap and the influence of avoiding flow separation in front of a Gurney flap by fairing.

flows, as the classical experiments with cylinders would suggest, but they can be also observed in turbulent flows. A tell-tale sign for an absolute instability is the occurrence of a single peak in the spectrum of the fluctuations in or near the wake. The reversed flow in the wake is a property which enhances this instability [18]. Since trailing edges of airfoils are seldomly very sharp, we also expect such an absolute instability in the wake of an (even slightly) blunt trailing edge.

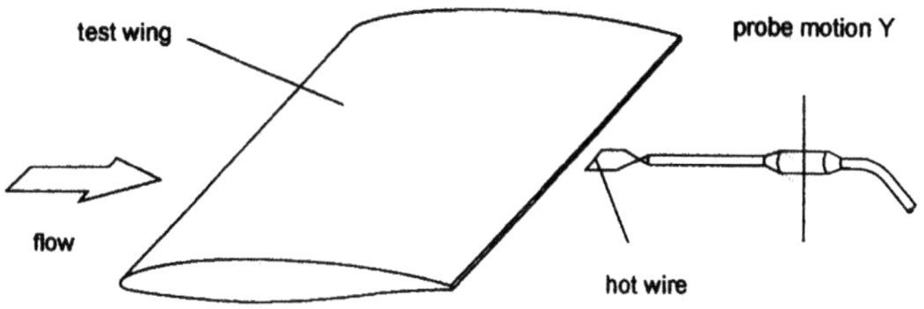

Fig. 8. Test arrangement for wake measurements

Our first aim is, therefore, to demonstrate experimentally, that a single frequency does exist in the fluctuations downstream of the trailing edge, and then devise trailing edge modifications to suppress those periodic fluctuations.

The easiest way to measure velocity fluctuations is offered by the well-established hot wire anemometry.

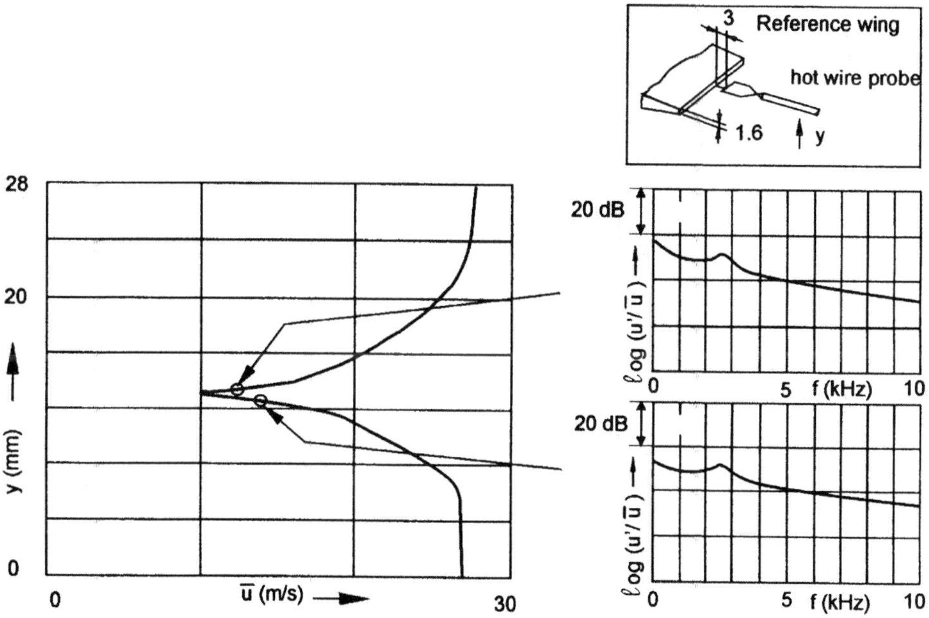

Fig. 9. Reference wing with finite trailing edge thickness of 0.33% chord. Data collected at a distance 3 mm (0.6% chord) downstream of the trailing edge. Mean flow $\bar{u}$ distribution and u' fluctuation spectra were collected at two locations with steep mean flow velocity gradient. Airfoil HQ17, Re = 1×10^6, $\alpha = -1°$ = Angle with minimal drag, $C_D = 0.0105$; $C_L = 0.42$. Dimensions in mm.

The hot wire instrument which we have used was a DLR-HDA IIIf anemometer with a built-in analogue linearizer. The anemometer output signals were plotted directly with an X-Y-plotter in order to obtain the mean flow distribution downstream of the wing. In places of maximum mean velocity gradient the highest streamwise velocity fluctuations u' occurred. The u' fluctuation signals were Fourier-analysed and the spectra were plotted. Fig. 9 shows recorded data of the reference wing. Due to the finite thickness of the trailing edge of this wing, there is a small hump in the spectrum, signifying an (albeit weak) absolute instability, which is highlighted by its single resonance frequency $f_0 = 2.4$kHz. By the way, this hump vanishes for a very sharp trailing edge.

As Fig. 10 shows, the wake of a (2-D) Gurney flap is wider, indicating a higher drag. In addition, the u'-fluctuation levels are higher and the periodic constituent with the frequency $f_0 = 0.8$kHz is much stronger. The data are recorded at Re = 1 x 10^6. Similar data are obtained if the mean flow velocity is reduced to one half of its previous value, yielding a Reynolds number of 0.5 x 10^6. As expected, the peak frequency f_0 is also reduced to half of its previous value. If one calculates a Strouhal number with the frequency f_0, the Gurney flap height[1] and the mean flow velocity, one obtains a resonance Strouhal number of $S \approx 0.14$.

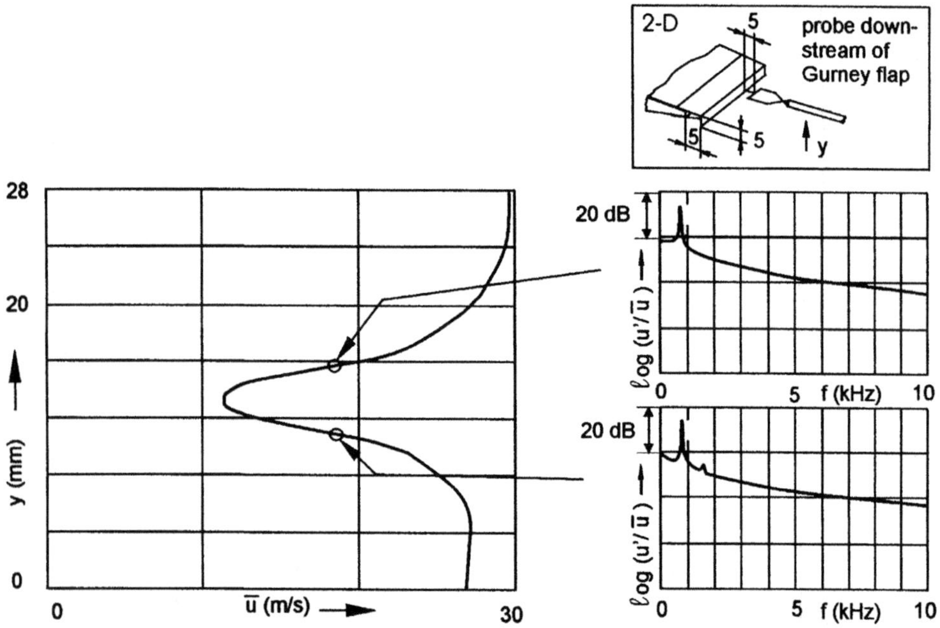

Fig. 10. Flow distribution and spectra in the wake of a 2-D Gurney flap with a height of 1% chord. Data collected at a distance of 5 mm (1% chord) of the trailing edge of the Gurney flap. Reynolds number and angle of attack as in Fig. 9. The Gurney flap height which we use throughout our instability investigations has a total height of 5 mm (1% chord) with an effective height of 3.4 mm (0.67% chord).

[1] It is conceivable that other reference lengths like, e.g., the momentum loss boundary layer (or wake) thickness may be a more appropriate reference length here.

4.2. Gurney flaps with slits

It is known from the Karman vortex street on cylinders, that a three-dimensional structure of the wake flow field effectively suppresses the absolute instability. On a cylinder, that can be achieved by applying a helical structure on its surface (i.e., the Screwton spiral) which can be seen quite often on industrial chimneys.

Obviously, the same approach cannot be literally transferred to our problem. However, as it will turn out, there are various possibilities to obtain a three-dimensional wake flow field. The one which we are considering first is the application of *slits* in the Gurney flaps. The effect of slits on the wake can be seen in Fig. 12. The hot wire data clearly show that the absolute wake

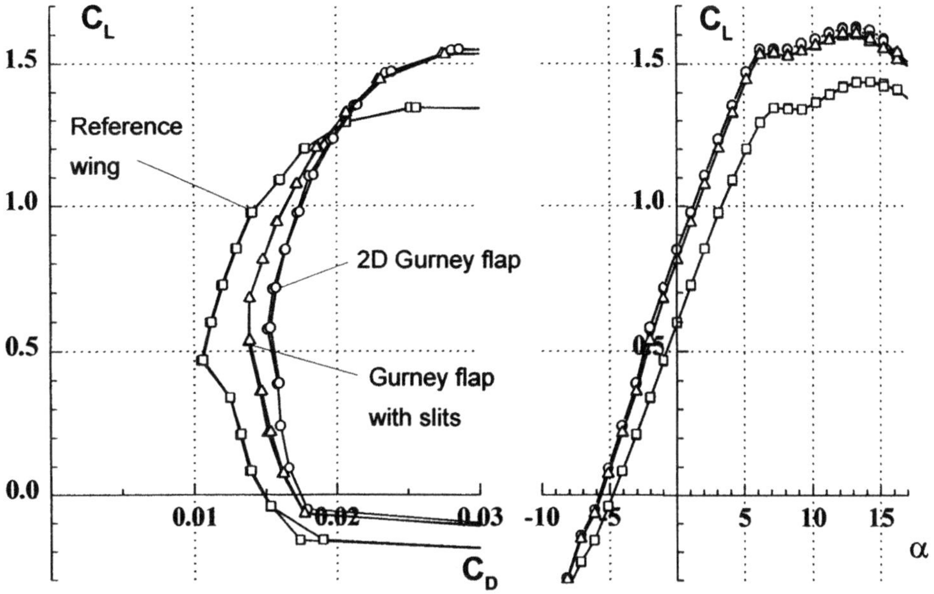

		h [%]	Δh [%]	L max	$C_{D\,min}$	$\varepsilon_{max} = \left[\frac{C_L}{C_D}\right]_{max}$
—□—	Reference wing	0.33	-	1.438	0.0105	69.6
—o—	2D Gurney flap	1	0.67	1.627	0.0151	63.6
—Δ—	Gurney flap with slits	1	0.67	1.605	0.0138	64.6

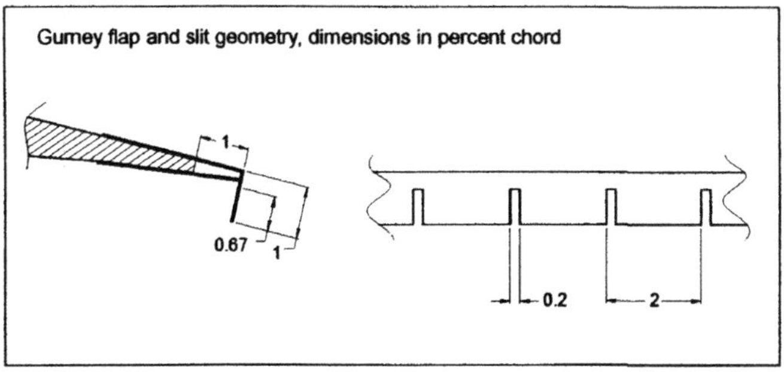

Fig. 11. Drag polars of a Gurney flap with slits. Re = 1×10^6

instability (i.e., the peak in the spectrum) has almost completely vanished. Drag polars can be seen in Fig. 11. It is obvious that the device drag is considerably decreased, by 28%. On the other hand, the bleed air through the slits causes the Gurney flap to appear smaller, which slightly decreases the gain in lift.

At higher angles of attack, however, the improvement due to the slits in the flaps in less pronounced. The hot wire data of the 2-D Gurney flap (not shown here) do not exhibit any

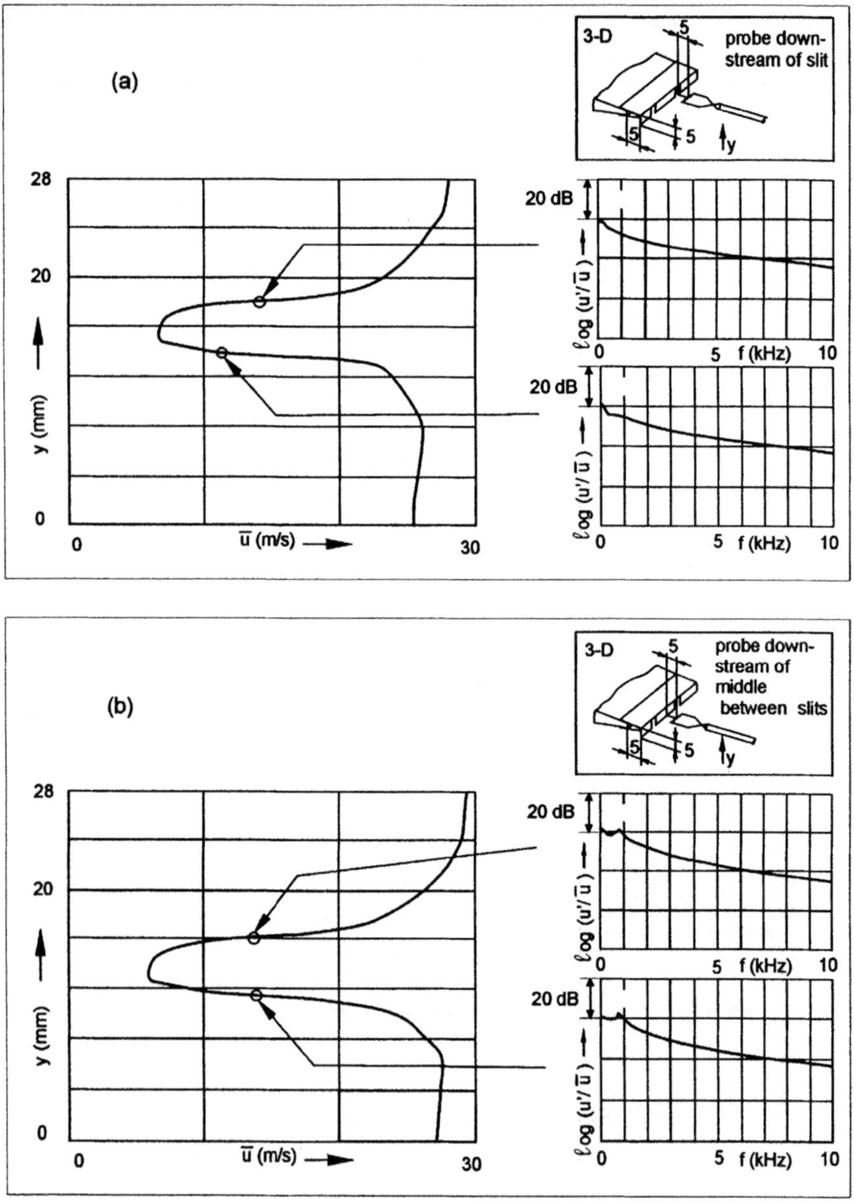

Fig. 12. Mean velocity distribution and spectra at 5 mm distance (1% chord) downstream of a 3D-Gurney flap (height 1% chord) with slits. Slit dimensions as shown in Fig. 11. Airfoil HQ17, Re = 1 x 10^6, α = -1°. Upper diagram downstream of slit (a), lower diagram downstream of middle between slits (b).

resonance anymore either. As a matter of fact, Koch [17] has already predicted that the absolute instability disappears for strongly asymmetric wakes. Obviously, for high angles of attack, the wake of an airfoil becomes strongly asymmetric. Therefore, one cannot reduce anymore the drag at high angles of attack by instability suppression in the wake. On the other hand, drag reduction is particularly desired at low angles of attack, and, as Fig. 11 shows, it is indeed achieved there.

Finally, we expect that, due to the elimination of the absolute instability of the wake, also induced mechanical vibrations of the wing and radiated noise are very likely to be reduced.

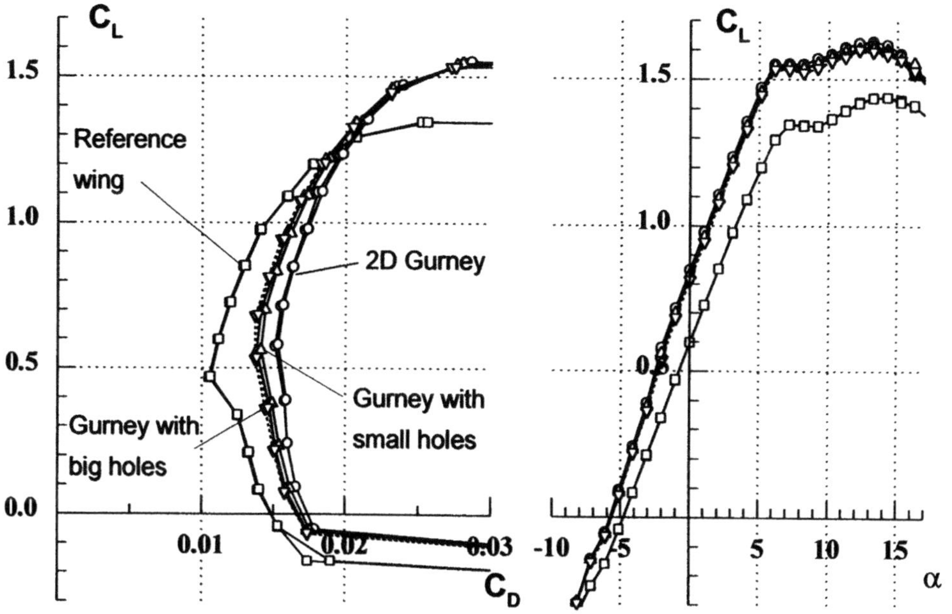

		h [%]	Δh [%]	$C_{L\,max}$	$C_{D\,min}$	$\varepsilon_{max} = \left[\frac{C_L}{C_D}\right]_{max}$
—□—	Reference wing	0.33	-	1.438	0.0105	69.6
—o—	2D Gurney flap	1	0.67	1.627	0.0151	63.6
—Δ—	Gurney flap. with small holes	1	0.67	1.611	0.0138	65.4
—∇—	Gurney flap. with big holes	1	0.67	1.599	0.0137	65.6

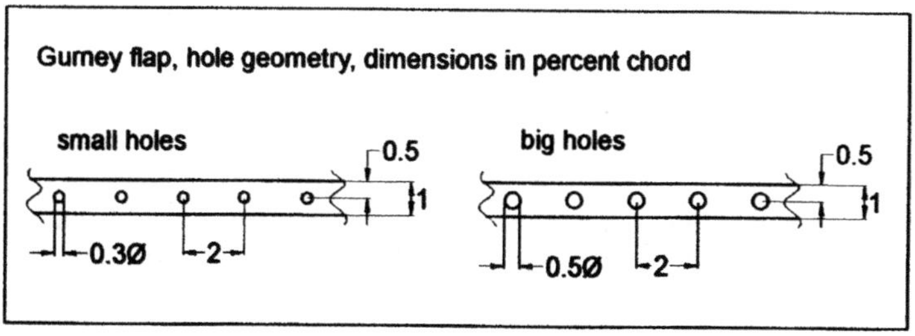

Fig. 13. Drag reduction by holes in the Gurney flap .

4.3. Gurney flaps with holes

One disadvantage of having slits in the Gurney flap is that the additional mechanical stiffness of the trailing edge with Gurney flap is no more available. Based on the idea that the reversed flow in the wake [5] is of crucial importance for the existence of an absolute instability [18] we considered *holes* in the Gurney flap as an alternative. This has the advantage that the mechanical stiffness and stability of the Gurney flap is maintained.

We have carried out tests with two sizes of holes, 0.3% chord and 0.5% chord in diameter. As Fig. 13 shows, both are effective. The smaller holes draw less bleed air and thus cause lower losses in lift. By the way, the same data as with the smaller holes are obtained if the bigger holes are placed at double the lateral spacing. According to our hot wire data (not shown here), the elimination of the wake instability is not so complete as with the slits, but it is significant enough to cause an equivalent drag reduction.

5. Acknowledgement

the various parts of this research were supported by different contributors. The measurements on a transonic wing were carried out together with Dr. E. Stanewsky, DLR Göttingen. The suggestion to carry out these measurements came from Dr. J. Mertens, DaimlerChrylser Aerospace Airbus, Bremen. This company together with the German Federal Ministry of Research and Technology (BMBF) provided the funding. The low-speed measurements were carried out in close co-operation with the Institute of Fluid Mechanics (Hermann-Föttinger-Institut) of the Technical University of Berlin. Funding was provided by the Volkswagen Foundation and by the German National Science Foundation (Deutsche Forschungsgemeinschaft) in the Special Research Activity (Sonderforschungsbereich 557) on Control of Turbulent Shear Flows.

6. References

[1] Liebeck, R. H. 1978, Design of subsonic airfoils for high lift. Journal of Aircraft, Vol. 15, No. 9, Sept., pp. 547 – 561.

[2] Henne, P. A. 1990, Innovation with computational aerodynamics: The divergent trailing edge airfoil. In: Applied Computational Aerodynamics (Ed. P. A. Henne), Progress in Astronautics and Aeronautics, Vol. 125, AIAA, Washington

[3] Pailhas, G., Sauvage, P., Touvet, Y. & Coustols, E. 1998, Flow field in the vicinity of a thick cambered trailing edge. Paper presented at the 9th International Symposium on Applications of Laser Techniques to Fluid Mechanics, 13 – 16 July, Lisbon, Portugal.

[4] Coustols, E., Pailhas, g. & Sauvage, P. 1999, Scrutinizing flow field pattern around thick cambered trailing edges: Experiments and computations. Paper presented at the 4th International Symposium on Engineering Turbulence Modelling and Measurements, 24 – 26 May, Corsica, France.

[5] Sauvage, P. 1995, Ph.D. Thesis, École nationale Supérieure de l'aeronautique et de l'espace, Toulouse.

[6] Ross, J. C., Storms, B. L. & Carrannanto, P. G. 1995, Lift-enhancing tabs on multielement airfoils. Journal of Aircraft, Vol. 32, No. 3, May/June, pp. 649 – 655.

[7] Kentfield, J. A. C.. 1994, Theoretically and experimentally obtained performances of Gurney flap equipped wind turbines. Wind Engineering, Vol. 18, No. 2, pp. 63 – 74.

[8] Stanewsky, E. 1981, Wechselwirkung zwischen Aussenstroemung und Grenzschicht an transsonischen Profilen. Dissertation, Technische Universität Berlin.

[9] Bechert, D. W., Stanewsky, E. & Hage, W. 1999, Windkanalmessungen an einem Transsonik-Fluegel mit stroemungsbeeinflussenden Massnahmen. Teil 1: Polaren, Teil 2: Druckverteilungen. DLR-IB 223-99 C05; IB 92517-99/B3-1 & 2.

[10] Bechert, D. W., Bruse, M., Hage, W., Meyer, R. 1997, Biological surfaces and their Technological Application . Laboratory and Flight Experiments on Drag Reduction and Separation Control, AIAA-Paper 97-1960.

[11] Rae, W. H. Jr. & Pope, A. 1984, Low-speed wind tunnel testing, Wiley, New York.

[12] Boermans, L. M. M. 1995, Wind tunnel data of the HQ17/14.38 airfoil, Delft University of Technology, Faculty of Aerospace Engineering.

[13] Althaus, D. 1996, Niedrig-Geschwindigkeits-Profile, Vieweg-Verlag, Braunschweig.

[14] Drela, M. 1996, "Xfoil" 6.8. An analysis and design system for airfoils. MIT, Dept. Of Aeronautics and Astronatics. E-mail: drela@henry.mit.edu.

[15] Bechert, D. W., Meyer, R., Hage, W., 2000, Drag Reduction of Airfoils with Miniflaps; can we learn from Dragonflies?, AIAA-Paper 2000-2315.

[16] Bechert, D. W. 1985, Excitation of instability waves, Zeitschrift für Flugwissenschaften und Weltraumforschung, Vol. 9, Nov/Dez, Heft 6, pp. 356 – 361.

[17] Koch, W. 1985, Local instability characteristics and frequency determination of self-excited wake flows. Journal of Sound and Vibration, Vol. 99, pp. 53 – 83.

[18] Huerre, P. & Monkewitz, P. 1985, Absolute and convective instabilities in free shear layers. Journal of Fluid Mechanics, Vol. 159, pp. 151 – 168.

Next Steps Envisaged to Improve Wing Performance of Commercial Aircrafts

Josef Mertens
University of Applied Sciences
FB6 Aerospace
Hohenstaufenallee 6, 52064 Aachen, Germany
Email: J.Mertens@fh-aachen.de
formerly:
DaimlerChrysler Aerospace Airbus GmbH
28183 Bremen, Germany

Summary

Several aerodynamic technologies were prepared which are to be introduced into the next generation of transport aircraft. This requires a technical solution guided by a well balanced design for aerodynamics, structure, systems and handling qualities. Here emphasis is on devices to control the trailing edge separation condition. It leads to an advanced wing which allows for adaptation to different flight conditions by local circulation control or spanwise redistribution of lift. Envisaged aircraft performance improvements are charcterized by design targets which combined offer important benefits in daily airline operation. Reference system are rather simple hinged flaps at the trailing edge of the wing. Further systems providing more flexibility, smaller size or better performances are under consideration.

Introduction

In several national and European programmes new technologies were prepared to improve the aerodynamic efficiency of future transport aircraft wings. Still laminarization is the most promising single aerodynamic technology. But here emphasis is on an advanced wing; it modifies and controls the trailing edge separation condition which determines wing circulation.

Airfoil performance is strongly related to control of circulation and type of the pressure distribution, especially for transonic and laminar airfoils. Trailing edge devices which modify circulation and pressure distribution, enable an extended performance range with respect to Mach number and cruise altitude. Shock position can be stabilized and off-design performance limits like buffet boundaries are extended.

For three-dimensional wings and complete aircraft, induced drag, interference drag, wing weight and handling qualities become important. They are influenced by spanwise lift distribution which can be altered by circulation control: for optimum aerodynamic performance during cruise and take-off or for weight optimal load distribution during manoeuvres and gusts. At approach, high drag is required which will be increased by aerodynamic devices along the wing span.

Reference solution for circulation control is a number of hinged flaps along the trailing edge called tabs, comparable to small ailerons. But those aileron like flaps must be integrated into the structure of the low speed flaps. This is a challenging task for structure and systems design including tab control. An optimum solution will only be achieved by simultaneous design of aerodynamics, structure and systems.
Other small devices like Gurney flaps, split flaps or trailing edge wedges offer very promising performance. To predict their aerodynamic performance under various design conditions, more in depth studies are required. Interdisciplinary design optimization will follow, including the evaluation of the total aircraft benefits.
Knowledge on those devices and their design is built up in several programmes in Europe. But it will require a concentrated approach to introduce them into a better performing product.

Operating Environment for Commercial Aircraft

Aircraft design consumes about 15 years; then, a large number of aircraft must be sold to recover the development costs and the investments for production. To become a successful programme, an aircraft family should be produced for more than 30 years. Adding an aircraft's operational life of about 20 years, the life cycle of an aircraft programme spans a human life. During this period, the operating environment can change dramatically. Therefore, a successful aircraft programme must be able to react to those changes and to provide variants which require only minor modification costs.

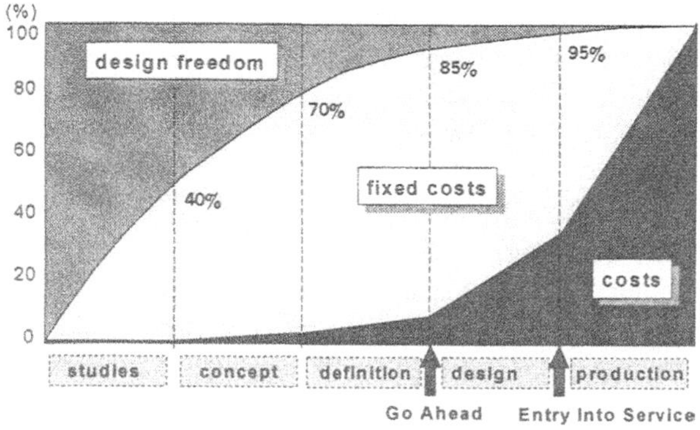

Figure 1: Key buying factors

Today, the main buying factors for airlines are operating costs and aircraft price (see fig. 1). But nearly as important are performances (e.g. payload-range) and further features which often guide the purchase order. But soon, environmental acceptability will become even more dominant, particularly noise and emissions; this is a common view of airlines and manufacturers.
The manufacturer has to respect the customer requirements at costs as low as possible. Mostly the aircraft's total programme costs are fixed by very early decisions, when only few details of the project are known (fig. 2). Therefore, those decisions must be carefully prepared,

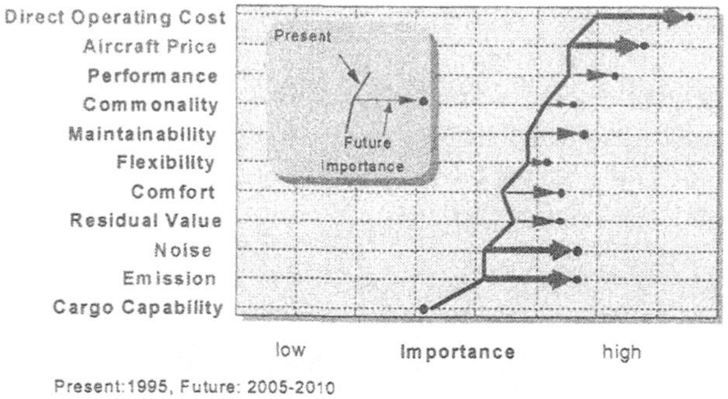

Figure 2: Cost of project development

e.g. by early technology programmes. With the advanced multifunctional wing, future aircraft programmes gain increased flexibility to better react on new requirements which only later on become obvious. It will also be possible to cure late detected design imperfections.

Concept of the Advanced Multifunctional Wing

Main component of an aircraft is the wing. It enables flying and determines flight performance. If the aircraft is to become more efficient and provide additional performance, the capability of the wing must be increased. Therefore, the DA-concept of the advanced multifunctional wing equips the trailing edge with adaptive elements (fig. 3), so called camber tabs. They are developed in the national technology programme "Process Chain High Lift with Multifunctional Control Surfaces (Pro-HMS)" and will improve the performance and application range of the classical high lift elements.

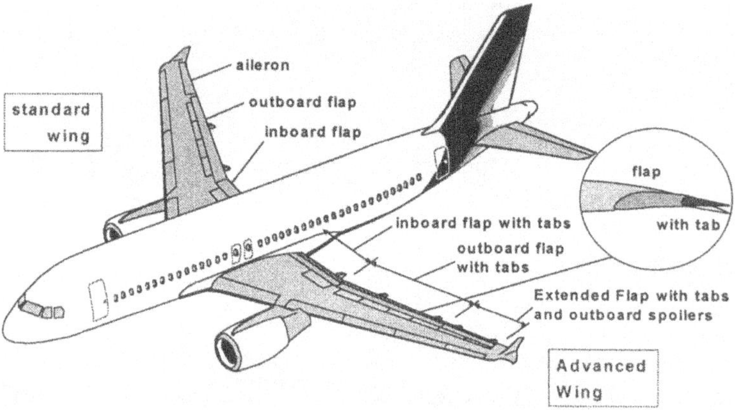

Figure 3: Basic concept of the advanced wing

Design targets characterize the envisaged improvements for the whole flight regime from take-off to landing. Performance and handling qualities are improved, design critical loads reduced or critical cases defused. Here, the most important functions of the tab application are briefly introduced (fig. 4).

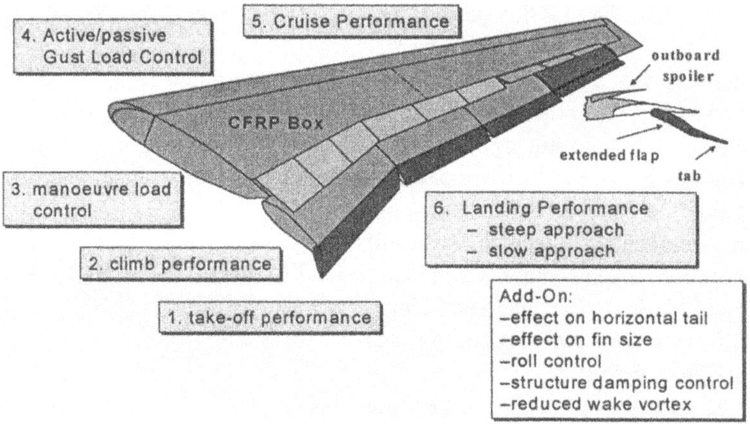

Figure 4: Target functions for the advanced wing

Improved performance for take-off and slow approach

Figure 5: Tail strike risk for long and heavy aircraft

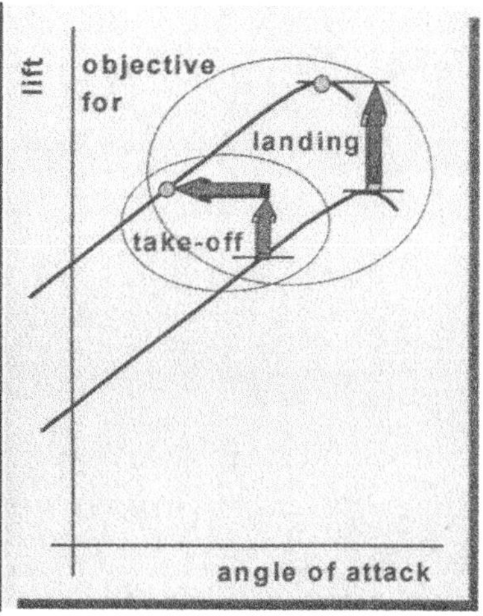

Figure 6: Low-speed targets

Aircraft cruise at high speed, but for take-off and landing they must be as slow as possible. This enables utilisation of airfields at high elevation or of small size, and it reduces wear, improves safety and reduces noise. Furthermore, heavier variants of the aircraft should be able to serve all important airports; this requires tail strike prevention for stretched versions. Figs. 5 and 6 illustrate those requirements.

Climb

Figure 7: Climb

During climb the aircraft must gain altitude as soon as possible and at the same time accelerate (fig. 7). Maximum climb angle is given by thrust/weight (T/W) and the aerodynamic performance L/D. Lift nearly equates weight; so by good approximation the climb angle γ is:

$$\gamma = T/W - 1/(L/D)$$

The better L/D or the lower drag, the steeper the aircraft can climb. By climbing steeper it reaches earlyer safety altitude and is higher over habitated areas, i.e. less noisy.

The heavier an aircraft, the more its total take-off distance (up to 50 feet altitude) is determined by the required climb angle. When the climb angle becomes the limiting condition (for heavier aircraft), smaller flap deflections are selected: it does reduce lift, but also reduces drag and improves L/D; the longer roll distance is more than offset by the steeper climb. (But still, total take-off distance rises for higher weight). Fig. 8 demonstrates the significant improvement of the aerodynamic low-speed performance by the multifunctional wing, especially for high lift coefficients.

First contribution comes from the additional fowler flap at the position of the "old aileron", the "Extended Flap". Additionally, the lift distribution can be optimized by the tabs. This is demonstrated in fig. 9, the upper part of fig. 8 showing the possible gains.

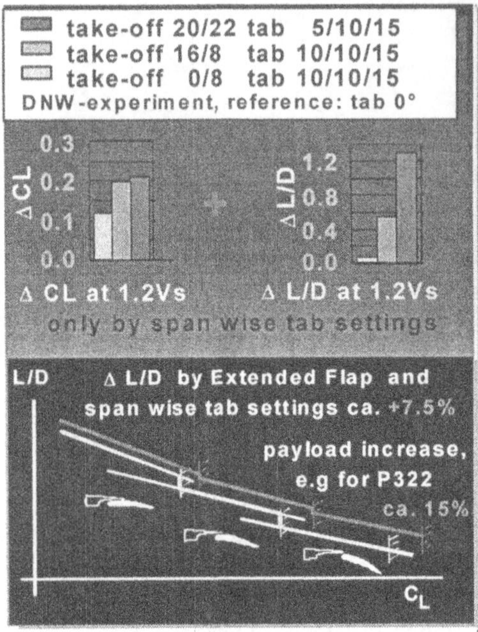

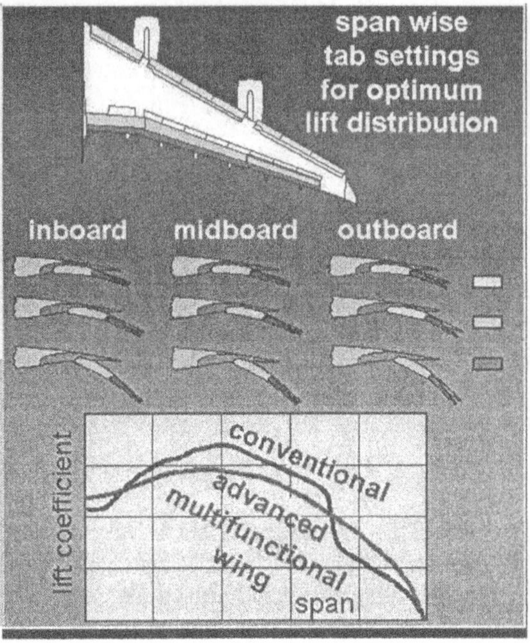

<u>Figure 8:</u> High-lift improvement by tabs

<u>Figure 9:</u> Tabs for Take-off and climb

A conventional wing generates low-speed lift mainly inboard where the flaps are positioned, often with a cut-out behind the engines. But where the spanwise lift changes, induced drag originates. Strong lift variations generate high induced drag. Using tabs, the best lift distribution can be selected over the whole span for all flap settings.

For large flap deflections, the flaps extend in the engine jet (fig. 10), producing additional jet interference drag (ΔC_{Dj}), particularly bad for take-off. Here, the tabs can be deflected a bit upwards (a so called "thrust gate") to minimize this drag (fig. 11).

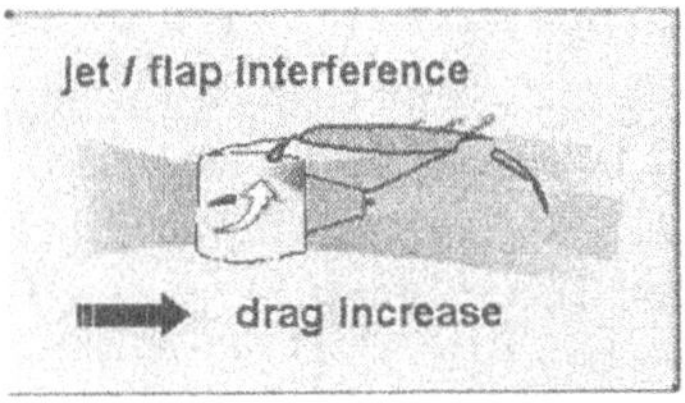

<u>Figure 10:</u> Jet interference

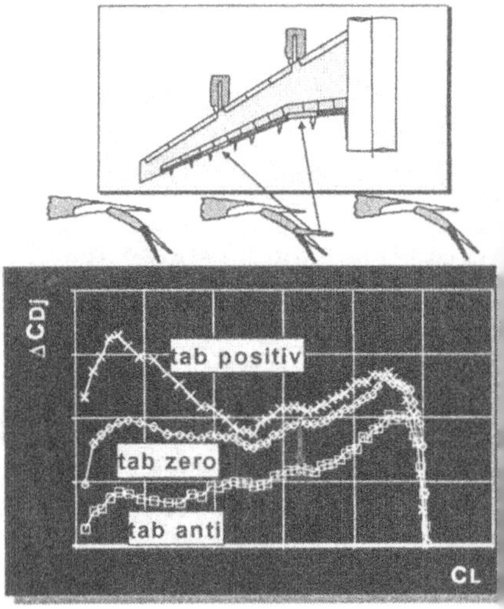

<u>Figure 11:</u> Thrust gate

Manoeuvre load control

In cruise flight, drag should be minimized. For a given span, minimum lift dependend (induced) drag is generated for elliptic (transonic nearly elliptic) lift distribution. Applying the same lift distribution for the weight defining load cases (e.g. 2.5 g manoeuvre),

large bending moments are generated in the inboard wing. Here the tabs and spoilers of the advanced wing enable a significant redistribution of the aerodynamic load: at the large inboard wing the tabs deflect downwards to generate more lift, whereas outboard lift is reduced by negative camber or spoiler deflections (fig.12).
This generates additional drag; but manoeuvres are short. Therefore the fuel consumption does not matter whereas the weight reduction is important.

Gust load control

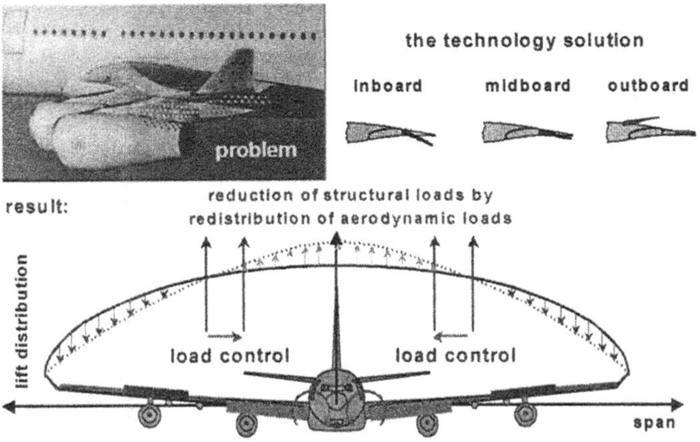

Figure 12: Reduction of structural loads by redistribution of aerodynamic loads

Similarly to the manoeuvre case, the aerodynamic loads can also be redistributed for gusts to reduce the inboard loads. But in contrast to a manoeuvre which is intentionally initiated (e.g. to avoid a collision or for a pull-up), gusts occur surprisingly. The load redistribution including the additional drag must be commanded as a precaution, or a gust sensor has to be used to detect the gust soon enough.
Precautionary load redistribution we call passive gust load control. It can be applied only, when additional drag does not harm, e.g. during descent; but often, gust loads become dominant during descent. For passive control, the load stays redistributed during the whole critical flight phase.
When using a gust sensor, it is possible to react directly against the incoming gust by defletion of tabs and spoilers. If an up-gust is detected, the tabs (and spoilers) on the whole wing deflect upwards to destroy as much lift as the gust would induce. Such a gust sensor could be positioned at the aircraft's nose (vane, pressure taps) or optically detect the gust some distance in front of the aircraft. The american bomber B2 already reduces gust loads significantly during low level flight using flush sensors at the nose and deflecting the trailing edge flaps. For civil application, such an active gust load limiting system has to demonstrate a very high reliability.
In down gusts, usually no critical structural loads occur. But when the 0 g level is passed, not secured people can be thrown against the ceiling, suffering injuries and even death. Those "inflight accidents" generate yearly costs of about 100 M$ for the airlines worldwide, mostly by injuries of flight attendants. Down gusts can be largely compensated by downward (positive) deflection of the tabs, or at least positive g-levels can be maintained. Because this is no certification requirement, such a system would be treated as "comfort improvement".
The benefit of load control can be reduced weight or improved aerodynamic performance at fixed weight, e.g. by increased span.

Performance control

During cruise, an aircraft flies at different altitudes and different speeds, depending on load, weather conditions or air traffic control requirements; those altitudes often deviate from the optimum conditions called design point. E.g. a long range aircraft carries about 40% of the take-off weight as fuel which mainly is burned during flight. Therefore, an aircraft cannot only be designed for the design point but rather for the whole cruise regime. Therein the performance must only slightly deviate from optimum. Thus, the wing is allways a compromise to cover the whole design space.

With flaps -here the tabs- we can adapt the wing to various flight conditions; so we can design it closer to the design point to gain performance. But now, for each tab deflection we have another optimum wing (a VC-wing = Variable Camber wing).

Fig. 13 shows wind tunnel results of two comparable research wings, one with fixed camber, the other with VC: The VC-performance in the design point was improved by 1.5% due to the narrow point design. But because optimum C_L and buffet boundary were increased, the wing area could be reduced by 8% with respect to the reference wing. This resulted in a further 2% reduction of the total aircraft drag due to lower friction drag. So total improvement at the design point was 3.5%. Additionally, the cruise regime was significantly expanded.

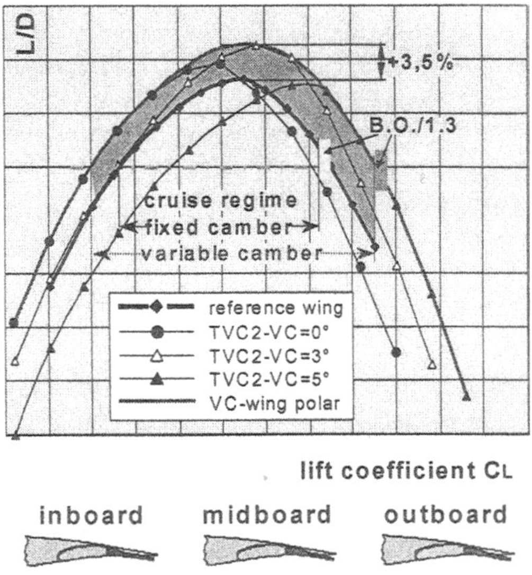

Figure 13: Cruise flight improvement

Steep descent

Maximum certificated *cruise* speed M_{MO} is often defined by a (possibly 20 s undetected) speed increase during a (7°) dive. But if an automatic system increases drag when passing M_{MO} (or v_{MO}), e.g. with a tab deflection as for manoeuvre load control, the aircraft accelerates more slowly and reaches only a reduced M_D or v_D. Thus, the loads at M_D resp. v_D are reduced.

Another case is an emergency descent (e.g. due to rapid decompression in the cabin), when the aircraft has to reach an altitude with sufficient pressure as soon as possible. The maximum vertical speed v_v at a given speed limit v_D (at idle thrust) results as

$$v_v = v_D / (L/D) .$$

The lower the L/D (i.e. the higher the drag D), the more rapidly the aircraft can descent. Here also a tab deflection as for manoeuvre load control can produce additional drag and (besides the increased sink rate) can prepare the aircraft for the loads during pull-up.

For *landing* we want to touch-down slowly in order to use short runways, save brakes and wheels, touch-down safely and reduce noise. In the future noise must be reduced even more;

this requires a steep approach to be still higher over the habitated areas. Also, smaller aircraft may fly above the glide path of the large ones to avoid the dangerous wake vortices (fig. 14). During approach, the descent angle is by good approximation

$$\gamma = 1 / (L/D) - T_i / W$$

T_i beeing the idle thrust to be maintained during approach. The lower L/D or the higher drag D, the steeper is the approach. Today, aircraft approach on a 3°-glide slope. To cover external disturbances, the possible glide path must be undisturbed about 5.5° to 6°. With the advanced multifunctional wing we want to approach at 5° to 6°, i.e. undisturbed we need about 7.5° to 9°. This requires a drag increase of about 40%. Fig. 15 shows measures to increase drag, investigated in the wind tunnel. Most efficient were strongly deflected tabs inboard with outboard spoilers in front of the Extended Flap.

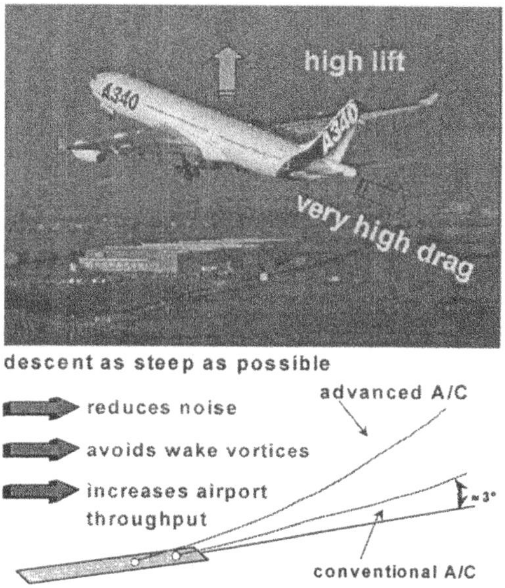

Figure 14: Steep approach

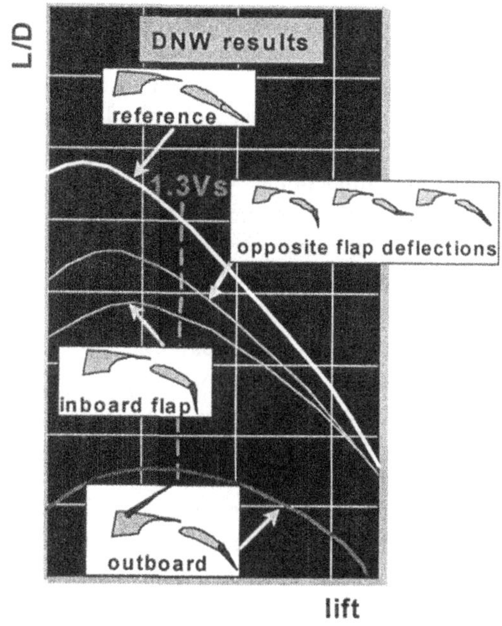

Figure 15: Wind tunnel results for drag increasing measures

To enable a safe go-around, it must be possible to reduce this additional drag rapidly at nearly constant lift. Tabs and spoilers are hinged flaps which can be turned rapidly and reliably. The lost drag at the inbord tabs is nearly compensated by the retracted outboard spoilers. Thus, this is an ideal means for steep approach.

The wing lift produces a wake vortex behind the wing with strong downwinds. To maintain existing separation distances to following aircraft, the wake vortices behind large aircraft must dissipate within a short distance. Fig. 16 shows a wind tunnel investigation with the (right hand) end of the down wind field, where vortex strength is

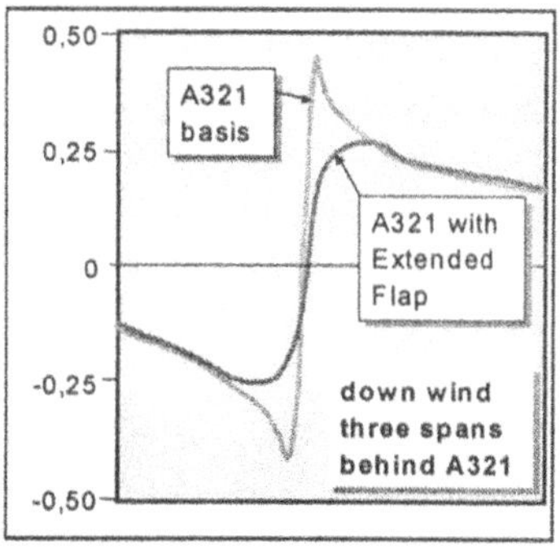

Figure 16: Downwind field

concentrated. Vortex strength is significantly reduced by spanwise differentiated tab deflections; this is a hint to rapid vortex dissipation.

Improved roll control

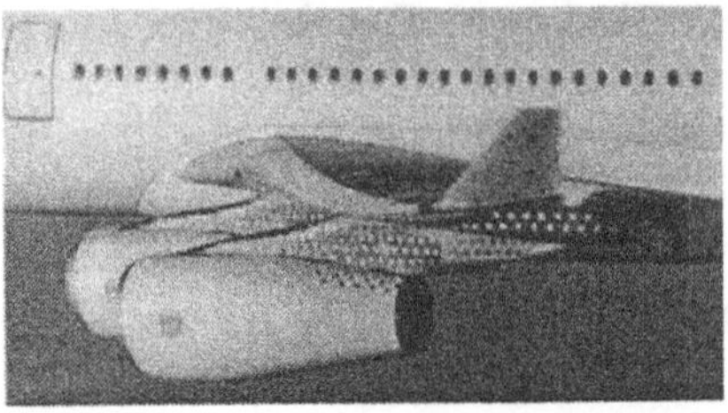

Figure 17: Reduced aileron efficiency by aeroelastic effects

Conventional aircraft use ailerons at the outboard wing for roll control supported by spoilers (see fig. 3). The advanced multifunctional wing has tabs over the whole span and additional outboard spoilers; this provides strongly improved roll control capability.
Especially on aircraft with slender wings, the efficiency of outboard ailerons -although having a long lever arm- is reduced by wing bending and torsion at high speeds (fig. 17). Finally aileron reverse occurs: the aircraft turns left for a right aileron command.
When scaling up an aircraft linearily, the mass rises with 3rd power, the lifting surfaces with 2nd, the inertias with 5th but rudder and aileron effectiveness only by 3rd power. Therefore, roll control for large aircraft becomes marginal, especially at low speeds. Here, the advanced wing with its multifunctional control surfaces offers significant improvements (fig. 18).
Midboard and outboard, the tabs deflect downwards on the up-wing, upwards on the down-wing. On the down-wing, additionally spoilers can be applied. Inboard tabs can be deflected downwards on both wings to compensate for spoiler lift losses. This is important near the ground, when immediately after lift-off or before touch-down reaction on side-gusts is required. If at this moment the aircraft looses a significant part of lift, the fuselage tail could strike the ground.

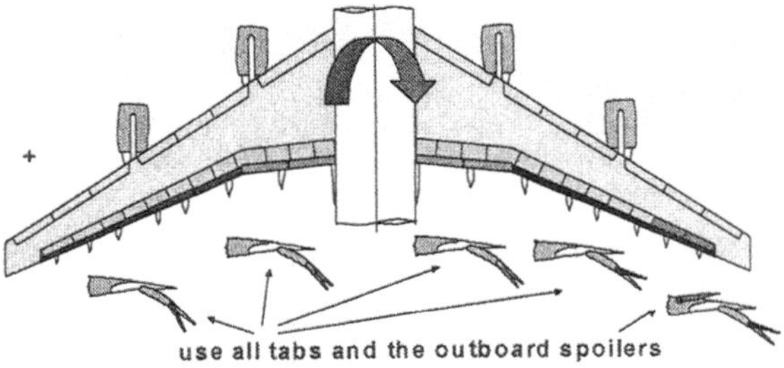

Figure 18: Roll control with the advanced wing

Trailing Edge Alternatives

Investigations have shown that small trailing edge devices (called Mini-TED = Mini Trailing Edge Devices) can have surprisingly high efficiency. Best known is the Gurney flap, a small normal fence at the trailing edge of about 1% chord. Comparable elements were investigated for local control of gaps (fig. 19).
It turned out that the efficiency of Gurney flaps is better than that of a 10% camber tab (fig. 20). Due to the small dimensions, forces are small (but high specific loads) and require only small actuators with low weight. Small

dimensions enable extremely high deflection speeds for control of sudden or high frequency loads (e.g. active gust or flutter control). But hitherto there exists not yet a realized structure or actuation concept nor is the aerodynamic data base sufficient for the whole design range.

Conclusion

Aerodynamic trailing edge elements were proposed and the wide application range was demonstrated. Besides the present basic concept with camber tabs along the whole span, future possibilities with Mini-Trailing Edge Devices were mentioned.

At DA, technology development is performed for aerodynamics, the suited CFRP-structure, actuation elements and the control system for the complex wing. But all single components will only be able to provide their full potential when beeing incorporated into a new aircraft design, optimally incorporating all benefits.

In Pro-HMS, technologies for this purpose are developed. Main elements are tested in ground tests to prepare a proof of concept flight test to follow.

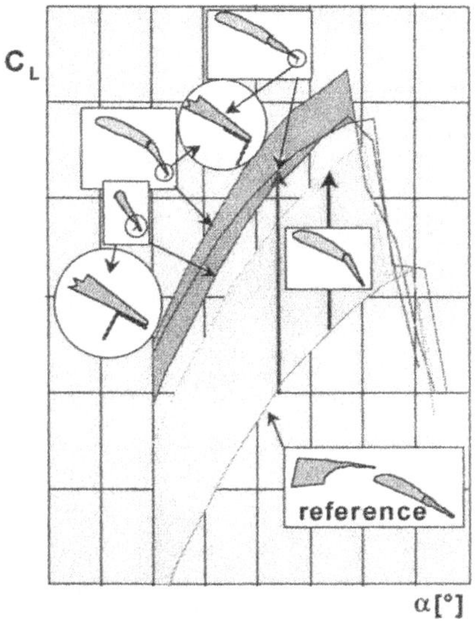

Figure 20: Mini-TED at high lift

Acknowledgement

These technologies were developed in the technology programmes "RaWid" and "HAK" of DaimlerChrysler Aerospace Airbus, partially sponsered by the German ministry for research and technology (BMBF). The ongoing programme "Pro-HMS" of DaimlerChrysler Aerospace Airbus is partially sponsored by the German ministry for economy and technology (BMWT).

dimensions enable extremely high-deflection speeds for control of rudder or high-frequency loads (gust alleviation or flutter control). But failure there exists not yet fully realized structure or reduction ... as the aerodynamic data base sufficient for this whole design range.

Session 8
Future Prospects

INDUSTRIAL PERSPECTIVES OF DRAG REDUCTION TECHNOLOGIES

Roland Gerhards, Joachim Szodruch
EADS Airbus GmbH
Kreetslag 10, D-21129 Hamburg, Germany

Summary

Competition in the global aircraft market forces the airlines to reduce cost considerably. This in turn requires from the manufacturer a reduction in development time and cost. Consequently, the manufacturer has to take the right decisions today for investments in cost effective technologies for the products of tomorrow which are successful, efficient and respond exactly to the variety of future customer demands and transportation system requirements.

One major part of these requirements is the significant reduction of fuel burn, where a 50% decrease is envisaged for the next 20 years. A large part is contributed by aerodynamic improvements divided into lift dependent drag reduction and friction drag reduction. Engine manufacturers are working on SFC reduction. Furthermore, materials and structural technologies as well as improved systems contribute to weight reductions of the aircraft and the engine. However, almost all of these technologies are only successful when considered on an interdisciplinary basis.

Introduction

In only about twenty years Airbus Industrie has been able to build up a full aircraft family utilizing advanced technologies as much as necessary and beneficial to our customer and ourselves. This concept of high quality and technological advanced aircraft made us a serious competitor in the world market and also ensures the competitiveness of Airbus products in the longer term future. The increasing economical constraints forces all involved especially also the airframe manufacturer to reduce cost. That certainly must be one of the major goals in the future, however, that can and must only be one part of our common efforts. On the technological side it is mandatory to maintain our standard, and future products, derivatives or all new aircraft must make the most beneficial use of advanced technologies as required by the market in terms of performance, comfort, infrastructure aspects, regulations or environmental concerns.

One of the ongoing technological developments is the reduction of aerodynamic drag because this has direct impact on fuel reduction, which is a major topic due to the influence on many other requirements (i.e. DOC, emission, performance, etc.). However, the drag reduction may not be seen as a single technology, but always in the general context with other industrial aspects, like manufacturer targets or customer acceptance. Additionally new technologies have to be integrated in a new or enhanced product with its own development process and timescale.

Driving Factors

For the analysis of the industrial aspects of new technologies - and in this paper especially drag reduction - it is necessary to understand the whole development process of a new aircraft. Initially it is worthwhile for the manufacturer to consider what presently are believed to be the main design drivers, which define the aircraft requirements. At present and in the mid term future these are- beneath other aspects - the following four main items: Economy, Growth, Congestion and Ecology. Fig. 1 shows exemplary a rough timeline for the next 50 years with some topics and events concerning these main drivers.

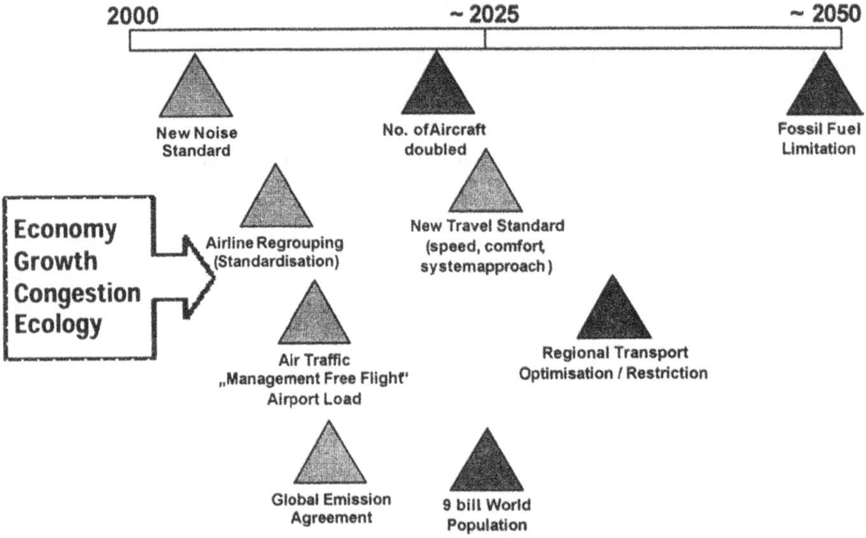

Fig. 1 Design drivers in the next 50 years

First, the aeronautic industry is an essential contributor to the worldwide social, economic and industrial development. In the year 2010 this industry alone will have more than 30 mio. employees (direct and indirect) worldwide generated by 1,25 billion passengers (business and leisure travel) with a profit of 1.740 billion US$ (before tax). And for example one slot at Frankfurt Airport creates 750 jobs direct at the airport and 1.650 jobs in the region. On the airline side the direct operating cost have to be decreased due to the competition and the yield decline as already observed presently. The A3XX needs to have 15-20% lower DOC for being accepted by the airlines for instance and so every new design has to improve the economics as one of the major goals also in the future.

Fig. 1 also indicates a doubling of numbers of aircrafts in the next 20-25 years. The average growth rate is predicted at 5% per year, which means the air traffic is doubling in the next fifteen years. Through larger aircraft and a higher productivity, the increase of aircrafts is only at a rate of 3.1% per year. Therefore, a new design has to cope with restrictions and new requirements caused by this additional traffic. One aerodynamic task is for example to reduce the negative effects of non-optimal flight levels caused by congestion on specific routes.

And finally the ecological aspects with respect to new aircraft design will be of significantly higher importance. These requirements are mainly driven by

- the growing public attention to environmental compatibility of new products
- the market response by airlines, airports and leasing companies
- and last but not least political considerations and proposals for restrictions, regulations and possible taxes.

The additional design parameter "environmental compatibility" is focused primarily on external noise, emissions, production, maintenance procedures and eventually even the question of recycling. The first two items are the most important ones because the economical effects are already visible today on a few airports through higher landing fees and emission taxes (e.g. CO_2-tax).

Development Process

A new product definition process starts after market analysis indicating a window of opportunity. Intensive scenario based studies (Fig. 2) for envisaged markets and regions follow to observe program viability under a number of different assumed developments of future economy, infrastructure, environmental legislation and further elements in the entire air transport system. A task not too easy, given the principal unpredictability of the future, however careful observation and analysis of trends can be done. From this, technical requirements for the aircraft must be identified. Then technologies have to be developed, that best respond to the requirements. Finally, those technologies that best serve the purpose will be decided to make their way into the aircraft. A continuous monitoring and controlling of the process makes several iteration steps necessary during the product definition.

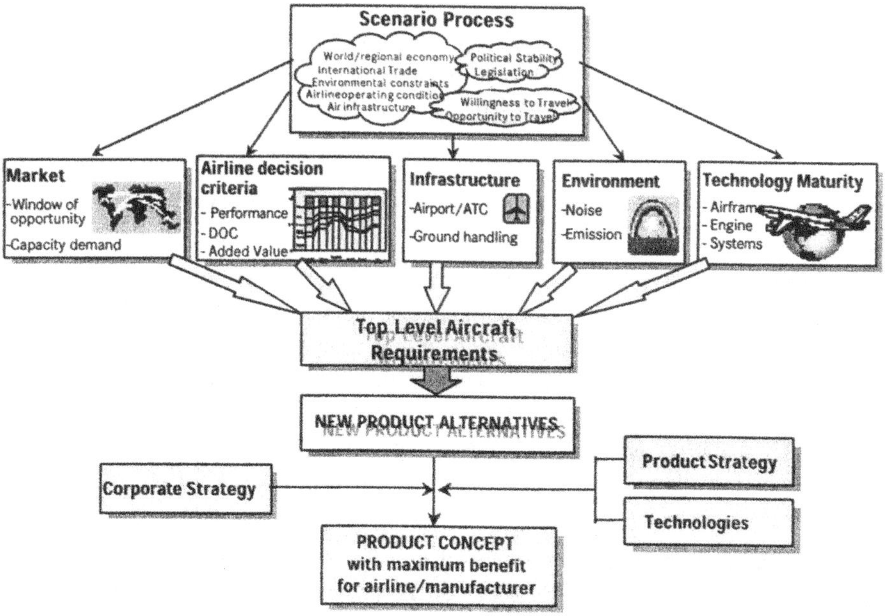

Fig. 2 Product development process

Although creativity is something one can hardly measure the timing for technical work and specifically validation is crucial. A typical life cycle of a commercial aircraft (Fig. 3) indicates the importance for the manufacturer to design an aircraft, which is able to fulfill major customer requirements during an in-service period of more than forty years. The figure shows also the timeframe of about 10 to 15 years before Go Ahead of a new program in which the research and technology integration takes place in relation to the development phase of a new product.

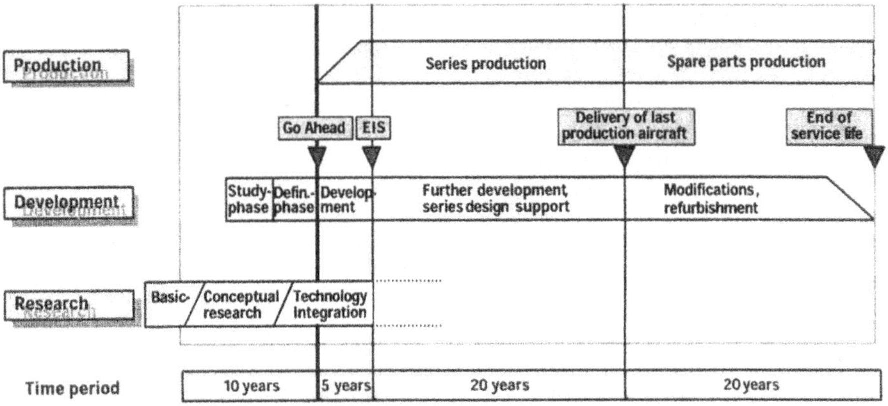

Fig. 3 Typical life cycle of a commercial aircraft

Technological Challenges

World air traffic experienced in the past a growth of approximately 5-8% per year but at the same time due to the quest for improving operating economics, fuel consumption was considerably reduced. Considering that the average flight efficiency can be expressed by

$$\frac{TripFuel}{Distance} \approx \frac{SpecificFuelConsumption}{MachNumber} \times \frac{Weight}{AeroEfficiency}$$

then using Airbus Industrie's products as a reference over the past twenty years the trip fuel has been reduced by about 37%.

Assuming twenty years of continuous, focused and efficient research it is estimated that another 50% net reduction potential in fuel burn is achievable. A large part (~33%) is contributed by aerodynamics divided into 11% lift dependent drag reduction (variable camber, smart wing, higher aspect ratio, reduction of vortices, wingtip devices) and 22% friction drag reduction (shock-boundary layer control, laminar flow, turbulence control). Engine manufacturers are working on SFC reduction (~22%). Furthermore, materials and structural technologies as well as improved systems contribute to weight reductions of the aircraft and the engine (~8%). Up to 4% improvement will be reached through new or advanced configurations.

Despite the breakdown in single disciplines, it is in the end the interdisciplinary, concurrent adaptation of all these technologies, which might contribute to the 50% fuel savings mentioned above. In Fig. 4 this multidisciplinary approach is shown for the laminar flow area. Additionally the three different research steps (basic research with technology feasibility studies, applied research with technology development and application-oriented research with validation and demonstration) are mentioned together with a sketch of a simplified cost distribution.

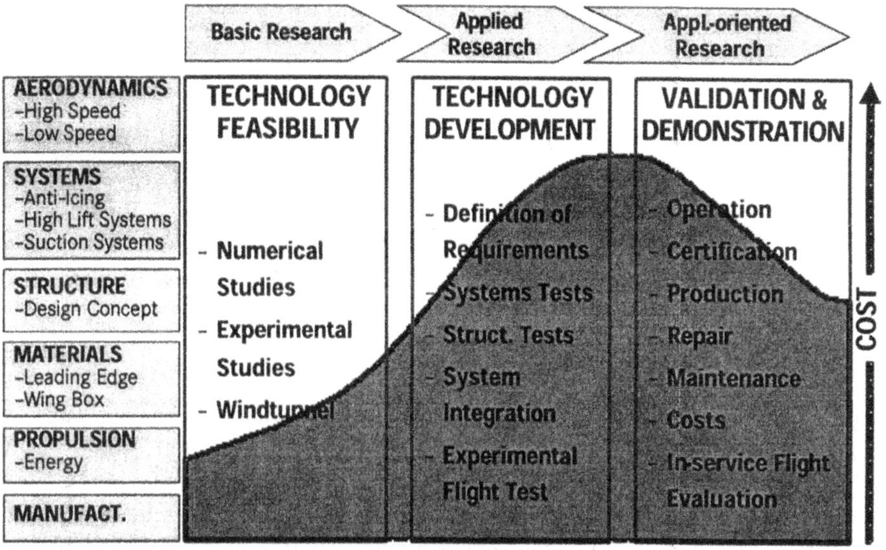

Fig. 4 Multidisciplinary approach: laminar flow

A considerable research budget, time and resources are required in order to implement these technologies and make them economically viable. So even under optimistic conditions, like continuous and harmonized European research actions, the next ten years will very likely provide only 16% to 20% of fuel burn reduction. In combination with the aerodynamic activities, other technologies must be advanced as well in order to find interdisciplinary optimized solutions to the various challenges described above. It is most important concerning all technology work to be on the market at the right time with the appropriate product.

Interdisciplinary Design

A key element is the careful observation and analysis of trends in the aircraft acquisition proceeding of the airlines. The aircraft selection process is influenced by many individuals within the airline organization. Although the concept of the new product has to provide a superior value to the decision makers in the airline, it is obvious that the ideal airplane, suiting everybody's requirements and expectations, does not exit. A purchase decision is, similarly to the design objectives of the manufacturer, at best a compromise between diverging interests. Market success is linked to the question how well a project responds to long term customer requirements.

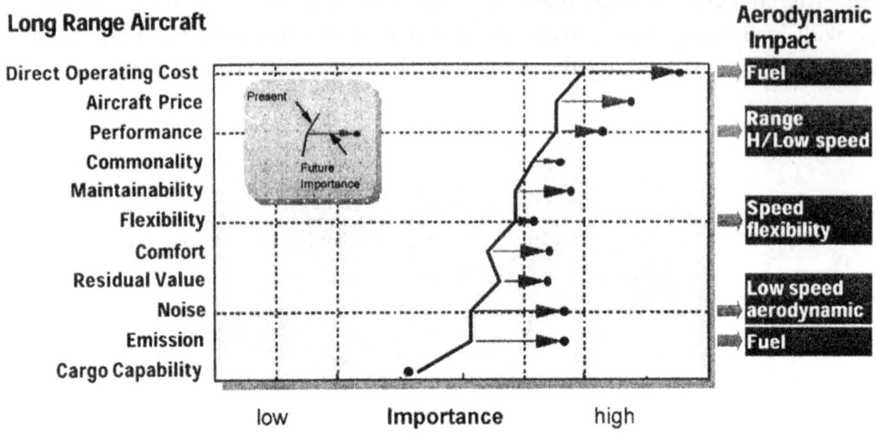

Fig. 5 Key Buying Factors

For the product evaluation only aircraft related Key Buying Factors will be considered, factors related to the manufacturer without any direct reference to the aircraft characteristics are excluded. These "Key Buying Factors", the combination of DOC, commonality and Added Values, determine the competitiveness of an aircraft. The selection of these factors is a result of an inquiry amongst 106 IATA airlines, operating approximately 80% of today's world jet fleet (Fig. 5). Particular attention has been given to criteria, which will dominate the airline market in the coming decade. On average, more than 50% of an airline's decision is determined by factors other than economics and commonality. Based on Fig. 5 the aerodynamic impact from Key Buying Factors can be deducted (Fig. 6).

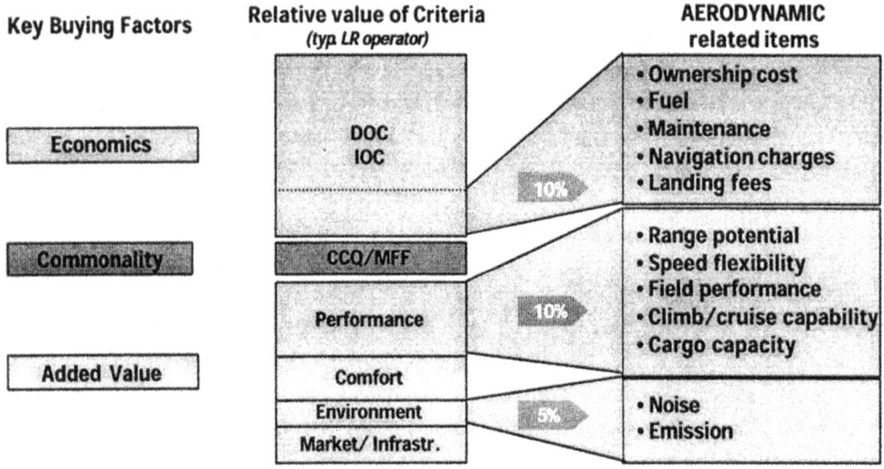

Fig. 6 Aerodynamic related items determine about 25% of the Key Buying Factors

Impact on Aircraft Design

On the basis of the above-mentioned figures one can extrapolate the potential fuel burn reduction over the next decades. However, only 50 to 60% of those potential improvements may be realized in the end due to restrictions in detailed design. The potential technologies have to be integrated in the aircraft configuration, which will lead to a rearrangement of the main aircraft components. One more advanced configuration not yet fully studied and balanced in the design is shown in Fig. 7.

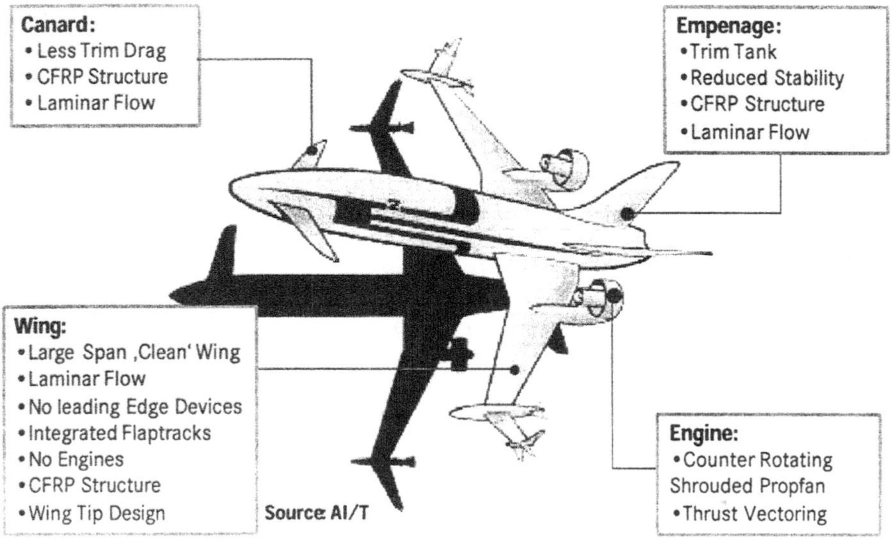

Fig. 7 Next generation 200-seater aircraft configuration

This new generation aircraft is characterized by:

- **Wing**: A large span "clean" CFRP wing without engines, slats and flap tracks and with natural laminar flow. A moderate sweep for a slightly reduced design speed. Reduced aerodynamic noise.
- **Engines**: Counter rotating shrouded propfan engines with an aft fan for optimum integration at the rear wing. Reduced engine noise.
- **Tailplane**: A laminar flow CFRP T-tail for rear engine installation. Trim tank in horizontal tail. Reduced stability.
- **Canard**: The weight concentration at the rear end of the fuselage moves the wing aft and the payload creates a forward CG position. A canard for this specific configuration could compensate the trim drag as long as the canard downwash could be separated form the wing and the engine intake.
- **Noise**: Footprint of 75 dBA within the airport boundaries.

Conclusion

The civil aircraft industry is steadily growing in size and in economic importance, a prediction supported by all market analysts. The requirements for future civil aircraft based on:

- Increased competition among aircraft manufacturers on one hand and among airlines on the other hand with emphasis on economics
- Growing importance of operational and infrastructural aspects
- Increasing stringency of environmental aspect, i.e. emission and noise
- Safety requirements

For the development of an aircraft this translates into

- Optimizing the development process and value stream (reduction of time and costs)
- High degree of interdisciplinary design work starting already in early phases of technology activities up to entry into service.

References

[1] Szodruch, J., Hilbig, R.: Building the Future – Aircraft Design for the next Century, AIAA-98-0135, 1998.

[2] Szodruch, J., Meller, F.: Enhanced Methodology for Aircraft Evaluation and Technology Identification, CASI, Montreal, 1999.

Session 9
Turbulent Drag Reduction Methods

CURRENT STATUS AND PROSPECTS FOR TURBULENT FLOW CONTROL

Clyde Warsop

BAE SYSTEMS Advanced Technology Centres - Sowerby
PO Box 5, Filton, Bristol, BS12 7QW, UK

Summary

This paper reviews some of the recent developments in and future prospects for turbulent boundary layer flow control. It addresses turbulent flow control from the point of view of both skin friction drag reduction and flow separation control. The fundamental aspects regarding the evolution and propagation of turbulent boundary layers is described. The state of the art with regard to passive flow control technologies such as riblets and sub-boundary-layer vortex generators is presented and addressed from the point of view of engineering application. Finally, an overview of recent research into the potential for the active control of turbulent boundary layers using emerging technologies such as MEMS is given. In addition to addressing aspects of fluid mechanics the critical importance for research into these technologies being of a multidisciplinary nature is emphasised together with the need for the involvement of the airframe and system integration engineer at an early stage in concept evolution.

1. Introduction

Due to their size and operating speeds the majority of commercial and military aircraft in service today are dominated by flows that result from the presence of turbulent boundary layers, which generally cover the majority of the aircraft's surface. It is well known that although, compared to laminar boundary layers, turbulent boundary layers give rise to significant skin friction drag penalties they do result in a reduced susceptibility to flow separation. Many attempts have been made in the past and are currently being undertaken to develop technologies with which to promote significant regions of laminar flow over the lifting surfaces of large aircraft. However, these technologies are unlikely to be ready for full-scale industrial deployment for a number of years due to the complexities associated with systems integration and the current relatively low price of fuel. Even when laminar flow aircraft become a commercial reality it will still be necessary for wings to be designed so that a turbulent boundary layer exists over the rearward part of the upper surface in order to achieve a satisfactory pressure recovery without susceptibility to flow separation. It therefore appears that turbulent boundary layers will feature on all commercial and military aircraft for the foreseeable future.

The desire to minimise drag (both skin friction and pressure) and to control flow separation in order to improve the performance of high lift and other trailing edge control devices is providing a driver for increased research activity in this field. Over the last 20 years or so our knowledge about the evolution and propagation of turbulent boundary layers has increased significantly. This has been made possible through the advent of new experimental techniques and the development and use of computation tools such as Large Eddy Simulation (LES) and Direct Numerical Simulation (DNS). Developments of new manufacturing technologies such as Microfabricated Electro-Mechanical Systems (MEMS) and solid state actuator technologies such as piezo-electric materials and shape-memory alloys has also led to the possibility for active flow control to be realised at both a macro and micro scale. MEMS technology, for example, offers the potential for the large-scale active control of coherent flow structures within the boundary layer. This could lead to the reduction of skin friction drag or the postponement of flow separation through the use of 'smart skins' capable of detecting and reacting to the state of the local boundary layer.

Despite over a century of intensive research, turbulence remains largely an enigma that is analytically unapproachable yet practically very important. The mysteries of turbulence are only now being solved by the use of physical and numerical experiments, which is a far-from-trivial task at the high Reynolds numbers of practical interest to the aerospace engineer. Controlling a practical turbulent flow to achieve a desired effect such as drag reduction, lift enhancement or noise reduction is a very difficult task. Passive control methods, while always preferable, are generally limited in their utility. Brute force suppression, or taming, of turbulence via active, energy consuming control devices is always possible, but the penalty for doing so often exceeds any potential savings. The challenge is to actively achieve a desired effect with a minimum of energy expenditure by utilising the natural instabilities within the fluid structures to amplify control inputs.

2. The Nature of Turbulence

Shear flow turbulence is neither homogeneous nor isotropic. It is dominated by a quasi-periodic sequence of large-scale structures often referred to as *turbulent coherent structures* (Fig. 1). Coherent structures are not only quasi-periodic, but are different in size and shape depending on the location of these structures within the flow. Furthermore, the coherent structures are born, grow and die within the boundary layer, evolving in both space and time. The precise dynamics involved in the turbulence activities are far from clear. Qualitatively, according to the generally accepted school of thought[1], the process starts with pairs of elongated, counter rotating, stream-wise vortices buried within the near-wall sub-layer region (Fig. 2). These vortices are often referred to as 'hairpin' vortices and arise in the first instance during the latter, non-linear stages of transition from laminar to turbulent flow. These hairpin vortices exist within a strong shear layer and induce low and high-speed regions between them. The low speed regions, close to the wall, termed *streaks*, grow downstream and develop inflectional velocity profiles. At the same time, the interface between the low and high-speed fluid begins to oscillate, signalling the onset of a secondary instability. The low speed region lifts up away from the wall as the oscillation amplitude increases and the flow rapidly breaks down (*bursts*) into a completely chaotic motion. In the sequence of turbulent activities within the boundary layer, there are two important events for energy production, called *sweeps* (or inrushes) and ejections (*bursts*). Over 80% of the turbulent kinetic energy production occurs during these events. The process described above is self- regenerating resulting in the continuous cyclic propagation of near near-wall hairpin vortices, streaks and bursts. There is still much speculation concerning the relationship between the stream-wise vortices and the streaks and significant further research in this field is required.

Large peaks in turbulent wall-shear stress are produced between a pair of counter-rotating longitudinal vortices as the high momentum fluid is brought down towards the wall during the near-wall burst events. These shear stress peaks give rise to the large increases in skin friction drag associated with a turbulent boundary layer compared to one that is laminar (typically an order of magnitude greater). Most of the activity in turbulent drag-reduction and separation control, both passive and active, relies on the manipulation by suppression, enhancement or modification of coherent turbulence structures.

3. Passive Turbulence Control

Riblets

The study of riblets for turbulent drag reduction began in the late 1970's at the NASA Langley Research Centre[2]. For typical triangular riblets, skin friction drag is reduced by up to 8%. Although riblets are quite tolerant to misalignment of the flow drag reduction is completely lost when the yaw angle exceeds about 30°. The net skin friction drag reduction is almost linearly proportional to the amount of coverage of riblets over the body surface, suggesting that the turbulent drag reduction by riblets is a predominantly local phenomenon. Recently, research at the DLR in Berlin[3] has made considerable improvements in drag reduction by optimising the shape of riblets. As a result, drag reductions of as much as 10% or more can now be achieved using more optimum geometry.

The drag reduction performance of riblets has been studied at high Reynolds and Mach numbers by many researchers and both parameters have been found to have little influence. The effectiveness of riblets in turbulent boundary layers under longitudinal pressure gradients has also been investigated. At Delft. Experiments have shown that drag reductions of up to 13% can be obtained using riblets in a boundary layer under adverse pressure gradient. There is still much debate about the exact mechanism by which riblets achieve a reduction in turbulent skin friction drag. However, all hypotheses are based on the control of the near-wall turbulent coherent structures. One school of thought is that the riblets act as small longitudinal fences impeding the lateral movement of the stream-wise vortices during the near-wall burst events. This causes the bursts to occur prematurely over the riblet surface leading to a reduction in the burst duration and intensity. Other researchers (Choi, Pollard) have found evidence for riblets shifting the origin of the turbulent boundary layer further from the wall. This reduces the effect of downwash during the near-wall burst events.

Although the most optimistic drag reductions for riblets lie in the region of 10% practical considerations probably limit the value that can be obtained to between 5% and 8%. However, skin friction drag for a large commercial transport aircraft is of the order of 40% of the total. Therefore even covering half the aircraft surface with riblets could amount to total drag reductions of the order of 3%. Resulting in significant fuel savings over the lifetime of the aircraft.

The practical application of riblets to an aircraft has been achieved through the use of special adhesive films such as those developed by the 3M company. Both Boeing and Airbus have tested riblets the most recent study being the application of riblet coatings to approximately 60% of the surface of an Airbus A340 in collaboration with Lufthansa/Cathay Pacific Airlines. The results have suggested an overall drag reduction in the region of 2% and that some of the issues related to contamination and degradation of the riblet material were less serious than originally envisaged. These latest tests have concluded that with current fuel prices and the cost of installing and maintaining the riblet coating the performance benefits are marginal. Any further significant rise in oil prices could, however, make the technology economically viable.

New approaches using three-dimensional riblet shape are also currently being investigated by a number of researchers[4].

Vortex Generators

Vortex generators (VGs) are devices that have always been considered as a means of fixing an unforeseen problem encountered during flight test. They have been considered synonymous with poor initial aerodynamic design. In general, VGs are used to delay flow separation occurring on various parts of an aircraft in order to reduce separation drag improve maximum lift or to solve a control problem. Pearcy[5] described the general aerodynamic principles associated with conventional VGs in 1961. Conventional VGs are typically vanes arranged in co- or counter-rotating arrays having a height at least equal to that of the local boundary layer. Since they are only generally required during short phases of the flight envelope they constitute a significant source of parasitic drag during cruise. Although VGs feature on most, if not all, Boeing aircraft Airbus philosophy has always been to refine the design of the wing so that it can operate without aerodynamic add-ons. The driver behind this philosophy has always been the desire to minimise excrescence drag and to have an efficient, clean, well-engineered, advanced technology design.

Although VG technology has advanced little over a number of decades there has recently been a renewed interest shown in this technology. This has come about for a number of reasons: Firstly, experimental and numerical evidence is beginning to show that, with good design and careful consideration of the fluid dynamics aspects, VGs having heights considerably less than that of the boundary layer height can be as effective as much larger devices. For example, studies at NASA[6] and at DERA/BAE SYSTEMS[7] suggest that device heights may only need to be 0.2 of the boundary layer height. Much reduced device sizes lead to significantly reduced parasitic drag; Secondly, there has been renewed interest in applying flow control devices at the early stages of design in order to arrive at more optimised designs with regard to wing size. For example there is renewed interest in improving the performance of high lift systems. Research is ongoing in the USA and the UK to study the use of sub-boundary layer vortex generators to obtain simpler, lower cost, lower weight flap systems. Another potential area of application for VGs is the control of shock/boundary layer interaction. Preliminary experimental and numerical work to investigate novel VG configurations has proven the viability of such applications. Further work is required to understand the impact of VGs on the aircraft design optimisation process. In order to accomplish this properly design/analysis tools are required to be able to examine the effects of VGs on the aerodynamic performance of three-dimensional configurations in a design environment. Methods must be accurate, reliable, and quick to apply. CFD has shown itself to be capable of modelling the effects of VGs on a boundary layer but as yet most studies have been restricted to simple two-dimensional flows and simple arrays of VGs[7]. Full Reynolds Averaged Navier-Stokes (RANS) simulations of entire wings complete with VG arrays are still a long way from practical realisation due to the extremely large grid sizes required to model the very small VGs. Attempts have been made[8] but the computational costs are excessive. A simpler approach is required to enable the main effects of VGs on a boundary layer to be modelled in either integral methods or in full RANS simulations. Such models must take account of the effect of the influence of the VG on the boundary layer within pressure gradients and perhaps be capable of coping with some level of flow separation. While sub-boundary layer VGs will have reduced parasitic drag influence.

There is still interest in having VG configurations that can be switched on and off according to the prevailing flow conditions. This could result in the total elimination of any adverse effects that may be experienced during either cruise or takeoff for example. In this field research has been going on for a number of years to study the effects of air-jet VGs, either continuous blowing[9] or pulsed[10]. Other possibilities are the use of solid-state actuators employing piezoelectric or shape-memory alloys to deploy VG arrays according to the prevailing flow conditions. Significant further research and development effort is required to bring such technologies to maturity.

4. Active Turbulence Control

Active turbulent flow control is a currently emerging field of technology and can be divided into two areas: The first involves a marriage of evolving smart materials and structures with variable, quasi-steady geometry. Examples include active vortex generators, pneumatic/actuated strake- forebody vortex control and "mission adaptive wings". The second field is highly speculative and is based on the vision of applying highly integrated smart materials to controlling highly non-linear dynamic systems with the objective of controlling coherent turbulence structures to effect drag reduction or separation control. MEMS is one of the current high profile technologies in this area. Due to space limitations within this paper it is only possible to mention a very small fraction of the activities in the field of active flow control. It is therefore only intended to present a few examples in order to draw a few key conclusions.

Microblowing

The microblowing technique[11] is a new technology being developed at the NASA Glenn Research Centre (US Patent No. 5,803,410). Turbulent skin friction drag reduction is obtained blowing a very small amount of air vertically through specially designed porous surfaces. Fundamental experiments have demonstrated turbulent skin friction drag reductions of between 50% and 70% on a flat plate at subsonic speeds and reductions of up to 80-90% at supersonic Mach numbers. The technique is purported to work by reducing the gradient of the velocity profile at the wall. However, it is suggested that the fundamental reasoning could well be connected to the displacement of near wall coherent structure activity away from the wall. A flight evaluation of the micro-blowing technique is to be conducted at the NASA Dryden Flight Research Centre on the F15B flight test fixture in 2000. It is hoped to demonstrate the technology at aircraft scale Reynolds and Mach numbers under real atmospheric conditions. It is proposed to use ram-air as the source of blowing.

MEMS

At the present time MEMS and turbulence control are seen as the "Holy Grail" of fluid mechanics. The term "Microfabricated-Electro-Mechanical Systems" (MEMS) embodies an area of technology in which sensors, actuators and electronics may be manufactured on a common substrate using integrated circuit fabrication techniques and compatible bulk and surface micro-machining processes. MEMS devices, including micro-sensors and micro-actuators, are attractive because they can be made small (characteristic dimensions may be between a few microns and a few millimetres), be produced in large numbers with uniform performance, include electronics for complex functionality, and be inexpensive. Their application in everyday use is increasing, for example, low cost micro-accelerometers used in the triggering mechanisms for automotive airbags. One potential application for MEMS technology is the control of fluid flows through the active manipulation of the coherent structures that develop in the near-wall region of a boundary layer.

Research into the application of MEMS for flow control has been ongoing in the USA for a number of years. Significant activity exists at a number of US universities where financial support from the Government via the AFOSR, among others, is being directed[12]. Recent experiments and simulations in the USA have demonstrated the fundamental feasibility of active boundary layer control. A Group at Stanford has developed actuator arrays using a combination of micromachining technologies together with mesoscale assembly techniques. Researchers at CalTech have designed, fabricated and evaluated microfabricated shear stress sensors, intended to resolve both temporally and spatially the small-scale streaks in the near-wall region of a turbulent boundary layer. These sensors have been used to obtain the correlations required to design real-time flow control logic which is part of a MEMS-based neural network system that is to be evaluated for active turbulent shear-stress control.

Until recently little coordinated research to evaluate the application of MEMS for flow control has taken place within Europe. In view of this a three-year, CEC supported "basic research" project, AEROMEMS, was set up at the end of 1997 to assess, quantify and demonstrate the viability of employing MEMS for a medium term aerospace flow control application

The flow control application to be addressed was selected with regard to the objective of ensuring the industrial requirements of practicality, cost and commercial benefit. The active control of a turbulent boundary layer to reduce skin friction drag is perhaps the most ambitious of all flow control objectives. However, this application would require huge arrays of sensors and actuators with all the attendant interconnection and processing problems. For many reasons it is unlikely to be commercially realistic to coat large areas of an aircraft's surface with high densities of MEMS devices for many decades. For nearer term applications it is therefore necessary to restrict considerations to those flow control applications which can make use of small numbers of actuators and sensors. One potentially nearer term application is the use of MEMS to suppress flow separation.

The detailed objectives of the project are to:

- Identify the basic design requirements of the appropriate MEMS hardware and the control strategies required for a successful commercial application.
- Demonstrate under laboratory conditions as representative of full-scale conditions as possible, the application of MEMS to the control of flow separation on a wing.
- Demonstrate that the concepts have the potential for viability with regard to the demands of commercial application (Physical robustness in an aircraft operating environment, maintainability and repairability).
- Identify if the hardware technology could be matured for industrial application within the medium term and to propose an outline plan for the commercial exploitation programme.

As the project draws to a close MEMS sensor and actuators have been developed and are about to be tested in a wind tunnel experiment to demonstrate separation delay on a high lift system. Simulated MEMS actuators have already been experimentally shown to effect separation delay on the demonstration model to achieve a 0.1 increase in maximum lift coefficient.

Concluding Remarks

This paper has described a wide range of both passive and active turbulent flow control technologies. Some have been with us for many years while others are in their embryonic state relying on modern advances in the understanding of fluid mechanics and the development of analytic/numerical tools and materials technology. Although there are important activities going on to investigate the application of new technologies such as MEMS for active flow control it is unlikely that any will be mature for application, other than as a component of very simple systems, within the next decade or two. However, there is still great potential for the application of passive or semi-active technologies and the renewed interest in vortex generators is one example of this.

Whether or not any of the flow control technologies will see a widespread application on a production aircraft in the future depends on two important factors:

- Does it work?
- Does it make practical economic sense?

The first question is probably the easiest criteria to address. In the first instance any new technology has to be demonstrated experimentally at large scale under relevant Reynolds and Mach numbers. It must then be proven to operate in the presence of real world conditions and shown to enhance or improve a valuable performance metric. Any system must be demonstrated to be manufacturable, robust, reliable, maintainable and inspectable. Any performance side effects must be acceptable

The second criterion is more difficult to assess and the answer may change with evolving worldwide economic market and political conditions. To be viable any flow control application must have a favourable overall cost/benefit. The initial design/integration cost combined with the cost of energy expenditure during operation must be outweighed by any benefits obtained. Issues related to the legal/regulatory standpoints of product liability, safety, and environmental/acoustic pollution also need to be resolved.

Although many flow control technologies have been identified and researched at the basic level for many decades few have ever reached maturity and full-scale deployment on a commercial product. It could be argued that the basic research community is not sufficiently aware of the practical issues of implementation and that in some cases non-application useful research is carried out while in other areas the research is not carried far enough to allow technological evaluation. With the limited research investment available it is becoming essential that the research community work more closely with the application community to develop practical technologies in the most efficient manner. This requires all involved to work in a multidisciplinary environment to develop the basic tools and understanding and then to progress this towards large-scale demonstration and evaluation. It is important that unworkable concepts are filtered at the earliest possible opportunity in order to avoid unnecessary effort being directed at hopeless causes. Early in the assessment process it is essential that industrial studies be made to identify the potential benefits and practical implications of any new flow control technology. In order to do this industry requires robust, rapid tools with which to undertake aerodynamic analysis.

Turbulent flow control should not just be considered as a fix to solve a problem or a means of improving even further the performance of an already optimised design. Consideration should be given to employing turbulent flow control early in the design optimisation process. A commercially better design may be obtained by the use of flow control to recoup performance losses associated with simplifying other aspects of the design to reduce manufacturing costs, system complexity or structural weight (e.g., reduced sweep, thicker wings, smaller, simpler high lift systems).

Research into turbulent flow control within Europe is still strong and despite the much greater activity going on in the USA is still able to compete in the international arena. However, it is fragmented and would benefit from the greater cohesion that could be afforded by focussed research initiatives involving the basic research community and industry across a broad range of disciplines.

References

1. Kim, J.J., Study of Turbulence Structure through Numerical Simulations: The Perspective of Drag and Reduction, in AGARD Special Course on Skin Friction Drag Reduction, AGARD-R-786, March 1992.

2. Walsh, M.J., Drag characteristics of V-groove and transverse curvature riblets, in Viscous Drag Reduction, Washington, DC: AIAA.168-184, 1980.

3. Bechert, D.W. et al, Experiments on drag-reducing surfaces and their optimisation with an adjustable geometry, J. Fluid Mech. 338:59-87, 1997.

4. Sirovich, L. Karlsson, S, Turbulent drag reduction by passive Mechanisms, Nature 388: 753-755, 1997.

5. Pearcey, H.H., Shock induced separation and its prevention by design and boundary layer control, in Boundary layer Control, its Principles and Application, Vol 2, Lachmann, G.V. Ed, 1961.

6. Lin, J.C. et al., Separation Control on High-Lift Airfoils via Micro-Vortex Generators, J Aircraft Vol. 31 No. 6, 1994.

7. Ashill, P.R., Sub Boundary Layer Devices to Control the Flow on Swept Wings, RAeS Aerodynamics Conference 2000, London, 17-18 April 2000.

8. Yu, N.J., Computational Simulations of a Commercial Airpkane Configuration with Vortex Generators, RAeS Aerodynamics Conference 2000, London, 17-18 April 2000.

9. Walsh, M.J., Drag characteristics of V-groove and transverse curvature riblets, in Viscous Drag Reduction, Washington, DC: AIAA.168-184, 1980.

10. Tilmann, C.P. et al, Characterisation of Pulsed Vortex Generator Jets for Active Flow Control, RTO AVT Symposium Braunschweig, Germany, 8-11 May 2000.

11. Hwang, D.P., An experimental Study of Turbulent Skin Friction Reduction in Supersonic Flow Using a Microblowing Technique, AIAA-2000-0545, Dec. 1999.

12. McMichael, J. M., Progress and Prospects for Active Flow Control using Microfabricated Electro-Mechanical Systems, AIAA-96-0306, Jan. 1996.

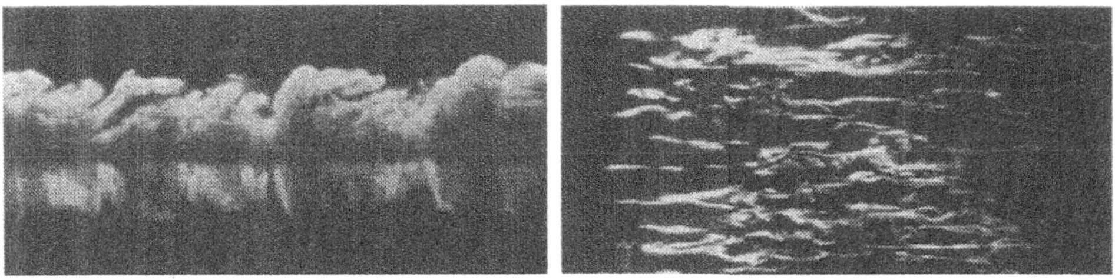

Fig. 1 Coherent Structures in a turbulent boundary layer (Left: Large scale eddies, Right: Near wall streaks).

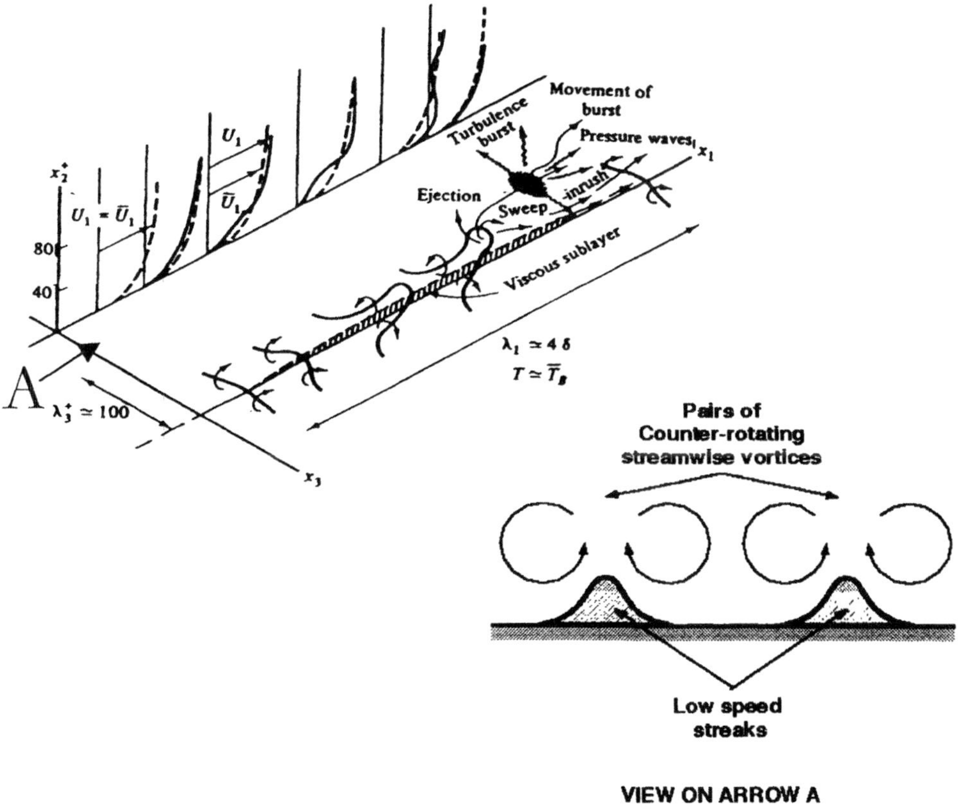

Fig. 2 Schematic depiction of the generation of near-wall turbulence.

Yaw Angle Effects on Optimized Riblets

W. Hage, D.W. Bechert, M. Bruse[1]

German Aerospace Center (DLR), Institute of Propulsion Technology,
Dept. of Turbulence Research,
Mueller-Breslau-Str. 8, D-10623 Berlin, Germany,
e-mail: wolfram.hage@dlr.de
[1]German-Dutch wind tunnels (DNW),Bunsenstr.20, D-37073 Göttingen, Germany

Summary:

The drag reducing effect of riblets, has been demonstrated previously both experimentally and theoretically. Several aspects of the application of riblet film on an long range aircraft are considered. The aim of the present investigation is to check experimentally the effects of an angle of yaw on the performance of riblets. For this purpose, various riblet geometries were tested experimentally. With an increasing misalignment to the flow direction the drag reduction ability of the riblets deteriorates. This effect is related to the geometry of the riblets, e.g., a riblet geometry with trapezoidal grooves and a tip angle of the ribs of $\alpha = 45°$ is less affected by misalignment as compared to other geometries.

Keywords: riblets, drag reduction, yaw effects

1. WALL SHEAR STRESS REDUCTION

Drag reduction of a moving vehicle can be achieved by wall shear stress reduction and, in some cases, by separation control. This paper will focus on turbulent wall shear stress reduction. Obviously, laminar flow offers the lowest attainable wall shear stress. It has been introduced successfully on glider aircraft wings. On commercial aircraft, however, the problems associated with the implementation of laminar flow on swept transonic wings are at present not yet completely solved. However, even with laminar flow being maintained on a major percentage of the wings, most of the aircraft surface will still be exposed to turbulent flow. Thus, turbulent wall shear stress reduction is obviously important in all scenarios of future aircraft design.

1.1. The mechanism of drag reduction

More than a decade ago, it had become clear that a turbulent boundary layer on a surface exhibiting longitudinal ribs can develop a lower shear stress than that on a smooth surface (Walsh 1980 & 1990, Nitschke 1984). In the following, the drag reducing effect of this ribbed structure is outlined. In order to obtain generally valid data, quantities are defined in dimensionless wall units. For instance, the spacing s between two ribs will take the form $s^+ = s \cdot u_\tau/\nu$. This is a Reynolds number defined with the velocity $u_\tau = \sqrt{\tau_0 / \rho}$, where τ_0 is the reference wall shear stress of a smooth surface exposed to the same flow conditions as the ribbed surface. ρ and ν are density and kinematic viscosity, defined as usual. In normal circumstances, u_τ assumes a value of a few percent of the free-stream velocity.

The strong exchange of momentum in a turbulent boundary layer is produced by high speed flow lumps approaching the surface ("sweeps") and by low speed flow moving away from the surface ("ejections") into the high speed regime of the flow. Regions of impinging flow

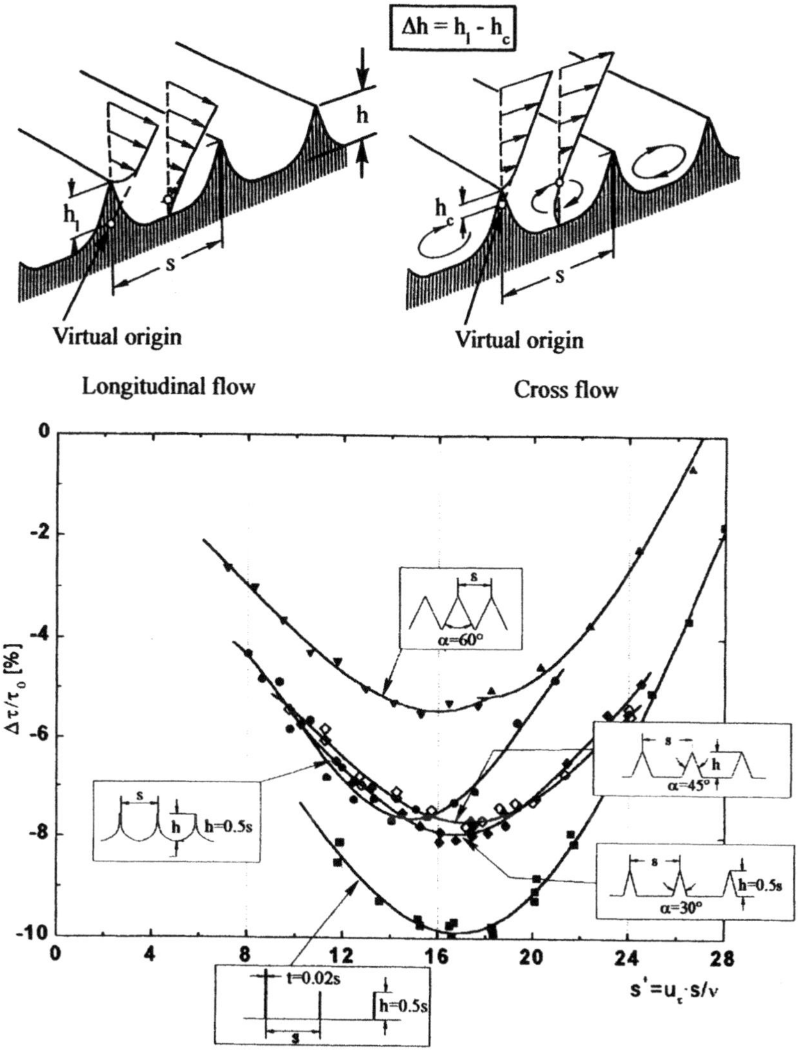

Figure 1: Longitudinal and cross flow on a ribbed surface (above), drag reduction performance of various rib geometries (below)

("sweeps") mostly occur in regions of elevated instantaneous pressure, whereas "ejections", i.e., where the flow moves away from the surface, correspond mostly to regions of lower pressure. This exchange of fluid normal to the surface generates the enhanced shear stress of a turbulent flow because high speed flow is decelerated efficiently when it is swept towards the surface. By contrast, such an exchange normal to the surface does not occur in a laminar flow where the streamlines are essentially parallel.

It is also obvious that local events such as "sweeps" and "ejections" require fluid motion in the lateral z-direction. Thus, hampering w-velocities in the z-direction will reduce momentum transfer and skin friction. It is plausible that that can be achieved with ribs on the surface which are aligned in the mean flow direction. On the other hand, it is known that surfaces exhibiting protrusions higher than about $y^+ \approx 5$ do increase the wall shear stress (Schlichting 1979). However, for protrusions smaller than $y^+ \approx 3\text{-}5$, ribs or other roughnesses are imbedded in the viscous sublayer. In this layer, very close to the wall, any fluid behaves like a highly viscous fluid. Therefore, it is admissible to calculate the flow around very small ribs with a viscous theory. Under viscous flow conditions, it turns out that the ribbed surface appears as a smooth surface located at a virtual origin, see Figure 1. However, the location of the origin, i.e., its elevation above the groove floor depends on the flow direction. For cross flow on the ribs, the virtual origin is closer to the rib tips than for longitudinal flow. The difference between these two heights, Δh, we call the "protrusion height difference". The existence of this difference has interesting consequences: If a flow motion is generated by a fluid lump in a plane at a height y^+ above the surface, the fluid lump experiences a higher resistance if it moves laterally than if it moves in the longitudinal direction. In this way, the cross flow is hampered by the ribs, as desired, and thus ribs do indeed reduce momentum transfer and shear stress. Therefore, for an optimization of the shear stress reduction, we have to select a ribbed surface which exhibits a maximum difference of the virtual origins for longitudinal and cross flow. As the theory shows (Luchini et al.; 1991), the maximum height difference is obtained for very thin blade-like ribs. The rib height should be $h \geq 0.6$ s, where s is the lateral rib spacing. For this value of the rib height, the maximum difference of the elevation of the origins, or the protrusion height difference is $\Delta h_{max} = 0.132$ s. Experiments in the Berlin oil channel (Bechert et al.; 1992) with an adjustable blade rib height (Bechert et al.; 1997) have shown that the optimal rib height is actually slightly lower, at $h = 0.5$ s. With this surface, a turbulent shear stress reduction of 9.9 % below that of a smooth surface has been achieved (Bechert et. al 1997).

Figure 1 shows the data of optimal blade rib surfaces as compared to riblets with triangular and semicircular cross-section. Clearly, it is difficult to manufacture surfaces with blade ribs. Therefore, we devised wedge-like ribs which still produce an impressive wall shear stress reduction, see Figure 1. Ribbed surfaces such as the ones shown in Figure 1 perform well within a certain s^+ range, which corresponds to a particular velocity range. In Figure 1, $\Delta\tau$ is the difference between shear stress τ on the ribbed surface and τ_0 on a smooth reference surface, i.e., $\Delta\tau = \tau - \tau_0$. Negative values of $\Delta\tau/\tau_0$ refer to drag reduction and positive values to an increase of drag.

2. TECHNOLOGICAL APPLICATIONS ON AIRCRAFT

At present, riblet film is on its way to implementation on long-range commercial aircraft. In the following, the application of riblet film on an Airbus A340-300 is considered. On long-range aircraft the fuel costs contribute perceptively to the direct operating costs and the fuel weight exceeds by far the payload. The contribution of the skin friction to the total drag of this aircraft is about 48 %. This skin friction contribution can be reduced with the optimized riblets with trapezoidal grooves by 8.2 %. If the whole aircraft could be covered with riblet film, that would yield a total drag reduction of the aircraft of about 4 %. For various reasons, not the whole surface of the aircraft can be coated with riblet film: Damage by dust does not permit to cover the leading edges of the wings and the vicinity of the landing gear. In addition, at the leading edges of the wings, the riblet film would interfere with the de-icing system and it would be incompatible with the laminar flow there. Locations where fuel and/or hydraulic fluid may come in contact with the plastic riblet film should also be avoided. Obviously, the windows cannot be covered either. Thus, only about 70 % of the aircraft can be coated with riblet film. In addition, in some places, riblet spacing and alignment will be suboptimal.

On the other hand, some of the roughnesses on the aircraft surface will be covered and thus be smoothened by the film. Moreover, a reduction of the wall friction ensues a slightly thinner boundary layer which, in turn, causes a reduction of the form drag in the rear part of the fuselage. This means that a 3 % reduction in total drag of the aircraft is probably achievable.

The weight of the riblet film is in the order of the weight of the paint which it replaces, i.e., 100-250 kg, depending on the percentage of the surface being covered. Additional technological issues, like durability, UV radiation tolerance, etc., have been solved in the meantime. An additional advantage of the riblet film application is the possibility to coat the aircraft in steps parallel to other mechanical or maintenance work. Furthermore, if the riblet film replaces the paint, the time required for painting and the time for attaching the riblet film do not differ significantly. Moreover, the producer of riblet film (3M, St. Paul, Min., USA) has developed a riblet film which can be removed more conventiently than paint. The perspective of having an aircraft with a dirt-repellent film surface, (which was observed since the riblet film was attached to a long-range aircraft), may be an additional attractive item in terms of maintenance costs.

Currently, the share of the fuel costs is about 30 % of the direct operating costs. Thus, one would save about 1 % of the direct operating costs by a 3 % reduction of the fuel consumption (which would be attributable to a 3 % total drag reduction). More important, however, is that about 2.4 tons of fuel can be replaced by additional payload. Therefore, the latter could be increased by 5 %.

2.1. Yaw angle effects on different riblet geometries

Due to the technical constraints mentioned above, only 70% of the aircraft surface can be coated with riblet film. Therefore, it is desirable to obtain the maximum performance of this riblet film coating. Previous work has shown that it is essential to align the riblet film well with the flow pattern on the aircraft in order to obtain an optimal drag reduction (Viswanath 1999, Gaudet 1989).

It is important to know the deteriorating effect of misalignment because deviations in flow direction do occur on the fuselage at varying angles of attack of the aircraft. in addition, for practical reasons in connection with the application of the riblet film, an optimal alignment may not be achieved everywhere on the surface of the aircraft.

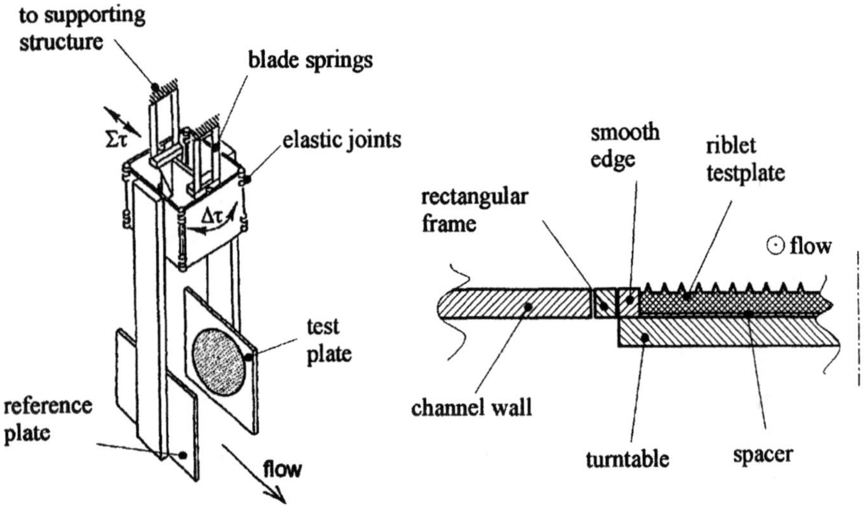

Figure 2: Experimental setup: differential shear stress balance (left), sectional view of test plate (right)

With the present experiments, the influence of the yaw angle on riblets with different riblet geometries is examined. Four different riblet designs, which are differing in the shape of the valleys between the ribs were tested: V-grooved riblets (developed by M.J. Walsh, NASA Langley), one riblet geometry with semicircular grooves between the ribs and two different optimized trapezoidally grooved riblets (see Figure 1).

The experiments were carried out in the Berlin oil channel. The various riblet test plates were mounted on a turntable, which was inserted into a rectangular test plate. (see figure 2) The angle of attack of the riblet test plate to the flow was varied during the experiments by up to ± 90°. With the differential shear stress balance we compare directly the forces which occur on the test plate and on a smooth reference plate. Thus we obtain a very high measurement accuracy of ± 0.3% (Bechert et al.1992). Figure 3 shows the drag data from the experiment. For a better comparison, the <u>losses</u> of the drag reduction are plotted versus the angle of yaw. It is worth mentioning, that these losses are based on the maximum drag reduction of the related riblet geometry, e.g., a loss in drag reduction of 10% means that the absolute drag reduction for the triangular riblets is reduced from 6% to 5.4% and that that of the trapezoidally grooved riblets (tip angle $\alpha = 30°$) is reduced from 8.2 to 7.4%.

With increasing yaw angle, one observes a degraded performance of the riblet surface. It turned out, that the sensitivity to the crossflow depends strongly on the riblet geometry. The trapezoidally grooved riblets ((a) and (b) in Figure 3) are less sensitive to misalignment as compared with the triangular or semicircular grooved riblets ((c) and (d) in Figure 3).

The geometry which is the least sensitive to misalignment is the one with trapezoidally grooved riblets with a tip angle of $\alpha = 45°$. For smaller yaw angles of up to 10°, the losses in drag reduction seem to be independent of the chosen nondimensional lateral riblet spacing s^+. This latter statement is valid for the s^+-range between 13 and 19, i.e., near the optimum of the riblet performance.

A similar behavior is found for trapezoidally grooved riblets with a tip angle of $\alpha = 30°$, exhibiting a slightly inferior resistance to misalignment. Compared to the data of the trapezoidally grooved riblets with a tip angle of $\alpha = 45°$, the sensitivity to misalignment depends more strongly on the chosen s^+. Furthermore, the losses in drag reduction are slightly

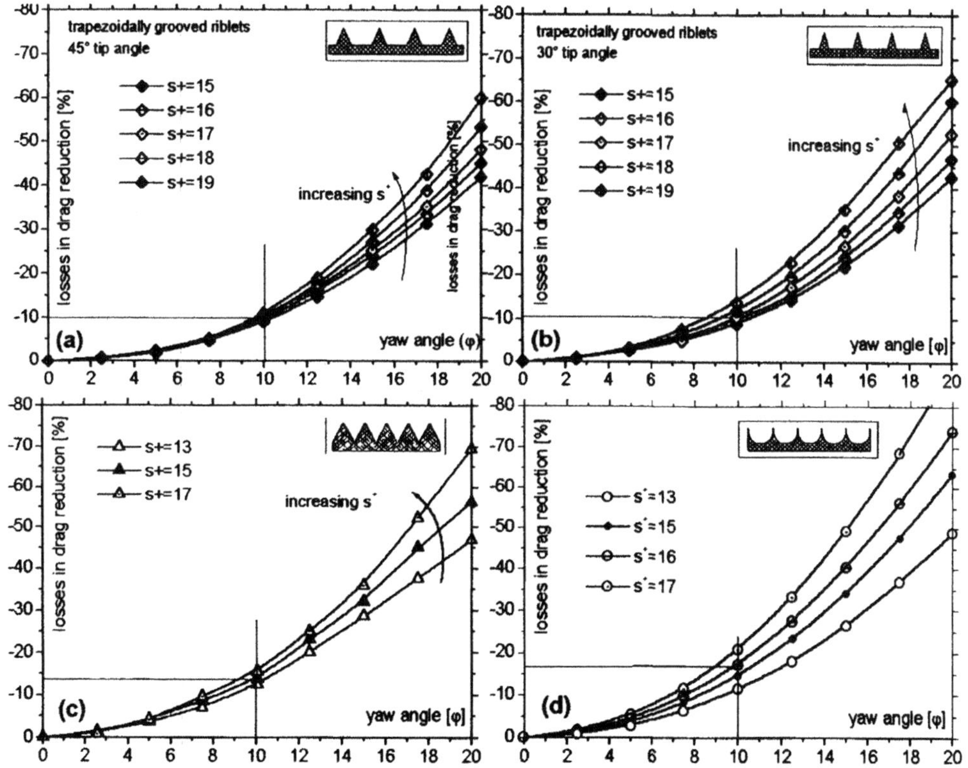

Figure 3: Drag data of various riblet geometries (a) trapezoidally grooved riblets, tip angle 45° (b) trapezoidally grooved riblets, tip angle 30° (c) triangular grooved riblets (d) semicircular grooved riblets.

higher than those of the geometry with a tip angle of $\alpha = 45°$.

The third riblet geometry with the triangular grooves is more effected by misalignment than the latter ones, the fan of data curves is shifted upwards to higher losses. Even at smaller angles of yaw, losses in drag reduction do occur. These losses start at lower angles of yaw as compared to the trapezoidally grooved riblets. These losses do also depend on the selected s^+: with increasing s^+, a lower drag reduction occurs.

This behavior is even more prominent when one is looking at the data of the semicircular grooved riblets. Even for small angles of yaw larger losses in drag reduction are found, and again they are related to the selected s^+.

The fact, that for smaller s^+ the drag reduction is less sensitive to misalignment might be an important aspect when the appropriate rib spacing is selected. If the chosen rib spacing s^+ is smaller than that of the maximal drag reduction, a larger tolerance to misalignment is obtained. On the other hand, for a rib spacing s^+ being larger than that at which the maximum in drag reduction occurs, the riblets will be more affected by misalignment.

Two geometric quantities seem to be responsible for the sensitivity to misalignment: the ratio of rib height to rib spacing h/s and the shape of the cross section of the valley between two ribs. For a given ratio of $h/s = 0.5$ of the two trapezoidally grooved riblets the one with the larger tip angle of $\alpha = 45°$ performs a little better if being compared to that with a tip angle of the ribs of $\alpha = 30°$. With a growing ratio h/s of triangular shaped riblets and of riblets with the semicircular grooves with $h/s = 1$ and $h/s = 0.7$, respectively, an even higher sensitivity to misalignment occurs. It seems also to be important how the floor of the valley between the ribs is shaped. The riblets with the triangular grooves do perform slightly better in terms of insensitivity to misalignment as compared to the semicircular grooved ones, although they have the larger depth of the valley between the ribs. Obviously, the shape of the triangular groove is very different from that of the semicircular one. It has the shape of a narrow notch while that of the latter one has a comparable broad semicircular floor. One likely reason for the sensitivity to misalignment of the different rib geometries is the phenomenon, that the fluid in the valleys between the ribs tends to slosh. Sloshing is a lateral movement of fluid in between the ribs, which produces a motion normal to the wall. This additional momentum exchange is responsible for additional losses in drag reduction. This particular type of sloshing was first described by Clark (Clark 1989) and was made visible with a low Reynolds number flow visualization. It is plausible, that the appearance of sloshing is more likely for broader valleys. This is also found for the two trapezoidally grooved riblets (a) and (b) in Figure 3, where the latter ones with the smaller tip angle of $\alpha = 30°$(which, of course, means better drag reduction, see Figure 1) shows a larger tendency to loose that drag reduction ability under the influence of a non-parallel flow.

REFERENCES

(Bechert et al., 1997) Bechert, D.W.; Bruse, M.; Hage, W.; van der Hoeven, J.G.Th. & Hoppe, G.; "Experiments on drag-reducing surfaces and their optimization with an adjustable geometry"; J. Fluid Mech., 338, pp. 59-87.

(Bechert et al., 1992) Bechert, D.W.; Hoppe, G.; van der Hoeven, J.G.Th. & Makris, R; "The Berlin oil channel for drag reduction research"; Exp. in Fluids, 12, pp. 251-260.

(Clark1989) Clark, D. G. Boundary Layer Flow Visualization Patterns on a Riblet Surface, QMC-EP-1081, Faculty of Engineering, Queen Mary College, University of London 1989

(Gaudet 1989) Gaudet, L.; "Properties of Riblets at Supersonic Speeds", Applied Scientific Research, Vol. 46, No. 3

(Luchini, et al., 1991) Luchini, P.; Manzo, F. & Pozzi, A.; "Resistance of a grooved surface to parallel flow and cross-flow"; J. Fluid Mech., 228, pp. 87-109.

(Nitschke, 1984) Nitschke, P.; "Experimentelle Untersuchung der turbulenten Strömung in glatten und längsgerillten Rohren"; Max-Planck-Institut für Strömungsforschung, Göttingen, Bericht 3/1983.(1983). Trans.: "Experimental investigation of the turbulent flow in smooth and longitudinally grooved tubes. NASA TM 77480 (1984).

(Schlichting, 1979) Schlichting, H.; "Boundary layer theory"; Transl.: J. Kestin, 7th Edition, McGraw-Hill, New York.

(Viswanath, 1999) Viswanath, P.R.; "Riblets on Airfoils and Wings: a Review"; AIAA paper 99-3402

(Walsh, 1980) Walsh, M.J,; "Drag reduction of V-groove and transverse curvature riblets". In: "Viscous flow drag reduction" (ed. G.R. Hough), Progress in Astronautics and Aeronautics, Vol. 72, pp. 168-184, AIAA, New York.

(Walsh, 1990) Walsh, M.J.: "Riblets", In: "Viscous drag reduction in boundary layers" (eds. D.M. Bushnell & J.N. Hefner), Progress in Astronautics and Aeronautics, Vol. 123, pp. 203-262, AIAA Washington.

Effects on the Resistance and on the Separation of V Shapes Passive Manipulators in a Turbulent Boundary Layer

Rossella Monti [1], Sergio De Ponte [2], Eugene Levich[3]

[1] Terr@A, H.R.C.E., Dept. IIAR, Politecnico di Milano, pza L. Da Vinci 32, 20133 Milano, Italy.

[2] Politecnico di Milano – Dip. di Ingegneria Aerospaziale via Lambruschini,I-20158 Milano, Italy.

[3] Constellation 3D Inc., 230 Park Avenue, suite 453, New York, NY 10169 & Orlev Sci. Com. LtD, div. Ormat Ind. LtD, p.o.b. 68, 81100 Yavne, Israel.

Summary

In the present investigation surfaces with passive manipulators randomly distributed are studied both in an equilibrium turbulent boundary layer with slightly favourable pressure gradient and in decelatered flows. The main aim of the work is to evaluate the effectiveness of the manipulators for drag reduction in a turbulent boundary layer and their behaviour in the limiting strong adverse pressure gradient that characterizes the conditions for maximum pressure recovery. The manipulators consists of a pseudo random 3D Vshaped roughness whose geometry and distribution are function of the Reynolds Number. The drag variations was estimated by direct comparison with measurements conducted with smooth boundaries. What we found is a significant stable drag reduction (up to a maximum of 30%) in a narrow range of Reynolds Numbers, and a small effect on separated flow.

Introduction

The researches collected in the last 20 years show that it exists the possibility to reduce the skin friction drag in turbulent flows. Nevertheless it seems that we are still far from an industrial technological solution. Big efforts were put on riblets technology and it was demonstrated in a large number of experiments that it is a successful method to reduce drag. Neither this, at present, any real application does not exist at an industrial scale. Probably because they are expensive in comparison with the benefit and rather delicate from a structural view point (Carpenter, 1977). In general terms, the methods of reducing drag are based on the idea of controlling the bursting. The key question is what cause the bursting. There are several different models and as a consequence, several methods of controlling turbulence generation. The manipulators we tested are based on the idea that the energy dissipation in turbulent flows strongly depends on the non-linear interaction of some particular coherent structure near the wall. This is described as the main reason for intermittency by Levich (1987) and Levich and Malomed (1994). The structures in the boundary layer were identified (Sirovich et al., 1990) in the transient region between the linear distribution of the velocity and the logarithmic profile. Physically this region, called "buffer zone", is extremely thin; nevertheless it plays a fundamental role in the energy exchange among the different zones of the flow field.

Theoretical studies (Moffatt, 1969, 1975, 1985; Levich et al., 1991) demonstrated that the energy flux is delicately correlated with the flux of helicity H from the large scales of motion to the dissipative scale. In physical terms the flux is possible only if a gradient exists. It is known from the experience that in bounded turbulent flows the energy flux is mainly directed from the outer region to the wall. It seems also that the energy flux correlates with the helicity flux (Levich, 1996). In the center region of the flow, the energy is transmitted to the large scale eddies from the mean flow. These eddies are themselves unstable and decompose on smaller eddies. Referring to Kolmogorov, the decomposition of vortices into smaller ones is always accompanied by growth of vorticity or enstrophy. Such growth occurs by the mechanism of vortex lines stretching, well known in hydrodynamical turbulence. This cascade process has the additional effect in turbulent boundary layer of bringing the fluctuations into closer and closer contact with the wall, while their vortex lines are more and more stretched.

Approaching the wall, as expected and as verified numerically by Shtilman et al. (private communication, 1995), the helicity density $h=\omega \bullet v$ growths, reaching its maximum approximately in the buffer zone.

The exchanges of helicity and energy among the different regions of motion are accelerated in a turbulent flow by the so-called "burst activity". The bursts are associated with concentrated injections of helicity in the outer region accompanied by mass transfer.

The burst activity, which is a typically intermittent phenomenon, is due to the non-linear interaction among a variety of coherent structures in the buffer region (Robinson, 1991). In particular Sirovich et al. (1990) showed that the Reynolds stresses depends mainly on the interaction between two structures: the rolls (which are the primary vortices) and the oblique propagating waves (weak oblique plane waves propagating downstream at an angle of about 40-60°).

From a topological viewpoint, the phenomenon of intermittency is connected with the mechanism of the helicity transfer from the large scale structures to the smaller ones (Levich, 1987; Levich and Malomed, 1994; Levich, 1996). In fact, the rolls are quasi-steady counter-rotating vortices and are typical helical structures with high energy content (nearly 80% of the turbulent energy is stored in the rolls) (Schoppa and Hussain, 1998). When bursts occur the roll structures are destroyed with subsequent release of high energy and helicity content. The bursts occurrence is often accomplished by negative helicity spectrum (Levich, 1996). Roughly speaking in the physical space, this means a transfer of helicity to the wall i.e. an instantaneous vorticity transfer to the wall. In other words, the turbulence activity tends to redistribute the vorticity in the boundary layer in similarity as it happens during the transition to turbulence (albeit on a longer time scale). It was observed that 95% of the vorticity has moved closer to the wall, much of it very close indeed (Passoni, 1999).

Furthermore, as reported by Dracos et al. (1989), in conjunction with the bursts occurrence, the Reynolds stresses exhibit a strong intermittent behaviour. Their peaks occur when the non-linear interaction between the rolls and the propagating modes determines the destruction of the most energy bearing structures, i.e., the rolls themselves. This means that a possible way for the reduction of the energy dissipation is the destruction of the topological coherence between the wall structures.

Because of the previous considerations, the drag reduction can be achieved by two different approaches: the excitation of few selected modes, thus forcing the coherence of the rolls by superimposed boundary conditions as usually done in the "riblets" technology (Luchini et al., 1991) or, as in this new perspective, destroying the rolls themselves (Levich, 1996). This last method means the application of a proper random disturbance in order to avoid or to limit stochastically the storage of energy in the near wall longitudinal structures. Handler et al. (1993) demonstrated in direct numerical experiments that artificial randomisation of the phases of appropriate velocity harmonics (responsible for the coupling of rolls and oblique waves) results in reduction of bursting and consequently in a drag reduction up to a 60%; an effect comparable to dilute polymer solutions in wall bounded flows. When bursts occur under such conditions the peak values of the Reynolds stresses are much lower and statistically the interval of quiescence between them is much longer. Ostensibly because the energy content of the rolls has been minimised, thus reducing the average shear stresses at the wall.

On the other hand, it was shown that arbitrary disturbances of all other groups of modes results in no effect or into drag increase. Therefore it is possible to conjecture that probably all methods of drag control in turbulent flows are connected with disturbance of the natural coherence of major propagating wave packets in turbulent boundary layer.

From the engineering viewpoint this problem is not easy to solve and different solutions were proposed. Among these dynamical systems that generate disturbances as indicated by theoretical considerations and numerical simulations. At the same time it appeared that the enhancing of the instability of the rolls could be done quite easily by laying a specific random pattern on the wall surface that enforces the decomposition of the rolls through generation of propagating wave packets with random phases. Based on the above concepts then it appeared the possibility to control turbulence by passive means. Specific patterns acting as protrusions along the solid walls were designed by Orlev Sci. Comp. LtD (patents: Sirovich et al. 1993, 1994, 1996, 1997) and tested in the framework of a contract with Brown University (Rhode Island, USA).

Practically the pattern consists of a pseudo random 3D V shaped roughness whose geometry and distribution are functions of the Reynolds Number. As a result, their distribution over the surface generates oblique waves with random phases, impeding the formation of the rolls and their interaction with the natural coherent-phase oblique waves. Experimental tests with the 3D V pattern were conducted by Orlev Sci. Comp. Ltd in the framework of a contract with Brown University (USA) in a fully developed turbulent channel flow. The experiments showed that the energy losses due to skin friction can be reduced up to 26% (Sirovich and Karlsson, 1997). The head losses were measured by means of pressure drops in a range of Reynolds Number: 15.000<Re<40.000, with Re based on the half height of the channel. In the same experiments it was verified also that using a different distribution of the patterns, i.e. an aligned distribution, in a range of Re 24000-30000, the C_f can be increase of about 50%. A peculiarity in fact of this technical solution is that it is equally suitable for drag reduction and drag increase, depending on the application.

These preliminary results show clearly that the V-shapes technology may be more efficient than every other known method.

Basing on the previously encouraging results, a new series of experiments were undertaken in order to test the 3D V technology in an equilibrium turbulent boundary layer with slightly favourable pressure gradient. This is with the aim of testing the technology in conditions closer to the industrial applications and to identify the criteria for its optimisation depending on the boundary conditions.

Test Series

In particular the following test serie were run:

1) shear stresses measurements (deduced from velocity profiles) with smooth boundaries at different streamwise positions. These measurements are the term of comparison with the analogous measurements conducted with manipulators. Moreover they established the quality of the measurements and of the flow and the accurancy of the post process method. The good agreement with NACA 63-006 data with the present measurements is surprising and probably somewhat fortuitos but it demonstrates at least that our flow is without any unusual properties. In addition, the natural growth of the boundary layer were checked plotting streamwise both the pressure and the integral parameters. Both show an excellent agreement with a linear behaviour as expected in a turbulent boundary layer, with a R squared of about 0.995.

2) the same investigations described at point 1 for smooth boundaries were repeated with manipulators for drag reduction at the same transverse sections.
3) some shear stresses measurements reversing the plates, i.e. orienting the V shapes downstream instead of upstream, were also done in the same range of Reynolds Numbers of the previous series, measuring the velocity profiles at the inlet and outlet of the test section.
4) some rough turbulent power spectra at a given x station, at different distance from the wall both with manipulators and smooth condition.
5) velocity profiles in reattaching conditions.

The manipulators

The manipulators consisted in rows of "V"shaped protrusions from al flat surface, 0.2 mm in high, 60° apex angle, equally spaced in the row and also the rows were equally spaced in streamwise direction. On contrary, the stagger between the rows was "random", according to a gaussian distribution.
At first, three equal plates 0.3 x 0.3 m were obtained by electroerosion on thick (10 mm) aluminium alloy plate, and later on the plates were used as mould for manufacturing longer elements covering other 1 x 0.3 m aluminum plates with shaped epoxy resin.

Test Facility

The preliminary tests were conducted in a boundary layer wind tunnel of 0.3 m width, 0.1 m height and 1.5 m long test section. It was soon recognised that the length was unsufficient and a new tunnel of much long (6 m) and high (0.4 m) test section was built. Probably the manipulators require a rather significant length to produce some significant effect
The tunnel was open circuit and the blower placed downstream.
Pressure taps were placed all along the tunnel to control longitudinal pressure gradient.
Velocity profiles were measured by total pressure probe and wall pressure taps. The probe was driven by a transversing gear and all was monitored and controlled by means of a small computer.
Of course, although the cross section area is constant, the boundary layer growth on the walls gives a negative longitudinal pressure gradient, linear as expected from boundary layer theory.
Furthermore, for evaluating the flow properties in a strong adverse pressure gradient, an insert was placed into the wind tunnel, producing a strong contraction and a following flow deceleration. At the minimum section a slot was placed, in order to blow the boundary layer and to attempt to prevent separation on the wall opposite to the measuring one.
The insert was designed according to the results of East and Sawyer (1979), looking for almost zero skin friction on one-dimensional model.
As told before, we attempted to control the flow by introducing a blowing slot. Here again, the longitudinal pressure gradient was monitored by means of 20 pressure taps along the flow, connected to a Scanivalve system.

Data reduction

Suspecting that, in presence of turbulence manipulator, the law of the wall may not be reliable, it was decided to evaluate the skin friction by the von Karman integral equation. It is sure that the effect of flow convergence due to the sidewall boundary layer is not negligible and in the absolute sense, it introduces a large error: it was therefore decided to compare any times the

results with the manipulators with the one of smooth surface in similar conditions. This was possible due to the modular design of the test section. Because the manipulators in the first stage were obtained by electroerosion on aluminum alloy plates, the reference "smooth" surface was obtained flat with the same technique to get reasonably comparable data.
The possibility of using a "skin friction balance" was not feasible for technical reasons.
The probe was a tube of 1.6 mm outer diameter, flattened to 0.4 mm in direction normal to the wall, as common practice in wall measurements.
In order to use rather correctly the von Karman integral equation it is necessary to evaluate the streamwise derivative of the momentun thickness, which is always a delicate matter, because, although the momentum thickness is evaluated by integration on many crosswise stations, the accidental errors may reduce significantly the final accuracy when we make a numerical derivation. For this reason the derivative was evaluated by difference approximation on different lengths, with the advantage of getting local or more overall skin friction values.

Data results

It might be seen from figure 1 that there is a significant skin friction reduction in a definite Reynolds Number range, although the data dispersion seems to be rather large. This is the first result of this research that confirms in some way the duct flow results. The most important difference seems to be due to the fact that the Reynolds Number at which a significant skin friction reduction is obtained seems to be realted to the streamwise length in some way. Probably it is related to the boundary layer thickness, while in the duct flow the developed turbulent properties are streamwise constant. In fact by integration of the Karman equation with linearized integral parameters over the whole length of the tunnel we obtained a slight drag increase (+10%). Nevertheless, integrating on a shorter distance, a net drag reduction takes place at x=3m, in a well defined range of Re_θ=2500-4500, corresponding to U=10-19 m/s. This behaviour is repeated both choosing as upstream sections x=1m and x=2m, i.e. integrating the Karman equation over a length of 2m and 1 m, respectively. It means that only in certain range of Reynolds Numbers there is a positive effect. A further experiment was coducted reversing the test wall with the point of the "V" of the manipulators directed downstream instead of upstream (fig. 2). There was a strong drag increase, meaning that the effect of the manipulators is not simply a roughness effect.
Hot-wire power spectra has shown a slight decrease in turbulent kinetic energy in the inertial range.

Tests in adverse pressure gradient

Being encouraged by this results, it was attempted to investigate the effect of strong adverse pressure gradient, although, in this case, the sidewall boundary layer growth would have very large effects. In fact it was soon noticed that the accuracy and repeatibility of the tests was worst that in the preceeding cases and it woud be necessary to redesign a wider test section (at least three or four times wider) and, for boundary layer control, to swich from blowing to suction, as said later on. This is the aim of future research.
For evaluating the flow properties in a strong adverse pressure gradient, an insert was placed into the wind tunnel, producing a strong contraction and a following flow deceleration, At the minimum section a slot was placed, in order to blow the boundary layer and to prevent separation on the wall opposite to the measuring one.

Due to the unsufficient flow control, we got attached flow at the upper and lower wall, but backflow in the center of the tunnel, when we attempted the maximum deceleration.
The experiment was then conducted forcing separation by a variable step introduced into the flow and controlled by an accurate transversing gear. Increasing the step depth into the stream it was noticed that up to a size of the step of 10 mm both the flows (with and without manipulators) reattached downstream; at a size of 11 mm both flows did not reattach, while for the intermediate size of 10.5 mm the flow without manipulators reattached, but the one with manipulators remained separated.
It is not a definite answer to the question of the effect of the manipulators on separation, but in any case there was a reasonable experimental evidence that the manipulators would slightly anticipate the separation. This result is in agreement with the consideration that separation occurs at zero skin friction and any mean of drag reduction may have this kind of effects.

Conclusions

These results, when considered and interpreted in a proper way, can not be seen as a simple repetition and extension of the experiments at Brown University (Sirovich and Karlsson, 1997). If from one side there are results that reflect the intrinsic properties of boundary layer turbulence, from the other side some doubts remain that the methodology developed by Orlev Sci. Computing Ltd (owner of the technology) and used in the experiments at Brown University is not optimal or universal. The present study discovered essential differences between the constant Reynolds Number properties of a duct flow and the increasing Reynolds Number properties of a boundary layer.
The achieved results indicate that the theoretical principles (Levich, 1987; Levich and Malomed, 1994; Levich, 1996), on which the technology is based, are substantially correct even if some peculiar aspects of turbulence in a boundary layer had not been considered.
It leads therefore to the need of identifying the dependence of the mechanism of turbulent energy exchange on some new, probably external scale parameter to be incorporated in the design of the passive manipulators, which was not considered in the previous experiments in the channel flow conditions.

Acknowledgements

We wish to thank the JRC (Ispra) of the European Commission who supported economically and institutionally the experiments here described; Orlev Sci. Comp. LTD that in the framework of an agreement with the JRC allowed us to use the patents for scientific purposes; Constellation 3D Inc. for the encouragements and the economical supports.

References

Bechert D.W., Bruse M., Hage W., van der Hoeven J.G.T. and Hoppe G.. Experiments on drag reducing surfaces and optimisation with an adjustable geometry. J. Fluids Mech. Vol. 388, pp. 69-87.
Bechert D.W., Hoppe G, Reif W.E. 1985. On the drag reduction of shark skin. AIAA Paper 85-0546.
Carpenter P., 1977. The right sort of roughness. Nature, vol. 388. August 1997.
Choi K.S.1987. Test of drag reduction riblets on a one-third scale racing yacht. Presented at Turbulent drag reduction by passive means, Conf. 15-17 Sept. , London, UK Royal Aeronautical Society.
Dracos, Kholmyansky M., Kit E., Tsinober A.; 1989. Velocity gradients measurements in turbulence behind a grid. Proc. IUTAM Symp. on Topological Fluid Mechanics, Cambridge (Cambridge Univ. Press).

East L.F, Sawyer W.G ; 1979. An investigation of the structure of Equilibriun Turbulent Boundary Layer AGARD CP 271, **Den Haag, 24 – 26 sept 1979**

Handler R.A., Levich E., Sirovich L.; 1993. Drag reduction in turbulent channel flow by phase randomization. Phys. Fluids A 5 (3), March 1993.

Levich E., 1987. Certain problems in the theory of developed hydrodynamical turbulence. Physics Reports (review section of Physics letters) 151, Nos 3&4 (1987) 129-238.

Levich E., Shtilman L., Tur A.V; 1991. The origin of coherence in hydrodynamical turbulence. Physica A 176, pg 241-296.

Levich E., 1996. New developments and classical theories of turbulence. Int. J. Modern Physics B, vol 10, pg. 2325-2392.

Levich E., Malomed B.; 1994. Model of boundary layer turbulence. Progress in Turbulence Research, ed. by H.Branover and Y.Unge, pp 28-56 (AIAA, Washington, DC).

Levich E., Sirovich L., 1995. Enginneering Turbulence Control. Final Report 1994/1995. ORLEV Sci. Comp. LtD, Min. of Energy and Infrastructure, Div. Res. and Dev. July 1995.

Luchini P., Manzo F., Pozzi, 1991. Resistance of a grooved surface to parallel and cross-flow. J. Fluid Mechanics, 228, 87.

Moffat H.K.; 1969. J.Fluid Mechanics, 35, 117-129.

Moffat H.K.; 1975. Magnetic Field Generation in Electrically Conducting Fluids, Cambridge Univ. Press.

Moffat H.K.; 1985. J.Fluid Mechanics, 159, 359.

Monti R. 1996. Coerenza, Intermittenza e Resistenza dei Moti Turbolenti. XXV Congr. Naz. Idr. E Costr Idr., Torino.

Monti R., 1999. Drag reduction in a turbulent boundary layer. Report EUR 18966 EN, European Commission.

Murakami Y., Shtilman L., Levich E.. 1992. Ruducing drag by phase juggling. Phy. Fluids A ,4 (8), August 1992.

Nieuwstadt F.T.M., Wolthers W., Leijdens H., Krishna Prasad K. and Schwarz-van Manen A.. 1993. The reduction of skin friction by riblets under the influence of an adverse pressure gradient. Experiments in Fluids 15, 17-26.

Passoni G. ; 1999. Detailed Investigation of the velocity field in the vicinity of solid wall by DNS. Report JRC-EC Ispra, Italy.

Ponti A., 1999. Studi sperimentali su manipolatori di turbolenza. Master Thesis, Politecnico di Milano, Milano, Italy.

Robinson S.K.; 1991. Coherent Motion in the turbulent boundary layer. Annual Rev. Fluid Mechanics, vol. 23, pp 601-639.

Sawyer W.G. and Winter K.G., 1987. An investigation of the effect on turbulent skin friction of surfaces with streamwise grooves. In Turbulent drag reduction by passive means, Conf. 15-17 Sept. , London, UK Royal Aeronautical Society.

Schoppa W., Hussain F., 1998. Genesis of longitudinal vortices in near wall turbulence. Meccanica n.5, vol.33.

Shtilman L., Polifke W., 1989. On the mechanism of the reduction of nonlinearity in the incompressible Navier Stokes equation. Phy. Fluids A 1, 778.

Sirovich et al. 1991. Propagating structures in wall-bounded turbulent flows. Theoretical and Computational Fluid Dynamics, Springer-Verlag, pg 307-317.

Sirovich L., Ball K.S., Keefe L.R.; 1990. Plane waves and structures in turbulent channel flow. Phys. Fluids A 2 (12), December 1990.

Sirovich L., Karlsson S. 1997. Turbulent drag reduction by passive mechanisms. Nature, vol. 388, 21 Aug. 1997.

Sirovich L., Levich E., Bronicki L.Y., Karlsson S. 1996. Method of and Apparatus for controlling turbulence in boundary layer and other wall-bounded fluid flow fields. USA. Patent n. 08/766,380, Dec. 1996.

Sirovich L., Levich E., Bronicki L.Y., Karlsson S. 1997. Method of and Apparatus for controlling turbulence in boundary layer and pother wall-bounded fluid flow fields. Europe Patent n. 97309935.1, Dec. 1997.
Sirovich L., Levich E., L.Y. Bronicki; 1993. Method and apparatus for controlling turbulence in a wall bounded fluid flow, US patent N. 5.263.793.
Sirovich L., Levich E., L.Y. Bronicki; 1994. Method and apparatus for controlling turbulence in a wall bounded fluid flow, US patent N. 5.362.179.
Sirovich L., Zhou X.; 1994. Dynamical Model of wall-bounded turbulence. Phys. Rew. Letters. vol.72, n.3.

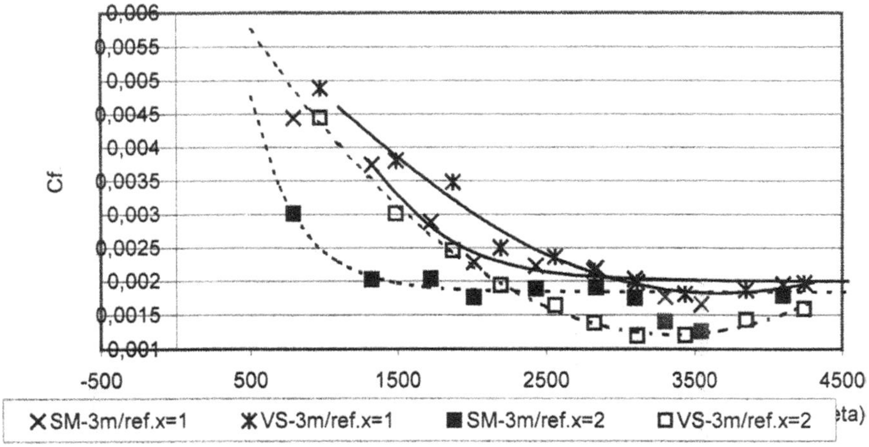

Fig. 1. Skin friction coefficient versus Re_θ at X=3m from the inlet, estimated integrating the von Karman equation over distances Δx=1m and Δx=2m, both with drag manipulators (VS) and smooth boundaries (SM).

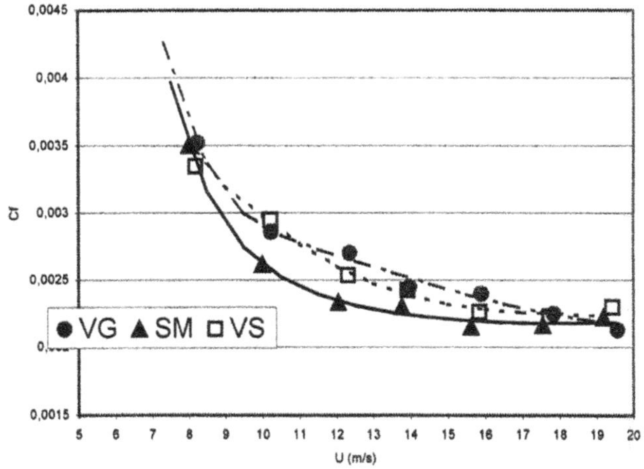

Fig. 2. C_f versus the centerline velocity U. C_f is calculated integrating the Karman equation between the abscissas X=1m and X=5m . VG-turned V shapes plate; SM-smooth palte; VS- V shapes plates

Session 10
Separation Control & Induced Drag Reduction

Active Control of a Laminar Separation Bubble

K. Augustin, U. Rist, S. Wagner
Universität Stuttgart, Institut für Aerodynamik und Gasdynamik
Pfaffenwaldring 21, D-70550 Stuttgart, Germany

Summary

In the present paper an active control mechanism for the control of laminar separation bubbles on airfoils is investigated by means of direct numerical simulation and linear stability theory. Boundary layer instabilities excited by periodic oscillations are utilized to control the size and length of the separation bubble and to make it finally disappear when desired. Unlike traditional vortex generators a sensor-actuator system based on this method will be adaptive to the respective flow conditions and will cause no additional undesired drag.

Introduction

Operated in a low Reynolds number regime the drag characteristics of laminar airfoils at "off-design" conditions can be considerably deteriorated by *laminar separation bubbles (LSB)* due to the inability of the laminar boundary layer to overcome the stronger adverse pressure gradient caused e.g. by a higher angle of attack or an extended trailing edge flap. When the laminar boundary layer separates from the airfoil, laminar-turbulent transition may occur and the turbulent boundary layer may reattach to the airfoil because of the increased momentum transfer towards the wall, thus forming a *laminar separation bubble*. To improve the behavior of the airfoil in such an "off-design" situation active control should set in to avoid the occurrence of these bubbles. We suggest to use controlled excitation of boundary layer perturbations upstream of the bubble. These disturbance waves become amplified by instability mechanisms within the laminar boundary layer and in the separation bubble itself which leads to an upstream shift of the transition and the now turbulent boundary layer does not separate. As a constraint the upstream shifting should be achieved by a minimum of disturbance amplitude A_0 respectively of disturbance energy introduced into the flow, but the necessary amplitude A_0 can only be minimized by choosing the proper wavelength λ_x, frequency β and of course the location and length of the disturbance strip exciting sinusoidal disturbance waves.

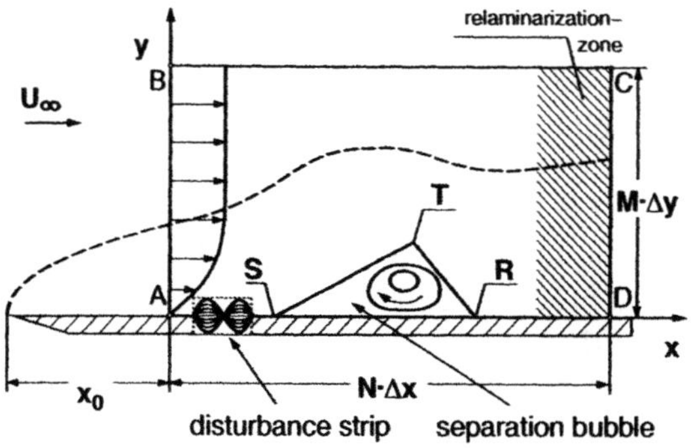

Figure 1 Schematic overview of the integration domain of the DNS with LSB

An indicator for the effectiveness of this control is the location of the point of transition within the LSB as shown in figure 1 where "S", "T" and "R" mark separation, transition and reattachment, respectively. The physical investigation of the laminar separation bubble phenomenon by means of *linear stability theory (LST)* and *direct numerical simulation (DNS)* is used to estimate a control method capable to avoid separation in the "off-design" case without deteriorating the design properties of the airfoil.

Based on the considered method an actuator-sensor concept will be designed, where the LSB is detected by sensors and an upstream actuator driven by a controller (figure 2) excites the boundary layer disturbances.

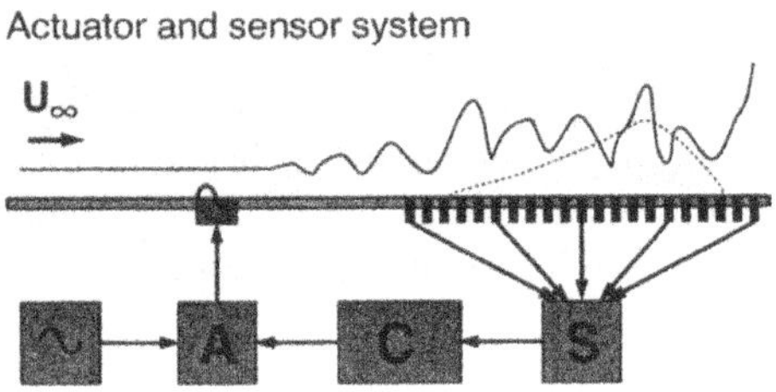

Figure 2 Sensor (S)-actuator (A) concept with controller (C) and signal generator (~)

Numerical method

The implemented 3D DNS method [2] solves the complete *Navier-Stokes* equations for incompressible flow in a vorticity-velocity formulation

$$\frac{\partial \underline{\omega}}{\partial t} - \mathrm{rot}\,(\underline{v} \times \underline{\omega}) = \frac{1}{Re}\,\Delta\underline{\omega} \tag{1}$$

$$\text{with} \quad \underline{v} = (u,v,w) \quad \text{and} \quad \underline{\omega} = (\omega_x,\omega_y,\omega_z)$$

within a rectangular integration domain (A-B-C-D in figure 1) over a flat plate.

All spatial coordinates are non-dimensionalized by a reference length $\tilde{L}$ and velocities by the freestream velocity $\tilde{U}_\infty$, where $\tilde{}$ denotes dimensional variables. This leads, together with the kinematic viscosity $\tilde{\nu}$, to the following definitions for the vorticity vector $\underline{\omega}$ and the frequency β

$$\underline{\omega} = -\mathrm{rot}\,\underline{v} \quad , \quad \beta = \frac{2\pi\,\tilde{f}\,\tilde{\nu}}{\tilde{U}_\infty} \cdot 10^5. \tag{2}$$

Fourth-order accurate finite differences are used on the equidistant numerical grid in streamwise (x-) and wall-normal (y-) direction whereas a spectral *Fourier* ansatz in spanwise (z-) direction is applied. The explicit time integration is realized by a fourth order, four step *Runge-Kutta* scheme. Once the three vorticity components are obtained, three Poisson equations for the remaining velocities u, v and w have to be computed. Due to the spectral ansatz in spanwise direction the Poisson equations for the u and v velocity reduce to ordinary differential equations (ODE) and only the v-equation has to be solved iteratively by a line relaxation method accelerated by a multigrid scheme [9].

At the inflow boundary steady *Falkner-Skan* or, at zero pressure gradient, *Blasius* velocity profiles are prescribed. In a relaminarization zone [3] upstream of the outflow boundary the unsteady vorticity components are damped to steady state values to avoid non-physical reflections and as a consequence the unsteady velocity components $\underline{v}'$ vanish exponentially. At the wall the no-slip condition is applied except for a disturbance strip where 2D and 3D perturbations can be introduced into the flow by periodic suction and blowing.

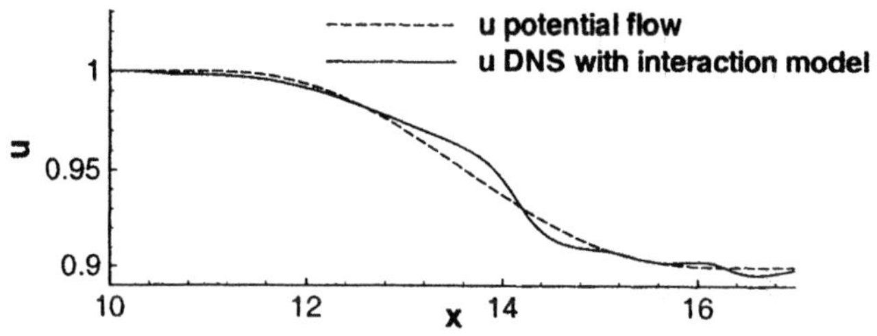

Figure 3 u-velocity distribution at the free-stream boundary

To take into account the strong boundary layer displacement effects of the LSB a viscid-inviscid interaction model is implemented into the multigrid algorithm for the v-velocity component. At run time a velocity distribution which allows for the typical boundary layer displacement effects of a LSB evolves from the initially prescribed potential flow. For a detailed summary of the interaction model and boundary conditions refer to [6]. Figure 3 shows the initial potential distribution of the u-velocity (dashed line) and the resultant distribution considering the effects of the LSB (solid line).

Once the DNS has yielded a reasonable mean flow the development of excited boundary layer perturbations can be studied with less effort by using the *linear stability theory (LST)* [10]. However, the LST provides reasonable results in the linear regime of the disturbance development only, i.e. upstream of the transition ("T") and can not cover variations of the mean flow due to the strong influence of different disturbance amplitudes A_0 on the size of the LSB. Applying the LST the spatial amplification rate α_i of discrete fluctuations specified by their streamwise wavenumber α_r and disturbance frequency β in

$$v(x,y,t) = \hat{V}(y)\ e^{(\alpha x - \beta t)} \quad \text{with} \quad \alpha = \alpha_r + i\,\alpha_i \tag{3}$$

can be determined by solving the *Orr-Sommerfeld* equation. Values of $\alpha_i < 0$ mean amplification and $\alpha_i > 0$ damping of the disturbance amplitude. The disturbance amplitude A can thereon be obtained by integrating the amplification rate α_i in streamwise x-direction and multiplying these values with an initial disturbance amplitude A_0.

Disturbance development

To check the accuracy of the DNS and LST results, comparisons to experiments at the laminar wind tunnel at the Universität Stuttgart [5] and at the KTH Stockholm, Sweden [1] have been performed and good agreement of the results has been achieved.

A comparison of the amplification curves $A = u'_{max}(y)$ for the fundamental disturbance mode (α_r,β) of a typical "midchord-bubble" case at $Re_{\delta_1} = 1722$ at the inflow boundary, considered in greater detail in [8], in figure 4 a) confirms the good agreement between LST and DNS, "S" and "T" again mark separation and transition. Both curves evidently coincide until close to transition which shows that the LST is a proper tool to investigate the disturbance development in flows

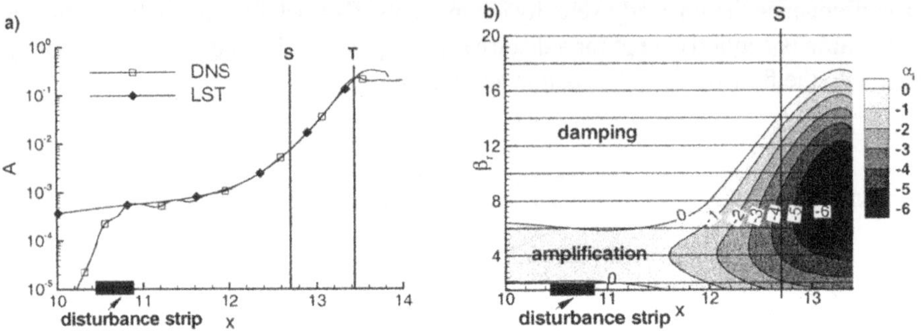

Figure 4 Amplification curves as a result of LST and DNS a) and stability diagram of LST showing amplification rate α_i b)

with LSB. In figure 4 b) the corresponding linear stability diagram $\alpha_i = \alpha_i(x,\beta)$ reveals a narrow band of amplified frequencies β only upstream of the LSB, comparable to a *Blasius* flow, whereas the amplified frequency band broadens further downstream towards the separation line.

For that reason, a suitable disturbance mode for influencing the LSB has to be carefully selected within the amplified frequency band with respect to the streamwise position of the disturbance strip x_0, its length x_S, wavenumber α_r, wave propagation speed Ω of a travelling wave and initial amplitude A_0. Langhoff [4] investigated a variety of these parameters in *Blasius* flow where the wall-normal velocities v_0 at the disturbance strip followed the prescription

$$v_0(x,t) = \underbrace{V_0}_{\substack{\text{initial}\\ \text{amplitude}}} \cdot \underbrace{\sin(B\,t)}_{\substack{\text{amplitude}\\ \text{modulation}}} \cdot \underbrace{\sin(\alpha_r\,x - \Omega\,t)}_{\substack{\text{travelling}\\ \text{wave}}} . \tag{4}$$

Depending on the choice of the frequency B for the oscillations at the disturbance strip and of the wave propagation speed Ω different kinds of wall forcing are possible: travelling waves, travelling oscillating waves and standing oscillations. Their ability to excite a *Tollmien-Schlichting* (TS-)wave for a given frequency $\beta = B + \Omega$ is compared in figure 5. It turns out that the boundary layer is most receptive to forcing with travelling waves ($\Omega = \beta$ and $B = 0$). Note that results for two different disturbance frequencies β are shown in figure 5.

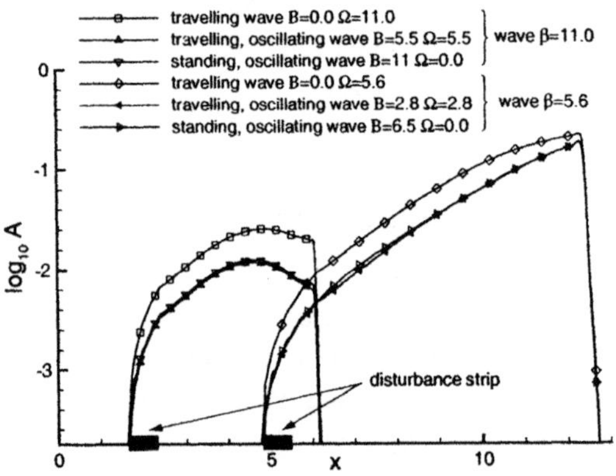

Figure 5 Amplification curves for different waveforms and disturbance frequency

For optimal efficiency the position of the disturbance strip for each frequency β should be located in the vicinity of the point where each wave becomes unstable according to LST (so called *branch I* of the neutral stability curve). Only in this case a maximum of integral amplification A can be achieved. At all other positions either the amplitude is damped before the amplified region is reached or the amplified range is to short.

Control of LSB by boundary layer disturbances

An estimation of suitable dimensional disturbance frequencies $\tilde{f}$ for maximum amplification is quoted in table with respect to different free stream velocities $\tilde{U}_\infty$.

Table 1 Estimation of required dimensional frequencies $\tilde{f}$ with respect to the free stream velocity $\tilde{U}_\infty$

$\tilde{U}_\infty$	$\frac{m}{s}$	20	30	40	50
$\tilde{f}$	Hz	425	955	1700	2650

The LST allows to estimate the necessary initial disturbance amplitude A_0 to influence the LSB. By integrating the stability diagram (cf. figure 4 b)) in x-direction for every frequency and multiplying with the initial disturbance amplitude A_0 the amplification curves $A = A(x,\beta)$ can be obtained. However, if one specifies a disturbance amplitude A_1 at a position x_1 downstream of the disturbance strip x_0 the necessary initial amplitude A_0 for every frequency β at the disturbance strip to gain A_1 can be determined as shown in figure 6.

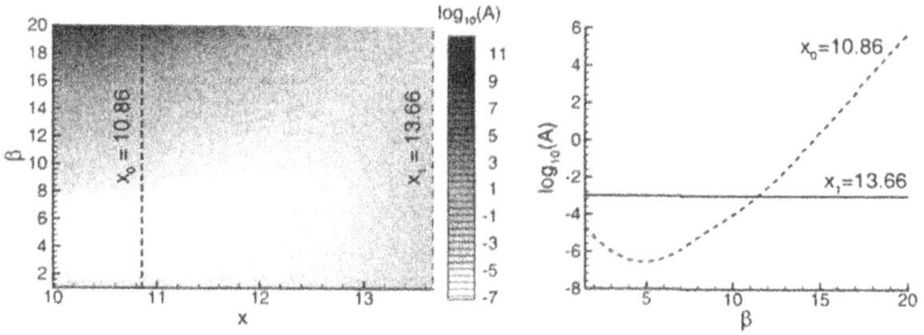

Figure 6 *Re*-normalized amplitude diagram of LST for amplitude $A_1 = 10^{-3}$ at x_1; a) overview and b) cross-section at x_0 and x_1

It is obvious that an appropriate frequency has to be chosen within the amplified frequency band, otherwise the disturbances are damped in such manner that the necessary amplitude A_0 has to be larger than the desired amplitude A_1 at x_1, which is obviously extremely inefficient. However, if $\beta = 5$ is selected for the disturbance a 1000-fold amplification can be achieved.

Results from simulations of the above mentioned “midchord bubble” with the disturbance amplitude V_0 at a frequency $\beta = 5$ ($B = 5, \Omega = 0$) changing from $V_0 = 10^{-12}$ to 10^{-6} over 120 disturbance cycles T, show that the laminar separation bubble is strongly sensitive with respect to the TS-wave. In figure 7 two results from this simulations at a disturbance level of $V_0 = 0.89 \cdot 10^{-8}$ and $2.51 \cdot 10^{-8}$ indicate that even a very small change, compared to the free stream velocity, has a lasting effect on the shape of the bubble.

Beneath the contour figures the corresponding skin friction ω_{zW} curves display the negative shear stress in the area of the LSB as well as the fluctuations of ω_{zW} downstream of the bubble due to vortex shedding in the vicinity of the reattachment line which is also visible from the

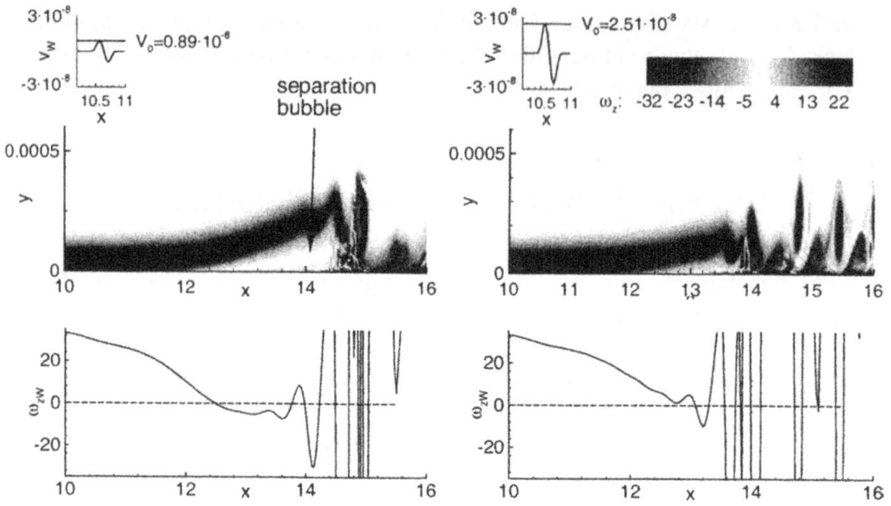

Figure 7 Instantaneous contours of spanwise vorticity ω_z and distribution of skin friction w_{zW} at two different disturbance levels V_0

figures above. The onset of the vortex shedding coincides with the location of transition ("T") which moves further upstream if the disturbance amplitude is increased. The curve of ω_{zW} on the right hand side shows almost random-like fluctuations at positions (e.g. $x = 14$) where the flow seems to be quite uniform in the left hand side picture. Because of these fluctuations the detection of LSB from instantaneous skin friction data must be avoided and one has to consider time averaged distributions of the skin friction.

Based on the present results a controlling mechanism (cf. figure 2) consisting of an oscillating piezoceramics actuator, which has already been build and tested [7], will be derived. For the present design the oscillation frequency signal for the actuator is provided by a signal generator and only the amplitude V_0 will be controlled. The controller itself will read the signal from the shear stress sensor and determines the size of the LSB by averaging the signal over a certain period of time, e.g. ten disturbance periods. The controller is necessary to avoid excessive disturbance amplitudes, to react on nonlinearities in the control loop, and to enable the mechanism to respond to changing flow conditions where a LSB might not be present.

Conclusion

A method that utilizes the instability mechanisms of a boundary layer flow with laminar separation bubble for amplification of proper disturbances up to an amplitude where an evident change in size of the bubble can be observed has been suggested and investigated by linear stability theory and direct numerical simulation. The simulations will lead to a sensor-actuator system based on wall shear stress sensors and a piezoceramics driven actuator capable of generating the desired disturbances, which is only activated while a separation bubble is present and otherwise generates no additional drag.

Acknowledgments

The financial support of the Deutsche Forschungsgemeinschaft (DFG) via the Sonderforschungsbereich 409 "Adaptive Strukturen im Flugzeug- und Leichtbau" at the Universität Stuttgart is gratefully acknowledged.

References

[1] Casper Hildings. Simulation of laminar and transitional separation bubbles. Technical report, Department of Mechanics, Kungliga Tekniska Högskolan (KTH, Royal Institute of Technology), Stockholm, Sweden, December 1997.

[2] Markus Kloker. Direkte numerische Simulation des laminar-turbulenten Strömungsumschlags in einer stark verzögerten Grenzschicht. Dissertation, Universität Stuttgart, 1993.

[3] Markus Kloker, Uwe Konzelmann and Hermann F. Fasel. Outflow boundary conditions for spatial Navier-Stokes simulations of transition boundary layers. *AIAA Journal*, 31(4):620–628, April 1993.

[4] Wolfgang Langhoff. Numerische Untersuchungen zur aktiven Grenzschichtstörungsanregung über schwingende Wandstreifen. Studienarbeit, Institut für Aerodynamik und Gasdynamik, Universität Stuttgart, Januar 1999.

[5] Ulrich Maucher, Ulrich Rist and Siegfried Wagner. Secondary instabilities in a laminar separation bubble. In Horst Körner and Reinhard Hilbig, editors, *Notes on fluid mechanics (NNFM)*, volume 60, pages 229–236. Vieweg Verlag, 1997.

[6] Ulrich Maucher, Ulrich Rist and Siegfried Wagner. Refined interaction method for direct numerical simulation of transition in separation bubbles. *AIAA Journal*, 2000.

[7] Jörg Müller. Konzeptstudie für ein aktives Oberflächenfeld zur Verhinderung laminarer Strömungsablösung. Studienarbeit, Bereich Flugzeugentwurf, Fakultät Luft- und Raumfahrttechnik, Universität Stuttgart, September 1999.

[8] Ulrich Rist. Zur Instabilität und Transition in laminaren Ablöseblasen. Habilitationsschrift, Universität Stuttgart, 1998. Shaker Verlag, 1999.

[9] Ulrich Rist and Hermann F. Fasel. Direct numerical simulation of controlled transition in a flat-plate boundary layer. *J. Fluid Mech.*, 298:211–248, 1995.

[10] Hermann Schlichting. *Boundary Layer Theory*, chapter Origin of Turbulence I, pages 459–483. McGraw-Hill, 1979.

Induced Drag Reduction with the WINGGRID Device

Ulrich* and Lucas La Roche, La Roche Consulting
Heilighüsli 18, CH-8053 Zürich, Switzerland, email: ularoch@attglobal.net

Summary

The WINGGRID is a nonplanar, multiple wake producing wingtip, which allows span loading control independent from wake interaction. Experimental verification and new design proposals of wing systems at subsonic speed are reported.

Recent theoretical and experimental work has cleared the way to understand more radical reductions of induced drag than possible with wing systems describable by the Munk stagger theorem, cf. [1]. These are wingtip configurations, that allow for span load control independent from wake interaction of the lifting wing system. Experiments and theoretical insights are treated in parallel to other configurations belonging to the same class of devices, i.e. the split-wing [2] and the SPIROID [8].

Paradigmata on limits of induced drag for lifting systems

In classic wing systems with linear interference of streamwise vortices, deviation from elliptic lift distributions towards rectangular distribution are limited, because of the interdependency between vortex-sheet linear interference and span load. Recent work (cf. [2], [4]) on non planar multiple wake systems show experimentally and theoretically, that the Munk stagger theorem (cf. [1]) is limited to wing systems with linear interference of the streamwise vortex-sheet. Munk's theorem and it's connected variation calculus for the determination of the minimum induced drag limit does not explain nonplanar multiple wakes with wake independent span load.

Smith's work [2] is especially enlightening due to its detailed theoretical and experimental investigation. It identifies the split wing-tip configuration with non-linear interference and clearly bigger span-efficiencies than the Munk theorem would allow for experimentally and by calculation using force-free wake calculus in the Trefftz plane. It also compares the split wing to conventional wings with an nonplanar wake obeying Munk's stagger theorem, such as investigated by C.D.Cone [7].

For assessment of possible reductions of induced drag the result of Spreiter & Sacks [3] is used, that only two basic parameters are necessary to characterize the induced drag of any wing-system (Munk or non-planar multiple wake).

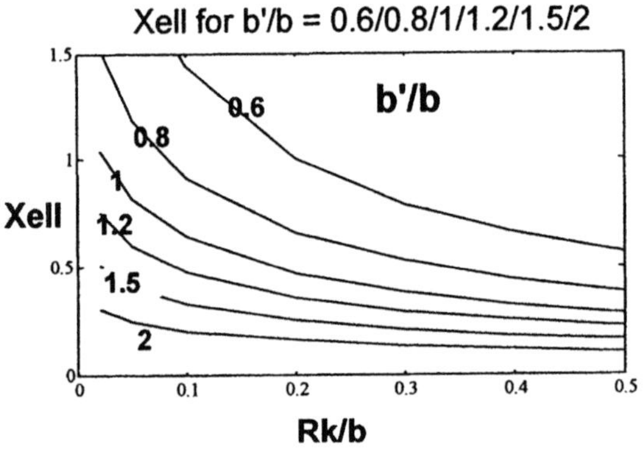

Fig. 1: Relative induced drag vs core radius Rk and vortex separation b'/b after [3]

These two parameters are the separation distance of the rolled up vortices b' and the radius Rk of the Rankine vortex-core. Independent choice of these two parameters allows for any reduction of induced drag.
Figure 1 shows Xell, the relative induced drag compared to an elliptic planform wing of the same aspect ratio vs the core radius Rk of the rolled up vortex. As parameter the separation distance of the vortices b' relative to the span b of the wing is used.
Table 1 summary of the two positions of *stagger theorem* and the *multiple wake* configurations

category	stagger theorem [1]	multiple wakes ([2], [4])
wake	streamwise vortices with linear interference of all lifting elements	Multiple wake on part span with nonlinear interference
airfoil elements	Superposition of arbitrary lifting lines – fields get represented by circumference	Different parts of span only coupled by circulation transfer
minimum drag limit	is defined by variation calculus in linear interference space – "constant downwash"	No limit defined – induced drag can be reduced to near zero with the right configuration
span loading	control of span loading is linked to streamwise vortices – elliptic distribution is optimal (planform)	Span loading control independent from wake interaction – e.g. true rectangular lift distribution possible

Flight-Model- and wind tunnel studies on WINGGRID and SPIROID

Asymmetric free flying model planes were used for configuration screening and wind tunnel tests for quantitative and configuration analysis with the most successful screening selections, cf. [5].

Two potential competitors in the new class „multiple wake" are the SPIROID and the WINGGRID, which resulted both in flyable asymmetric models, indicating the two conditions for stable flight, namely near rectangular span load and massive reduction of induced drag to be present.
The wind tunnel-tests verified the reduction of induced drag (apparent aspect ratio) and the rectangular lift distribution by force-moment measurements and independent smoke trail measurements, cf. [4].

Comparison of the slope representing the apparent aspect ratio of WINGGRID and SPIROID models to an elliptic reference wing show aspect ratio * e, where e is the span efficiency.

The wind tunnel tests based on force measurement and smoke trail verifications, show that the SPIROID and the WINGGRID are equivalent in effect and operate on the same principles.

Progrid 97 full scale tests WINGGRID

Based on the wind tunnel tests, the first fullscale WINGGRID was designed to fly on the testbed PROMETHEUS, a jet powered motor glider. The tests comprised glide path measurements of absolute L/D with GPS 8 channel and piezo-barograph. Span load measurements of the WINGGRID blades using strain-gauges did provide independent check on the WINGGRID in flight behavior.

- L/D polars results

Below a certain critical speed or above the equivalent Cl-value the WINGGRID starts to behave like an ordinary slit-wing, loosing the effect of induced drag reduction. As was learned in the wind tunnel experiments this critical speed of the „cutoff" is a function of the stagger angle used. It is a design parameter, [4]. For speeds above „cutoff" polar fitting consistently confirmed a span efficiency of around e = 2.

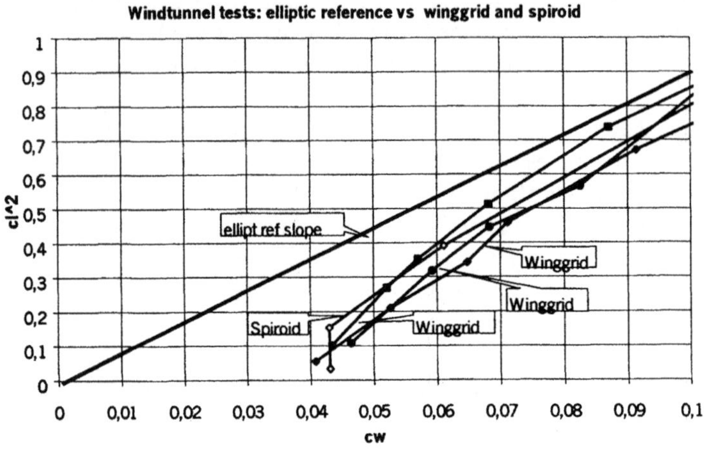

Fig. 2: Wind tunnel experiments: slope comparison of elliptic wing, SPIROID and WINGGRID show span efficiency of > 1.5 for the latter two

- Lift-Loads of WINGGRID blades measured on the testbed

Above the critical speed of > 45 m/s a loading of 200 N per winglet is reached, equivalent to a rectangular span load, below the load of lift typically is about 70% of this former value. The critical speed as measured turned out to be higher than expected from the extrapolation from the preliminary wind tunnel tests, this will be an ongoing concern in further developement.

- Load calculations of WINGGRID blades

Since above cutoff in supercritical speeds the blades produce lift individually, a basic vortex-superposition-calculation is sufficient to reproduce the resulting lift distribution, including mutual interference. Calculation and actual load measurement are found to be within 10% difference, cf. [4].

Idaflieg 99 full scale tests WINGGRID

L/D polar and smoke visualisation for control of downwash-geometry and span load and comparison of the testbed's two wing-configurations for direct evaluation of the drag-components were performed by Idaflieg/DLR.

- Idaflieg & DLR method

The identical testbed with WINGGRID has been measured with the well known Idaflieg/DLR method of comparison to a calibrated sailplane by the classic photographic measurement and also by a new differential GPS-method, cf. [11] and [12]. The low mean error of 2% for the photographic measurements allowed for direct comparison of two configurations of the testbed. One is the testbed with its 23m span Stemme S10-wing, the other with the base S10-wing and WINGGRID added with total span of 12m. The comparison allows to eliminate directly fuselage drag and to obtain the WINGGRID drag figures on an absolute basis. Important results of this full scale tests is confirmation of the findings of the 97 PROGRID tests for the span efficiency. A new result is the assessment of the additional drag the WINGGRID exhibits. It is firstly interference drag of the blades at base and tip and secondly profile drag due to lower Re-numbers on the smaller chords of the blades and drag of the interconnection body and endplates.

- *cl-cw* polars

If we analyze the *cl-cw* polars obtained, the cutoff already identified in the 97 tests is plainly visible as a discontinuity at Cl = 0.67 in fig. 3. A curve fit L/D= $1/(\pi*e*AR)*cl^2+cd0$ shows a span efficiency e of 2. A curve fit with L/D= $1/(\pi*e*AR)*cl^2+B*cl+cd0$ does also show e=2 for cl<0.67 in fig. 4.
Alternative explanations that could explain this anomaly as data scatter, laminar drag bucket or fuselage separation were not found to match the required magnitude in cw or location in terms of cl.

- L/D polars compared

The measured points of the L/D polars are fitted with the expected polars for e = 2 (supercritical with WINGGRID) and e = 1 (elliptic with /without parasitic drag of WINGGRID), see fig. 5. Again the result, that in supercritical speeds the full WINGGRID-effect results in e = 2 is plainly visible. Also visible is the influence of the additional drag of the WINGGRID-configuration used: interference drag and profile-drag. If verified on absolute values, cf. [9], this measured additional drag is within 10% of calculated values. Based on [9] and its cited literature it is possible that the interference drag can be eliminated essentially to zero by proper fillets and roundings.

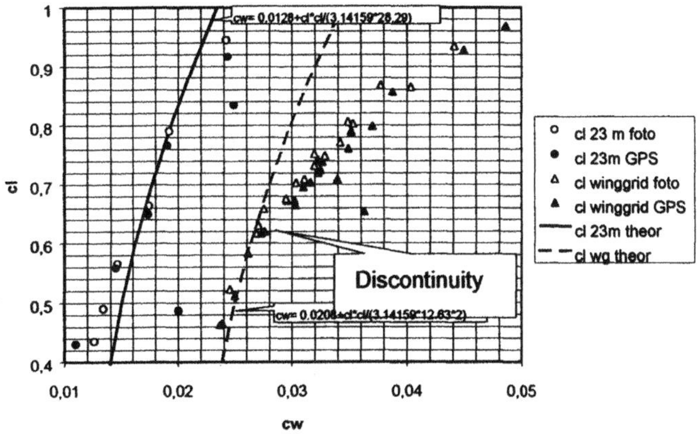

Fig. 3: Fit polars (L/D= $1/(\pi*e*AR)*cl^2+cd0$) for Idaflieg measurements show a span efficiency of 2 for the WINGGRID

Since no effort on interference drag reduction was spent on this test configuration a properly designed WINGGRID will exhibit almost negligible additional drag, because higher profile drag is mainly compensated by the overlap being less than 0.7 in most practical cases and interference drag can be eliminated by known means.

- Vortex-sheets visualizations

From the independent test with smoke bombs the experimental evidence can be inferred, cf. [5]:
-b'/b≈ 0.95, which means we have a near rectangular span load (elliptic has b'/b≈ 0.8).
-the smoke pattern taken with the WINGGRID endplate mounted camera shows indentations of individual vortex-sheets leaving the two foremost WINGGRID-blades, which is evidence for a multiple wake configuration as shown in fig. 6.

Summary of improvement potential and design of WINGGRID

As summarized in [6] the information from the early tests 97 and 99 Idaflieg/DLR permits to do reasonable prototype design with precalculation of performance. It is important to

recognize, that the speed of „cutoff", which with the testbed is situated at a cl-value below 0.7, is a key design parameter of the WINGGRID-configuration to be set.
In the following the main design features of the WINGGRID are summarized:

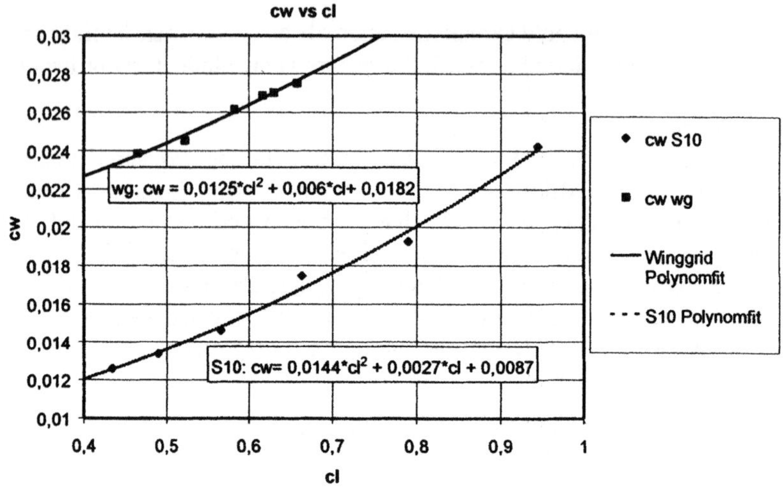

Fig. 4: Polar fit with linear element, L/D= $1/(\pi*e*AR)*cl^2+B*cl+cd0$, for cl <0.67 shows a span efficiency of 2 for the WINGGRID, i.e. $1/(\pi*e*AR)=0.0125$ with AR=12.6

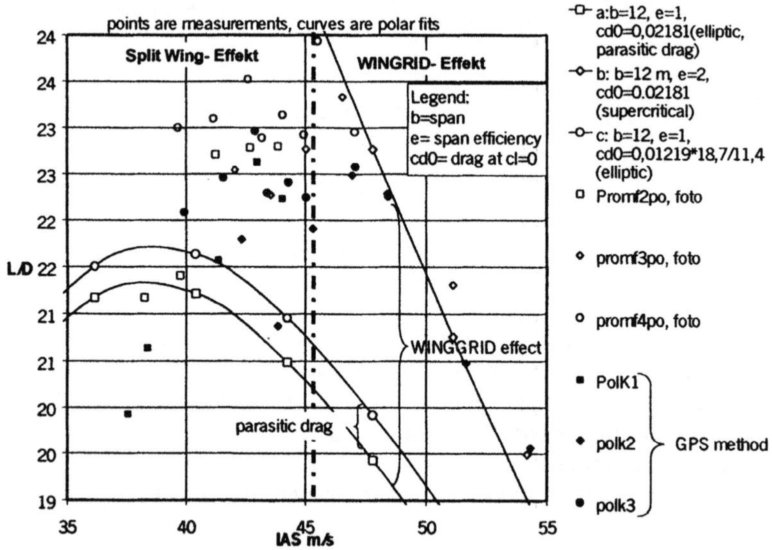

Fig. 5: L/D measurements compared to fit polars

There are five conditions for a WINGGRID to operate successfully, which have to be met as a logic AND.

- the devices span should not exceed a certain part of total span, say L2 < 50% of L.
- the device has to have a stagger angle of at least the maximum angle of attack (corresponding to lowest speed without stall) considered for the main wing.
- the devices grid winglets should be parallel.

- the device should have essentially the same lift per span as the main wing profile in 2-D flow would have at the attachment point (rectangular spanload).
- overlap of the grid winglets should be less than 1.

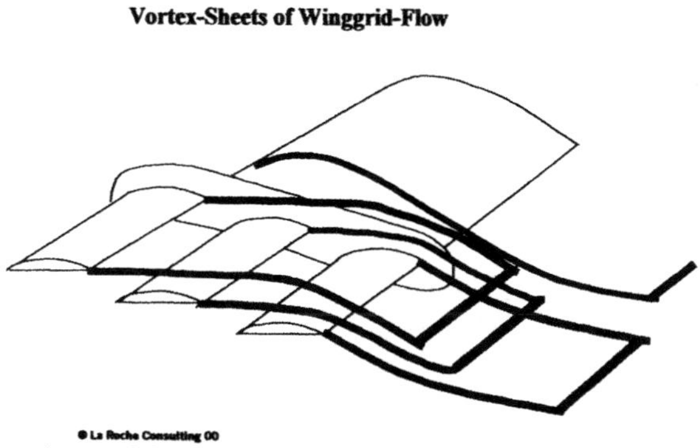

Fig. 6: Multiple vortex sheets of the non planar wake configuration WINGGRID. The vortex sheet of the main wing is split into several vortex sheets at the tip.

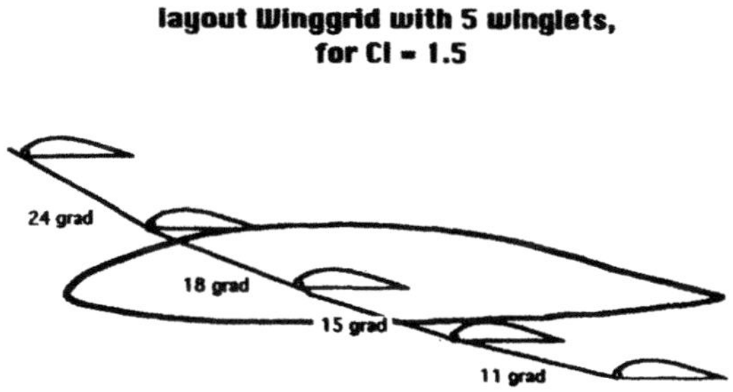

Fig. 7: WINGGRID design with individual stagger angles

- Stagger-angle design criteria

A design consideration worked on is treatment of the so called stagger angle, the rotation between the WINGGRID and the main profile chord necessary to obtain the nonlinear multiple vortex-sheets of the WINGGRID operation to reduce induced drag.
The example in fig. 7 demonstrates the present approach for the design of the WINGGRID stagger angle for the MCR01 project. The condition for stagger angle depending on Cl-value is set for each blade individually.

- Lift-distribution designs

The different designs of a WINGGRID regarding blade load are listed below:

a) all blades with the same angle of attack. This will lead to the foremost blade having more than 200% of lift then the last, but will guarantee on the other side correct angle of attack at all speeds.
b) as a) but with foremost blade corrected for a smaller angle of attack. This will alleviate the problem of very high lift for the first blade.
c) As b) but with chord of first blade made bigger. This allows to independently have a smaller local Cl-value.
d) For a preferential operational speed adjust all angles of attack in order to arrive at equal lift load for all blades

Potentials of improvements WINGGRID

- L/D comparison

L/D polars can be used to compare different wing designs. Of prime interest is the relation of induced drag to total drag as a function of span efficiency e, see also details in [4] and [6]. With the much increased span-efficiency available with the WINGGRID the L/D for aircraft designed to fly near the optimum relation of induced drag to total drag around 30 to 50% an improvement of 30 to 100% of L/D is within realistic expectations.

UL-Prototype MCR01

A project is in progress to bring the WINGGRID to production (in cooperation with DYN'AERO and ONERA). The aircraft main parameters are described in table 2.

Table 2: Target specification Data of the MCR01 motor glider-Version

Span: 9.8 m	TOW: 450 kg	Vmin 55 kmh	Vcruise 240 kmh
Vglide 90-140 kmh	Vmax 300 kmh	L/D_{max} 30	Wing Area 9.35 m^2

Three different wingtip-arrangements with identical span are tested:

- No wingtip/Classic wingtip/WINGGRID

Wing system with WINGGRID at transsonic speeds

The preliminary design shown is for demonstrating the main design challenges. Starting from the profiles, the critical mach numbers of the profiles were evaluated with cl-values used. With three blades and a relative span of 20% of the WINGGRID (relative to total half span) the resultant span efficiency will reach e = 2. It is of course essential to keep compressibility drag small.

- Shocks on WINGGRID-blades

Because of the small overlap-value we do not have effective overlap. Shock wave patterns will develop individually on each blade of the WINGGRID (as in transsonic compressor-bladings)

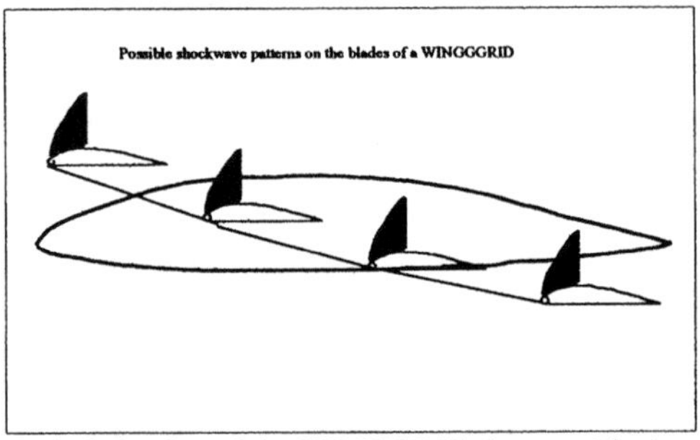

Fig. 8: Transonic behavior of the WINGGRID (supercritical profiles preferred)

- Preliminary design of transonic wing

The design shown in fig. 9 is an adaption of a „infinite sheared wing“ with spanwise rectangular lift distribution after [6]. No corrections are added for the critical parts of the wing, where the „infinite sheared wing“ model has to be substantially modified for its ending at the fuselage and WINGGRID junctions. The same applies to the individual blades of a WINGGRID.

- Sweep angles

The necessary sweep angles result from profile and cl-data of the WINGGRID for use in the preliminary design.

WINGGRID's would add 10 to 30% lift to drag improvement to transonic transport airplanes with shorter span, prolonging the lifecycle of the classic Cayley type aircraft.

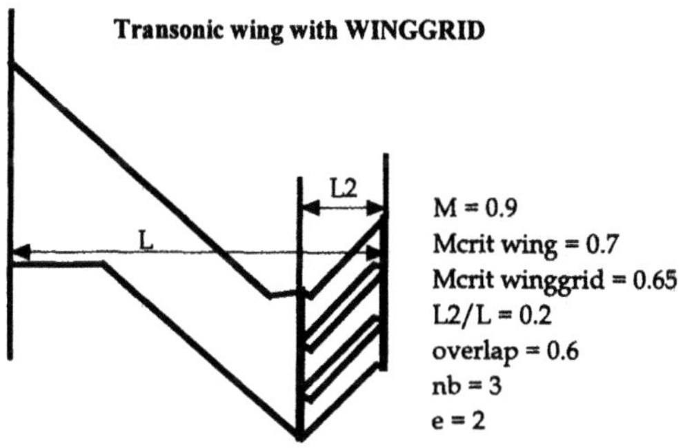

Fig. 9: Base design of transonic WINGGRID

Conclusion

With span efficiencies e > 2 possible there are profound consequences to be exploited. In order to get the most benefits of the WINGGRID one usually has to design a completely new wing since we have:

- Rectangular span load for maximum effect, which means no taper and no twist
- Rectangular spanload leads to Cl_{max} being equal to $Cl_{average}$, e.g. higher wing load for the same Cl_{max}
- Lift load on the WINGGRID is high, one has to take the structural consequences for this
- Smaller kinetic energy in the rolled up vortices in the wake

Acknowledgements

This work would not have been possible without enthusiastic and visionary cooperation of various organizations and people. We acknowledge and thank for:

- The initiative of Idaflieg to do the tests in 99 coached by Andre Jansen.
- The ontime cooperation of Dietmar Schmerwitz from DLR with test results [10].
- The preparation of the WINGGRID's by Ivo Stengele from ETHZ/ILS.
- The availability of the testbed from EFF represented by Thomas Bircher.
- The GPS measurement results provided by Gerko Wende and Stefan Ronig from IFF [11].

References:

[1] Munk Max M., *The Minimum Induced Drag of Aerofoils*, NACA Report 121

[2] Stephen C. Smith, *A Computational and Experimental Study of nonlinear Aspects of Induced Drag*, NASA Technical Paper 3598, Ames Research Center

[3] Spreiter J.R., Sacks A.H., *the rolling up of the trailing vortex sheet and its effect on the downwash behind wings*, Journal of the Aeronautical Sciences jan.51, p. 21 ff.

[4] La Roche U. et al., *WING-GRID, Development, Qualification and Flight Testing of a WINGGRID on a jetpowered testbed*, Proceedings ICAS 98 Melbourne (Australia),

[5] http://www.rhone.ch/WINGGRID, *the website of the WINGGRID*

[6] Küchemann D., *The Aerodynamic Design of Aircraft*, Pergamon Press, ISBN 0-08-020515-1

[7] Cone C.D., *The theory of induced lift and minimum induced drag of nonplanar lifting systems*, NASA technical report R-139 1962

[8] Procter P., *Winglet designs to cut fuel burn*, Aviation week&space technology, dec, 6 1994, p.3

[9] Hoerner S., *Fluid-Dynamic Drag*, Library of congress card number 64-19666

[10] Schmerwitz D. et al., *Messbericht Idaflieg/DLR 1999*, Flugsystemtechnik DLR-Braunschweig/IFF TU-Braunschweig/Idaflieg)

[11] http://www.idaflieg.de

THE LIFTING SYSTEM WITH MINIMUM INDUCED DRAG

Aldo Frediani*, Mario Chiarelli, Alberto Longhi, Enrico Troiani.
Department of Aerospace Engineering "L. Lazzarino", Pisa University
Via Caruso, 56100 Pisa (Italy)

Summary

The lifting system with minimum induced drag is a closed wing with a box shaped front view and was named Best Wing System by Prandtl; on the horizontal wings, the circulation results from the superposition of an elliptical and a constant distributions and, on the the side walls, it is butterfly shaped. Very recently, the Prandtl results (obtained by an approximate procedure) were confirmed by a closed form solution, briefly described in the paper. The concept of minimum induced drag can give rise to a family of aircraft, ranging from very small to very large (larger than the A3XX) aircraft and still compatible with the actual airport areas. These aircraft are named as PrandtlPlane, in honor of Prandtl. An example of a 600 seat category aircraft is presented.

1. Introduction

It is well known that, for given total lift and wing span, the induced drag of a biplane is lower than that of a monoplane and also that, among all the biplanes, there is one with minimum induced drag in which the lift is equal on the two wings and elliptically shaped. A triplane has an induced drag lower that of a biplane and an optimum triplane exists, in which the internal wing is less loaded than the external ones. In the optimum lifting system with $n+1$ wings, the induced drag is lower than the optimum n wing system; so, the lifting system with the minimum induced drag has an infinite number of wings. On each wing of the n-plane, the circulation distribution is symmetrical along the span; the total lift on a single wing varies according to the vertical position and is symmetrical with respect to the horizontal center line of the system. According to Prandtl, the infinite-plane is equivalent to a box shaped wing, in which the side walls are equivalent to tip vortex generators. A condition of optimum for this box system exists; in this case, the system was named as "Best Wing System" by Prandtl. The ratio between the induced drag of the Best Wing System and the optimum monoplane with the same lift and total span was calculated by Prandtl himself before 1920 and published as a NACA paper in 1924 [1]; in that paper, Prandtl used an approximate procedure, even though no detail was given of it. A closed form solution of the Prandtl problem was given in 1999, [2] and it will be briefly resumed in this paper. The problem is to assess the circulation function along the vortex line, which gives rise to the minimum induced drag for a given total lift. Owing to a theorem, the condition of minimum is verified when the drag induced by the free vortices is constant along the two horizontal wings and identically zero on the vertical side wings. Thanks to this theorem, it is possible to prove that the optimum lift distribution along the vortex line is the superposition of a constant and an elliptical distributions along the wings and, on the side walls, it is a butterfly shaped distribution, confirming the results of Prandtl. When the ratio h/b between the vertical distance of the wings and the wing span increases, the induced drag reduces and it is remarkable that the induced drag in the best wing system could be very lower than a that of a monoplane.

In spite of these results, the Prandtl intuition had no influence on the development of aeronautics; the biplane configuration was abandoned owing to the parasite drag due to struts and cables (necessary for improving static and aeroelastic stiffnesses of the wings). Starting from the '20s, all the aircraft were designed with metallic cantilever wings; so, the metallic box wing allowed to improve the bending an torsional stiffnesses and use the box as tanks. In the field of civil aviation, it is easy to observe that the bigger aircraft are derived from the smaller ones by simply improving the scale factors; owing to the limitations of the airfield areas, this process is limited to aircraft with a wingspan of 80 meters, the maximum compatible with actual airfields; therefore, the next generation aircraft following A3XX will be non conventional. The present paper shows a non conventional aircraft with a box shaped wing front view, based on the Prandtl's best wing system concept; in honor of Prandtl, it is named as PrandtlPlane. At the beginning of 20s, structural problems did not allow to build the proposed wing system; owing to the technologies and materials available to-day, the construction of the wings according to the best wing system concept seems to be possible, with a significant reduction of induced drag. Because the induced drag is proportional to the square of the load factor, the effects of maneuvers and gusts on the aircraft performances are less negative than a conventional aircraft; besides, with a proper design of the high lift devices, the drag of PrandtlPlane during take off and landing is significantly reduced; the same for emissions and noise close to the airports.

2. The induced drag for multiwing systems

We assume the classical hypotheses that the wing is a lifting line model ($\Gamma(y)$ is the intensity distribution) and that the horizontal component of the induced velocity can be disregarded with respect to the vertical component $w(y)$. With the usual meaning of the symbols, $\Gamma(y)$ satisfies the fundamental equation of the vortex line theory:

$$\Gamma(y) = \frac{1}{2} C_{L\alpha}(y)\, c(y) \left[V_\infty \alpha(y) - \frac{1}{4\pi} \int_{-\frac{b}{2}}^{\frac{b}{2}} \frac{d\Gamma}{d\eta} \frac{1}{\eta - y} d\eta \right]$$

and the problem is the following:

$$\left\{ \begin{array}{c} \min D_i\left(y, \Gamma\left(y\right)\right) \\ \rho V_\infty \int_{-\frac{b}{2}}^{\frac{b}{2}} \Gamma\left(y\right) dy = L \\ \text{geometry given} \end{array} \right\} \quad where \quad D_i = \rho \int_{-\frac{b}{2}}^{\frac{b}{2}} w\left(y\right) \Gamma\left(y\right) dy.$$

In the case of a monoplane with minimum induced drag, $\Gamma(y)$ is elliptical; in this case, the induced velocity is constant along the span and, besides $D_i = L^2/q\pi b^2$ (q is the dynamic pressure).

In the case of a biplane with $b_1 = b_2 = b$, it results: $L_1 = L_2 = L/2$, and the induced drag becomes: $D_i = \frac{L^2}{q\pi b^2}\left(\frac{1}{2} + \frac{1}{2}v_{12}\right)$ where: $v_{12} = \frac{1}{4}\ln\left(\sqrt{\frac{b^2}{h^2} + 1}\right)$.

By comparing an optimum biplane with an optimum monoplane, we have: D_i (biplane) $\leq D_i$ (monoplane). when $\frac{b^2}{h^2} + 1 \leq e^8$

We have, trivially, that $v_{12} \to 0$ when $h \to \infty$; in this case D_i (biplane) $= \frac{1}{2} D_i$ (monoplane).

The condition of minimum induced drag for a given lift and a given front view of a multiwing system, results from the two following theorems due to Munk.

Theorem 1. The induced drag of a lifting system is invariant when the wake vortices are translated along the the asymptotic speed direction.

Theorem 2 (of mutual induction): If two lifting elements are positioned on the same plane normal to the asymptotic speed, the drag induced by the first on the second equals that induced by the second on the first.

Owing to these theorems, if ε is the angle of an element of the vortex line with respect to the horizontal axis, the condition of minimum induced drag of the system becomes

$w_n = w_0 \cos\varepsilon,$ where w_0 is a constant.

3. The Best Wing System

We consider a box wing system, composed of two horizontal wings connected by two side wings at the tips, in the conditions of minimum induced drag; the problem is to determine the circulation distribution to be generated along the vortex line. Owing to the previous results, the induced velocity is identically zero on the side walls and is constant on the horizontal wings.

With reference to figure 1, the total induced velocities in the vortex segments are:

$$w\left(y_1^1\right) = \frac{1}{4\pi}\left[\int_{-b/2}^{b/2} \frac{d\Gamma_1}{dy_1}\frac{1}{y_1 - y_1^1} dy_1 + \int_{-b/2}^{b/2} \frac{d\Gamma_2}{dy_2}\frac{\cos^2\alpha}{y_2 - y_1^1} dy_2 + \int_0^h \frac{d\Gamma_3}{dz_3}\frac{\cos\beta\sin\beta}{h - z_3} dz_3 + \int_0^h \frac{d\Gamma_4}{dz_4}\frac{\cos\delta\sin\delta}{z_4 - h} dz_4\right].$$

$$w\left(y_2^1\right) = \frac{1}{4\pi}\left[\int_{-b/2}^{b/2} \frac{d\Gamma_2}{dy_2}\frac{1}{y_2 - y_2^1} dy_2 + \int_{-b/2}^{b/2} \frac{d\Gamma_1}{dy_1}\frac{\cos^2\alpha}{y_1 - y_2^1} dy_1 + \int_0^h \frac{d\Gamma_3}{dz_3}\frac{\cos\varepsilon\sin\varepsilon}{z_3} dz_3 - \int_0^h \frac{d\Gamma_4}{dz_4}\frac{\cos\theta\sin\theta}{z_4} dz_4\right].$$

$$w\left(z_3^1\right) = \frac{1}{4\pi}\left[\int_{-b/2}^{b/2} \frac{d\Gamma_1}{dy_1} \frac{\cos^2\beta}{z_3^1 - h} dy_1 + \int_{-b/2}^{b/2} \frac{d\Gamma_2}{dy_2} \frac{\sin^2\varepsilon}{z_3^1} dy_2 - \int_0^h \frac{d\Gamma_4}{dz_4} \frac{\cos^2\nu}{z_4 - z_3^1} dz_4 - \int_0^h \frac{d\Gamma_3}{dz_3} \frac{1}{z_3 - z_3^1} dz_3\right].$$

$$w\left(z_4^1\right) = \frac{1}{4\pi}\left[\int_{-b/2}^{b/2} \frac{d\Gamma_1}{dy_1} \frac{\cos^2\delta}{h - z_4^1} dy_1 - \int_{-b/2}^{b/2} \frac{d\Gamma_2}{dy_2} \frac{\sin^2\theta}{z_4^1} dy_2 + \int_0^h \frac{d\Gamma_3}{dz_3} \frac{\cos^2\nu}{z_3 - z_4^1} dz_3 + \int_0^h \frac{d\Gamma_4}{dz_4} \frac{1}{z_4 - z_4^1} dz_4\right].$$

Now, we introduce the nondimensional quantities $\eta = \frac{y}{b/2}$, $\mu = \frac{z}{h}$, $k = \frac{2h}{b}$, we apply the conditions of minimum induced drag and observe that, for symmetry, just two equations are independent; afterwards, we prove that [2] the lift distribution on the wings is the superposition of an elliptical and a constant part (Figure 2) and, then, the most general solution of the system is reduced to the following two cases: (A) the circulation is elliptical on the two wings and identically zero on the side walls and (B) the circulation is constant on the two wings and unknown on the side walls.

Case A. Owing to the hypotheses: $\Gamma_3 = 0$, $\Gamma_1 = \Gamma_{01}\sqrt{1-\eta_1^2}$, $\Gamma_2 = \Gamma_1$ and using polar coordinates: $\eta_1 = \cos\varphi_1$ e $\eta_2 = \cos\varphi_2$, the two equations of the system become

$$-\int_0^\pi \frac{\Gamma_{01}\cos\varphi_1}{\cos\varphi_1 - \cos\varphi_1^1} d\varphi_1 - \int_0^\pi \frac{\Gamma_{02}\cos\varphi_2\left(\cos\varphi_2 - \cos\varphi_1^1\right)}{\left[\left(\cos\varphi_2 - \cos\varphi_1^1\right)^2 + k^2\right]} d\varphi_2 = \text{constant}.$$

$$\frac{k^2}{4\pi}\left[\left(1-\mu_3^1\right)\Gamma_{01}\int_0^\pi \frac{\cos\varphi_1}{\left[(1-\cos\varphi_1)^2 + k^2\left(\mu_3^1 - 1\right)^2\right]} d\varphi_1 - \mu_3^1\Gamma_{02}\int_0^\pi \frac{\cos\varphi_2}{\left[(1-\cos\varphi_2)^2 + k^2\left(\mu_3^1\right)^2\right]} d\varphi_2\right] = 0$$

The closed form solutions of the integrals are given in [2].

Case B. We preliminary remark that the circulation function is continuous when passing from a wing to a side wing and viceversa; so, the circulation on top and bottom in the side wings is the same of that at the wing tips; we indicate this intensity as A_0, unknown. As for the circulation function along the side wings, we assume a cubic behaviour and, indicating as e the angular coefficient of the function in the middle of the side wall, we obtain

$$\Gamma_3(\mu_3) = A_0\left[-4(e+2)\mu_3^3 + 6(e+2)\mu_3^2 - 2(e+3)\mu_3 + 1\right].$$

The two equations in the case B are

$$\int_0^1 \frac{d\Gamma_3}{d\mu_3} \frac{\left(1-\eta_1^1\right)}{\left[\left(1-\eta_1^1\right)^2 + k^2(1-\mu_3)^2\right]} d\mu_3 + \int_0^1 \frac{d\Gamma_3}{d\mu_3} \frac{\left(1+\eta_1^1\right)}{\left[\left(1+\eta_1^1\right)^2 + k^2(1-\mu_3)^2\right]} d\mu_3 = \text{constant}$$

$$\frac{1}{4\pi}\left\{\int_0^1 \frac{-12(e+2)\mu_3^2 + 12(e+2)\mu_3 - 2(e+3)}{\mu_3^1 - \mu_3} d\mu_3 - k^2\int_0^1 \frac{\left[-12(e+2)\mu_3^2 + 12(e+2)\mu_3 - 2(e+3)\right]\left(\mu_3^1 - \mu_3\right)}{\left[4 + k^2\left(\mu_3^1 - \mu_3\right)^2\right]} d\mu_3\right\}$$
$$= 0;$$

the closed form solutions of the integrals were obtained in [2].

Now, collecting together the solutions of cases A and B, we obtain the following system of equations:

$$A_0\left[\arctan\left(\frac{k}{1-\eta_1^1}\right)\left(\frac{12(e+2)\left(1-\eta_1^1\right)^2}{k^3} - \frac{2(e+3)}{k}\right) + \arctan\left(\frac{k}{1+\eta_1^1}\right)\left(\frac{12(e+2)\left(1+\eta_1^1\right)^2}{k^3} + \frac{-2(e+3)}{k}\right) + \right.$$
$$\left. - \frac{6(e+2)\left(1-\eta_1^1\right)}{k^2}\ln\left(\frac{\left(1-\eta_1^1\right)^2}{k^2 + \left(1-\eta_1^1\right)^2}\right) - \frac{24(e+2)}{k^2} - \frac{6(e+2)\left(\eta_1^1+1\right)}{k^2}\ln\left(\frac{\left(1+\eta_1^1\right)^2}{k^2 + \left(1+\eta_1^1\right)^2}\right)\right] + \pi[-\Gamma_{01} +$$

$$-\Gamma_{01}\frac{A}{A^2+k^2} + +\Gamma_{01}\frac{A}{A^2+k^2}(\sqrt{\frac{2}{(A^2+k^2)(\sqrt{V}-a)}}G - \frac{2k}{(A^2+k^2)(\sqrt{V}-a)}F + \frac{1}{2(4+k^2)}R)] = 2\pi bw$$

$$\frac{A_0}{4\pi}[24\,(e+2)\,\mu_3^1 + \ln\left(\frac{1-\mu_3^1}{\mu_3^1}\right)^2\left(6\,(e+2)\,(\mu_3^1)^2 - 6\,(e+2)\,\mu_3^1 + (e+3)\right) +$$

$$\ln\left(\frac{4+k^2(\mu_3^1-1)^2}{4+k^2(\mu_3^1)^2}\right)\left(\frac{6(e+2)\left(4+k^2(\mu_3^1)^2\right)}{k^2} - (e+3)\right) + (12e+2)k^2\,(1+\mu_3^1)$$

$$[-\frac{1}{k^2} - \frac{\mu_3^1}{2k^2}\ln(\frac{4+k^2(\mu_3^1-1)^2}{4+k^2(\mu_3^1)^2}) - \frac{2}{k^3}(\arctan\left(\frac{k}{2}(\mu_3^1-1)\right) - \arctan\left(\frac{k}{2}\mu_3^1\right))]]+$$

$$\frac{k^2}{4}[(1-\mu_3^1)\Gamma_{01}\sqrt{\frac{\sqrt{4+k^2(1-\mu_3^1)^2}-(1-\mu_3^1)}{2k(1-\mu_3^1)\left(4+k^2(1-\mu_3^1)^2\right)}}(I_2 - H_2\sqrt{\frac{k(1-\mu_3^1)\left(\sqrt{4+k^2(1-\mu_3^1)^2}-(1-\mu_3^1)\right)}{2\left(4+k^2(1-\mu_3^1)^2\right)}})+$$

$$-\mu_3^1\Gamma_{01}\sqrt{\frac{\sqrt{4+k^2\mu_3^{12}}-k\mu_3^1}{2(4+k^2\mu_3^{12})k\mu_3^{12}}}(I_1 - H_1\sqrt{\frac{k\mu_3^1\left(\sqrt{4+k^2\mu_3^{12}}-k\mu_3^1\right)}{24+k^2\mu_3^{12}}})] = 0$$

where the quantities I_1, I_2, H_1 and H_2 are rational functions depending only on parameter $k=2h/b$.

The problem of optimum is not influenced by the total lift; so, for simplicity sake, we put $A_0 + \Gamma_1 = 1$; moreover, the optimum circulation distribution (that giving minimum induced drag) depends on e. The optimum value of e can be obtained by a numerical procedure, by minimizing the error; we define as error the L^1 norm of the difference between the right and the left hand sides of the second equation and, then, we have the following problem of minimum (the functions f, g and h have proper expressions)

$$Min_e \int_0^1 \left|A_0\left[(e+2)\,f(\mu_3^1) + g(\mu_3^1)\right] + h(\mu_3^1)\right|\,d\mu_3^1;$$

the result is that the minimum occurs for $e = -2$ and the resulting circulation distribution along the sidewings becomes linear, that is, butterfly shaped, according to the predictions of Prandtl. Finally, figure 4 shows a comparison between the present and the Prandtl results; the difference is small in the range of small h/b values (of interest in the case of the PrandtlPlane) and larger for high h/b values.

4. The PrandtlPlane aircraft

The Prandtl results on the best wing system were published at the beginning of the history of aeronautics, but they had no influence on the development of aircraft design. Following the introduction into service of light alloys, by the end of 10s, in Germany, the cantilever wing became the standard solution in aeronautics; even though the concepts on induced drag were well known, the structural solutions, with the presence of struts and cables in the biplanes, did not allow to minimize the total drag. The main idea presented in this paper is that materials and technologies available to-day in aeronautics makes it possible to design wing systems with a very low induced drag, according to the best wing system concept. This concept can be applied to any category of aircraft because, owing to the first Munk theorem, the induced drag is independent of the sweep angles of the wings. Supposing to maintain the same total wetted surface of the wings (including the horizontal tail) and the same wing span of a conventional configuration, the thickness of the PrandtlPlane profiles are about the half of a conventional one; on the other side, the PrandtlPlane wings are overconstrained beam systems. The comparison between cantilever and PrandtlPlane wings with the same wetted surface and wing span, is not immediate; anyway, an optimized PrandtlPlane wing is very stiff with regard to the static loads, even though some problems could occur when dealing with the coupling of dynamic modes as far as flutter is concerned [3]. Recent research activities show that these problems could be solved, especially by a proper use of tailoring of composite materials. In order to compare a PrandtlPlane and a conventional aircraft, we consider as an example a 600 seat category aircraft with about 500 tons of Maximum Take Off Weight. The proposed configuration is an aircraft with a 3 deck fuselage, two for passengers and the lower one for goods and luggage. The lifting system is made of two swept wings at different levels, connected at their tips by means of two vertical wings, so that the front view is box-shaped; in order to prevent flutter, the aft wing is connected to the bottom fuselage and the rear wing to the upper fuselage (but the opposite is also possible). The distance between the wings allows us to obtain a h/b ratio, for which the induced drag reduction is about 25 %. The main landing gear is positioned on central fuselage, inside proper fairings. The aircraft is powered by four engines, two under the aft wings and the other two at the rear fuselage; the position of the centre of gravity is subjected to small variations during cruise, with positive effects on optimization. Figure 3 shows an artist's view of the proposed configuration. The aerodynamic optimization can be obtained by a proper use of computer codes; owing to the complexity of the geometry proposed, these codes have been written for the purpose. The aerodynamic configuration of an aircraft is recursively analyzed with the aid of Computational Fluid Dynamic (CFD) codes, until satisfactory configurations are obtained. Thus, many different configurations can be analyzed and the simplicity of generating and modifying the aircraft geometry becomes fundamental. A new approach was developed [4], based on a software which was written in order to connect the following commercial programs in a proper way: an advanced CAD package, a pre- and post-processor and a CFD code. To start with, a domain is created with suitable dimensions with respect to the aircraft. In the second step, the volume of the aircraft is "excavated" in this solid domain; thus, the computational volume between the outer surface of the prismatic domain and the inner surface (that of the aircraft) is created. In the third step, the computational volume is meshed using the facilities of the CAD program. The codes used are: PTC-Pro/Engineer , MSC-Patran and Fluent . Using this parametric CAD code, it is possible to change the aerodynamic configuration of the aircraft by modifying the values of a few parameters; the properties relevant to boundary surfaces, as inlets, outlets, far-field conditions are then added by using the Patran code. Preliminary results using the Euler code show that the aft wing is free of shock waves and that the rear wing is critical at the wing root; it also appears that the design solution of the fuselage in that region has an important effect. Figure 5 shows a typical CFD result.

The high lift devices are designed and positioned on the whole wingspan, in such a way to maintain the "Best Wing System" configuration during take off and landing as well; so, the needed engine power in take off and landing is reduced and the same happens as far as vorticity, noise and emissions are concerned.

The results obtained with the aforementioned method, show that the lift coefficient is very high both during cruise and take off and landing and, besides, the aircraft is always stable with respect to stall conditions. So, take off and landing occur with a relatively small rotation on the runway; moreover, lift is always positive on both the wings. In the previous procedure, the generation of the initial geometry of the aircraft is one of the most time-consuming tasks; to shorten this time, a new automatic procedure was developed. The kernel of this approach is the same CAD code which, in this case, is used as main controller of the CFD (Fluent) and the FEM (MSC-Nastran) packages, running on different machines; the three packages are used in a network to obtain a multidisciplinary optimization of the aircraft configuration in which the aerodynamic and structural meshes are generated in an automatic way. This approach has proved to be efficient for generating conventional aircraft, but less for a PrandtlPlane, owing to the high number of parameters involved. To overcome this disadvantage, a new and faster approach is now under development by using the Matlab software as a programming language, in order to generate the geometry of the aircraft and the surface mesh; once the mesh is generated, it can be easily transferred into a commercial CFD and FEM software. With this new software, a PrandtlPlane configuration can be generated in less than one minute with a Personal Computer, compared to many hours needed before.

The results obtained show a high aerodynamic efficiency and, also, that the wings are stiff as for static deformations, but flutter could be more critical owing to the possible coupling between pitch rigid body and bending-torsional modes [6].

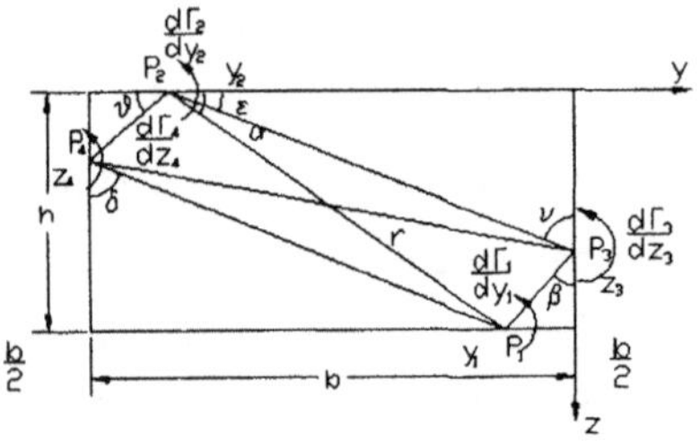

Figure 1

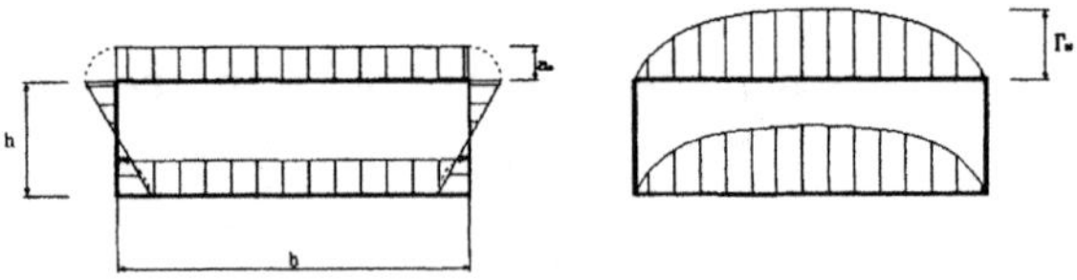

Figure 2

Figure 3

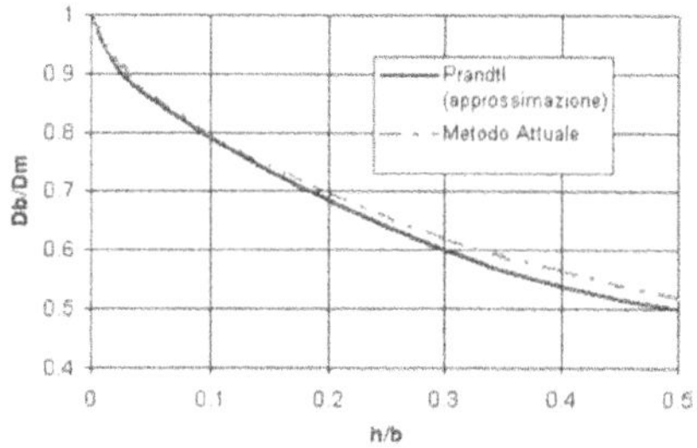

Figure 4

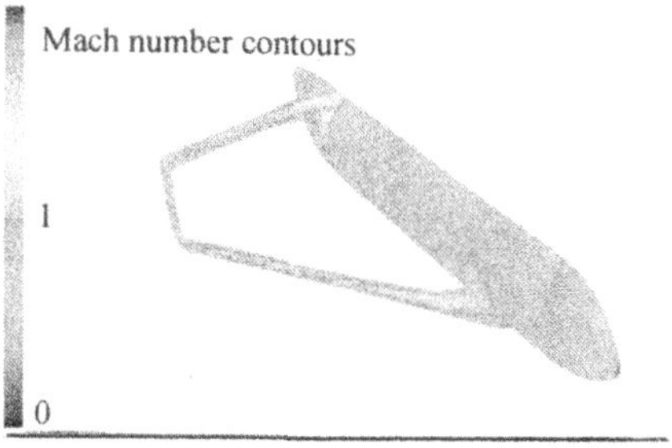

Figure 5

References

[1] L.Prandtl, Induced Drag of Multiplanes, NACA TN 182, (1924).

[2] A. Frediani, G. Montanari, M. Pappalardo, Sul problema di Prandtl della minima resistenza indotta di un sistema portante (in Italian), Proceedings of the XV Congress of AIDAA, Turin 15-19 November, (1999), pp.267-278.

[3] Frediani A., Chiarelli M., Longhi A., D'Alessandro C. M., Lombardi G., Structural Design and Optimization of the Lifting System of a Non-Conventional New Large Aircraft, CEAS International Forum on Aeroelasticity and Structural Dynamics, Rome, (1997).

[4] A. Frediani, A. Longhi, A mesh generation Procedure for Optimization in computational Fluid Mechanics, L'Aerotecnica Missili e Spazio, Vol. 78, N.1-2, (January-June 1999), pp 8-13.

[5] A. Longhi, M. D'Alessandro Caprice, E. Troiani, Aeroelastic Analysis of a New Large Airliner with a Non-Conventional Configuration, 24th MSC European Conference, Naples, (1997).

Poster Session 1

Assessment of the e^N Method as a Transition Prediction Tool for Zero Pressure Gradient Flows with and without Boundary Layer Suction

J.E. Ellis, S.A. Walsh & D.I.A. Poll
Cranfield College of Aeronautics
Cranfield, Bedfordshire, MK43 0AL
United Kingdom

Summary

This paper compares theoretical predictions for the onset of transition to turbulence with experimental results observed in a 2-D zero pressure gradient boundary layer (typical unit Reynolds number 2 million) subjected to suction through a perforated strip extending 20cm in the streamwise direction, for suction coefficients in the range 0.001 to 0.01. In the zero suction case experimental observations agreed well with theoretical predictions obtained by using the e^N method with a transition criterion of N=9. However, with suction applied this criterion failed to predict transition onset in the post-suction region. The discrepancy is attributed to disturbance effects not accounted for in linear stability theory, made apparent, either directly or indirectly, by the application of boundary layer suction.

Introduction

The design of a hybrid laminar flow control (HLFC) system requires a detailed understanding of the laminar boundary layer development, including both its mean profile and its stability characteristics. Critically, there is a requirement to accurately predict the location at which laminar flow breaks down to be replaced by turbulence; a process know as transition.

Currently transition onset is predicted either by experiment or using computational methods. Although experiments can accurately measure the location of transition, the high cost of flight tests and the difficulty in scaling wind tunnel data make this an impractical design tool. Alternatively, there are relatively inexpensive computational methods which model the growth of the disturbances which cause transition. For two-dimensional flows, linearised theory is used to determine the growth rates of small amplitude disturbances. This theory accurately represents the physical situation and the predicted disturbance growth rates agree well with experimental observations. However, transition occurs physically when the small amplitude disturbances begin to grow in a non-linear fashion; this is, by definition, beyond the scope of linearised theory and as a result the method cannot be used directly to predict transition. To overcome this, researchers have produced an empirical correlation between experimentally determined transition onset and theoretical disturbance amplification levels. Thus, the correlation provides an estimate of transition location; this being said to occur at a critical upper value of disturbance amplification. This convenient augmentation of linearised stability theory has resulted in the semi-empirical e^N method which is the current state-of-the-art design tool.

The aim of the present paper is to assess the performance of the e^N method for 2D flows incorporating discrete suction strips with large suction coefficients followed by a region of subsequent boundary layer regrowth in a near neutral pressure gradient. Experimental data has been produced to provide a physical benchmark against which the computational results can be assessed.

Experimental method

Several series of wind tunnel experiments were performed to provide test-data against which the e^N method results could be assessed. Both non-suction and suction test cases were considered. The limited streamwise extent of the suction strip and its location aft of a small adverse pressure gradient region required the use of suction coefficients larger than those typically considered. For HLFC design, the operational trade-offs between suction extent and suction rate are not yet clear and, in any case, are likely to depend on the precise application and other external constraints. In effect, the current experiment modeled one extreme of the HLFC concept, and afforded an opportunity to assess the e^N method for suction rates approximately one order of magnitude larger than normally considered. The results provided experimental data which related transition location to boundary layer suction rate and free stream conditions. Scaling effects were avoided by using the e^N method to model the exact conditions of the experiment.

The test-bed wind tunnel model was a deep flat plate with an elliptic nose section, as illustrated in Figures 1 and 2. The model was static pressure tapped along its length. Suction could be applied to the model by means of a permeable strip aft of the nose.

The model was mounted horizontally between the side walls of a low speed, low turbulence, closed return wind tunnel. The working section was 0.5m x 0.5m. The maximum free stream speed was 45 m/s and the free stream turbulence was less than 0.1% across the speed range. At 465mm (as measured along the surface) downstream of the leading edge was a perforated titanium suction surface which extended 200mm in the streamwise direction. The suction surface was interchangeable which allowed surfaces with different perforation diameters to be tested. In the current experiments two surfaces were tested: the first had a 0.0625mm perforation diameter and the second had a 0.125mm perforation diameter. The two surfaces gave the same results within the expected scatter of the experiment. During the non-suction cases the perforated surface remained installed on the model.

The perforation pattern was geometrically similar for both surfaces tested. Each row of perforations was at constant chordwise distance. Adjacent holes in the same row were 10 diameters apart. Adjacent rows were also 10 diameters apart. To avoid holes lining up in the streamwise direction each successive row was staggered by 3 diameters. This produced a 'slanted grid' perforation arrangement constructed from a repeating pattern of parallelograms. A simple geometrical calculation gives the open area / total area ratio equal to 0.79%. This value is independent of hole size.

The suction was provided through four plenum chambers. Each extended 50 mm in the streamwise direction and covered the whole span. Unless otherwise stated all tests can be considered to use only the first chamber, between x=465mm and x=515mm.

During all tests, outflow was avoided through the inactive plenum chambers by simply shutting off the valve to that chamber. This precaution was sufficient as the external static pressure did not vary significantly over the chambers (see Figure 3) removing any problems of partial inflow and partial outflow on the same chamber.

The surface static pressure distribution is given in Figure 4, where an arbitrary chord length of 1600mm has been chosen. After the pressure peak over the nose region, the pressure gradient becomes nearly zero. During testing only the region $0<x/c<0.45$ was actively measured as this was the region of preliminary influencing the transition process. Zero suction transition onset typically occurred at around $x/c=0.4$. If values of C_p at $x>0.45$ were required for input into the stability calculations, then values were calculated by extrapolation as shown in Figure 4. This is a reasonable assumption as the wind tunnel roof lining had been specifically shaped to maintain a constant static pressure downstream.

A single hotwire was used to detect turbulence in the boundary layer. This was mounted on a 2D traverse, which allowed movement in the chordwise direction, x, and normal to the surface, y. The traverse can be seen in Figure 2. The spanwise position remained fixed at the midspan. Attaching a flattened Pitot to the traverse allowed the boundary layer profiles to be measured in both the suction and non-suction cases.

Tests provided a series of transition onset Reynolds numbers and suction rates. Combined with the velocity profile data described previously, the onset locations were then used as benchmarks against which to compare the computational boundary layer solutions and stability calculations described in the following sections.

Boundary layer computations

For the present exercise all boundary layer calculations were performed using a typical finite difference boundary layer code. The code, designated CBLF, had been validated extensively prior to the current study and is known to provide standard velocity profiles to five significant figure accuracy [1]. The boundary layer is computed by solving the thin shear layer equations using a two point finite difference Keller Box method [2].

Typical boundary layer output is shown in Figure 4 where displacement thickness development is given for three test cases. In all cases the freestream conditions and the pressure distributions are identical. Heat transfer is zero, but constant velocity surface transpiration (suction) occurs between $465mm<x<515mm$ ($0.291<x/c<0.322$). Upstream of the suction location, the results from all three cases are coincident. Downstream, the zero-suction case continues its natural development, while in the other two cases the boundary layer is thinned by suction. In the post-suction region, boundary layer re-growth occurs.

For these particular test cases experimental data were available from previous work [3]. This provides some validation of the code for modeling boundary layer suction cases, the small discrepancies within the limits of experimental scatter. Demonstrating agreement between theory and experiment is particularly important if the boundary layer data are to be used for a subsequent stability analysis. Clearly a stability analysis based on physically unrepresentative boundary layer data cannot reasonably be expected to produce physically relevant results.

Stability calculations

The e^N method has been employed to predict transition to turbulence in the laminar two-dimensional boundary layer. The method uses linear stability theory, based on the Orr-Sommerfeld equation, more details of which can be found in Mack [4]. Its implementation is the numerical solution of this equation, with stability data is usually presented in terms of N, a convenient non-dimensional measure of disturbance amplification, defined as the natural logarithm of the amplitude ratio A_1/A_0.

For the present work, stability data was obtained using a typical Orr-Sommerfeld solver, designated MELISSA. MELISSA solves the incompressible Orr-Sommerfeld equation, spatially, using a shooting method with a fixed step size Runge-Kutta integrator, a Newton Raphson search procedure and Gram Schmidt orthonormalisation to control error growth.

Typical linear stability theory output in shown in Figure 5. In the experiment, a Reynolds number of 2.04×10^6 produced transition onset at x/c= 0.675. The intersection with the stability envelope corresponds to an N factor of 9.6. This N factor at which transition onset is observed experimentally is designated N(xt/c).

Results

By measuring the transition location experimentally, the prediction process can be reversed to produce an assessment of the method. The N factor calculated at the location of experimental transition is designated N(xt/c), with the e^N method relying on N(xt/c) being approximately constant in all cases. Correlations between transition location and N factor have produced mixed results, with wind tunnel experiments usually producing lower transition N factors than those experienced in flight. Having performed stability calculations for flight test data from a Fokker F100 NLFC glove, Schrauf et al[7] reported Tollmien-Schlichting dominated transition N factor averages of about 21, which reduced to about 14 when curvature effects were included.

For three-dimensional flows the issue is further complicated by cross-flow instability and attachment line effects. However, for two-dimensional flows, N=9 seems to have been generally adopted, although without any conclusive validation. Flow control applications, particularly cases involving boundary layer suction, have yet to be validated to the stage where any empirical criterion can be used without caution. This section details the validation work performed in the present investigation, treating non-suction cases and suction cases separately.

Experimental transition data were obtained over the Reynolds number range $0.5 \times 10^6 < R_c < 4.5 \times 10^6$. As expected, these data show a forward movement of the transition onset location with increasing Reynolds number. Figure 6 compares the experimental results with lines of constant N factor. The close agreement with the N=9 criteria is encouraging, predicting the transition onset location to within 0.1c of the measured value. It should be noted that the N=9 criterion was originally based on experiments with non-porous surfaces, while the current arrangement incorporated a perforated strip. The presence of the perforated surface seems to have no effect on the N=9 criterion, which suggests that the perforations do not introduce significant disturbances into the flow. However, this may not be the case for larger perforation sizes, increased porosity, or for situations where outflow exists.
The current validation of the N=9 transition criteria for the non-suction cases is in good

agreement with other established publications in the field, verifying that this implementation of the e^N method is essentially comparable to previous work. This increases the credibility of the computational methodologies employed and provides a platform from which the flow control cases can be investigated.

When boundary layer suction is applied to a disturbed laminar flow, the effect is to produce local disturbance damping and a reduction in N factors for all frequencies in the suction region. Figure 7 shows a typical stability output. Figure 8 shows the values of N(xt/c) in the suction cases. There are four single-chamber suction cases with transpiration velocities ranging from 0.034m/s to 0.227m/s; in these cases N(xt/c) ranges from 3 to 8. In addition, there is a four-chamber suction case with N(xt/c) values ranging from 0 to 4. Clearly, all these results are significantly less than the N(xt/c)=9 criterion inferred from the non-suction data.

More importantly, very large errors are produced if the N(xt/c)=9 criterion is used for transition prediction. This is illustrated by Figure 9 which compares the theoretical and experimental results for the v_w=-0.034 m/s suction case. Experimental transition onset locations are typically 0.2c to 0.4c upstream of the calculated N=9 line. Taking a specific example, at $R_c=3.1\text{x}10^6$, the experimental transition onset location is about x/c=0.4; an assumption of N(xt/c)=9 predicts a transition location close to x/c=0.65 and a corresponding error of $\Delta x/c=0.25$. For larger suction regions (the four chamber case) where the N(xt/c) values are much lower, the errors increase. As the error approaches and exceeds 0.5c it is reasonable to conclude that some form of correction will be required if the method is to continue to be used as a practical design tool. No suitable correction, modification or quantitative correlation between a correction factor and suction was found during this study.

The widely scattered results in the suction cases bring into question the applicability of the e^N method for flow control applications of this type. It should be noted that the suction rate considered in the experiment are larger than those considered for production NFLC configurations to date, but still well within the range useful to the designer.

The cause of the discrepancy could be related to either the experiment or the computations. It is believed that both the computations and the experiment were performed correctly. Good agreement was observed between the experimental and computational boundary layer profiles and integral parameters, even in the post-suction region where the very thin boundary layer made experimental measurement very difficult. The stability code had already been extensively validated[1] and the stability analyses were performed carefully and repeatedly to test the sensitivity of suction distribution, pressure distribution and node spacing; no influences were found that could drive the discrepancies observed in Figure 8. For severe suction cases (high C_q, all four chambers active) the post-suction region developed from very small boundary layer thickness and the stability analysis was found to be comparable to the Blasius case, as expected. Overall, no problems were observed or suspected that could explain the gross variation of the transition N factors. It was therefore concluded that experiment was producing an adverse effect not modeled within the computations.

Two main candidates were identified for this effect, both involving disturbance introduction or growth not modeled by non-linear theory.

Firstly, it is possible that when a large N factor is achieved, subsequent suction will not reduce disturbance levels in the manner predicted by linear theory. In experiments performed on the same wind tunnel model as that used in the current study it was found that the efficacy of suction applied a short distance upstream of natural (non-suction) transition was greatly reduced[3]. This effect might be related to the pathological cases described by Atkin and Schrauf[8] where experimentally observed transition occurs downstream of the maximum N factor, often in a region where linear theory predicts that disturbances are being damped. If linear theory were to break down prior to transition onset, this would have an important implications for the design specification of the suction system. We would not expect this phenomena to be apparent for monotonically increasing N factor cases, but for HLFC test cases the effect could be significant.

Secondly, it is possible that the application of suction through discrete holes is causing new disturbances to be introduced into the boundary layer. It is documented that suction through perforations produces trailing vortex structures which form around a perforation and propagate downstream. For large perforation diameters (typically 1mm) these trailing vortices have been observed experimentally with the aid of smoke-visualisation[5] and LDV equipment[6]. While trailing vortex structures have not been observed for the very small diameter perforations used in the current experiments, it seems reasonable to assume that the same physical effect remains, albeit on a smaller scale.

Other work performed on the wind tunnel model used in the current study has shown that the disturbance effect is dominated by perforation diameter[3]. For large perforation diameters, typically 2mm, suction acts as a boundary layer trip, this effect gradually becoming less severe as perforation diameter is reduced. Extrapolating to very small perforation diameters the disturbance effect is likely to be much reduced, possibly to the case where it can no longer trip the flow, only limit the extent of its development in the laminar state. This would explain the reduced values for N(xt/c) with the trailing vortices, either directly or indirectly, increasing the disturbance level in the boundary layer and causing transition upstream of the theoretical predictions.

Suction conditions in the current experiments were of the correct order of magnitude to suggest some adverse effect of transition using the criterion produced by Ellis and Poll [9] although the current situation is slightly different as they attempted to define the more severe oversuction condition where transition moves forward of the natural (no-suction) transition onset point.

Conclusions

Comparison of the stability output with experimental transition data showed good agreement with the N=9 criterion for the non-suction cases. However, for the suction cases agreement was poor, with no clear onset criterion and transition N factors ranging from 0 to 8. This unexpected difficulties in correlating suction results may have been encountered for the first time in the current experiments because of the relatively large suction rates used, or they may be related to the pathological test cases previously observed on European programmes.

It seems that the addition of boundary layer suction has increased the physical complexity of the experiments, such that linear stability theory is no longer representative and transition is observed upstream of the theoretical predictions. From this we can suppose that disturbance levels in the experiment were larger than those modeled by the theory and that the causal mechanism was related, either directly or indirectly, to the application of suction. This would suggest the need for caution when using the e^N method for this type of design, and further investigation into the potentially adverse effects of suction through a large array of discrete perforations.

Nomenclature

c	nominal model chord (c=1.6m)
C_p	pressure coefficient
N	N factor of a disturbance
$N(x_t/c)$	N factor at experimental transition location
R_c	$=U_\infty c/\nu$
U_∞	free stream velocity
v_w	wall transpiration velocity (negative value indicates suction)
x	distance measured along the surface of the model from the nose
x_t	experimentally determined location of transition onset
ν	kinematic viscosity

Subscripts

w	wall
c	chord
t	transition onset

References

[1] Walsh S. A. "Surface transpiration effects on low speed flows" University of Manchester Ph.D. Thesis 1997

[2] Keller H. B. "A new difference scheme for Parabolic problems" Numerical Solution of Partial Differential Equations II, Academic Press, New York, 1970

[3] Ellis J. E. "Laminar and laminarising boundary layers by suction through perforated plates" University of Manchester Ph.D. Thesis 1996

[4] Mack L. M. "Boundary Layer Stability Theory" AGARD-R-709 June 1984

[5] Goldsmith J. "Critical laminar suction parameters for suction into an isolated hole or a single row of holes" Northrop Aircraft Inc. Report no. BLC-95 1957

[6] MacManus D., Eaton J., Barrett R., Rickards J., Swales C. "Mapping the flowfield induced by a HLFC perforation using high resolution LDV" First European Forum on Laminar Flow Technology, Hamburg, Germany 1992

[7] Schrauf G., Perraud J., Vitiello D., Lam F. "Comparison of boundary layer transition predictions using flight test data" Journal of Aircraft p891-897 Vol.35 No.6 November-Decemeber 1998

[8] Atkin C. J., Schrauf G. H. "Progress in linear stability methods for design applications" European Congress on Computational Methods in Applied Sciences and Engineering, Barcelona, September 2000.

[9] Ellis J. E., Poll D. I. A "Laminar and laminarising boundary layers by suction through perforated plates" Second European Forum on Laminar Flow Technology, Bordeaux, 1996

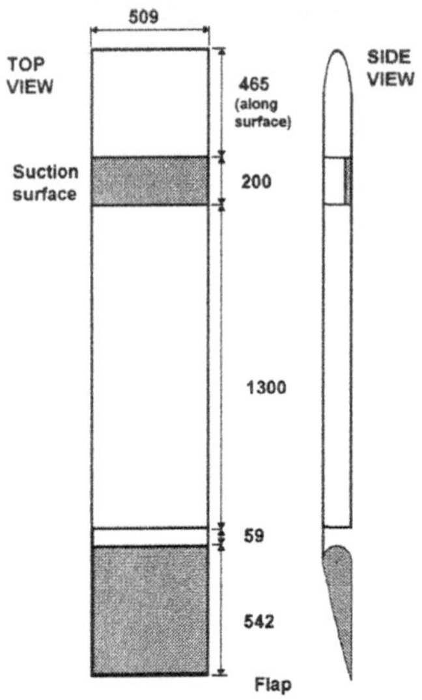

Figure 1 Sketch of wind tunnel model

Figure 2 Wind tunnel model as mounted in working section. Probe and traverse system visible

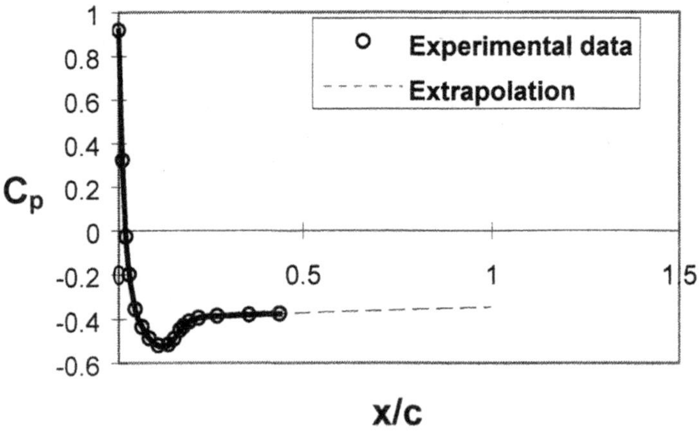

Figure 3 Experimental surface static pressure distribution over the model

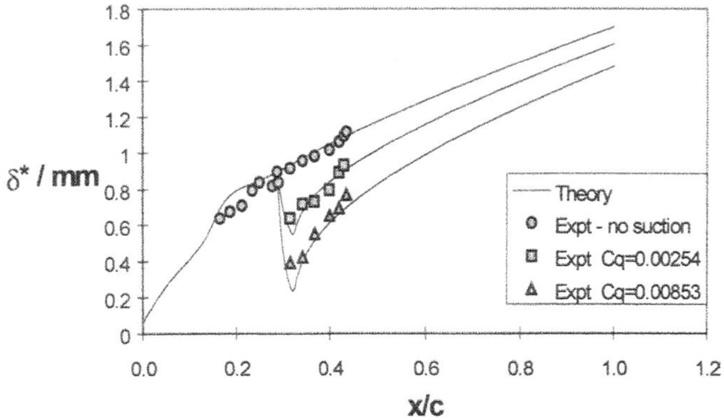

Figure 4 The variation of displacement thickness with surface location, in suction and zero-suction cases $Rc=2.47x10^6$ v_w=-0.06904m/s and v_w=-0.23185m/s

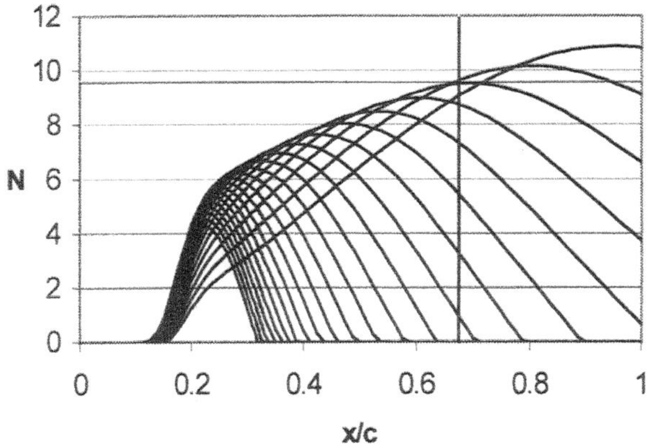

Figure 5 Typical N factor plot for a zero-suction case (R_c=2.04x10^6); Experimental xt/c=0.675; N(xt/c) = 9.62

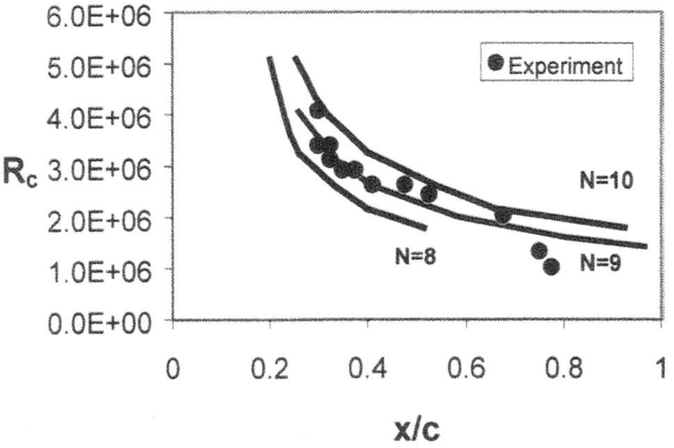

Figure 6 Experimental transition data shown against lines of constant N factor

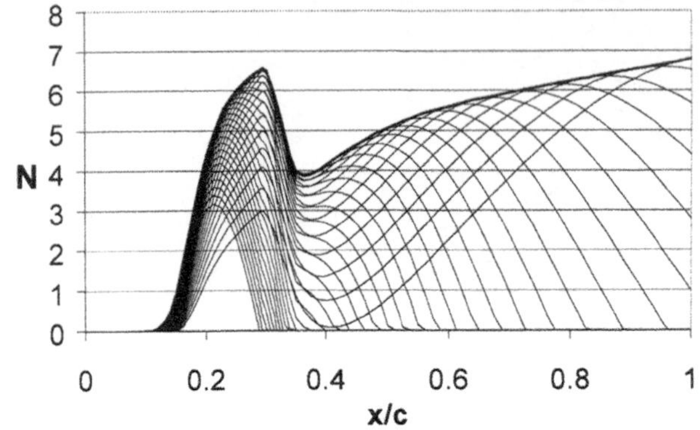

Figure 7 Typical N factor plot for a suction case (R_c=2.04x10^6 v_w=-0.068 m/s) Experimental xt/c=0.725; N(xt/c) = 5.60

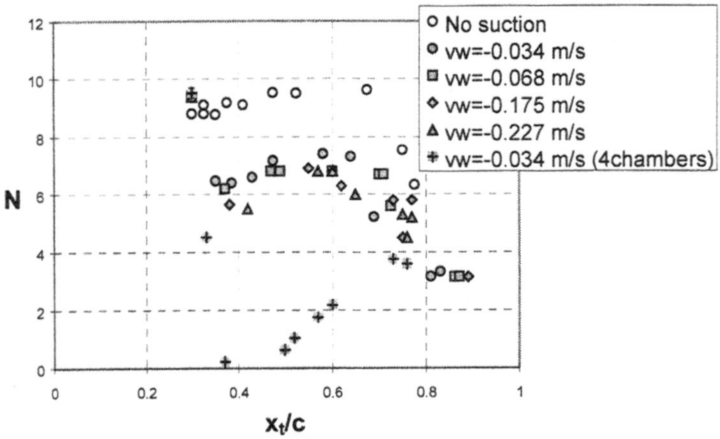

Figure 8 N factors existing at experimentally observed transition onset location for suction cases

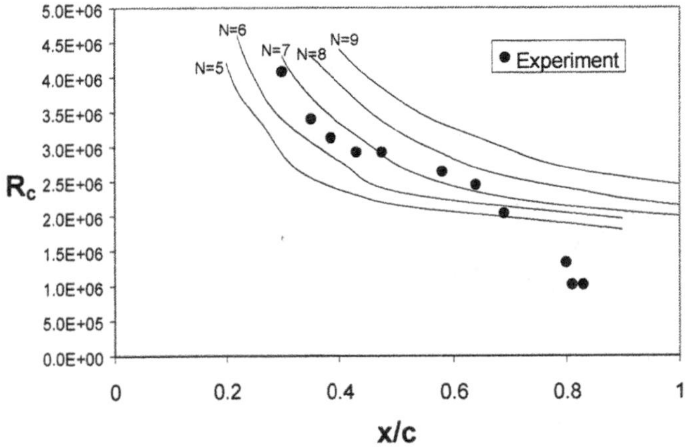

Figure 9 Experimental transition data shown against lines of constant N factor for the v_w=-0.034 m/s suction case.

Active Control of Nonlinear Disturbances in 2-D Boundary Layers

C. Gmelin, U. Rist, S. Wagner
Universität Stuttgart, Institut für Aerodynamik und Gasdynamik
Pfaffenwaldring 21, D-70550 Stuttgart, Germany

Summary

The purpose of the current investigation is the development of techniques which allow the active control of disturbances occurring in transitional boundary layers. Due to the inability of techniques following the wave-superposition principle to damp nonlinear high-amplitude disturbances we favour an approach called *ω_z-control*. It is shown, that the direct feedback of instantaneous flow signals from the wall is well suited to drive the active control without additional control loops between sensor and actuator. In our case the spanwise vorticity (ω_z) or its first streamwise derivative on the wall is sensed, multiplied by a certain factor A and prescribed as a new boundary condition at the wall with some time delay Δt. Investigations using Linear Stability Theory (LST) show a strong dependence of achieved damping rates from the phase between $\omega_{z,w}$ and wall-normal velocity v_w. With a properly adjusted phase even in very unstable base-flows (e.g. with adverse pressure gradient) a remarkable transition delay can be achieved. With aid of Direct Numerical Simulation (DNS) different transition scenarios have been simulated to confirm the LST-results and to study the active control of high amplitude disturbances in stages of transition where nonlinear interactions have already set in.

1 Introduction

In times of emerging fuel costs there is a strong interest in concepts to actively control the laminar-turbulent transition with the goal to reduce skin friction. In contrast to using the wave superposition principle [1] new approaches appeared, directly modifying the modal structure of the influenced disturbances. One of these methods is the *ω_z-control* which is, in contrast to wave superposition methods, able to reduce even nonlinear disturbances near the final transition to turbulence. In this case no kind of "anti-disturbance" is overlaid but wavenumber and amplification rate of existing eigenfunctions are changed significantly (cf. section 3).

Two different base flows are considered: a Blasius and a Falkner-Skan boundary layer with adverse pressure gradient (Hartree parameter is $\beta_H = -0.1$). Besides extensive investigations using LST the behaviour of nonlinear disturbances leading to a K-breakdown in a Blasius boundary layer is examined using spatial DNS (cf. section 5).

2 Control Principle

To damp disturbances in boundary layers we use the feedback of instantaneous signals of the spanwise vorticity measured at the wall. These signals are multiplied by an amplitude factor $|A|$ and are prescribed as a v-boundary condition after a time delay Δt at the wall which is necessary to produce the phase shift Θ shown in Fig.1 a).

Investigations (see section 3) using LST showed, that control is most effective for the case of $\Theta = \frac{\pi}{2}$, i.e. a phase shift of 90 degrees which can also be achieved by setting v_w proportional to $\frac{\omega_z}{x}$. To eliminate the inherent instability of the altered procedure a modified control algorithm was developed:

$$v_w = \frac{\partial \omega_{z,w}}{\partial x} \cdot \frac{|\omega_{z,w}|_{max}}{|\frac{\partial \omega_{z,w}}{\partial x}|_{max}} \cdot |A| \,. \tag{2.1}$$

The maxima in eqn. 2.1 are each calculated over approximately one wavelength up- and downstream. This algorithm is able to produce a control signal proportional to the first derivative of $\omega_{z,w}$ whose amplitude is always lower than $|A| \cdot \omega_{z,w}$. A detailed discussion of the underlying physical mechanisms can be found in [3]. In contrast to approaches based on the superposition of disturbances with opposite phase working with linear transfer functions between sensor and actuator like the FIR-filter concept (Fig.1 b), for details see [4]), no downstream transport of information has to be modeled here. Furthermore, there is no need of an error sensor located far downstream to tune control parameters, it represents a pure feed-forward approach.

3 Results of the Linear Stability Theory

To study the behaviour of linear instability waves under the influence of active ω_z-control investigations using LST were undertaken. Therefore, the boundary conditions at the wall for the Orr-Sommerfeld (and Squire equation) had to be changed (index w denotes wall properties) to

$$\begin{aligned} v_w &= A \cdot \omega_{z,w} \\ \text{with } A &= |A| \cdot e^{i\Theta} \quad , \end{aligned} \tag{3.2}$$

where $|A|$ is the amplitude factor and Θ is the phase difference between v_w and $\omega_{z,w}$ similar to the time delay used in the simulations. Fig.2 shows the influence of active control on the unstable area in the stability diagram of the two scenarios we looked at. For the Blasius case (a) already small amplitudes are sufficient for a strong damping effect. If the amplitude factor $|A|$ is larger than approximately $5 \cdot 10^{-5}$ for a phase angle of $\Theta = 0$ no unstable area is present any more. For the Falkner-Skan base flow with adverse pressure gradient (b) even with large amplitudes only small decay of the disturbances is achievable. Only a proper phase shift of $\Theta \approx \frac{\pi}{2}$ yields a remarkable damping effect due to the stronger instability of the Falkner-Skan base flow. Detailed investigations concerning the dependence of the eigenvalues on the phase angle (Fig.3) show that even for the Falkner-Skan case attenuation can be reached only for a very narrow Θ-range (hatched areas in Fig.3 a) and b)). This leads directly to the necessity to adjust Θ to approximately 90 degrees and to the development of the modified control algorithm already mentioned above. Although the introduction of a phase shift in the Blasius case is not obligatory to achieve positive effects even in this case active control can be improved by applying the new control law.

4 Numerical Method

To investigate the behaviour of nonlinear waves in the Blasius-boundary layer a number of calculations using Direct Numerical Simulation (DNS) were undertaken. All simulations were performed in a rectangular integration domain with the spatial DNS-code developed by Konzelmann, Rist and Kloker [5][6][7]. The flow is split into a steady 2D-part (base flow) and an unsteady 3D-part. The x-(streamwise) and y-(wall-normal) directions are discretized with finite differences of fourth-order accuracy and in the spanwise direction z a spectral Fourier representation is applied. Time integration is performed by the classical fourth-order Runge-Kutta scheme. The utilized variables are normalized with $\tilde{U}_\infty = 30\frac{m}{s}$, $\tilde{\nu} = 1.5 \cdot 10^{-5}\frac{m}{s^2}$ and $\tilde{L} = 0.05m$ ($\tilde{}$ denotes dimensional variables):

$$x = \frac{\tilde{x}}{\tilde{L}} \quad , \quad y = \frac{\tilde{y}}{\tilde{L}} \quad , \quad z = \frac{\tilde{z}}{\tilde{L}} \quad , \quad t = \tilde{t} \cdot \frac{\tilde{U}_\infty}{\tilde{L}} \quad ,$$

$$u = \frac{\tilde{u}}{\tilde{U}_\infty} \quad , \quad v = \frac{\tilde{v}}{\tilde{U}_\infty} \quad , \quad w = \frac{\tilde{w}}{\tilde{U}_\infty} \quad , \quad Re = \frac{\tilde{U}_\infty \tilde{L}}{\tilde{\nu}} = 10^5 \quad ,$$

where u, v and w are the components of the unsteady velocity disturbances.
This leads to the dimensionless frequency $\beta = \frac{2\pi \tilde{f} \tilde{L}}{\tilde{U}_\infty}$, where $\tilde{f}$ is the frequency in $[Hz]$ and the dimensionless spanwise vorticity $\omega_z = \frac{\partial u}{\partial y} - \frac{\partial v}{\partial x}$.

5 DNS Results for Nonlinear Disturbances

As a test case for the effect of the ω_z-approach on disturbances with large amplitude a typical K-breakdown scenario (dotted lines in Fig.5) is used where a fundamental mode (1,0) with large amplitude and a stationary 3D-disturbance (0,1) (the first index denotes multiples of the frequency β, the second multiples of the basic spanwise wave number $\gamma = 20$) are excited initially. Because of nonlinear interactions the 3D-mode (1,1) arises and falls in resonance with the fundamental 2D-mode (modes (1,0) and (1,1) share the same wave number from $x \approx 3.4 \ldots 4.0$). The other modes shown appear due to higher order nonlinear combinations. They demonstrate transition to turbulence by generation of higher harmonics and the mean flow distortion (0,0). When the strongly amplified 3D-waves have reached the amplitude level of the fundamental mode, saturation sets in and transition to turbulence takes place. Applying ω_z-control to this scenario two main control effects can be distinguished: direct damping of nonlinear disturbances and the affection of the resonant behaviour. The first is comparable to the linear case where ω_z-control is able to directly damp Tollmien-Schlichting(TS)-disturbances. When active control is applied, an immediate transport of fluctuation energy from disturbance to base flow sets in, reducing the amplitude of the present disturbances [3].

In opposition to more traditional approaches like FIR-filters active feedback of instantaneous flow data for active control is better suited to control transition even in cases when strong nonlinear interactions have already taken place. In the present concept no downstream transport of information has to be modeled. Therefore, the main reason for the failure of linear approaches is removed, the control doesn't have to "know" anything about the processes happening further upstream. Furthermore, the new concept doesn't require strictly wave like disturbances, a point which is very important in regions close to transition when the large-amplitude 2D-wave is strongly deformed due to nonlinear interactions and the growing 3D-waves, reaching similar amplitude levels.

In Fig.5 a) the effect of ω_z-control with optimized phase, using the modified control algorithm (2.1) is shown compared to the uncontrolled K-breakdown case. Control is applied immediately before the final breakdown to turbulence. Looking at the phase velocities of the modes (1,0) and (1,1) (Fig.6) one can see that phase-synchronization and therefore fundamental resonance has already set in. Anyhow, a strong damping effect is visible, the fundamental 2D-mode is reduced to an amplitude-level below 1% compared to the uncontrolled case and the accompanied 3D-modes follow after a short delay leaving large stationary modes (0,1) and (0,2). Fig.5 b) shows the damping capability compared to four different cases using a FIR-filter and therefore linear superposition of anti-phase waves to reduce nonlinear disturbances. Here the bars at the bottom of the picture denote the position of the control-strips (Index CS) of the different runs where disturbances with opposite phase are introduced. The FIR-cases have been discussed in detail in [4]. The achievable amplitude reduction of the FIR-concept is becoming worse in the vicinity of nonlinearity close to transition. When the control-strip is positioned at $x_{CS} = 2.42$ almost "linear" reduction is achievable whereas control fails completely when applied at $x_{CS} > 3.6$, in regions where disturbances are no longer wave-like. On the other hand, the ω_z-control is able to damp even those nonlinear 2D- and 3D-disturbances occurring just before transition sets in. With regard to the instantaneous flow data presented in Fig.4 it is possible to suppress the appearance of transitional structures like high-shear layers or Λ-vortices with control applied approximately one to two wavelengths upstream (Fig.4 a) and b)). Comparing wall-normal velocity v and pressure p at the wall (Fig.4 c)) a proportionality of v_w to $\frac{\partial p}{\partial x}$ can be distinguished, similar to investigations using compliant walls to control transition [2], implying similar control-mechanisms for both approaches.

The second effect mentioned above, the altering of the resonant behaviour can hardly be separated from the direct attenuation. From Fig.6 it is obvious, that both resonant modes (1,0) and (1,1) do not share the same wave number and phase speed any more immediately under the influence of control ($x > 3.46$) long before direct attenuation can be recognized ($x \gtrsim 3.7$, Fig.5 a)). This indicates a disruption of resonance forced by active control due to the altered wave speeds of the dominant modes.

6 Summary

With the aid of Direct Numerical Simulations (DNS) it was possible to develop a simple, yet effective control algorithm to actively control the laminar-turbulent transition occurring in a 2D boundary layer. It combines two main effects: the direct attenuation and a reduced resonance according to an altered phase velocity of the involved modes. Calculations using Linear Stability Theory (LST) show a strong dependence of the resulting wave number and amplification rate on the chosen amplitude and phase difference between ω_z (sensed) and v_{wall} (stimulated).

It is shown that this approach works very well even close to transition where the boundary layer instabilities have reached a highly nonlinear stage. Investigation concerning the application of the ω_z-approach to more complex scenarios like wave packets eg. are under the way. Further research has to show how far transition can be shifted downstream and whether a complete relaminarisation of the flow is possible using this approach.

References

[1] M. Baumann and W. Nitsche. Investigations of Active Control of Tollmien-Schlichting Waves on a Wing. In R.A.W.M. Henkes and J.L. van Ingen, editors, *Transitional Boundary Layers in Aeronautics*, volume 46, pages 89–98. KNAW, Amsterdam, North Holland, 1996.

[2] C. Davies. *Evolution of Tollmien-Schlichting Waves Over a Compliant Panel.* PhD thesis, University of Warwick, 1995.

[3] C. Gmelin and U. Rist. Active control of laminar-turbulent transition using instantaneous vorticity signals at the wall. *submitted to Phys. Fluids*, 2000.

[4] C. Gmelin, U. Rist, and S. Wagner. Investigations of active control of wave packets and comparable disturbances in a Blasius boundary layer by DNS. In Hilbig Nitsche, Heinemann, editor, *Notes on Numerical Fluid Mechanics*, volume 72, pages 194–201. Vieweg-Verlag, Braunschweig, 1999. STAB-Symposium, Berlin, Germany, 1998.

[5] M. Kloker. *Direkte Numerische Simulation des laminar-turbulenten Strömungsumschlages in einer stark verzögerten Grenzschicht.* Dissertation, Universität Stuttgart, 1993.

[6] U. Konzelmann. *Numerische Untersuchungen zur räumlichen Entwicklung dreidimensionaler Wellenpakete in einer Plattengrenzschicht.* Dissertation, Universität Stuttgart, 1990.

[7] U. Rist and H. Fasel. Direct numerical simulation of controlled transition in a flat-plate boundary layer. *J. Fluid Mech.*, 298:211–248, 1995.

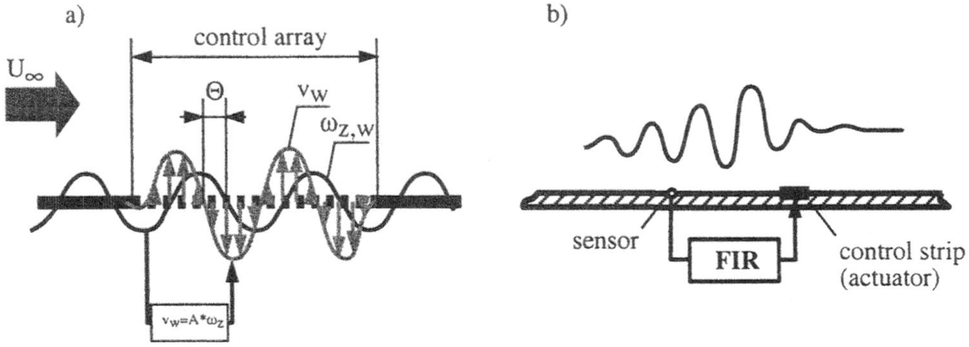

Figure 1 Control principles: a): ω_z-control. b): active control via FIR-filter

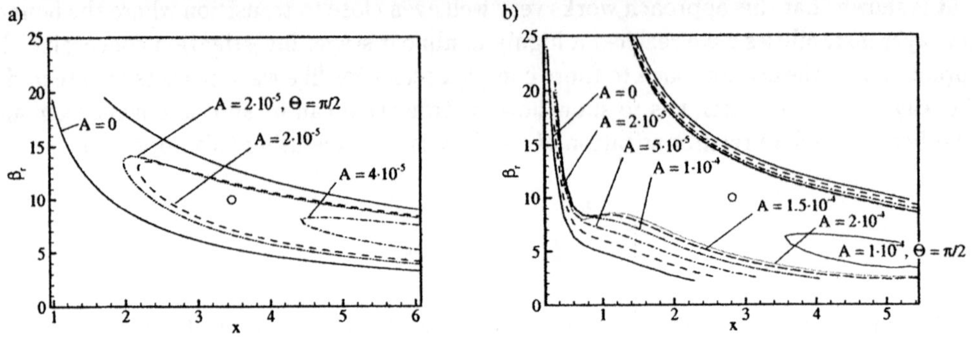

Figure 2 Curves of zero amplification of the investigated base flows.
a): Blasius b): Falkner-Skan base flow with Hartree-parameter $\beta_H = -0.1$. Free stream velocity was $U_\infty = 30\frac{m}{s}$ and x was made dimensionless with $L = 0.05m$. The small circles mark the position in the x/β-diagram of the calculations shown in Fig.3

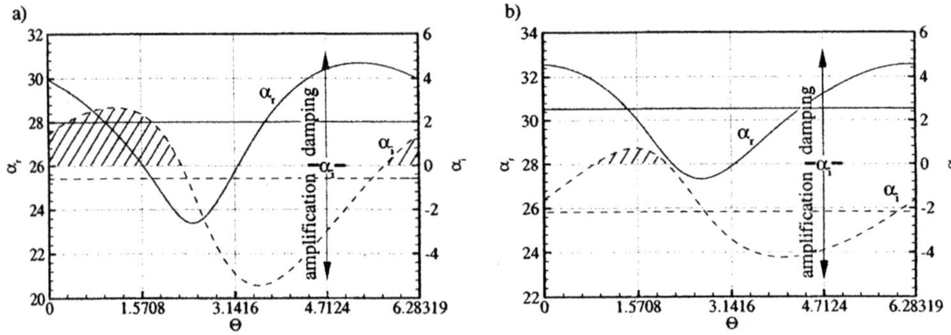

Figure 3 Eigenvalues for a variation of the phase angle Θ between $\omega_{z,w}$ and v_w for a given amplitude-factor of $A = 1 \cdot 10^{-4}$. a): Blasius b): Falkner-Skan. The circles in Fig.2 mark each position in the stability diagram. The horizontal lines (solid and dashed) show the uncontrolled values of α_r and α_i.

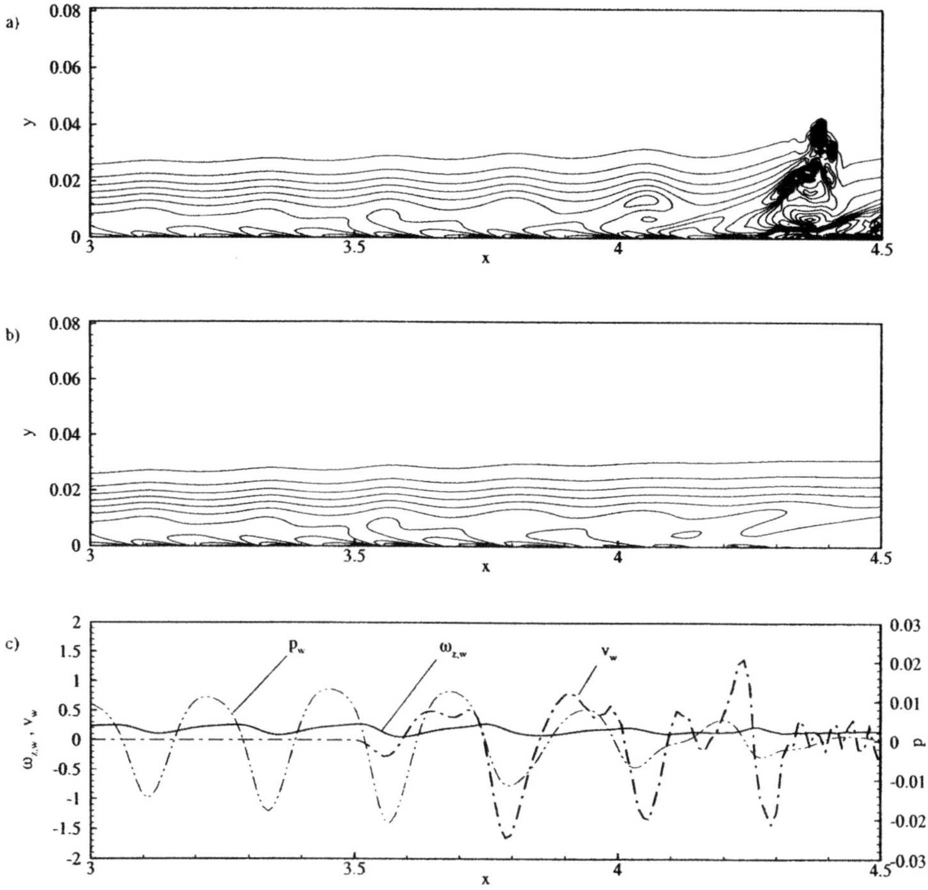

Figure 4 K-breakdown scenario. a): Instantaneous ω_z-iso lines (peak plane) of the uncontrolled case. b): same as (a) but with ω_z-control applied for $x > 3.46$ ($|A| = 2 \cdot 10^{-4}$, optimized phase). c): Distribution of ω_z, v and pressure p at the wall belonging to b).

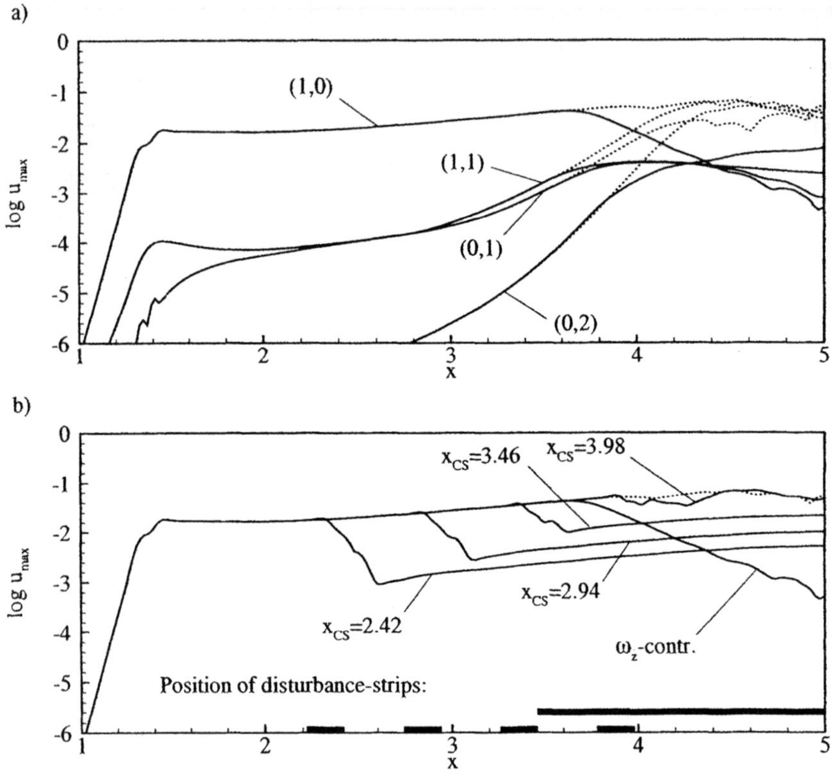

Figure 5 K-breakdown scenario, u_{max}-Amplitudes vs. x. a): Comparison of the most important modes with (solid lines) and without ω_z-control (dotted lines). Control parameters: $A = 2 \cdot 10^{-4}$, $x_{control} > 3.46$, optimized phase b): Same scenario as (a), Modes (1,0) controlled via FIR-filter at four different locations compared to the ω_z-control and the undisturbed case

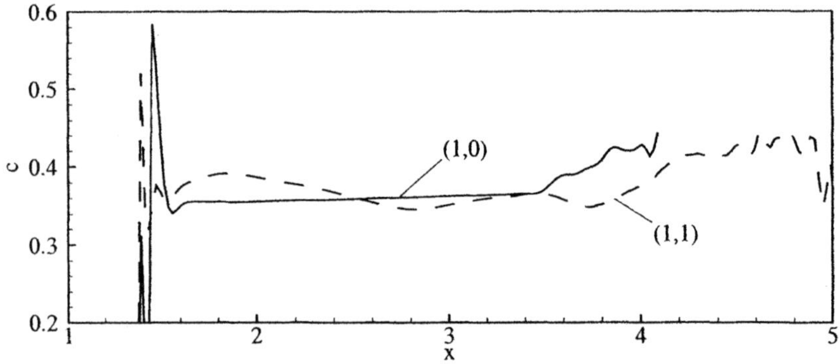

Figure 6 Phase velocities of the resonant modes (1,0) and (1,1). K-breakdown scenario with active control, same control parameters as Fig.5 a)

CONTROL OF THE COHERENT VORTICAL STRUCTURES OF A BOUNDARY LAYER

V. V. Babenko
Institute of Hydromechanics. National Academy of Sciences of Ukraine.
8/4, Zheliabov str., 02057 Kyiv, Ukraine.

SUMMARY

The various kinds of coherent vortical structures are arise and interleave in conformance with determined order at formation of a homogeneous boundary layer. The features of regularity of development and sequence of these structures are analyzed on basis of the results of physical experiment. The following hypothesis is offered: the similar vortical structures arise under certain conditions in near wall area of turbulent boundary layer. The combined method of drag reduction, which is based on simultaneous using of several methods of action on coherent vortical structures of a boundary layer, has been elaborated. The theory of boundary layer receptivity of different disturbances and the methods of efficient friction drag reduction due to control of coherent vortical structures of the boundary layer was worked out.

INTRODUCTION

Recently, ordered coherent vortical structures (CVS) have been detected in the transitional and turbulent boundary layers (BL). The study of the laws of their formation and development allows us to suggest a hypothesis: with proper control of these coherent structures, both the magnitude of hydrodynamic drag reduction and the intensity of heat and mass transfer can be regulated. Moreover, it is quite enough to influence not the whole of boundary layer, but a small region where the kernel of the coherent vortical structures are formed. It implies a need to influence only the disturbing motion. Hence, the energy expenditures necessary for control of coherent structures is an order of magnitude smaller than the energy required for traditional methods of boundary layer control.

1 MEASURING METHODS OF BOUNDARY LAYER PARAMETERS

The measurements were carried out in a low turbulence hydrodynamic bench. In experiments various methods of visualization, including a tellur-method, thermo- and laser anemometry were applied [1].

A low-turbulence hydrodynamic test bench comprises special equipment and devices.

Simultaneous photography of velocity profiles in vertical and transverse directions and also of longitudinal small tellurium jets allow the development of an exciting spatial-temporal picture of the velocity field during different stages of transition. Analysis of profiles U(z) yields the characteristic points along y and z axes where the boundary layer kinematics parameters are measured with LDA and DISA.

2 DEVELOPMENT OF DISTURBING MOTION IN A TRANSITIONAL BOUNDARY LAYER

The results of the visualization of the CVS in a transition a boundary layer are illustrated well in Fig. 1 which depicts the flow structure at different stages of transition. The whole process can conditionally be divided onto the following stages:

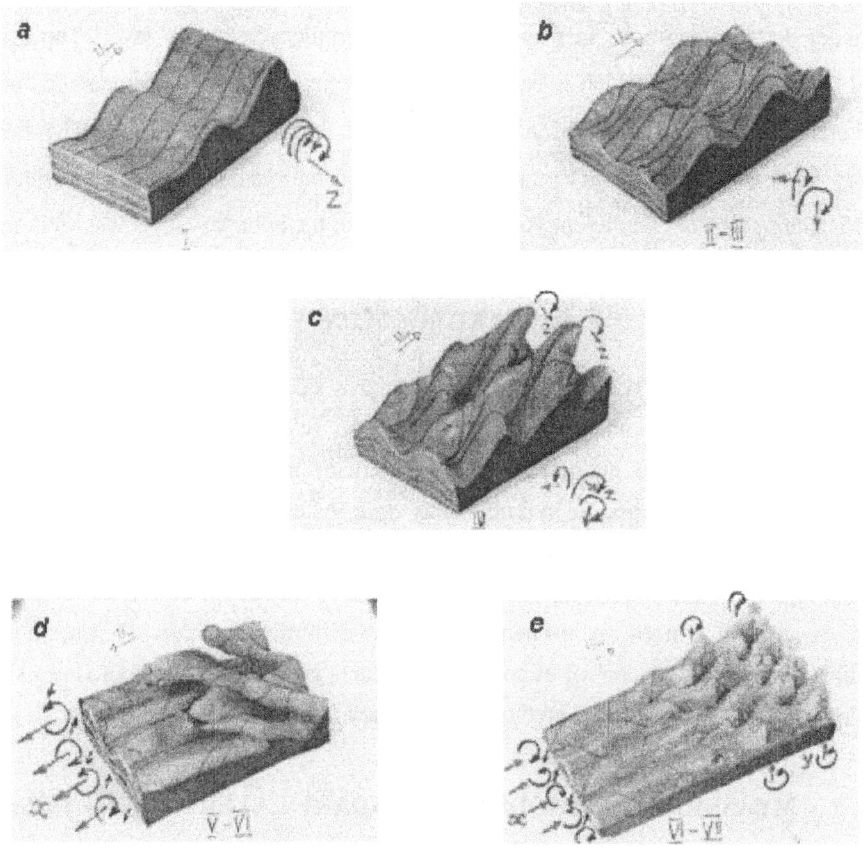

Fig. 1. Model of transition stages of laminar boundary layer to turbulent one: a - stage I, b-II-III, c-IV, d- V-VI, e-VI-VII-th stages of transition.

I. Pure laminar flow in a boundary layer ups the critical Reynolds number.
II. Amplification of plane Tollmien-Schlichting waves.
III. Deformation of this wave in plane XZ from the point of view of an observer who moves with speed of wave propagation velocity.
IV. Deformation of the wave in plane YZ, wave amplitude becomes a periodic function of Z with "crests" and "troughs" in this direction.
V. Amplification of deformed wave, which causes stretching of "crests" forwards and up wards and generation of longitudinally oriented "pin-like" (Λ-like) vortices. Appearance of turbulent spots.
VI. Development of vortical system with counter – rotating neighboring longitudinal vortices.
VII. Change of shape and intensity of vortexes; transition from straight-line to zigzag-like trajectory of vortices in XZ-plane.
VIII. Intensive development of turbulent spots and generation of turbulent boundary layer.

Depending on external conditions, the appearance of these stages can occur with different speed. Merging of the stages or development of the turbulent boundary layer may occur just after the detachment of the heads of Λ-like vortexes (after VI-th stage).

3 COHERENT VORTICAL STRUCTURES IN A TURBULENT BOUNDARY LAYER

Kline has formulated 8 types of CVS in a canonical boundary layer

1. Low-speed bands in the field of $0<y^+<10$.
2. Ejection of a low-speed fluid from a wall.
3. Driving high-speed fluid to a wall.
4. Vortical structures of several prospective types.
5. Forced interior shear layers in near wall area $y^+<80$.
6. Near-wall "pockets" precisely observed like areas of marked particles used in some methods of visualization.
7. Large (delta-like) ruptures of a longitudinal velocity.
8. Large-scale movements in exterior layers (for boundary layers: with convexities, super layers and deep hollows, completed fluid of free stream).

Kline explicitly has analyzed all these types of CVS. It is known, that on a flat plate the majority of turbulence (~70% or more) in the inner layer is generated via CVS mechanisms, and the minority (~30% or smaller) is generated by fluctuations in the outer layer.

Still there is no clear understanding of the spatial character of the coherent structures and their dependence both on each other and on coherent structures in the outer part of the boundary layer subjected to disturbances of different type from the outer and inner boundaries. We can the disturbances "deteriorating factors" in comparison with the idealized conditions of an undisturbed flat plate flow. The basic idea of the present investigations is based on results of experimental research that have allowed us to put forward the hypothesis that the viscous sublayer of a turbulent boundary layer can be represented as a quasi-laminar boundary layer. In

1980 we put forward the hypothesis that in the viscous sublayer of a turbulent boundary layer the development of coherent structures happens approximately in the same way as in a transitional boundary layer with influence of there deteriorating factors [2].

In 1983 Blakwelder has published work, in which put the analogy between flows in transitional and turbulent boundary layers [3].

4 METHOD OF RESEARCH OF RECEPTIVITY OF BOUNDARY LAYER TO VARIOUS DISTURBANCES

It is convenient to portray real disturbances with a complex form as combinations of simple wave or vortical movements of a different kind, frequency and amplitude. Thus, the concept of receptivity includes the interaction between of various disturbances brought in from the outside and result of such interactions. It is permit to present the boundary layer as a dynamic system with its own frequency spectrum, reacting to forcing in a manner depending on both the resonant and forcing mode. It is convenient to portray such interaction as follows. It is necessary to study the structure and character of a disturbance in a boundary layer in the absence of any significant external influences, and then to see how the development of the disturbance changes in the presence of different, of finite external forcing.

A study of the development of transition and the influence of final planar or vortical disturbances at each stage have allowed us to make two basic conclusions. First, the process of interaction of disturbances in a boundary layer can be presented as a superposition of "frozen", so-called natural structures with the structures of the forcing disturbances. Secondly, the interaction of natural structures with there introduced from the outside has a resonant character depending on the nature of the structures and the amplitude-frequency-wave characteristics.

All above-mentioned concerns apply not only to transitional but also to turbulent boundary layers. Experiments have shown that after ejection in viscous sublayer stages of disturbance development like there in a transitional boundary layer are present. Therefore, the considered categories of receptivity are also applicable to the study of the parietal area of a turbulent boundary layer.

5 METHODS OF INFLUENCE ON COHERENT VORTICAL STRUCTURES OF A BOUNDARY LAYER

Except the idea of structural character of disturbance development at different stages of transition, the idea of influence the frequency characteristics of ordered vortical structures lays in the base of control methods development. As we said before, fluid particle has dual property of its motion: it moves along a definite trajectory and has a spectrum of oscillations during motion where one can separate the definite energy-carrying frequencies for each stage of transition or for every type of coherent vortical structures.

In accordance with these two approaches the methods of influence the coherent vortical structures have been elaborated. As the principal factor (third approach or principle of influence), the influence energy must be of the same order as the energy and dimension of disturbing motion at corresponding stage of transition.

In table 1 there are having been developed methods consisted of seven groups: mechanical (I), dynamic (II), kinematics (III), electric (IV), acoustical (V), combined (VI) and active ones with feedback system (VII). Let consider mechanic methods. Oscillating surface has to contribute to ordering or damping the corresponding vortical structures. Oscillated my be the part of the surface, for example forepart.

The most perspective are combined methods, which use simultaneously the combination of different considered methods and/or combination with active methods of control.

Some of methods considered have been studied experimentally.

6 EXPERIMENTAL INVESTIGATION FLOW OF LOCALIZED CAVITIES

Experiments were carried out in the hydrodynamic channel of 1 m length and 0.02 m×0.2 m cross section [4].

At the bottom of working section were installed two cavity. At realization of experiments cavity placed under different angles in relation to a stream. That allowed visualizing the flow either of two cavities simultaneously or their interaction. In order to investigate effect of cavity geometry variation used was special construction with flexible membrane.

Visualization was carried out with colored streams.

Visualized pictures obtained and data for profiles allow suggesting the following picture of flow inside cavity. At small velocities flow a vortex with rotation axis in transversal direction OZ arises in cavity. Place of vortex location depends on the cavity shape and on steeples of its forward and lateral slopes.

In cavity of regular spherical shape, axis of rotation lies near center being closed by ends on lateral slopes. Maximum velocity in plane cavity will be on the axis and minimum one on the sides because of the constant value of $\omega \cdot G$, where ω - angular velocity and G - square of the vortex cross section. Motion remains vortical one outside the cavity too, but axis of its rotation is turned and directed along the flow OX.

When changing shape of cavity the vortex some changes its axis of rotation. So, in ellipsoidal cavity with sharp forward edge, the vortex axis lies nearer to front border and takes the arched shape with elongated somewhat forward and up butt-ends. Two longitudinal vortexes outside the cavity have the opposite rotation.

In cavity of irregular ellipsoidal shape, the vortex takes Ω -shaped shape. Forward part of the vortex lies near back edge of cavity. Then motion occurs along streamline rotating however, round OX axis. Two are rotating to meet one another vortical systems are situated much closer than that in cavity of regular spherical.

Increase of free stream velocity results in the fact that bursts/ejection's begin to take place from the cavity. The main factor which stimulates the bursts is disturbing influence arising during development of instability in initial boundary layer.

7 COMBINATION METHOD OF A DRAG REDUCTION

As is visible from a table 1, perspective the application of a combination method of a drag reduction is. The effectiveness of each method of action on CVS is based on influence of a method on physical process of development of disturbance. Thus each method influences mainly any characteristics of development of disturbance. The advantage of each separate method is in combination method of drag reduction used, and in aggregate effectiveness thus will increase. For optimum application of a combination method it is necessary to determine on what characteristic of development of disturbances moving each separate method of action on CVS influences. It is interesting to mark that in a living nature in majority the combination methods are applied. Let's to give some examples of combination methods of a drag reduction.

8 CONCENTRATED AND DISTRIBUTED FORMING OF LONGITUDINAL VORTICAL SYSTEMS ON ELASTIC PLATES

Since the tested elastic plates had high absorbing characteristics, then change of flow nature in boundary layer was observed only when introducing concentrated three-dimensional disturbances by means of vortex-generators V4 [5] which had such dimensions: height - $15{\cdot}10^{-3}$ m, length - $18{\cdot}10^{-3}$ m.

Distributed forming of longitudinal vortical disturbances of step λ_z=0.012 m was realized on plate in the same manner as on rigid standard: on the plate surface in longitudinal direction there were glued thin wires (fig. 2). In that case vortex-generation is stimulated due to different thickness of the boundary layer in Z direction.

Thus, introduced concentrated or distributed three-dimensional disturbances were stabilized by elastic plate. Therefore, as on rigid plate at small ε too, there were used the methods for increase of intensity of disturbing motion.

One of them consists in heating the ribs of stiffness (wires) by means of electric current. Due to thermal-insulating properties of elastic material, heat flux was directed into boundary layer increasing convective motion.

Another way of introducing into boundary layer three-dimensional disturbances consisted in manufacturing elastic plates which exterior layers had specified order. As velocity U_∞ grew, increased dynamic loads caused in elastic material a response owing which into boundary layer from below there were "introduced" longitudinal disturbances, like by means of heated wires.

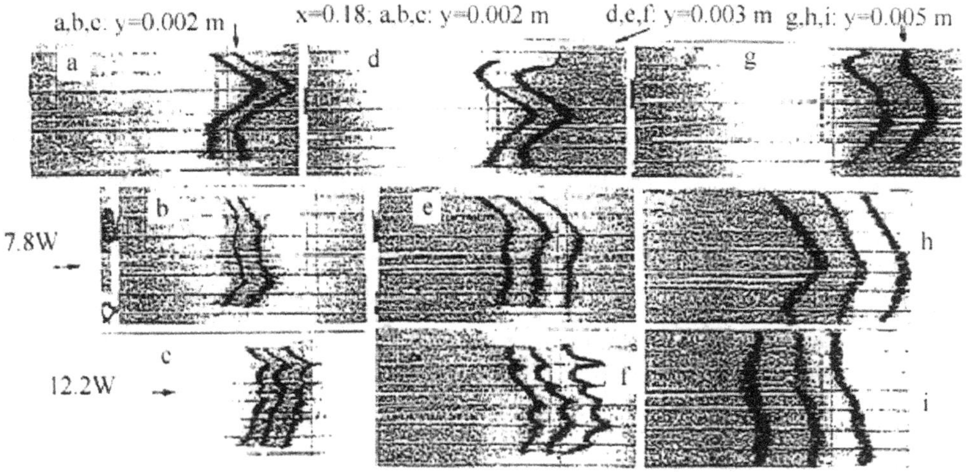

Fig. 2. Visualization of U(z) at U_∞=0.06 m/s over a plate 7a (x=2.18 m) at y=0.002 m (a-c), 0.003 m (d-f), 0.005 (g-i) without heating (a, f, g), with heating of longitudinal edges at consumed electrical capacity 7.8 W (b, e, h) and 12.2 W (c, f, i).

The second type of a combined method consists of joint application of a sword-like tip and injection of polymer solutions.

The third type of a combined method consists of joint influence to a boundary layer of polymer solutions and elastic surface.

9 MEASUREMENT OF INTEGRAL CHARACTERISTICS OF A BOUNDARY LAYER

The measurements of longitudinal average and fluctuation components velocity in the boundary layer on elasticity surfaces and at injection in it of water solutions of polymers are being made by means of laser anemometer. It is registered, the thickening of the viscous sublayer of a turbulent boundary layer, have decrease in longitudinal fluctuation a velocity and hear-wall shear velocity.

These results call be explained by changing of the form and sizes of the vortices in near-wall region as well as by change in the cascade process of vortex dissipation in the boundary layer.

Taking into account the similarity of the results in using the elastic plates and polymer additions, it is possible to expect, the similarity in the interaction mechanisms.

The measurements of friction drag on the elastic plate were being made at injection in a boundary layer a polymer solution through one nose slot (fig. 3).

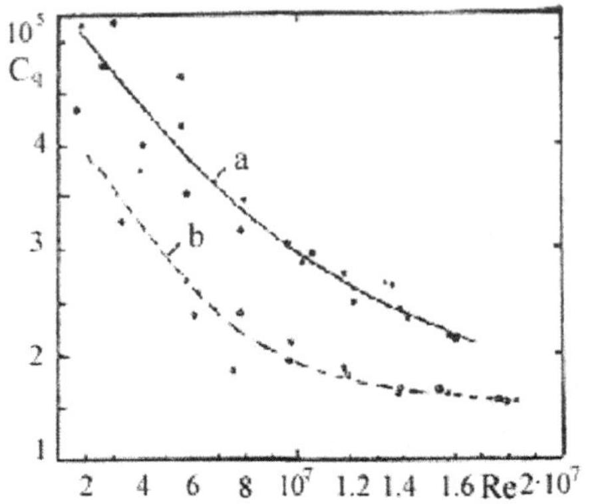

Fig. 3, a. Capacity coefficient of polymer solution of different Reynolds numbers. C_q=Q/(US), Q=V/t – capacity of polymer solution.

The used elastic coating was of polyurethane PY-3B of $3\cdot10^{-3}$ m thickness.

The measurements of friction were being made in hydrodynamic channel at velocities of towing 2.0-22.0 m/s.

On the rigid and compliant surfaces 2 series of experiments A and B were made. They differed with the amount of the polymer solution, injected in the boundary layer (fig. 3).

The outcomes of measurements of a drag reduction are shown on fig. 3. Curves 1, 3 correspond to the capacity C_q(Re) according to law "B", and the curves 2, 4 - correspond to law "A".

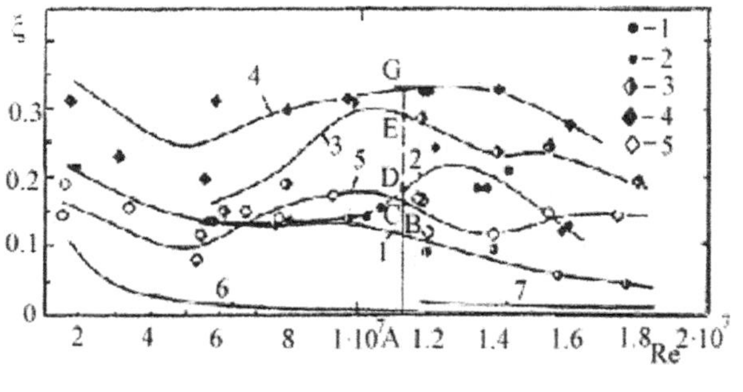

Fig. 3, b. Coefficient of reduced drag along the Reynolds number; in solid injection of polymer additions in the boundary layer on rigid (1, 2) and on elastic (3, 4) surfaces; 5 – without Polymer addition on elastic surface. For 1, 3 the law of capacity PEO is denoted by B; for 2, 4 – A on figure 5; 6, 7 – relative root-mean-square error of measurements in experiments of drag coefficient by Reynolds number – $\sigma_{\Delta C_q}/C_x$.

It is a remarkable, that the effects of the elastic surface and of the polymer are added. With allowance to experiment error, AC+AB=AE and AC+AD=AG.

10 THE EXAMPLES OF PRACTICAL APPLICATIONS

Because of method of control CVS and combination method of a drag reduction were developed various design solutions for interior and exterior problems flow of fluid.

The method for calculation of a boundary layer kinematics parameters at injection of liquids with different density (Dr. E. Shkvar)

The mathematical model of multiphase turbulent boundary layer with injection the system of liquids with different properties is developed. The boundary layer is consider as inhomogeneous flow and its phases are consider as mixing-up liquids during flow development.

This model is based on boundary layer type system of governing equations, which include continuity equation, longitudinal impulse transfer equation and transfer equations of concentration for each phase in flow. With the aim of modeling the turbulent properties the differential low-Reynolds k-ε model has been used. The especial attention was given by determination of coefficients of diffusion and local significance of moving fluid density.

For modeling the values of these coefficients the following hypothesis has been adopted: the local significance of diffusive coefficients and density are consider as partial value depending of concentrations of each phase from all ones, which are present in considered point of flow.

For solving partial differential equations system with the boundary conditions the effective marching method with second order accuracy has been elaborated. The general advantage of proposed finite-difference scheme and computational method is without-iteration two-step procedure, which allows quickly running along flow direction and simultaneously solve mentioned above system of linked differential equations. This methodology and program realization of this approach does not have any limitation of the number of injected liquids and order of their localization in flow.

11 CONCLUSION

The new approach and understanding of a problem of interaction of various disturbances in a boundary layer at various stages(phases) of its development (problem of a receptivity of a boundary layer) is offered. Is carried out of broad experimental researches of a receptivity by a boundary layer of flat and three-dimensional disturbances, and also their various combinations. Because of obtained results of physical researches of a boundary layer are developed a method of control CVS and combination method of a drag reduction. The appropriate experimental researches of these methods are carried out and the methodology of a design solution of various applied problems is developed.

ACKNOWLEDGEMENTS

The author wish to thank International Science Foundation J. Soros (Grant Number of Principal Investigators UAW000 and UAW200), President «Cortana Corporation» K. J. Moore (Contract with Institute Hydromechanics NAS Ukraine 1995-1997) for Sponsoring much of works described in this papers.

REFERENCES

1. L.F.Kozlov, V.V.Babenko. Experimental investigations of the boundary layer /Kiev, Naukova dumka. - 1978. - 184 p.
2. V.V.Babenko On interaction of flow with elastic surface // Mechanism's of turbulent flows. - M. - Nauka. - 1980. - P.292-301.
3. R.F.Blackwelder. Analogies between transitional and turbulent boundary lauers / The Physics of Fluids, - 1983, v.26, 10. – p.2807-2816.
4. V.V.Babenko Formation of vortex structure on curvilinear surfaces and semi-spherical cavities // 11-th International Couette-Taylor Workshop, P.103, July 20-23, 1999, Bremen, Germany.
5. V.V.Babenko Experimental definition of the stability of longitudinal vortical disturbances // Ing.-Phys. J. – 1986, N2. - p.201-207.

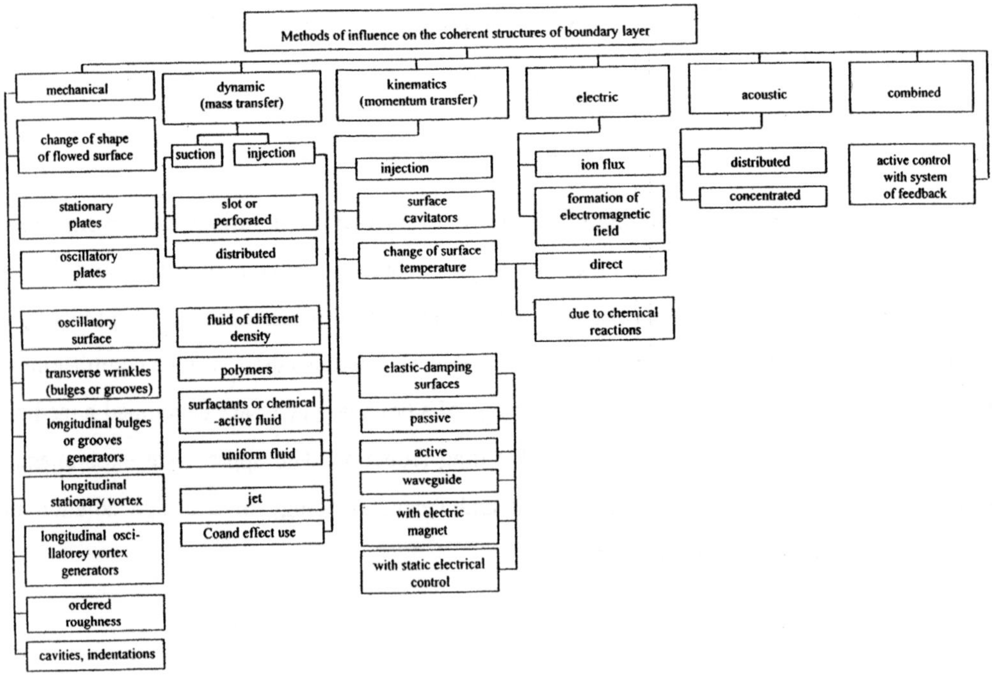

Novel Approaches To Combat Insect Contamination on HLFC Wings

D. O' Donoghue[†], T.F. O' Dwyer[†], T. Young[‡] & T. Pembroke[†]

[†] *Dept. of Chemical & Environmental Science, University of Limerick, Castletroy, Co. Limerick, Ireland.*
[‡] *Dept. of Mechanical & Aeronautical Engineering, University of Limerick, Castletroy, Co. Limerick, Ireland.*

SUMMARY

The aeronautical industry is faced with the problem of fuel inefficient performance due to drag. Drag reduction can be brought about by application of Hybrid Laminar Flow Control (HLFC) technology. The feasibility of this is threatened by natural roughness adhering to leading edge wing sections (ice, insect, dirt, dust). Current opinion identifies fluid exudation systems as having the greatest potential for alleviation of this problem. Presented are results of an investigation into the modification of such fluid systems to incorporate enzymes into the active matrix. Six proteolytic enzymes from the serine and cysteine protease families were identified for analysis. Assays of activity in solvent systems ranging from entirely aqueous to 80%v/v fluid point depressant were conducted using azoalbumin as substrate. In general, activity was diminished with increasing concentrations of ethylene and propylene glycol although to varying degrees. Enzymes showing greatest resilience were investigated further for activity in the presence of corrosion inhibitors and fluorosurfactants. Within each type of additive, the inhibition of enzyme activity varied according to chemical species and concentration. Overall testing allowed identification of those compounds and concentration which would be favoured for use in conjunction with enzymes for insect contamination systems.

1 Introduction

Although aeronautical fuel efficiency improvements have been achieved and are expected to continue into the future, fuel consumption is also expected to increase [1,2]. This is important as aviation fuel usage represents the second largest polluting transport mode after vehicular modes and for the foreseeable future, kerosene based fuels are expected to remain the industry's dominant energy source [3]. Hence, dramatic improvements in fuel efficiency are essential. While Hybrid Laminar Flow Control (HLFC) systems show promise, they are prone both to surface roughness and pore blockage by adherands [4]. Therefore, before widespread implementation of such systems, means of reducing the adhesion of roughness elements need to be addressed. Of particular concern is insect debris [5]. Efforts at alleviating such contamination have received attention over the years although none have been incorporated into current aircraft systems [6-9].

1.1 Potential New Fluids for Assessment

Fluid based systems containing glycols have shown promise in flight contamination alleviation [10-13]. Such systems may also help remove pre-existing contamination, thereby reducing the severity of flight maintenance regimes. To help bring about this, biological additives, known as enzymes, that decompose proteins on wing surfaces, may prove useful to insect contamination alleviation fluids. The enzyme as a cleaning aid is a principle applied daily in cleaning agents for household and industrial uses [14-17]. Application of enzymes to HLFC surfaces as cleaning aids would simply broaden the already common use of these materials in the public arena.

Applying such a concept to HLFC contamination mitigation systems requires adoption of enzyme screening processes. In the present study, a range of serine and cysteine proteases were selected for investigation in solvent compositions representing the transition from completely aqueous to predominantly organic phases. The effect of such a transition and of organic co-solvent type on protein degradation were of interest. The co-solvents of choice are used in the aeronautical industry as de-/anti-icing agents and are proposed to be the dominant component of novel wing contamination alleviation solutions. This paper presents some results on a selection of possible enzymes screened from both protease families.

2 Materials & Methods

2.1 Materials

α-Chymotrypsin (E.C. 3.4.21.1, from bovine pancreas), trypsin (E.C. 3.4.21.4, from bovine pancreas), Subtilisin Carlsberg (E.C. 3.4.21.14, from *Bacillus licheniformis*), Papain (E.C. 3.4.22.2, from *Papaya latex*), bromelain (E.C. 3.4.22.32) and Proteinase K (E.C. 3.4.21.64, from *Tritirachium album*), were purchased from Sigma (Dublin, Ireland) and used without further purification. 1,2-ethanediol (hereafter ethylene glycol), 1,2-propanediol (hereafter propylene glycol) trisma base, sodium hydroxide, benzotriazole and 2-Butyne-1, 4-diol were all purchased from the same source. Flourosurfactants used were donated by DuPont (Belgium). All other materials used were supplied by the CES Department, University of Limerick.

2.2 Buffer Systems

Buffers used contained from 0 to 80%v/v ethylene or propylene glycol. All buffers were prepared by mixing in the appropriate ratio, volumes of water based and glycol based buffer. Each was 50mM Tris-HCl modified to a pH of 7.5 for use at 40°C. Mixing of buffers was found not to cause appreciable deviation from the desired pH.

2.3 Enzyme & Substrate Preparation

Stock enzyme solutions were prepared at the appropriate concentration (e.g. 1mg/ml Subtilisin, 10mg/ml α-Chymotrypsin) in aqueous buffer. Aliquots of these were diluted into mixtures of aqueous and non-aqueous buffers such that the final enzyme working solutions had a known concentration of non-aqueous solvent. Fresh azoalbumin substrate was prepared to a final concentration of 10mg/ml in appropriate buffer systems to ensure a constant concentration of non-aqueous solvent per assay.

2.4 Enzyme Analytical assays

Assays were conducted following the method of Sarath [18]. For each enzymatic reaction, 250μL of appropriate substrate solution was spiked with 150μL of enzyme working solution pre-incubated at 40°C. Assays were run for 30 minutes and terminated with 1.2ml of 10%w/v TCA. Samples were centrifuged at 8000g for five minutes. For each assay, 0.6ml was added to 0.7ml 1.0M NaOH and colour developed was read at 440nm. Using the same method but with varying substrate concentrations (0.01176mM to 0.09408mM), kinetics assays were performed over 10 minute periods for each enzyme.

One unit of enzyme activity is defined as that amount of enzyme required to cause an absorbance change of 1.0 at 440nm in a cuvette of 1.0cm path length, under the conditions of the assay (*Note: Where shown in graphs values of U/mg are Abs@400nm/mg enzyme used in assay*).

3 Results and Discussion

3.1 Comparative Activity Assays

For use in cleaning agents, especially in environments such as the leading edge of aircraft, it is important that the enzyme can degrade a broad range of protein types. This can be measured using enzyme kinetics. The kinetics of azoalbumin degradation was studied in aqueous buffer for the six enzymes selected within this study. The enzyme turnover number was then used to evaluate the comparative activity of each enzyme and the potential each had for HLFC wing protection systems. Figure 1 shows k_{cat} data for each enzyme. Clearly, Subtilisin Carlsberg has the highest turnover number. It is observed to be 69 times more effective than trypsin in aqueous buffer under test conditions. This, and the order of the results, was not unexpected given the general lack of bond specificity of this enzyme compared with that of trypsin (see Appendix II of Ref 18).

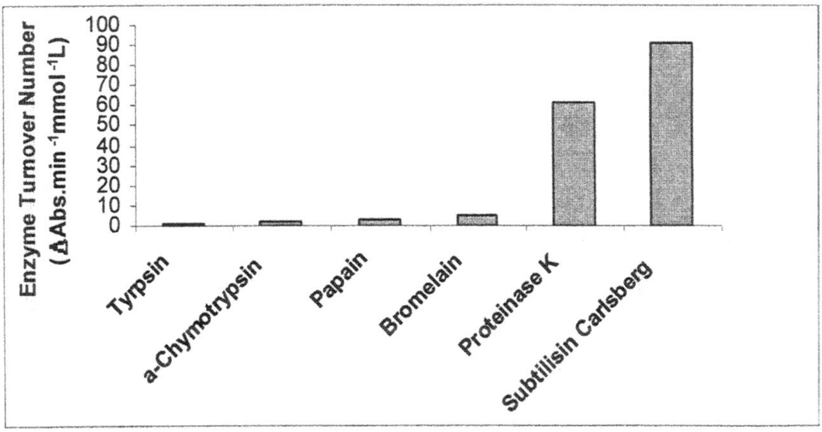

Figure 1: Enzyme Turnover Numbers for Azoalbumin Degradation in aqueous buffer, pH 7.5 (50mM Tris-HCl); Temp.40°C, Substrate Concentration Ranged from 0.01176mM to 0.09408mM azoalbumin).

3.2 Protease Activity In Glycol Solutions

The effect on enzyme activity of varying organic co-solvent concentration is shown in figure 2 for ethylene glycol and in table 1 for both glycols used. Both solvents are used as aeronautical de-/anti-icing agents. Each is a recognised cryosolvent and has been used as protein perturbants in conformation studies suggesting their mild nature toward biocatalysts [19,20]. A downward trend in activity was observed with increasing organic co-solvent concentration for all enzymes although the onset and severity of this loss was unique to each enzyme investigated. Subtilisin Carlsberg retained the highest activity (relative to aqueous activity) at all concentrations in excess of 40% v/v organic co-solvent while Proteinase K was next most resilient. Activity was negligible for all enzymes in co-solvent systems lacking deliberate water concentrations (as % v/v).

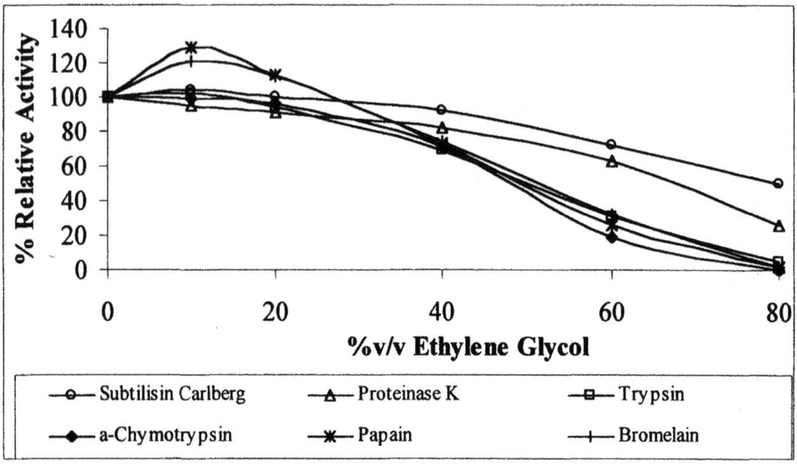

Figure 2: % Relative Activity of Proteases with Concentration of Co-solvent Ethylene Glycol; Assays were conducted in triplicate with enzyme blanks at 40°C, pH 7.5 (50mM Tris-HCl).

Table 1. Variation of Protease activity with co-solvent type and concentration (Assay conducted at 40°C, pH 7.5 (50mM Tris-HCl). % Relative Activity is that absorbance at 440nm in glycol fluids relative to that detected in aqueous buffer.

	% Relative Activity									
%v/v Glycol Fluid	**10**		**20**		**40**		**60**		**80**	
Enzyme	EG	PG	EG	PG	EG	PG	EG	PG	EG	PG
Subtilisin Carlsberg	103.78	110.41	99.55	100.50	92.29	88.43	71.94	67.71	49.77	24.36
Proteinase K	95.00	95.66	90.85	95.69	82.10	72.95	63.02	44.66	25.91	15.45
Trypsin	102.44	94.20	93.89	78.17	68.85	42.75	30.69	21.98	5.34	12.21
α-Chymotrypsin	99.22	93.52	96.24	79.27	71.11	34.97	18.78	3.24	0.00	1.55
Papain	128.84	117.19	113.10	101.39	72.61	48.74	26.26	7.30	1.95	1.13
Bromelain	120.77	108.74	112.32	96.67	74.40	57.58	31.64	14.78	2.27	0.00

Some enhancement of enzyme activity was observed at low concentrations of organic co-solvent. This has been observed in other enzyme/ water /organic co-solvent systems up to 30%v/v organic co-solvent in the reaction mixture [21]. In the context of enzymes as additives for HLFC insect contamination alleviation systems, these results are largely meaningless as the enzymes showing enhanced activity do so at unacceptably low concentrations of fluid point depressant and have aqueous reaction turnover numbers much lower than the enzymes Subtilisin Carlsberg and Proteinase K. At higher concentrations of organic co-solvent, the lack of water could be argued to be responsible for the loss of activity. This is however, unlikely to be the sole causal agent as the water concentration never dipped below 20% v/v of any reaction mixture. Physicochemical parameters of the solvent systems might offer a more rational explanation for the progressive loss of activity observed. Increased viscosity with reduced water content and the difference between the viscosity of the two glycols could be providing a physical barrier to enzyme-protein interactions.

3.3 Time Dependent Stability of Proteases in Glycol Fluids

From baseline testing results described above, Subtilisin Carlsberg and Proteinase K underwent further investigation to characterize their potential as additives in proposed insect contamination alleviation fluids. To gauge their usefulness as components in such products, lifetime of usefulness studies were conducted under conditions deemed to mimic those found in the anticipated work environment. Enzyme solutions were stored in PVC containers at room temperature in darkness until required. Tests were conducted over a twenty eight-day period on batches of both enzymes to determine the effect of long term storage and glycol concentration on enzyme residual activity. Results are presented in figures 3 and 4.

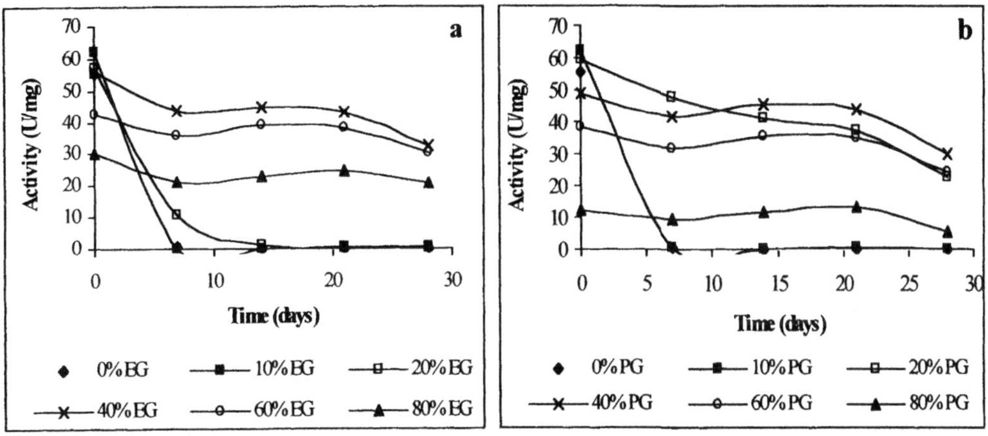

Figure 3: Subtilisin Carlsberg Activity (U/mg) as a function of Time and Glycol Concentration (%v/v); **a** Ethylene Glycol; **b** Propylene Glycol (Test conditions: Temp. = 40°C, pH 7.5 (50mM Tris-HCl buffer, variable glycol concentration)

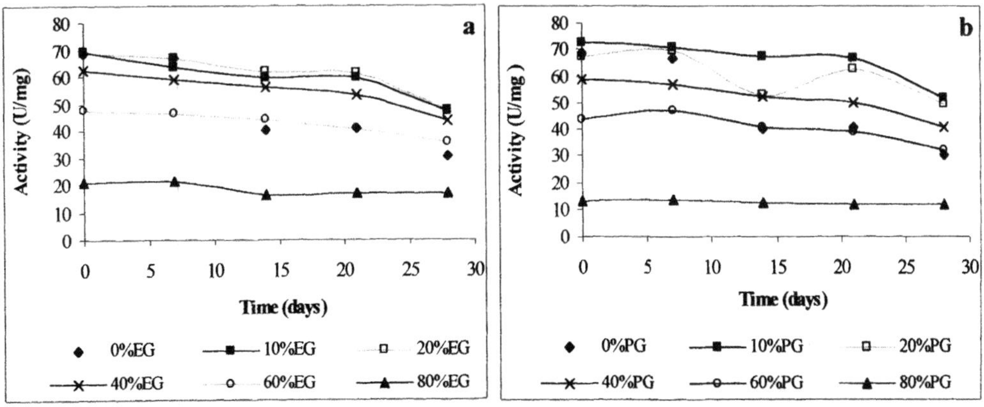

Figure 4: Proteinase K Activity (U/mg) as a function of Time and Glycol Concentration (%v/v); a Ethylene Glycol; b Propylene Glycol; (Test conditions: Temp. = 40°C, pH 7.5 (50mM Tris-HCl buffer, variable glycol concentration)

Propylene glycol was observed to offer each enzyme marginally higher resistance to autoproteolysis (a kind of self destruction of enzymes) than ethylene glycol. This was the case over the first two weeks of testing, whilst on four weeks the results suggested a reversal in the trend for Subtilisin Carlsberg under test conditions of 60 and 80%v/v glycol co-solvent used.

Linear regression analysis was conducted on each data set to identify the enzyme tendency to undergo autoproteolysis as a function of organic co-solvent concentration. The rationale was that the less stable the enzyme the more negative the slope of the linear regression line. Results are shown in table 2. With the exception of Subtilisin Carlsberg in 60% v/v ethylene glycol, propylene glycol solutions had the greatest stabilising effect. Propylene glycol is already used in the preparation of liquid enzyme products [22]. Results also reveal that Subtilisin is particularly prone to autolysis in the aqueous range. This indicates that any aircraft cleaning agent would require glycol concentrations in excess of 40% v/v to have an appreciable shelf life. This is convenient as it provides a sufficient glycol concentration for combinational contamination alleviation and icing protection measures. Distinguishing between organic co-solvents for use in conjunction with the enzymes tested by this means allows a qualification of the results from initial testing. These results tend to suggest that using propylene glycol in HLFC contamination alleviation systems is better than ethylene glycol. However, the higher activity of all enzymes in fresh solutions of ethylene glycol, coupled to the good stabilising effect of ethylene glycol in mixed systems suggests a compromise is necessary.

Table 2: Regression Analysis Output of Enzyme Activity According to Solvent Composition (Slope Units are $\Delta Umg^{-1}/day$; EG = ethylene glycol; PG = propylene glycol; ND = not determined)

%v/v-Glycol	*Subtilisin Carlsberg* EG-Slope	*Subtilisin Carlsberg* PG-Slope	*Proteinase K* EG-Slope	*Proteinase K* PG-Slope
0	ND	ND	-1.4638	-1.4638
10	ND	ND	-0.7052	0.6514
20	-1.7576	-1.1824	-0.6724	0.6224
40	-0.6838	-0.5214	-0.6176	0.601
60	-0.3171	-0.361	-0.4276	0.4429
80	-0.2095	-0.1295	-0.1886	0.0662

3.4 Corrosion Inhibitors & Fluorosurfactants

Aircraft de-/anti-icing fluid compositions often use corrosion inhibitors to protect against metal ageing and oxidation. Surfactants are also added to aid fluid wetting rates over wing surfaces. In consequence, fluorosurfactants (Zonyl type, DuPont) were investigated for their effect on enzyme activity in otherwise favourable solution compositions. Figure 5 shows the effect of each surfactant concentration tested on the activity of Subtilisin Carlsberg in either 40 or 80% v/v ethylene glycol buffer. In general, deactivation was minimal and limited to Zonyl FSA and FSK. Zonyl FSK strongly deactivated the enzyme, perhaps due to a surfactant denaturing effect but also due to the effect of acetic acid (a carrier fluid for this surfactant) on the buffering capacity. With Zonyl- FSK after testing the pH was found to be significantly reduced in both 40%v/v and 80%v/v ethylene glycol solutions. This was unfortunate in the context of the surfactant as a wetting aid for enzyme containing insect contamination alleviation fluids as the surfactant showed excellent wetting characteristics.

Figure 6 shows the effect of the corrosion inhibitors on the activity of the enzymes Subtilisin Carlsberg and Proteinase K in the presence of 40 and 80% v/v ethylene glycol. Across the range of inhibitors used, the activity was nominally affected relative to the control tests. Whilst for the most part, each enzyme was observed to have a higher relative activity in 80% v/v glycol fluids, the actual activity is considerably less for an equal amount of added enzyme solution in this glycol concentration compared to a 40% v/v solution. Hence a tradeoff between enzyme activity and solution glycol concentration may again be required for contamination mitigation systems containing enzymes. These data again point to the utility of the enzyme Subtilisin Carlsberg as a possible supplementary additive such new fluids currently under investigation.

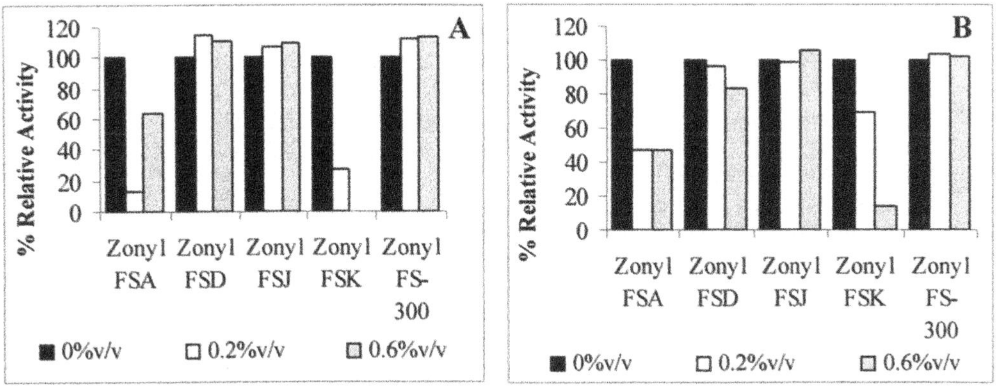

Figure 5: Effect of Zonyl Fluorosurfactant Type & Concentration on Subtilisin Carlsberg Activity at 40°C and pH 7.5 (50mM Tris-HCl Buffer); A = 40%v/v Ethylene Glycol; B = 80%v/v Ethylene Glycol, Substrate = 10mg/ml Azoalbumin in aqueous buffer (50mM Tris-HCl, pH 7.5), Assay Temp. = 40°C.

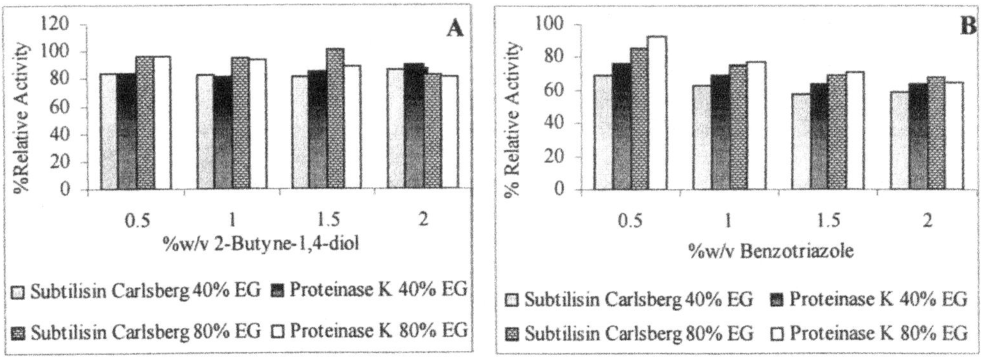

Figure 6: Corrosion Inhibitor Effect on Protease Activity in 40 and 80%v/v Ethylene Glycol. **A** = 2-Butyne-1, 4-diol, **B** = Benzotrialzole. Test Conditions: Temp. = 40°C, pH = 7.5 (50mM Tris-HCl, Substrate = 10mg/ml Azoalbumin glycol containing buffer containing corrosion inhibitor at concentrations 0 to 2.0%w/v)

4 Conclusions

The application of enzymes as additives for insect contamination mitigation in fluids with anti-icing capability on HLFC wings shows promise. Of six enzymes tested Subtilisin Carlsberg proved the most resilient to organic co-solvents and was effectively stabilised in both ethylene and propylene glycol. Propylene glycol was better at stabilising the enzyme but effected a reduced activity compared to that in ethylene glycol. Therefore, in the selection process for a solution containing both freezing point depressant and enzyme, a trade-off between the type of glycol fluid used and the concentration of enzyme used may be required. The fact that glycol fluids at intermediate concentrations allowed a high degree of enzyme activity and retained the solution in the liquid state at subzero temperatures indicates that a reduced glycol concentration may be more beneficial to contamination mitigation systems than a concentrated one. The effect of other potential product additives was found to be minimal and in some cases additive to the activity of Subtilisin Carlsberg.

5 References

1. Lewis, JS & Niedzweiki, RW; Chapter 7 in Aviation & the Global Atmosphere, Penner, JE, Lister, DH, Griggs, DJ, Dokken, DJ & McFarland, M (eds.), Cambridge University Press, 1998.
2. Jeziorski, A; " *Exhausting Issues*"; Flight International, 7th-13th June, 1995, pp.114.
3. Ellis, JH, Harris, NRP, Lister, DH & Penner, JE, Introduction to Ref 1 above.
4. Humphreys, B; pp. Proceedings of 1st European Forum on Laminar Flow Technology, Hamburg, Germany, March 16-18, 1992.
5. Barrett, RV, Rickards, J, & Eustace, R; pp. Proceeding of 2nd European Forum on Laminar Flow Technology, Bordeaux, France, June 10th – 12th, 1996.
6. Coleman, WS; in Boundary Layer and Flow Control, Vol. II, Lachmann, GV (ed.), Pergammon Press Ltd., pp. 682-747, 1961.
7. Wortmann FX; Report No.: NASA-TT-F-15454; pp.13., April, 1974.
8. Peterson, JB & Fisher DF; NASA CP 2036; In its CTOL Transport Technology Conference, Langley Research Centre, Hampton, Virginia, Feb 28- Mar 3, pp 357-363, 1978.
9. Yi, O.; Eiss, NS.; Wightman, JP.; REPORT NO.: NASA-CR-183041; pp.117; Sep. 1988
10. Croom, CC & Holmes, BJ; Report No.: SAE Paper 850860, General Aviation Aircraft Meeting and Exposition, Wichita, KS, Apr.16-19, pp.10, 1985.
11. Wagner, RD., Maddalon, DV, & Fisher, DF;; presention from the 16th Congress Of The International Council of Aeronautical Sciences, Jerusalem, Israel, August 28th -September 2nd, 1988.
12. Maddalon, DV and. Braslow, AL; NASA TP-2966; March 1990.
13. Moeken, B, Wolrath, W, & Horstmann, KH; Dornier Post No. 4, 1992, pp. 38-39
14. Koki Horikoshi; FEMS Microbiology Reviews, vol. 18, 1996, pp. 259-270
15. Greene, RV, Griffin, HL & Cotta, MA; Biotechnology Letters, vol. 18, no. 7, July, 1996, pp. 759-794
16. Banerjee, UC, Kumar Sami, R, Azmi, W & Soni, R; Process Biochemistry, vol. 35, 1999, pp. 213-219
17. Singh, J, Vohra, RM & Sahoo, DK; Biotechnology Letters, vol. 21, 1999, pp. 921-924
18. Sarath, G, de la Motte, RS, & Wagner, FW; Chapter 3 in Proteolytic Enzymes: A Practical Approach, Beynon, RJ & Bond, JS (eds.), IRL Press, 1989
19. Coll RJ and Fink AL; Cryobiology, vol. 24, 1987, pp.332-344
20. TE Creighton; "*Proteins Structure and Molecular Properties* (2nd Ed); WH Freeman & Company, New York, 1993.
21. Bicknell, R & Waley, SG; Biochemistry, vol. 24, 1985, pp.6876-6887
22. Godfrey, T & West, S; "*Industrial Enzymology (2nd ed.)*"; Stockton Press, New York, 1996.

Evolutionary Search Algorithms for Boundary Layer Control

W. MacCormack[†] O.R. Tutty[*] E. Rogers[‡] P.A. Nelson[†]

[†]Institute of Sound and Vibration Research
[*]School of Engineering Sciences
[‡]Department of Electronics and Computer Science
University of Southampton
Southampton SO17 1BJ, UK

Summary

In order to achieve an overall reduction in energy consumption when using suction to delay the transition from laminar to turbulent flow, it is necessary to apply the suction in an efficient if not optimal manner. Here an investigation is conducted into the use of distributed surface suction with multiple suction panels. Cost functions based on minimising the suction effort while maintaining transition in a fixed position and minimising the total energy consumption of the system are employed. Evolutionary methods, Genetic Algorithms and Simulated Annealing, are used to perform the optimisation. Realistic suction distributions are produced. In general better behaviour was found with the Genetic Algorithm than with Simulated Annealing, which was prone to premature convergence.

Introduction

It is well established that suction applied on the surface of a body can delay transition from laminar to turbulent flow, and thereby reduce drag. However, the ultimate objective, minimisation of the total net drag, will be achieved only if the energy saved in the drag is greater than that required to drive the suction system. This paper presents recent results on the optimisation of suction systems for hybrid laminar flow control, including results using a cost function based on the total energy consumption of the system.

The basic approach used consists, in effect, of monitoring the state of the flow together with the automatic control of suction applied through the surface of the body. This programme involves both algorithm development and verification and wind tunnel based experiments. Experimentally, pressure measurements are used to determine the position of transition in a boundary layer, and a feedback control loop is then used to maintain transition at a desired location with minimum power consumption by controlling the suction flow rates on suction panels.

Previous work has demonstrated that this overall strategy is technically feasible both in simulation studies [1, 2] and by wind tunnel based experiments [3, 4]. This work employed gradient based controllers, which used a local linear model constructed by standard linear system identification tools around selected operating points. For the case of flow over a flat plate these controllers performed well. In somewhat more complicated cases, however, they did not, e.g. when a non-zero pressure gradient is applied to a flat plate [2].

The gradient based approach fails because it can be incompatible with the basic physics of the flow. In this paper new cost functions are introduced for which no analytic (one step) solution algorithms exist. Instead, iterative search techniques must be used. Here the use of stochastic optimisation techniques is considered. In particular, Genetic Algorithms and Simulated Annealing are used and the performance of the resulting control/optimisation schemes analysed.

The calculations reported below use air at standard conditions, with kinematic viscosity of $1.5 \times 10^{-5}\ m^2/s$, and a free stream velocity of 20 m/s. All lengths will be in metres and velocities in m/s, and where convenient these units will be dropped.

Computational Modelling

The flow past the body is calculated using the interactive boundary layer equations with Blasius flow upstream. The grid is uniform in the axial direction and stretched in the vertical direction, so that the majority of the points are placed close to the wall. All derivatives are approximated with second order accuracy, using centered differences for derivatives in y-direction and three-point-backward differences for derivatives in x-direction. The finite difference equations are solved at each streamwise position using Newton's method, and the edge velocity is determined by performing Gauss-Seidel sweeps over the flow field. The effect of the interaction of the boundary layer with the outer inviscid flow is most pronounced at the ends of the suction panels where the boundary condition at the wall is discontinuous, and the solution with interaction is much smoother than that without interaction.

It is assumed that the transition from laminar to turbulent flow is preceded by the linear amplification of Tollmien-Schlichting waves, rather than through a bypass mechanism. Hence the disturbance is governed by the Orr-Sommerfeld equation. The e^N-method is used to relate the point of transition to the amplification of the disturbance waves. The position of transition is predicted as the first point at which N exceeds a critical value N_T which can vary according to the operating conditions of the flow. In the present case the value of N_T is determined by comparison with the experiments of Nelson, Wright & Rioual [3]. This gives $N_T = 4.3$, corresponding to a free-stream turbulence level of approximately 0.5%.

Cost Functions

Consider first the design requirement for a nacelle, where the transition position must be located at a pre-specified position while the suction effort is minimised. This can formulated as

$$\textit{minimise} \qquad \phi_s = K_1 \sum_{i=1}^{N} v_i^2 \tag{1}$$

$$\textit{subject to} \qquad x_T - d = 0 \tag{2}$$

where there are N suction panels, v_i is the (uniform) suction velocity on panel i, x_T is the transition position, d is the desired transition location, and K_1 is a constant used to provide a convenient scaling for ϕ_s. Here ϕ_s is used to estimate the effort made by the suction system. It is not, however, a true measure of the energy used by the pumps in the suction system (in the original experiments the effort for each pump was $O(v_i^p)$ where p was between two and three), but should vary monotonically with it, and therefore minimising ϕ_s will minimise the effort.

Note that the panels are weighted equally, which in effect assumes that all the panels have the same length. A weighting allowing unequal lengths could easily be included in (1).

The problem defined by (1) and (2) forms a constrained optimisation/optimal control problem. In [1, 2] this problem was solved using a gradient based controller (optimiser) which worked well for simple cases, but did not work well in more complex cases, and in certain cases, would not converge. In order to use (1) and (2) for iterative search techniques, it is necessary to define a cost function incorporating both (1) and (2). One possible form is

$$\phi = \phi_s + K_2 \mid x_T - d \mid \tag{3}$$

where K_2 is a positive constant. In the calculations reported below, K_1 is fixed at 10^4 and K_2 is varied to adjust the balance between the suction cost and the penalty on the error in the transition position.

As an alternative to the constrained optimisation problem defined by (1,2), suppose now that the aim is to minimise the power consumption of the system as a whole, based on the design requirements for a wing or an aerofoil. A model has been developed [5] which takes into account various factors giving a nonlinear cost function in the form

$$W_T = W_{wd} + W_{pump} + W_{ml} - W_{thr} \tag{4}$$

where W_T is the total power consumption of the system, W_{wd} is the power consumed in overcoming the wake drag, W_{pump} is the power consumed by the suction system, W_{ml} is due the momentum loss from withdrawing fluid from the main stream, and W_{thr} is from the thrust gained by expelling this fluid back into the flow in an optimum manner. Details of the model can be found in [5].

Optimisation Methods

The suction distribution was adjusted to minimise the cost functions (3) and (4), using a Genetic Algorithm (GA) [6] or Simulated Annealing (SA) [7] both in standard form. The GA has binary encoding, roulette wheel reproduction, single point crossover with a probability of 0.9, and mutation using a probability of 0.05. The SA uses the Metropolis algorithm to update its current model, i.e. a new model m_{new} is generated by perturbing the current model (changing one of the suction flow rates). If $\phi(m_{new}) \leq \phi(m_{cur})$ then the perturbed model is accepted and replaces the current model m_{cur}. If, however, $\phi(m_{new}) > \phi(m_{cur})$ the new model is not always rejected, but is accepted with a probability

$$P_{accept} = \exp\left(-\frac{\phi(m_{new}) - \phi(m_{cur})}{T}\right) \tag{5}$$

where T is the temperature at the current iteration. The temperature schedule used is

$$T = T_0(0.99)^k \tag{6}$$

where T_0 is a constant and k is the iteration number. Here $T_0 = 0.1$ is used. With (5) and (6) the probability of accepted a model with a higher cost decreases as the iterations continue.

Results - Flat Plate with Fixed Transition Position

A two panel flat plate will be used as a test case. In general terms, with the fixed transition position cost function, this models the original wind tunnel experiments [3, 4]. The suction panels are placed at $0.28 \leq x \leq 0.48$ and $0.58 \leq x \leq 0.78$. For this configuration the natural transition position, i.e. that with no suction, is at $x \approx 0.93$, downstream of the trailing edge of the second panel. This configuration has the advantage that the relationship between the suction flow rates and the transition position can be mapped completely, and hence the solution can be determined for any given case. With this particular configuration transition can be delayed until $x_T \approx 1.66$, and bounds are taken for the suction velocities which allow the full range of possible transition delay, setting $0 \leq v_i \leq 0.015$. As a first trial of the solution routines, a coarse 16×16 discretisation was taken, i.e. $v_i = (i-1) * 0.001$, $i = 1 \ldots 16$, with a population size of 10 for the GA, a desired transition position of $d = 1.4$, and $K_2 = 10$ in (3). With these parameters, the optimum on the grid has $(v_1, v_2) = (0.003, 0.002)$ with $x_T = 1.402, \phi = 0.155$ and $\phi_s = 0.130$. Both the GA and SA found the optimum, but the GA was considerably more efficient, taking 38 function evaluations (combined solutions of the flow, stability and transition prediction routines) compared with over 160 using SA.

Runs were also performed using a fine $2^{16} \times 2^{16}$ discretisation. With the pseudo-random number generator used in this study, this discretisation gives in effect a continuous distribution. The optimum with $d = 1.4$ is $(v_1, v_2) = (0.00125, 0.00246)$ with $\phi_s = 0.0761$. The algorithm was set to terminate on the iteration in which the 600th function evaluation was performed. The results for the GA with $K_2 = 10$ and a population of 20 are shown in figure 1. In this figure, and others for the GA, symbols show the sample space and lines the lowest cost solution attained so far. The final values are $(v_1, v_2) = (0.00127, 0.00245)$ with $x_T = 1.400$ and $\phi_s = 0.0761$, i.e. in effect the optimum. Other values of K_2 and the population size were investigated, but the values used for figure 1 gave the best results.

Results for this problem with the SA are shown in figure 2. In this and other SA figures, symbols/thin lines show the current model and lines/thick lines the lowest cost solution attained so far. Clearly, although the desired transition position has been found, the solution after 600 iterations is far from optimal. A large number of other options were tried for the SA, including other values of T_0 and K_2 in (3). The results were never better than those shown in figure 2, with either a worse solution or the algorithm taking much longer to approach the optimum. Convergence to a suboptimal solution was a common occurence. Hence it is concluded, that for this two panel problem, the GA is more reliable and efficient than the SA.

Consider now a flat plate with six suction panels 0.2 long, spaced 0.1 apart, placed at $0.3i \leq x \leq 0.3i + 0.2$, $i = 1, .., 6$, so that the first panel starts at 0.3 and the last finishes at 2.0. The natural transition position is still at 0.93, but with this configuration transition can be delayed to $x \approx 3.1$. It was found in [2] that for $d = 1.7$, 2.0 and 2.6, a Gradient Projection (GP) based controller/optimiser converges to and maintains a solution in less than 30 iterations. However, with $d = 2.3$ more iterations were required and the GP algorithm has trouble maintaining the solution due to the fact that within the tolerance of the algorithm the solution was not well defined. Even when the algorithm converged to a solution in terms of the transition position and suction cost, there were relatively large changes in the individual suction flow rates, with a trade off of effort between different panels with no significant effect on the position of transition or the suction cost.

For the GA an upper limit must be set on the suction flow rates. The same value as for the two panel case above could be used but this would be wasteful in that much of the time the algorithm would be searching in regions that are of no interest. A simple suction distribution, which is of practical interest, is to set the same suction flow rate on all panels. With this distribution it is found that although the transition position can be moved to its maximum value with sufficiently large suction, with suction as low as $v_i = 0.003$, $x_T \approx 2.9$, and with $v_i = 0.005$, $x_T \approx 3$. Accordingly, the suction flow rates were limited to $0 \leq v_i \leq 0.005$.

The GA was run with 2^{16} possible choices on each panel, $K_2 = 1$, and a population of 20. The results are shown in figure 3. The final values were obtained on iteration 31, after approximately 600 function evaluations. They give $x_T = 2.30$ and a suction cost of $\phi_s = 0.197$, which compares well with that from the 'best' case of 0.190 obtained with the GP algorithm [1, 2]. Note that due to the noise added to the suction flow rates in the GP algorithm, this difference is within the numerical uncertainty in the results.

The results for the SA for this case are shown in figure 4. The algorithm quickly converges on a solution but clearly this is much worse than that obtained by the GA (figure 3). The individual suction flow rates for the GA and SA at the end of the runs are given in table 1. Also included are values from GP algorithm and the suction cost. Although there is little difference in the total cost for the GA and GP, there are relatively large differences between individual suction flow rates, consistent with the large variability found in previous work [1, 2].

Table 1: Individual suction flow rates and suction cost for the different algorithms for the six panel flat plate case

panel	GA	SA	GP
1	0.00099	0.00160	0.00038
2	0.00155	0.00044	0.00180
3	0.00211	0.00055	0.00222
4	0.00168	0.00382	0.00296
5	0.00211	0.00301	0.00184
6	0.00214	0.00220	0.00174
ϕ_s	0.197	0.315	0.190

Results - Flat Plate with the Power Balance

Consider a flat plate 3.5 long with 12 suction panels each 0.1 long running from $x = 0.1$ to $x = 1.3$ with no gaps between the panels, and $0 \leq v_i \leq 0.015$. The GA was run allowing 16 flow rates on each suction panel, i.e. discrete changes of 0.001 in the v_i, and a population of 20. The run was terminated once 1200 function evaluations had been completed. The most expensive solution is that with no suction and transition at its natural position ($x \approx 0.93$). In the power balance the largest term by far is the wake drag which depends directly on the transition position. Almost any non-trivial distribution of suction will move transition a significant distance downstream, and therefore decrease the power consumption significantly. Hence the optimisation procedure in effect seeks to push transition as far downstream as possible with the minimum suction effort. In the final solution for this problem there is a small amount of suction ahead of the natural transition position, which delays transition, and sufficient suction on the last few panels to almost completely suck away the boundary layer and therefore push transition well downstream. The SA produced a solution with a similar suction distribution.

Although a realistic suction distribution has been produced, this configuration is not physically sensible as most of the panels are not used. Hence consider instead 12 panels starting at

$x = 0.8$, just ahead of the natural transition position and ending at $x = 2$. The results are shown in Figure 5. The initial population contains a good solution with $W_T = 23.0$ as compared to $W_T = 52.3$ with no suction. W_T drops to 19.8 by the end of the run. In the final distribution shown in Figure 5c, the suction applied on the first panel restabilises the boundary layer to an extent, with some suction on the central panels which delays transition and high suction on the last two panels which forces transition to $x = 3.13$. Again, this can be interpreted as a strategy to force transition far downstream with the minimum suction effort.

Results for the SA with the downstream configuration can be seen in Figure 6. The final suction distribution has a pattern which is broadly similar to that for the GA. The final cost and transition position are also similar ($W_T = 19.6$ and $x_T = 3.15$). There is however noticeable differences in the individual suction flow rates. In general it appears that suction effort can be traded off between panels with little effect on the gross results.

Conclusions

Simulations have been performed of boundary layer transition control using distributed suction with cost functions based on minimising the suction effort while holding transition in a fixed position, and minimising the total energy consumption of the system. Genetic Algorithm and Simulated Annealing techniques were used to solve the non-linear optimisation problems. It was shown that both approaches are feasible, and can be used where gradient based methods fail.

For the fixed transition/minimum suction case, it was necessary to define the constants in the cost function so that there was a suitable trade off between reducing the suction and penalising the error in the transition position. In general the Genetic Algorithm approach was better behaved in that it consistently found a good solution if enough iterations were performed. In comparison the Simulated Annealing algorithm was prone to premature convergence.

With the minimum energy criterion, physically realistic suction distributions were produced with a large number of suction panels. The results suggest that a reasonable general strategy is to place a judicious amount of suction ahead of and around the natural transition position to restabilise the boundary layer, with small amounts of suction downstream of this to further delay transition, and high suction towards the end of the panels to push transition as far downstream as possible, with the minimum suction effort. It is clear that an efficient suction system will require a good distribution of panels as well as a good distribution of suction on the panels. Work is continuing in this area on design optimisation methods which can choose the number and position of the panels as well as the suction distribution.

Acknowledgements

This work was supported by EPSRC through a ROPA grant.

References

[1] Tutty, O.R., Hackenberg, P. and Nelson, P.A. (2000a). Numerical optimisation of the suction distribution for laminar flow control. *AIAA J.*, **38**, 370-372.

[2] Tutty, O.R., Hackenberg, P. and Nelson, P.A. (2000b). Gradient projection based control and optimisation of the suction distribution for laminar flow control. Accepted for publication, *Proc. I. Mech. E. Part I: Systems and Control Engineering*.

[3] Nelson, P.A., Wright, M.C.M. & Rioual, J.-L. (1997). Automatic control of laminar boundary-layer transition. *AIAA J.*, **35**, 85-90.

[4] Wright, M.C. & Nelson, P.A. (1999). Four-channel suction distribution optimization experiments for laminar flow control. *AIAA J.*, **38**, 39-43.

[5] Rioual, J.-L., Nelson, P.A., Hackenberg, P. and Tutty, O.R. (1996). Optimum drag balance for boundary-layer suction. *AIAA J.*, **33**, 435-438.

[6] Goldberg, D.E. (1989) *Genetic Algorithms in Search, Optimization and Machine Learning*. Addison-Wesley, Reading, MA.

[7] Metropolis, N., Rosenbluth, A.W., Teller, M.N. & Teller, E. (1953). Equation of state calculations by fast computing machines. *J. Chemical Physics*, **1**, 1087-1092.

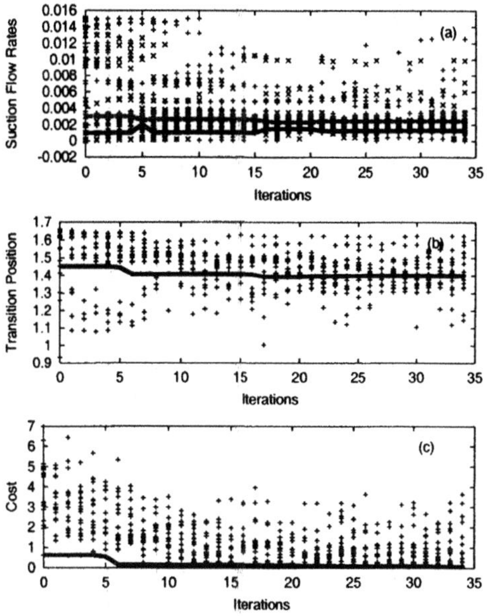

Figure 1: GA for two panels. (a) suction flow rates ($+ = v_1$, $\times = v_2$), (b) transition position, (c) total cost. The lines show the best case attained so far.

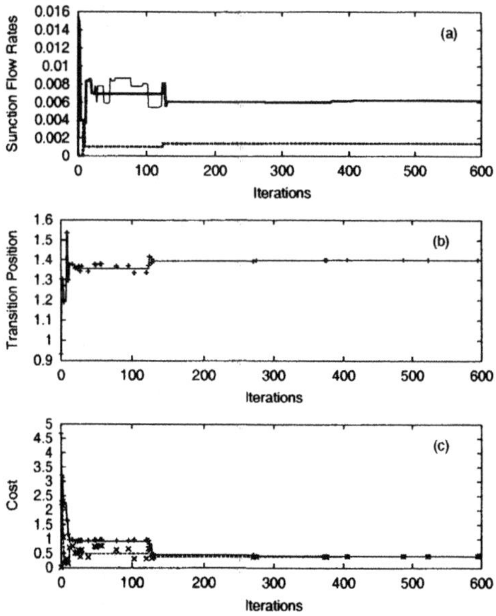

Figure 2: SA for two panels. (a) suction flow rates, (b) transition position, (c) total (solid/+) and suction ϕ_s (dashed/x) costs.

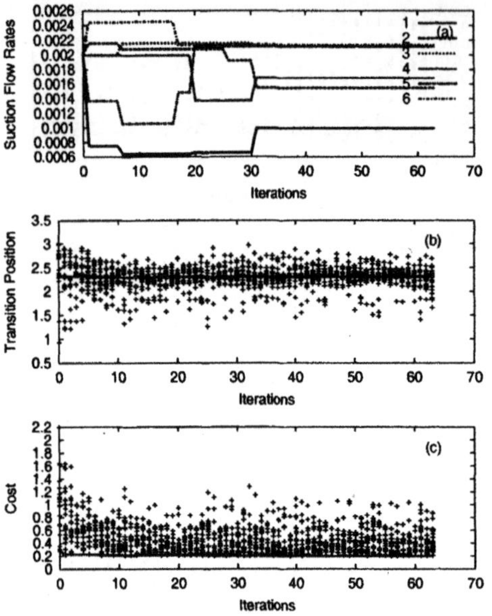

Figure 3: GA for six panels. (a) suction flow rates, (b) transition position, (c) total cost. The lines show the best case attained so far.

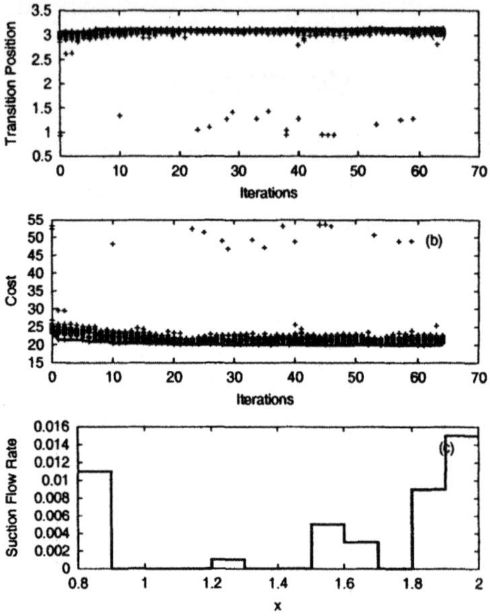

Figure 5: GA for 12 panels with the power balance. (a) the transition position, (b) the cost, (c) the final suction distribution.

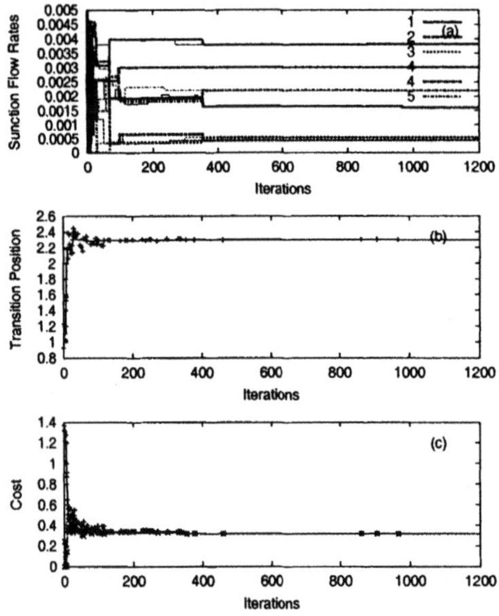

Figure 4: SA for six panels. (a) suction flow rates, (b) transition position, (c) total (solid/+) and suction (dashed/x) costs.

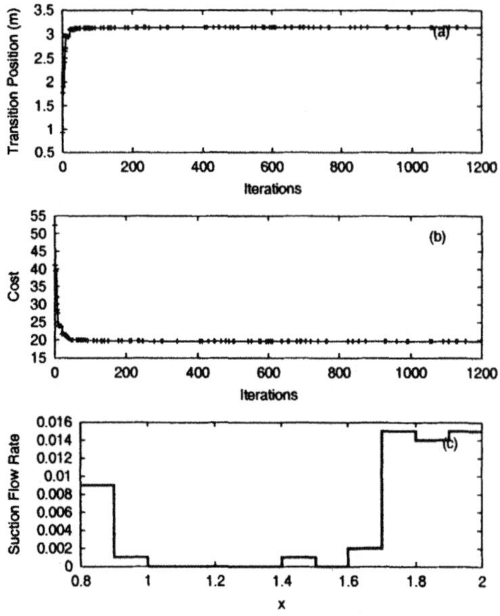

Figure 6: SA for 12 panels with the power balance. (a) the transition position, (b) the cost, (c) the final suction distribution.

Poster Session 2

Experimental Investigations on Active and Dynamic Instability Control of Separated Turbulent Wing/Cylinder-Flows

B. Gölling, U. Ch. Dallmann, H.-P. Kreplin

DLR - Institute of Fluid Mechanics, Section Transition and Turbulence, Bunsenstr. 10, D - 37073 Göttingen, Germany.

Summary

A novel concept of flow-separation control at high Reynolds numbers ($1.0 \times 10^5 \leq Re \leq 5.0 \times 10^6$) is under investigation in order to minimise drag and vortex shedding of wing-profiles and cylinder-flows which aims the excitation of large-scale spatial and small-scale temporal flow instabilities.

While at high-drag subcritical Reynolds numbers the laminar mean-separation lines appear straight, i.e. flow separation is two-dimensional under two-dimensional lateral boundary conditions, at low-drag supercritical Reynolds numbers the turbulent mean-separation is three-dimensional and large-scale spanwise-periodic wall-flow patterns appear (i.e. obtained by oil-flow visualisations). This regime is associated with laminar-turbulent transition due to small-scale spatio-temporal instabilities of the separated boundary layers and of the subsequently evolving separated shear- layers.

An active-stationary flow control is provided by spanwise-periodic continuous suction and an active-dynamic flow control is performed by spanwise- and temporally-periodic acoustic actuators which introduce large-scale spanwise non-uniformities at uncontrolled high-drag subcritical Reynolds numbers. Our preliminary results obtained by stationary and/or dynamic spanwise and azimuthal wall-pressure measurements exhibit the potential for considerable drag reduction by application of certain combinations of excitation Strouhal number, amplitude and spanwise wave-number.

Introduction

Most flow configurations in aerodynamics have to deal with the phenomenon of boundary-layer separation and subsequent vortex-flow development. At aerodynamically relevant conditions the flows around high-lift configurations or at the leeward side of fuselages or around any blunt bodies at angles of attack the separated flows are in general unsteady and three-dimensional. However, unsteadiness and three-dimensionality are inherent features also under nominally two-dimensional boundary conditions, for instance in separated flows around wing-profiles and cylinders in cross-flow. Laminar-turbulent transition phenomena are accompanied by coherent flow structures, yet not sufficiently understood flow instabilities, flow bifurcation and/or structural flow changes. Large-scale flow-separation structures, so called spanwise cells have been observed in surface-flow visualisations on high-lift airfoil configurations as well as on circular cylinders at high Reynolds numbers [1][2][3]. These non-uniformities have been addressed erroneously to effects of lateral boundary conditions [4]. However, large-scale spanwise almost-periodic flow-structures have been identified at supercritical Reynolds numbers, i.e. at $Re \approx 10^6$ in the flow around long-span circular cylinders and profiles. The observed large-scale spanwise pattern exhibits a wave-length of about 1.7 to 2.0 cylinder diameters or profile chord lengths. Dallmann and Schewe [2] conjectured that streamwise-oriented quasi-stationary vortices are associated with the observed large-scale cell-like surface-flow patterns and, should be present also in the separated near-wake flow region where also small-scale shear-layer vortices are inherent features. The exploratory numerical

simulations of Tamura et al. [5] confirmed these conjectures. Streamwise as well as spanwise coherent flow-structures appear at first in the wake and subsequently already at flow-separation as the Reynolds number is varied from the subcritical (laminar separation) via critical towards the transcritical (turbulent separation) flow regimes. Within the critical Reynolds number regime Schewe [6][7] in addition identified hysteresis effects on lift and drag. Hence, small disturbances should have substantial flow-control potential at about $Re \approx 3.5 \times 10^5$.
Different three-dimensional vortex structures are identified in the wake of circular cylinders at low Reynolds numbers between Re = 160 and Re = 500 [8] and small-scale vortices appear also in the turbulent shear layers in the separated near-wake flow region at $Re \approx 10^4$ [9][10][11].
During recent years flow control has received increasing interest with respect to various objectives: drag reduction, lift enhancement, mixing augmentation and flow-induced noise suppression [12]. In particular, most effective flow-control means has been substantiated wherever instabilities had be identified, for instance in laminar-turbulent boundary layer transition control. Various instability theories have contributed to a better understanding of structural changes of wake flows, however, most theories are restricted to rather low Reynolds number laminar or transitional flow separations [13][14]. Recent experiments [15][16] and direct numerical simulations [17] have indicated that the control of certain large-scale spanwise periodic vortical structures may lead to considerable drag reductions and/or damping of von Kármán vortex shedding at low Reynolds numbers. Large-scale spanwise-periodic flow instabilities have been recently discovered to be the result of a so called global flow-instability present in laminar separation bubbles [18][19]. This type of instability is independent of upstream introduced disturbances which otherwise undergo strong amplifications within the separation bubble and lead to laminar-turbulent transition [20] [21].
Our interest is focused on unsteady high Reynolds number flows in order to control spatial structural changes of the separated flows together with considerable jumps in drag and lift acting on the cylinder or wing-profiles whenever laminar-turbulent transition takes place in the separating boundary layers. Two hypotheses have emerged: Firstly, that those three-dimensional coherent structures can be excited as linear or non-linear instability modes of certain (still to be identified) transitional basic flows with boundary-layer separation and/or, secondly, that the turbulent separating flow regenerates ordered, large-scale spanwise almost-periodic flow structures. Our exploratory flow-instability controls of separated cylinder or wing profiles flows either directly excite the quasi-stationary spanwise almost-periodic flow regimes via stationary means and let transition to turbulence evolve freely or in addition affect the unsteadiness and turbulence generation within those large-scale coherent structures.
We perform the control on a circular cylinder in order to identify optimal response of the controlled flow, to minimise the total drag acting on a cylinder/wing-profile by simple variation of an actuator's location relative to the location of mean-flow separation. (However, in this first study flow control is applied only to one side of the cylinder.)
To reach aeronautically relevant Reynolds numbers the experiment is conducted in the DNW high-pressure wind tunnel (HDG) located at the German Aerospace Center (DLR) in Göttingen.
In the first part of our research we identify the various uncontrolled flow regimes by force measurements and surface oil-flow visualisations with respect to their three-dimensional, i.e. spanwise stationary and unsteady flow characteristics by mean-pressure as well as dynamic wall-pressure measurements. In the second part of the study active-stationary flow-control means is achieved by continuous suction of fluid through a spanwise row of neighbouring holes while an active-dynamic (unsteady) flow-control means utilises loudspeaker signals. These loudspeakers allow spanwise non-uniform or spanwise-periodic disturbance excitations in addition to streamwise-periodic control of flow separation and vortex shedding. The

efficiency of the applied flow-control means is given by the base-pressure coefficient, which indicates the behaviour of drag.

Experimental Set-up

The measurements are conducted in the high-pressure wind tunnel of DNW located at DLR in Göttingen. A circular cylinder model with a diameter of $D = 0.06$ m is mounted in a closed test section with the square size of 0,6 x 0,6 m^2, so that the length-to-diameter ratio is $L/D = 10$ and the geometric blockage ratio is $B = 10\%$. Different test sections and/or configurations are utilised to apply different flow-control means together with either force and mean wall-pressure measurements or mean and dynamic wall-pressure measurements, or to make oil-flow visualisations. By pressurising the air in the wind tunnel up to 50 bar (the maximum possible pressure is 100 bar) high Reynolds numbers from $Re = 0.1 \times 10^6$ up to $Re = 5.0 \times 10^6$ are obtained, while the wind tunnel reaches a maximum free-stream velocity of $U_\infty = 35$ m/s.
All cylinder models have a smooth surface with a mean roughness height of less than 10 μm with the exception of holes for the pressure sensors and the disturbance excitation chambers with a diameter of 300 μm.
A reference cylinder model is used in order to distinguish the behaviour of flows around different cylinders by comparing the force and base-pressure coefficients. The force measurements are made with a piezo-balance which is described in detail by Schewe [22]. The dynamic signals of the force components of drag and lift are measured by the balance and amplified by charge amplifiers. The Strouhal numbers are obtained by Fourier analysis of time signals of lift fluctuations. The base-pressure coefficients at different spanwise locations are measured with a piezo-resistive pressure-transducer via a scanner-valve system by Scanivalve.
A second cylinder model is used to conduct the active-stationary and active-dynamic flow-control means such that a row segment of 36 pressure chambers, evenly distributed along the span at a reference azimuth, is inserted into the model. The size of a single chamber is chosen in such a manner as to generate at least eight actuator points for an artificially excited wavelength of two cylinder diameters. All chambers inside the cylinder are covered by a thin metal plate with 9 small holes in the surface of the plate per each chamber. Every chamber is connected via flexible pressure-tubes with horn-driver loudspeakers for the active-dynamic flow-control means, or with an air pressure tank for the active-stationary flow-control means utilising the continuous blowing technique (not discussed in this paper), or with an air pressure reservoir located outside the tunnel in order to apply the continuous suction technique. In order to measure the mean-pressure coefficients along the span of the cylinder and the base-pressure coefficients, simultaneously, three rows of pressure sensors are located at relative azimuths of 55, 155, and 180 degrees and some further sensors are instrumented around the cylinder at the symmetric span position of the cylinder at $y_D = 0$. This second cylinder model is mounted on a spin table of the test section, so that the reference azimuth can be varied in a range of ±35 degrees. The mean-pressures are measured by piezo-resistive pressure-transducer via a scanner-valve system by Scanivalve and the dynamic pressures are detected by dynamic piezo-sensors by Kulite.

Uncontrolled separated cylinder flow

The behaviours of base-pressure, drag and lift coefficients as well as Strouhal numbers of the dominant separation modes in their dependence on Reynolds number are shown in Fig. 1a-d in the subcritical up to the transcritical Reynolds number regime of the separated cylinder-flow in good agreement with results obtained by Schewe [6].
The oil-flow visualisations in Fig. 2a/b show the spanwise behaviour of the separated cylinder-flow in the subcritical and in the *from supercritical up to transcritical regime*. As Schewe and Dallmann [2] discussed we reconfirm the appearance of spanwise periodic structures, called cells, in this *three-dimensional transitional regime*. Furthermore we obtain two separation lines

on both sides of the cylinder in the end of subcritical regime at Reynolds numbers above Re = 2.0 $^{x}10^{5}$, although in the beginning of the subcritical regime only one separation line on each side of the cylinder can be recognised. Following the argumentation of Dallmann et al. [2] this topological changing is necessary to obtain a rapid decreasing in drag in the critical regime.

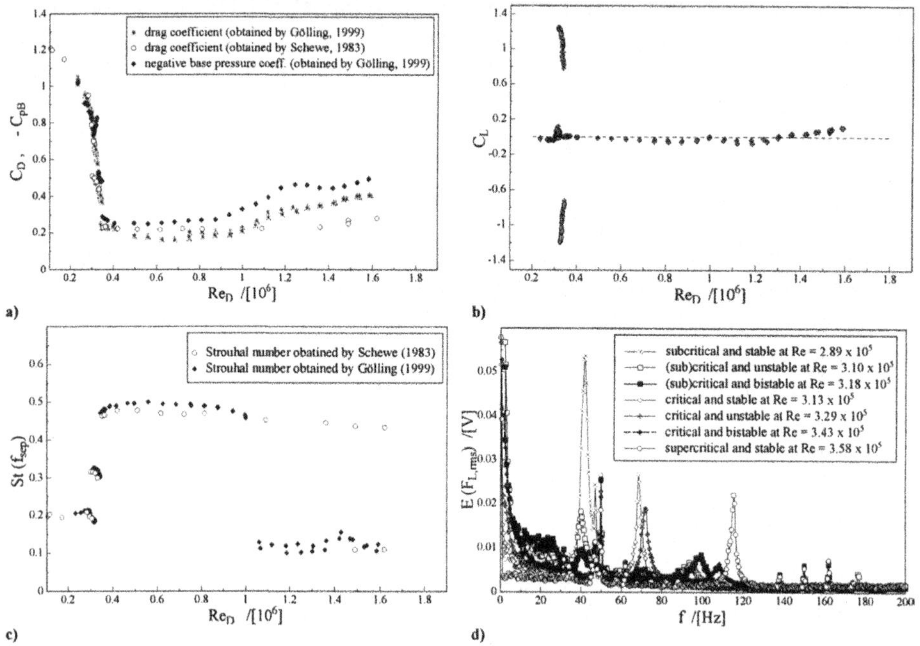

Fig. 1. Characteristic force coefficients, Strouhal numbers and spectra at the subcritical up to the transcritical Reynolds number regimes in comparison with results obtained by Schewe[6]: **a)** base-pressure and drag coefficients; **b)** lift coefficients indicating both asymmetrical flow states at critical Reynolds numbers, **c)** changes of Strouhal numbers of dominant separation mode calculated from experimental data by Fourier analysis of lift fluctuations and **d)** some power spectra of lift fluctuations according to the different flow states.

Because of the added roughness due to the oil-filament on the surface of the cylinder the appearance of spanwise almost-periodic large-scale coherent flow-structures should be slightly shifted to lower Reynolds numbers, in particular in the supercritical Reynolds number regime.

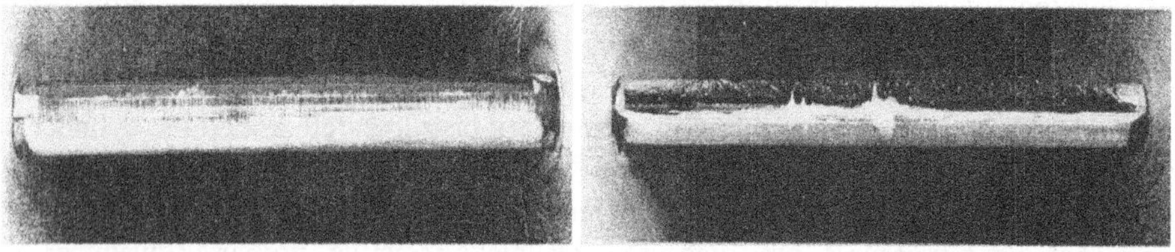

a) Oil flow visualisation at Re = 1.2 $^{x}10^{5}$ (**left**) with view on 45 deg. and at Re = 2.0 $^{x}10^{5}$ (**right**) with view on 90 deg. At the beginning of the subcritical regime exist only one separation line and at the end appears two separation lines on each side of the cylinder.

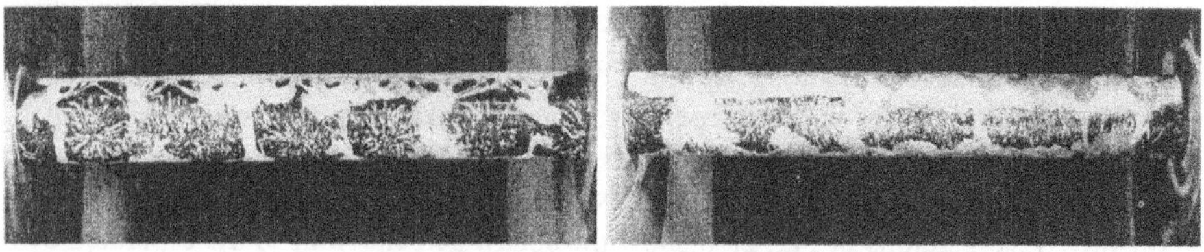

b) Oil flow visualisation at Re = 4.0 $^{x}10^{5}$ with view on 170 deg. (**left**) and at Re = 8.0 $^{x}10^{5}$ with view on 180 deg. (**right**). In supercritical up to transcritical regime spanwise periodic vortical flow structures dominate separation.

Fig. 2 Oil-flow visualisations for different Reynolds numbers according to different flow states: **a)** subcritical and **b)** from supercritical up to transcritical regime of cylinder-flow separation.

Therefore, the oil-flow visualisation in Fig. 2b (left) is actually conducted at a lower Reynolds number ($Re = 4.0 \times 10^5$) than its true Reynolds number of a smooth cylinder of roughly $Re = 5.5 \times 10^5$. The appearance of spanwise almost-periodic cells in the oil-flow patterns as well as in the patterns of the spanwise mean-pressure coefficients (here not shown), in particular with a spanwise wavelength of roughly two cylinder diameters in the early supercritical regime, is accompanied by a considerable reduction of the drag in comparison to the subcritical regime. The observations and information contained partly in Fig. 1 and Fig. 2 provide the motivation for the application of a spanwise-periodic excitation with a well-defined spanwise wave-number by the various flow-control means. The appropriate selection criteria of the excitation frequency (excitation Strouhal number) in case of combined spanwise-periodic and temporal-periodic acoustic excitation are discussed in the next paragraph.

Controlled separated cylinder flow

The motivation for checking different flow-control means in the high drag subcritical regime is to excite either a stable mode of spatial flow-instability which should affect the spanwise-periodic large-scale separated streamwise oriented vortex-flow structures and/or to stabilise a certain mode of the temporal flow-instability which corresponds to a certain shear-layer instability and/or to initiate laminar-turbulent transition with high-frequency excitation.
A characteristic parameter of effectiveness of the different flow-control means is the base-pressure coefficient, which is correlated to drag. We also choose the base-pressure coefficient enhancement relative to the base-pressure coefficient of the uncontrolled separated cylinder-flow to indicate control effectiveness. In the following we define the measure "***subcritical drag reduction***" by a Reynolds number averaged base-pressure coefficient enhancement:

$$\gamma_{sub} = \frac{1}{\Delta Re} \cdot \int_{Re_{sub}} -\frac{C_{pB,exc.} - C_{pB}}{C_{pB}} \, d\,Re ,$$

which reflects the effectiveness of flow-control means in the whole subcritical Reynolds number regime with $\Delta Re = [1.0 \times 10^5 , 3.1 \times 10^5]$.
The various flow-control actuators are positioned close to the mean-flow separation and only at one side of the cylinder. (Similar flow-control means applied to wing-profiles which would be effective also in an asymmetric manner.) Of course, in order to reach an optimum in drag reduction of a circular cylinder-flow both sides of the cylinder have to be involved in the flow-control means.
The scale of the spanwise periodicity of the excitation is given by the range of spanwise wave-numbers which are apparent in the oil-flow visualisations and spanwise mean wall-pressure coefficient distributions (here not shown) in the supercritical up to the transcritical regime of the uncontrolled separated cylinder-flow. These wave-numbers scale with the cylinder diameter. Hence, according to the aspect-ratio of the cylinders $L/D = 10$, a wave-number $k_L = 1$ means, that we excite a wavelength of about 10 D. Finally, we get the following wave-numbers: $k_L = 1$, $k_L = 2$, $k_L = 4$ (Fig. 2b left), $k_L = 6$ (Fig. 2b right), and $k_L = 8$. In particular, if we conduct a spanwise constant flow-control means, we define this wave-number by $k_L = 0$.
The excitation-amplitude of the different flow-control means is tuned until a sufficient and/or an asymptotic behaviour of base-pressure coefficient enhancement is obtained.
The additional parameter of the active-dynamic flow-control means, the time-periodicity of excitation, i.e. the excitation Strouhal number $St_{exc} = f_{exc} \cdot D/U_\infty$, is varied between $St_{exc} = 0.2$ and $St_{exc} = 6.0$. This has either the effect of triggering the periodic separated flow in the near wake downstream of the cylinder or possibly affect laminar-turbulent transition with a certain frequency that is much larger than the vortex shedding frequency of the uncontrolled wake-flow.

The results in subcritical drag reduction are given in Fig. 3. At first, we recognise (in Fig. 3 left) that with increasing excitation Strouhal number of the acoustic temporal excitation (beginning with $St_{exc} = 0.2$ and stopping at $St_{exc.} = 6.0$ - here summarised in $St_{exc.} \geq 1.5$) the *subcritical drag reduction* is also increased for every spanwise wave-number. In particular (shown in Fig. 3 right), spanwise-periodic excitations with even wave-numbers and with maximum excitation boundary conditions ("*max. cond.*") achieve best results in base-pressure coefficient enhancement with *subcritical drag reduction* of at least $\gamma_{sub} \geq 30\%$ while application of stationary active control realises a different behaviour. It seems that a well-defined combination of spatially and temporally periodic excitation of the separated shear-layer flow is most effective for separated cylinder- and wing-flows.

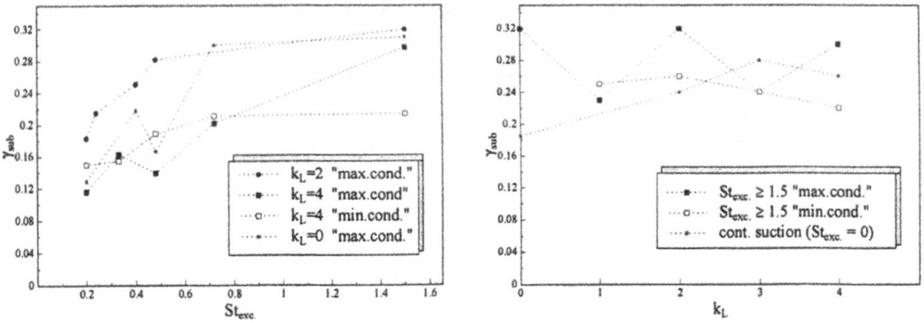

Fig. 3 ***Subcritical drag reduction*** γ_{sub} **Left:** in dependence on temporal excitation Strouhal number of different spanwise wave-numbers and different lateral excitation boundary conditions and **Right:** in dependence on spanwise wave-number of different temporal excitiation Strouhal numbers and different lateral excitation boundary conditions.

The spanwise mean-pressure coefficient distributions (obtained upstream of mean-separation, Fig. 4 top) indicate a critical mean-separation behaviour of the cylinder-flow with the onset of the flow control. In particular by instationary spanwise-periodic flow control means, the spanwise mean-pressure coefficient distribution follows the spanwise excitation-amplitude distributions at some subcritical Reynolds numbers.

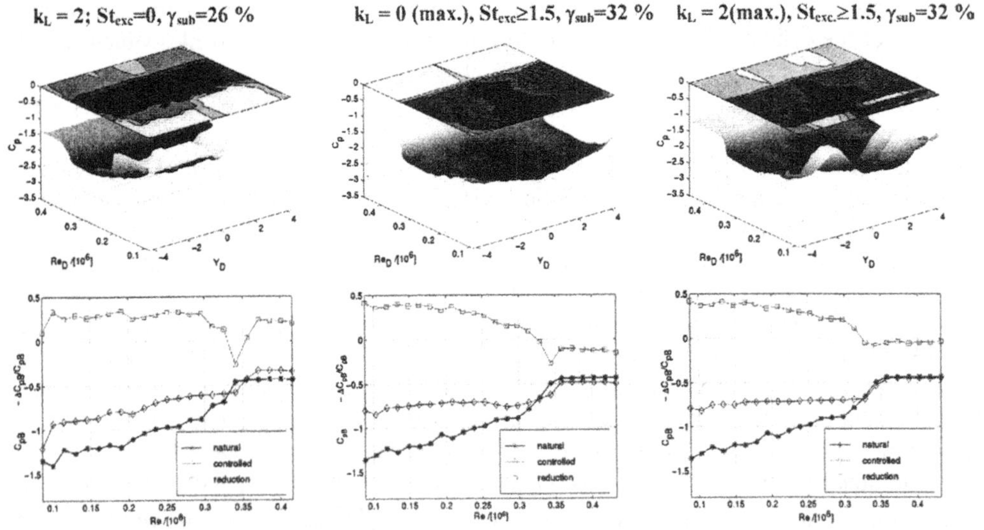

Fig. 4 Effect of different flow control means on spanwise pressure distribution (top) and base pressure coefficient (bottom) in the subcritical regime: Left: Active control by continuous suction with a wave-number of k_L = 2 ("*max.cond.*").resulted in a drag reduction of 26%; **Middle:** Active dynamic control by acoustic excitation with a spanwise wave-number of k_L = 0 ("*max. cond.*") and $St_{exc} \geq 1.5$ resulted in a drag reduction of 32% and **Right:** Active dynamic control by acoustic excitation with a spanwise wave-number of k_L = 2 ("*max. cond.*") and $St_{exc} \geq 1.5$ resulted in a drag reduction of 32%.

In case of active stationary flow control with a defined spanwise wave number of $k_L = 2$, the spanwise mean-pressure coefficient distributions at Reynolds numbers ranging from $Re \approx 2.2 \times 10^5$ up to $Re \approx 3.5 \times 10^5$ are similar to the spanwise mean-pressure coefficient distributions in the critical Reynolds number regime of the uncontrolled separated cylinder-flow, so that the laminar boundary-layer flow becomes transitional at lower Reynolds numbers (Fig. 4 left). Furthermore, the spanwise mean-pressure coefficient distributions for Reynolds numbers between $Re \approx 1.4 \times 10^5$ and $Re \approx 2.2 \times 10^5$ show no longer the quasi two-dimensional behaviour as obtained in the subcritical Reynolds number regime of the uncontrolled separated cylinder flow. However, the spanwise mean-pressure coefficient distribution does not indicate a direct excitation of spanwise-periodic large-scale vortical structures, although a spanwise-periodic disturbance distribution interacts with the laminar separating boundary-layer flow.
In case of spatio-temporal acoustic flow control with a spanwise wave-number of $k_L = 2$ ("*max. cond.*") and with $St \geq 1.5$ the spanwise mean-pressure coefficient distributions at Reynolds numbers between $Re = 1.0 \times 10^5$ and $Re = 1.7 \times 10^5$ indicate two mean-pressure coefficient maxima ($y_D \approx \pm 2$) and three mean-pressure coefficient minima - two at the lateral ends ($y_D \approx \pm 4$) and one at the symmetric span position of the cylinder ($y_D \approx 0$) (see Fig. 4 right). At Reynolds numbers between $Re \approx 1.7 \times 10^5$ and $Re \approx 3.4 \times 10^5$ the spanwise mean-pressure coefficient distributions show a similar behaviour like in the critical Reynolds number regime of the uncontrolled separated cylinder-flow and seems to be quasi two-dimensional up to the onset of laminar-turbulent transition taking place in the separating boundary layer.
Another example with a spanwise-constant excitation-amplitude and a temporal-periodic acoustic excitation of $St_{exc.} \geq 1.5$ is shown in Fig. 4 (in the middle). The spanwise mean-pressure coefficient distributions is immediately quasi two-dimensional and comparable with a spanwise mean-pressure coefficient distribution in the critical Reynolds number regime of the uncontrolled separated cylinder-flow.
These last two examples show, that although the spanwise wave-number of active-dynamic excitation are different, the same result in *subcritical drag reduction* of $\gamma_{sub} = 32$ is achieved by high-frequency temporal periodic excitation of flow-separation. This supports our second hypothesis, that the attenuation of turbulence by introduction of temporal disturbances in the separating boundary-layer flow or in the subsequently evolving separated shear-layer flow causes the evolution of large-scale spanwise periodic vortical flow-structures with $k_L > 0$ or, simply, turbulent spanwise nearly homogenous separation with $k_L = 0$.

Conclusions

This experimental investigation of separated circular cylinder-flows proves the effectiveness of purely spatial as well as spatially and temporally periodic flow-control means for drag reduction. The effectiveness is indicated by a considerable base-pressure coefficient enhancement at subcritical Reynolds numbers ($1.0 \times 10^5 \leq Re \leq 3.1 \times 10^5$), which describe the measure *subcritical drag reduction* γ_{sub}. While the active flow-control means by continuous suction reaches *reductions* of roughly $\gamma_{sub} \approx 25\%$, the active-dynamic flow-control means by acoustic excitation in particular at certain, even wave-numbers and high acoustic excitation Strouhal numbers of $St_{exc.} \geq 1.5$ causes *reductions* of at least $\gamma_{sub} \geq 30\%$. The spanwise mean-pressure coefficient distributions taken upstream of the controlled flow-separation, in particular in case of the active-dynamic flow control, indicate for some Reynolds numbers a spanwise modulated flow-separation behaviour according to the spanwise variation of excitation with a defined spanwise wave number. Although, it doesn't prove stringent the existence of spanwise-periodic large-scale structures at subcritical Reynolds numbers at the controlled side of the separated cylinder-flow. However, our preliminary investigations utilising excitation of temporal disturbances is likely to cause a critical transitional flow-separation behaviour at the controlled side of the cylinder-flow, similar to a (dynamic) turbulence generator, but attenuate also these spanwise variation in separation by spanwise variation of excitation. In contrast, the active-stationary flow control by spanwise-periodic continuous suction, i.e. without any direct excitation of boundary-layer turbulence, is less effective at subcritical Reynolds numbers and realise i.e. no spanwise-periodic pressure distribution.

Acknowledgement

The authors would like to thank Dr. G. Schewe of the Institute of Aeroelasticity for his technical support e.g. in using the piezo-balance, for the many discussions and his continuous support.

References

[1] Schewe G. (1994): Reynoldszahleffekte im Hochauftriebsbereich. Presentation at Symposium of AG STAB 1994, University of Erlangen, Germany.
[2] Dallmann U., Schewe G. (1987): On topological changes of separating flow structures at transition Reynolds numbers. AIAA 19th Fluid Dynamics, Plasma Dynamics and Laser Conference, June 8-10, 1987/Honolulu, Hawaii. AIAA-87-1266.
[3] Dallmann U., Herberg Th., Gebing H., Su W.H., Zhang Q.H. (1995): Flow field diagnostics: Topological flow changes and spatio-temporal flow structure. AIAA 33rd Aerospace Sciences, Meeting and Exhibit, January 9-12, 1995/Reno, NV; AIAA 95-0791.
[4] Norberg C. (1994): An experimental investigation of the flow around a circular cylinder: influence of aspect ratio. J. Fluid Mech., vol. 258, 1994, pp. 287-316.
[5] Tamura T. , Ohta I., Kuwahara K. (1990): On the reliability of two-dimensional simulation for unsteady flows around a cylinder-type structure. J. Wind Eng. Ind. Aerod., vol. 35, 1990, pp. 275-298.
[6] Schewe G. (1983): On the force fluctuations acting on a circular cylinder in crossflow from subcritical up to transcritical Reynolds numbers. J. Fluid Mech., vol. 133, 1983, pp. 265-285.
[7] Schewe G. (1986): Sensitivity of transition phenomena to small perturbations in flow round a circular cylinder. J. Fluid Mech., vol. 172, 1986, pp. 33-46.
[8] Brede M., Eckelmann H., Rockwell D. (1996): On secondary vortices in the cylinder wake. Physics of Fluids, vol. 8, 1996, pp. 2117-2124.
[9] Wei T., Smith C.R. (1986): Secondary vortices in the wake of circular cylinders. J. Fluid Mech., vol. 169, 1986, pp. 513-533.
[10] Wu J., Sheridan J., Hourigan K., Soria J. (1996): Shear-layer vortices and longitudinal vortices in the near wake of a circular cylinder. Experimental Thermal and Fluid Science, vol. 12, 1996, pp. 169-174.
[11] Mittal, R., Balachandar S. (1995): Autogeneration of three-dimensional vortical structures in the near wake of a circular cylinder. TAM Report No. 812, Universitiy of Illinois at Urbana-Champaign, November 1995, 50 p.
[12] Gad-el-Hak M. (1996): Modern developments in flow control. Appl. Mech. Rev., vol. 49, 1996, 365-379.
[13] Huerre P., Monkewitz P.A. (1990): Local and global instabilities in spatially developing flows. Ann.. Rev. Fluid Mech., vol. 22, 1990, pp. 473-537.
[14] Koch W. (1985): Local instability characteristcis and frequency determination of self-excited wake flows. J. Sound Vib., vol. 99, 1985, pp. 53.
[15] Owen J.C., Bearman P.W., Szweczyk A.A. (2000): The influence of streamwise vorticity on von Kármán vortex shedding. IUTAM Symposium on Bluff Body Wakes and Vortex-Induced Vibrations. BBVIV, 13-16 June 2000, Marseille, France.
[16] Darekar R.M., Sherwin S.J. (2000): Flow past a bluff body with a wavy separation line. IUTAM Symposium on Bluff Body Wakes and Vortex-Induced Vibrations. BBVIV, 13-16 June 2000, Marseille, France.
[17] Meneghini J.R., Bearman P.W. (1992): Numerical simulation of control of bluff body flow using a discrete vortex method incorporating viscous diffusion. IUTAM Symposium Göttingen/Germany 1992. Bluff-Body Wakes, Dynamics and Instabilities. Eds.: H. Eckelmann, J.M.R. Graham, P. Huerre, P.A. Monkevitz. Springer Verlag, pp. 257-262.
[18] Theofilis V., Hein S, Dallmann U. (2000): On the origins of unsteadiness and three-dimensionality in a laminar separation bubble. Phil. Trans. Roy. Soc. London A (to appear 2000).
[19] Barkley D., Gomes G. , Henderson R. D. (1998): Three-dimensional instability in flow over a backward facing step. J. Fluid Mech. (submitted 1998).
[20] Hein S., Thefilis, V., Dallmann U. (1998): Unsteadiness and three-dimensionality of steady two-dimensional laminar separation bubbles as result of linear instability mechanisms. DLR-IB 223-98 A39, Aug. 1998.
[21] Rist U., Maucher U., Wagner S. (1996): Direct numerical simulation of some fundamental problems related to transition in laminar separation bubbles. In Proc. Computational Fluid Dynamics Conf. ECCOMAS '96, 319-325.
[22] Schewe G. (1982): A multicomponent balance consisting of piezoelectric force transducers for a high-pressure windtunnel. Conf. Proc. Sensor & Systems 82, vol. 2, Pasadena (USA), 18-20.5.82; and Techn. Messen 49, p. 447 (German version).

DRAG REDUCTION IN AIRFOILS USING CONTROL DEVICES IN THE SHOCK WAVE-BOUNDARY LAYER INTERACTION REGION

N. Caballero
INTA, Fluid Dynamics Department, Ctra. Torrejón Ajalvir, Km 4.5, 28850 Madrid, SPAIN

Summary

The present work concerns the use of different techniques to control the shock wave boundary layer interaction phenomenon that takes place in transonic flows around airfoils at *off design* conditions. The idea behind the use of control mechanisms is to modify locally the flow in the vicinity of the shock wave in order to reduce total drag by means of reducing wave drag, and alleviate the shock induced separation that takes place at the foot of the shock when the Mach number or the angle of attack are far beyond the design point and causes the appearance of buffet.

1. INTRODUCTION

The aim of using control devices is to avoid the undesired effects that characterize the transonic regime at off design conditions, mainly the increase of total drag and the occurrence of buffet. The control devices can be broadly classified into three main categories. *Passive* ones are either based on use of a cavity placed underneath of the shock foot, separated from the outer flow by a perforated plate, or by placing a bump at the same location. Both techniques share the same physical principle: splitting the original shock thus reducing the total pressure jump. When cavities are considered, the induced blowing in front of the shock causes a boundary layer growth and as a consequence an increase in viscous drag. In order to weaken this last effect, a net mass suction trough the porous plate is enforced by adding extra power via the use of a pump (*active* control: power is externally supplied). Both the use of bumps and of active cavities lead to the reduction of wave drag with only marginal effects on the viscous drag. This occurrence indicates that blowing in front of the shock is the key physical ingredient behind the effectiveness of these techniques in controlling the shock wave boundary layer interaction region. A further improvement can be obtained by simultaneously controlling the boundary layer by using a suction slot upstream of the cavity/bump position pursuing a viscous drag reduction and/or the delay of the buffet onset. This last strategy, based on the use of a passive and an active device is usually termed as *hybrid* control technique. In the present work all of the mentioned methods are numerically investigated. It will be shown that properly selected devices can lead to both drag reduction and buffet delay.

2. NUMERICAL TOOL

The numerical simulations have been carried out using a 2D Navier-Stokes Reynolds averaged time accurate solver. The numerical discretization is based on the second order in time, implicit method of Beam & Warming [1] and on a second order central finite difference scheme in space. Turbulence closure is based on the Baldwin Lomax eddy viscosity model. Some modifications to the original model are needed to take into account the surface mass transfer and the longitudinal pressure gradient taking place within the control device regions. The effect of these new physical features have been modeled by means of a generalization of the *Van Driest's* [6] wall damping function, which in its original form reads like:

$$D = 1 - \exp\left(-\frac{y^+}{A^+}\right) \text{ being } A^+ = 26 \text{ and } y^+ = y\frac{u_\tau}{\nu}. \tag{1}$$

The value of A^+ has been modified to include the two mentioned effects. According to the work of *Bushnell et al.* [2] and *R. Bur* [3] , A^+ can be written as:

$$A^+ = 26[exp(11.8v_p^+) - \frac{p^+}{v_p^+}\{exp(11.8v_p^+) - 1\}]^{-\frac{1}{2}},\ v_p^+ = \frac{v_p}{\bar{u}_\tau},\ p^+ = -\frac{d\bar{p}}{dx}\frac{\nu}{\bar{\rho}\bar{u}_\tau^3} \tag{2}$$

being v_p^+ and p^+ the non-dimensional transpiration velocity and pressure gradient respectively. The calculation of v_p^+ and p^+ may cause problems in the region behind the shock where probably separation occurs ($\tau_w \leq 0$). To avoid this situation a different velocity scale has been considered in this zone: the friction velocity at the position where the absolute value of the shear stress reaches its maximum value.

The effects of the control devices have been incorporated by modifying locally the wall boundary conditions. The transpiration velocity through the perforated surface has been modeled using two different laws. Both consider the flow through a single hole as if it were alone: hole to hole interference is neglected. The first one under certain hypothesis assumes a developed laminar pipe flow through the hole. Then the velocity through the hole is a function of the pressure drop through the perforated plate. The second one is based on dimensional analysis and establishes a dependency of the Mach number at the hole with the pressure drop and the shear stress at the wall.

The first model used to predict the transpiration velocity along the perforated wall is based on the work of *Poll et al. [4]*. The basic assumption consists in considering the flow through an individual hole as a fully developed laminar pipe flow. Also the effects of compressibility are considered to be small enough to be neglected.

Then, the velocity profile at the exit plane is accepted to be parabolic and the pressure drop through the pipe can be written as:

$$\frac{\Delta P}{0.5\rho v^2} = 2.24 + \frac{64}{R_d}\frac{e}{d} \tag{3}$$

with v the velocity at the exit plane , e/d the ratio of plate thickness to hole diameter and R_d the Reynolds number based on hole diameter.

The latter relation can be worked out to give:

$$Y = [40.7 * X + 1.95 * X^2] \text{ with } X = \frac{\dot{m}}{\mu e} \text{ and } Y = \frac{\Delta P d^2}{\rho \nu^2}\left(\frac{d}{e}^2\right) \tag{4}$$

i.e., X proportional to the mass flux ($\dot{m} = v\rho\pi d^2/4$) , and Y proportional to the pressure difference (being ρ the density at the exit, μ the molecular viscosity and ν the kinematic viscosity).

Due to irregularities in manufacturing the perforated plates it is not possible to specify the diameter of an individual hole accurately, but rather it is possible to define a nominal diameter which can only be determined by a statistical survey of a large number of holes. Consequently, it is necessary to specify an equivalent diameter in equation (4) introducing a new factor k defined as: k= *effective* diameter/ *measured* diameter.

Then equation (4) becomes:

$$Y = \frac{1}{k}[40.7 * X + 1.95 * X^2] \; . \tag{5}$$

The second transpiration model [5] is based upon the work of Bohning and Doerffer. The main points of this new model are the introduction of the effect of tangential flow in the process of sucking fluid (tangential flow does not influence blowing) and the particular treatment of the plate porosity.

The proposed law for the transpiration velocity through a perforated surface is:

$$M_h = 1.2 \left(\frac{dP}{P_0} \right)^{0.555} \tag{6}$$

where M_h stands for the Mach number in the hole.

When tangential flow is present and suction takes place the proposed model is modified with the addition of a term that takes into account the presence of shear stress at the wall. In this case the velocity transpiration through a hole can be obtained from the expression:

$$\frac{dP}{P_0} = (M_h)^{\frac{1}{0.55}} \left[\left(\frac{1}{1.2} \right)^{\frac{1}{0.55}} + 25 \left(\frac{\tau_0}{0.5 \rho_h u_h^2} M_h \right)^{\frac{1}{1.52}} \right] . \tag{7}$$

As far as the porosity is concerned Bohning&Doerffer have suggested an expression for determining the so called effective porosity:

$$\frac{dP}{P_0} = 0.063 \frac{\dot{m}/A}{Por} \tag{8}$$

where $\dot{m}$ is the mass flow rate in $[Kg/sec]$, A the corresponding flow area in $[m^2]$, dP the pressure drop through the surface and Por the effective porosity in %.

The effective porosity is usually lower than the prescribed value given by the manufacturer of the plate and takes into account the possible irregularities that can exist. Once the effective porosity is determined its value has to be introduced in the computations that make use of the formula proposed by Bohning&Doerffer for the transpiration velocity.

3. USE OF BUMPS AS CONTROL DEVICES

In this section some aspects of the flow around airfoils with bumps as control devices are discussed. How these mechanisms act on the steady flow around the section wing and what the results are in terms of drag reduction are the main considered issues.

As a basic configuration we selected the DRA-2303 airfoil. The original shape has been modified with an asymmetric bump .0025c height located in the upper surface between $x/c = 0.5 - 0.68$.

For the computations it has been used a C-type grid with 300x70 points and the outer boundary located at ten chords distance from the airfoil.

The effect of the bump on the main flow has been investigated under the following flow conditions: $Mach = 0.68$, $Re = 18.8E6$ and transition fixed at 5% of the chord on both the upper and the lower surface.

In figure 1 the predicted pressure coefficient distribution corresponding to the original section is compared with the experimental data for different lift coefficients. Figure 2 shows the pressure distribution around the bump airfoil for the same angles of attack as in the former figure, (leftmost $\alpha = 1.0$, central $\alpha = 1.75$ and rightmost $\alpha = 2.25$). It is remarkable the good agreement between experimental and computed data.

The main effect induced by the presence of the bump is the split of the shock. At low incidences the flow undergoes a first compression when it reaches the bump and then expands until the shock wave occurs. When increasing the angle of attack the two compressions become closer until they coalesce and then the effect of the bump is negligible.

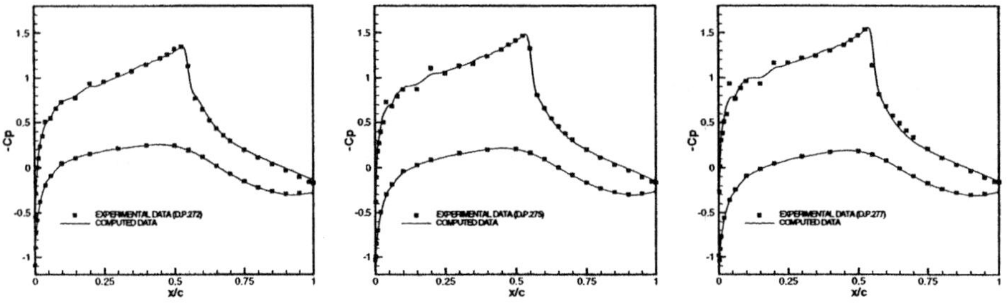

Figure 1 DRA-2303 DATUM: From left to right: $C_l = 0.61$, 0.72 and 0.79.

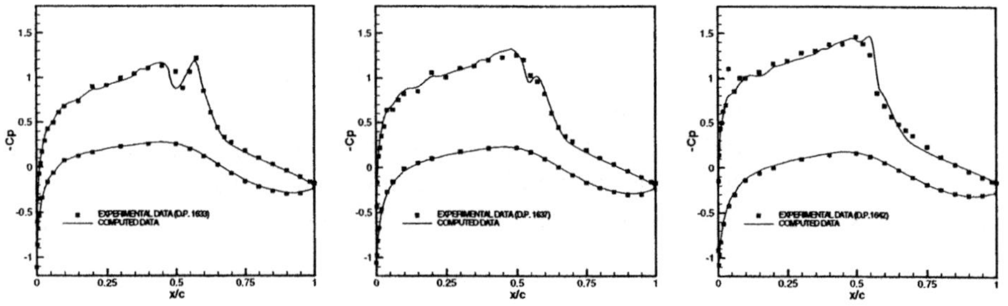

Figure 2 DRA-2303 WITH BUMP: From left to right: $C_l = 0.55$, 0.68 and 0.77.

4. USE OF CAVITIES AS CONTROL DEVICES

The use of cavities separated from the outer flow by a porous wall as control devices has been investigated for the laminar DRA-2303 airfoil. The cavity has been located in the upper surface from $x/c = 0.475$ to $x/c = 0.625$. The plate characteristics have been fixed to: thickness $1.0mm$, porosity 4%, hole diameter $0.076mm$ being the airfoil chord $635mm$. Experimental results are available for these tests.

In the next discussion the parameter C_q stands for the non dimensional mass flow rate through the porous plate: $C_q = \frac{q}{\rho_\infty U_\infty c}$, being q the net mass flow through the perforated plate and c the airfoil chord.

With the previously described cavity it has been considered the case of passive as well as active control. When comparing the computed results with the experimental data for any of those controls some discrepancies have been observed as far as mass flow rate and cavity pressure are concerned.

This is illustrated on the left part of fig. 3 where the pressure distribution for the passive case corresponding to $C_l = 0.6$ is plotted versus the experimental data. The two computed curves, obtained using Poll law model for the estimation of the velocity transpiration (similar results are obtained using Bohning&Doerffer law) , correspond to consider input data either the zero mass flow rate through the perforated plate or the corresponding experimental pressure coefficient in the cavity. According with these numerical results no passive control takes place when the pressure in the cavity is $(Cp)_{cav} = -0.946$ since some blowing occurs ($C_q = 0.0002$). In order to have passive control the pressure in the cavity should be $(Cp)_{cav} = -1.025$ much lower than the experimental value. Discrepancies can be explained with probably three dimensional effects in the experimental setup. In figure 3 right it is also plotted the non-dimensional computed transpiration velocity (the value of sound speed at infinity has been used for the non-dimensional variable).

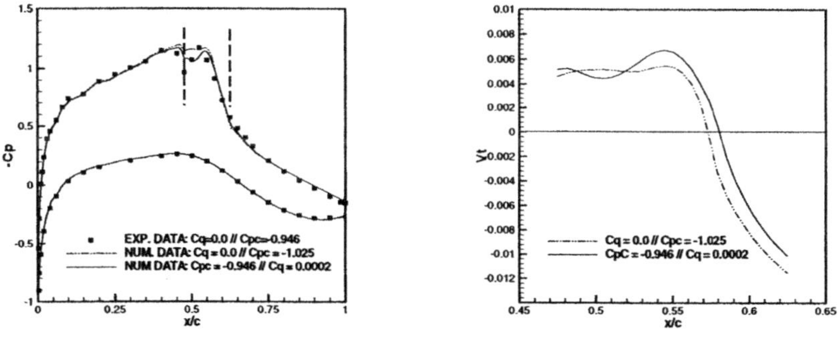

Figure 3 DRA-2303 WITH CAVITY: $C_q = 0.0$. Pressure and velocity distribution.

Figure 4 presents the results corresponding to active control (experimental $C_q = -.00009$). Similar comments as those done for the passive control can be made. Again the experimental pressure coefficient gives a net blowing and a slightly higher compression in front of the shock.

With respect to global coefficients it can be conclude that the use of cavities does not improve the performances of the airfoil. Experimentally and numerically it has been found a decrease in lift and a considerable increase in total drag (of the order 10-15when

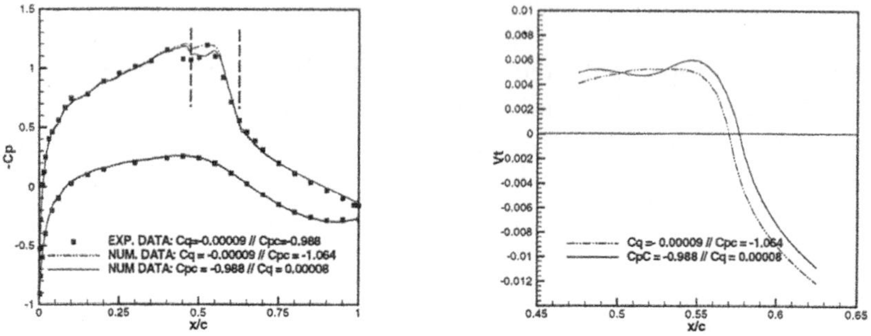

Figure 4 DRA-2303 WITH CAVITY: HIGH SUCTION. Pressure and velocity distribution.

the cavity is activated either as passive control or active one.

The investigations of the effect of hybrid control have been done adding a slot suction to the configuration already discussed. The slot of size 1%c has been located behind the cavity at the abscissas $x/c : 0.65 - 0.66$. The effect of the slot is mainly to suck part of the boundary layer, reducing its thickness and thus reducing viscous drag.

The simulations have been performed using two levels of suction through the slot and also with a zero net mass flow. The latter can be considered as passive control.

Some benefits have been observed when comparing with the case of the single cavity but still the net total drag is higher as the original airfoil total drag.

The left picture of fig. 5 shows the pressure coefficient distribution for the mentioned case that corresponds to the experimental data point 2291. The numerical results fits very well with the experiments. In the right picture the corresponding iso-mach contours are plotted. The flow structure does not differ much of that of the single cavity.

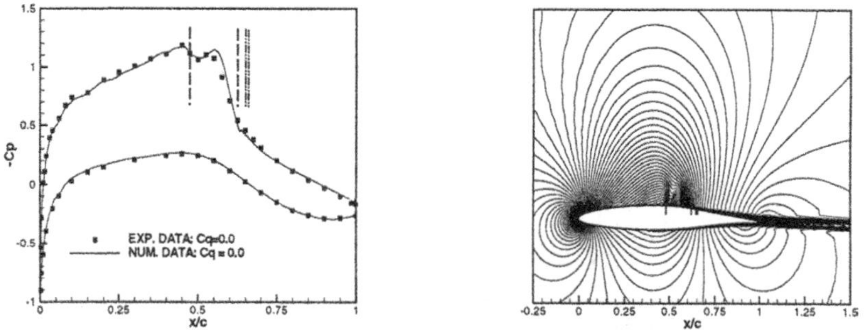

Figure 5 DRA-2303 + HYBRID CONTROL: $C_q = 0.0, C_l = 0.6$. Pressure distribution and iso-Mach contours.

5. UNSTEADY BEHAVIOR

One of the important issues of the use of control devices is the influence that such elements can have on the buffet onset. Some numerical investigations have been done to study the effect of suction slots in the buffet onset. The idea is to suck in front of the shock in order to alleviate or even eliminate shock induced separation.

To study the influence of these control devices in buffet we have followed the strategy of applying suction through a 1% size opened slot to a completely developed unsteady flow in the laminar DRA-2303 airfoil corresponding to an incidence of $\alpha = 4.8$.

The two parameters than can be freely chosen in order to obtain the desired effects are the position of the slot and the suction intensity.

In the developed unsteady flow, see fig. 6, the shock wave travels over the upper surface from a position of minimum lift upstream to a maximum lift position downstream.

For the low level of suction ($C_q = -7E - 5$) no relevant effect has been found with respect to buffet onset. The unsteady flow does not suffer any change and the buffet onset remains at the same limits as if it were without slot.

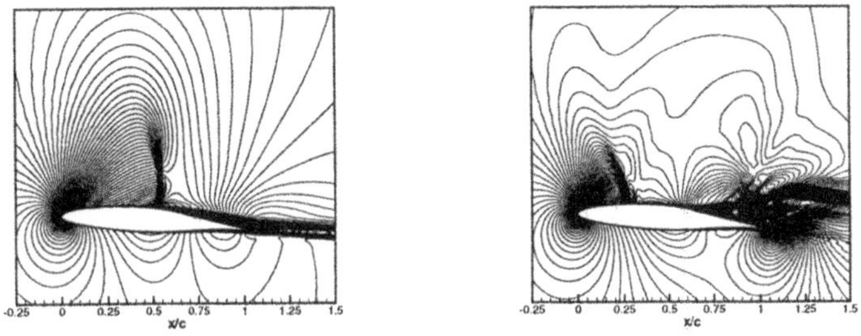

Figure 6 DRA-2303: Unsteady state: no control, $\alpha = 4.8$.

On the contrary when increasing suction ($C_q = -7E-4$) the situation changes drastically. In fig. 7 the lift coefficient history is plotted for the different slot locations.The first part of the three histograms corresponds to the no control case and it is followed by the control computed case. The results are quite relevant since it is predicted a completely absence of unsteadiness when the slot is located at $x/c : 0.54 - 0.55$.

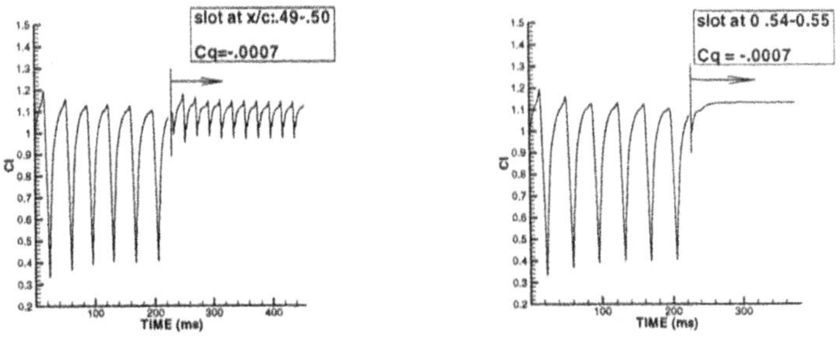

Figure 7 Histograms before and after application of control at two different slot positions: $\alpha = 4.8$.

6. CONCLUSIONS

Numerical investigations performed on an laminar airfoil have shown the possibility of controlling the shock wave boundary layer interaction phenomenon that takes place in the transonic regime when flying at off-design conditions.

From the various control devices investigated, it can be concluded that the use of properly designed bumps, may give excellent results in terms of total drag reduction. On the other hand, the use of cavities, in a passive fashion, gives very good results in terms of wave drag reduction but accompanied with an important increase in viscous drag. The different means of reducing the induced viscous drag, such as sucking fluid from the cavity or including suction slots, seem to be efficient only when very high suction intensity is applied and therefore they do not appear to be useful in wing applications.

With respect to the onset of buffet, the computations show a delay in its appearence when control mechanisms are used. The most outstanding results show how the developed buffet flow around an airfoil can even disappear when using an appropriate suction slot.

In any case, it is important to notice the good agreement found between numerical and experimental results. That shows, once more, the advantage of using numerical tools in order to predict the flow behavior and maybe to reduce the number of expensive tunnel tests.

References

[1] Beam R. M. and Warming R. F. "An implicit factored scheme for the compressible Navier-Stokes equations", *AIAA Journal* **16**, 1978. 393-402.

[2] Bushnell D.M., Cary A.M. and Harris J. E. "Calculation Methods for Compressible Turbulent Boundary Layers", NASA, SP-422,1977.

[3] Bur R. "Passive Control of a Shock Wave Turbulent Boundary Layer Interaction in a Transonic Flow", 1992 *Rech. Aérosp.*,**6**

[4] Poll D.I.A., Danks M. and Humphreys B.E. "The Aerodynamic Performance of a Laser Drilled Sheet", First European Forum on Laminar Flow Technology, Hamburg, March 1992. Paper 92-02-028

[5] R. Bohning and P. Doerffer "Hybrid and Active Control of the Shock-Wave Boundary-Layer Interaction and porous Plate Transpiration Flow", EUROSHOCK-II Final Technical Report, July 1999.

[6] Van Driest E.R. 1956 "On turbulent flow near a wall", *J. Aero. Sci. Journal* **23** 1007-1012.

[7] Baldwin R.S. and Lomax H. "Thin Layer Approximation and Algebraic Model for Separated Turbulent Flows", AIAA Journal, Vol 99, 1978.

Numerical Study of Separation Control by Movable Flaps

Markus Schatz*, Ulf Bunge, Holger Lübcke, Frank Thiele

Technische Universität Berlin, Hermann–Föttinger-Institut,
Müller-Breslau-Str.8, D-10623 Berlin, Germany

Summary

The effect of movable flaps on the upper surface of an airfoil is studied by an unsteady numerical RANS simulation of the flowfield around static flaps at different flap angles. The self-adjusting flap remains in an equilibrium position if the sum of all resulting aerodynamic forces is zero. In this case, the lift increases significantly compared to the clean airfoil. The numerical results, however, indicate that in equilibrium position, the flap does not yield the optimal performance. Additionally, the effect of an oscillating flap as a means of flow-control is studied to investigate the effect of different excitation frequencies.

Introduction

During the landing approach, the feathers on the upper surface of bird wings tend to pop up, which can be interpreted as a biological high-lift device (Fig. 1b). Experimental studies of Meyer and Bechert [1] and of Patone and Müller [2] show that feather-like movable flaps on the upper surface of a laminar airfoil can increase lift and reduce drag at high angles of attack, particularly in the post-stall region. Therefore, large scale flow separation is avoided and the reverse flow is divided into two closed regions. In experimental studies, this yields a gain of up to 18% lift without increasing drag. Additionally, free-flight tests revealed a favourable handling of sailplanes in stall condition.

a)

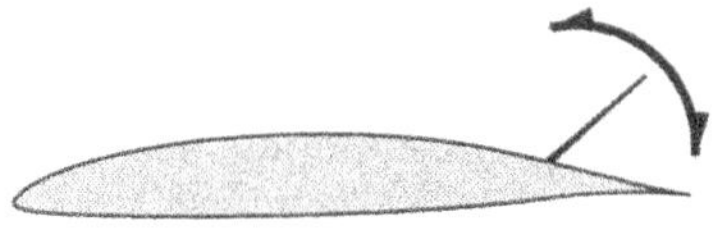

b)

Figure 1: a) Sketch of the movable flap on an HQ17 airfoil, b) lifted feathers of a landing Skua under high-lift conditions

The principal aim of this study is to obtain a better understanding of the effect and the flow physics of self-activated movable flaps. A flapped HQ17 airfoil at $Re = 10^6$ is investigated by a numerical simulation based on the RANS equations applying an advanced eddy-viscosity turbulence model. By comparing the results of the flow around a static flap at different flap angles, the angle at which the flap is in an equilibrium position can be found. The flowfield at this excitation is representative for the self-adjusting flap and can be compared to the experimental data.
Additionally, as a further step towards a real self-adjusting flap, the effect of an oscillating flap as a method of active flow-control is investigated. In order to find an optimum excitation, different frequencies are applied in the numerical simulation.

Computational Approach

Similar to the experiments, the numerical study considers an HQ17 airfoil (Fig. 1a). The numerical investigations were performed using a two-dimensional incompressible Finite-Volume code which solves the Reynolds-averaged Navier-Stokes equations. The two-dimensional method is fully implicit and of second order in space and time. In previous investigations of unsteady turbulent flows, improved one- and two-equation models exhibited the best performance. Therefore, turbulence effects are modelled by the LLR k-ω model [3] and results are compared to the Spalart-Allmaras model [4] and the standard Wilcox k-ω model [5]. For the static and oscillating flap cases however, only results obtained with the LLR k-ω model are presented here. Transition is fixed at positions taken from the experiments for all angles of attack.
For the simulation of the clean airfoil, a mesh with 27000 cells is used, and the grid for the flapped airfoil consists of 34000 cells. In the first cell, the non-dimensional distance to the wall remains below $y^+ = 2$ in both cases. To capture the geometry in the case of a moving flap, the numerical grid needs to be distortable. Starting with a basic grid at a low flap angle, the mesh around the upward-moving flap is deformed with a method based on an analogy to structure mechanics [6]. If necessary, a biharmonic smoothing algorithm by Yan [7] is applied to improve the mesh quality.

Results and Discussion

Clean HQ17 airfoil

Numerical investigations of the clean airfoil show that the flow becomes unsteady at high angles of attack with a vortex street developing in the case of $\alpha > 12°$. Consequently, all simulations must be performed with an unsteady method at a non-dimensional time-step of $\Delta t^* = \Delta t \cdot \frac{u_0}{c} = 0.01$.
The lift coefficients obtained by the LLR k-ω model are in good agreement with the experimental data. Results computed with the Spalart-Allmaras one-equation model and the Wilcox k-ω model, however, overpredict lift for high angles of attack (Fig. 2a). The difference in the models' performances arise from their ability to predict the separation point on the upper surface. The separation point can qualitatively be calculated by the LLR k-ω model (Fig. 2b), however, all detachment positions are slightly upstream compared to those obtained from the experiments.

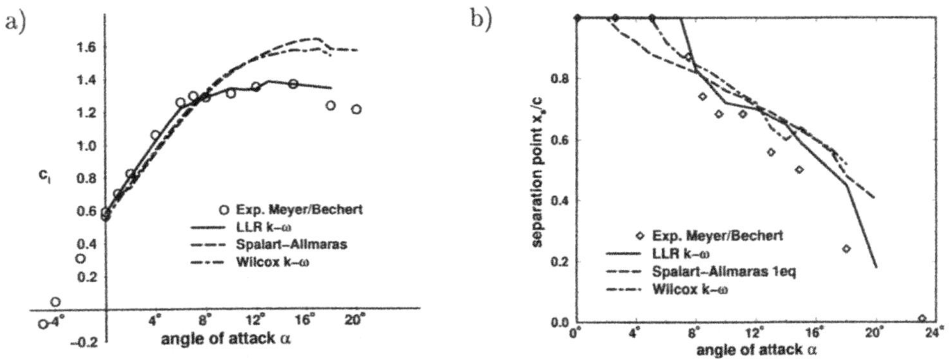

Figure 2: a) Lift coefficient for the clean airfoil, b) separation point

Static flap

A flap of 17.2% chord length is mounted on the airfoil at 80% chord. In the experimental study, Meyer and Bechert showed that with such a self-adjusting flap the maximum lift can be increased by about 18%.

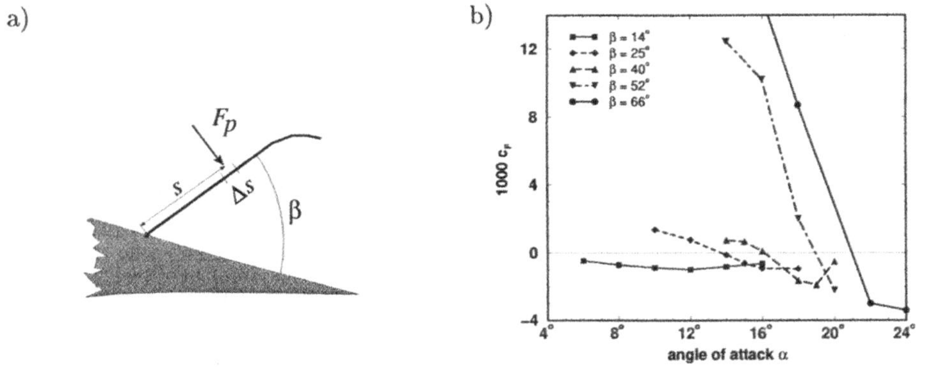

Figure 3: a) Integration of C_F, b) C_F for different flap-angles over the angle of attack

To find an optimum flap angle in the numerical simulation, the resulting aerodynamic forces on the flap are calculated for each angle. The integrated moment M can be represented by a non-dimensional coefficient C_F:

$$C_F = \frac{M}{\frac{1}{2}\rho u_o^2 c^2} \qquad \text{with} \qquad M = \sum_{flap} F_p \cdot s \quad = \sum_{flap} p \, \Delta s \cdot s \ , \qquad (1)$$

where F_p is the force due to pressure p, c is the airfoil chord, u_o is the reference velocity and s is the lever arm (Fig. 3a). Positive coefficients represent an upward movement. A self-activated flap will stay in its equilibrium position if the sum of all forces on the flap is zero and the coefficient C_F disappears.

Fig. 3b shows the flap coefficient over the angle of attack. For small flap angles β no equilibrium position can be found, and the coefficient remains negative for all angles of attack. In the case of large flap angles ($\beta \geq 25$), however, the equilibrium position can clearly be detected.

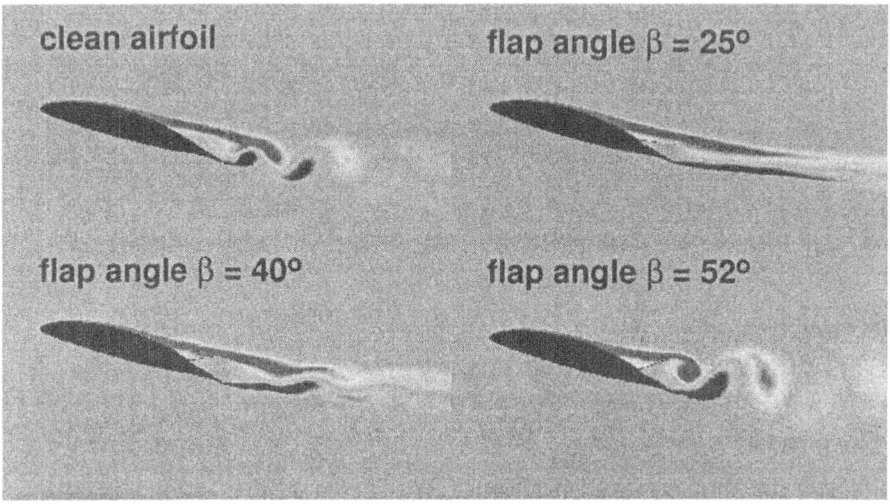

Figure 4: Comparison of different flap angles with the clean airfoil at $\alpha = 16°$, flowfield including detaching vortices is represented by the vorticity distribution

By introducing such a flap, the flow can be stabilized, and the maximum lift increases. In Fig. 4, the results of simulations at $\alpha = 16°$ for different flap angles are plotted. The comparatively strong vortex street, which appears behind the clean airfoil, is damped in the case of a low flap angle, and cannot be suppressed completely ($\beta = 25°$). In equilibrium position, the flap is just slightly touching the detached shear-layer ($\beta = 40°$). The vortex street is damped but its intensity exceeds that of the lower flap angle. In the case of too large angles, the flap works like a spoiler without any effect on the lift coefficient, generating additional drag ($\beta = 52°$). If the flap angle is too low, the flap remains inside the reverse flow region and shows less effect.
Compared to the clean airfoil, the maximum lift coefficient can be increased by more than 20% for the optimum flap angle ($\beta = 25°$ in that case). However, at the equilibrium position, it increases the lift by not more than 15%. The predicted lift coefficients agree fairly well with the experimental data.
Further investigations were carried out for different angles of attack. The lift coefficients for all investigated cases are shown in Fig. 5a. At the respective optimum flap angle, the lift coeffients are represented by an envelope curve. The resulting maximum lift is increased by about 20% over a wide range of angles of attack. In the case of the flap being in equilibrium, however, the lift remains about 5% below this maximum. This is due to the flaps tendency to adjust at too large angles.

The Strouhal number is defined by

$$St = f \cdot \frac{c}{u_o} , \qquad (2)$$

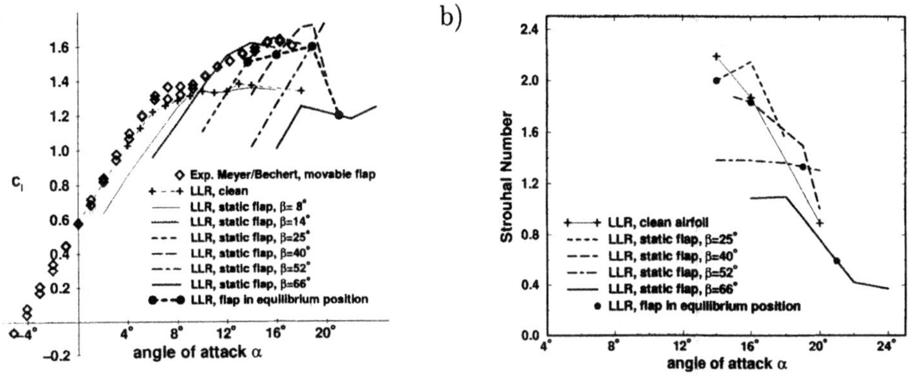

Figure 5: a) Lift coefficient for static flap at different angle, b) Strouhal number of detaching vortices

where f is a typical frequency of the detaching vortices. In Fig. 5b, Strouhal numbers for different flap angles are plotted over the angle of attack. For the flap in equilibrium position, the Strouhal number can obviously be compared with the clean airfoil. Therefore, the frequency of the detaching vortices is almost not affected by a self-adjusting flap.

Oscillating flap

Experimental investigations with movable flaps sometimes show high-frequency flap motions. In additional numerical simulations, the flap is oscillating around a mean flap angle similar to a flow-control device. To capture the effect of time-dependent flap angles, the mesh needs to distort at every time step, additionally the equations of motion must take the space conservation law into account. The flap motion is prescribed by:

$$\beta(t) = 38^\circ + 2^\circ \cdot cos\,(\frac{\pi u_o}{c} St \cdot t)\ . \tag{3}$$

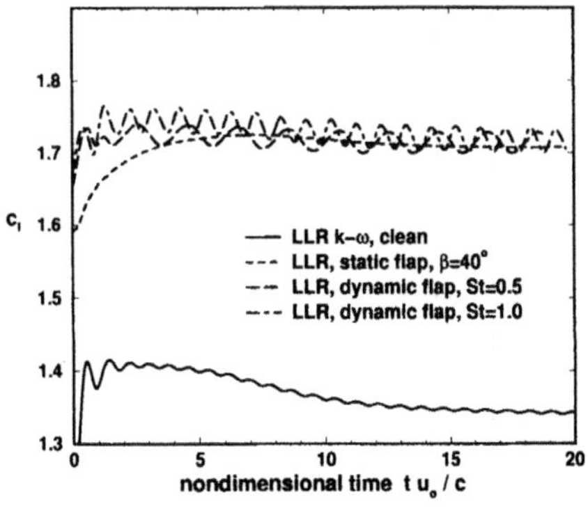

Figure 6: Unsteady lift coefficient over nondimensional time for different oscillation frequencies

It is found that the lift coeffients increase due to flap oscillation, however no additional gain is achieved compared to a static flap (Fig. 6); the drag is even larger than in the case of a static flap at the same flap angle. Different excitation frequencies do not significantly increase the lift coefficient.

Conclusions

Numerical investigations show that movable flaps can result in a 10% increase in lift and up to 10% reduction in drag. All important effects can be captured by an unsteady RANS approach, however, the most accurate results are obtained with the LLR k-ω turbulence model. For a static flap in equilibrium position, increased lift is predicted, however, the maximum lift corresponding to a flap at lower flap angle, can not be realized. In the equilibrium position, the self-adjusting flap is just touching the free shear layer. The Strouhal number of detaching vortices in the equilibrium position is similar to that in the clean airfoil case. Finally, the investigations show that flow control by oscillating flaps does not significantly improve the airfoil performance.

Acknowledgement

This research is funded by Deutsche Forschungsgemeinschaft under the umbrella of the Sonderforschungsbereich 557, 'Beeinflussung komplexer turbulenter Scherströmungen' at the Technical University Berlin.

References

[1] D.W. Bechert, R. Meyer and W. Hage, Aeroflexible Oberflächenklappen als Rückstrombremsen nach dem Vorbild der Deckfedern des Vogelflügels, DLR Institutsbericht 92517-96/B2, 1996.

[2] G. Patone and W. Müller, Aeroflexible Oberflächenklappen als "Rückstrombremsen" nach dem Vorbild der Deckfedern des Vogelflügels, Technical Report TR-96-05, Technical University Berlin, 1996

[3] T. Rung and F. Thiele, Computational Modelling of Complex Boundary-Layer Flows, 9th Int. Symp. on Transport Phenomena in Thermal-Fluid Engineering, Singapore, 1996.

[4] P.R. Spalart and S.R. Allmaras, A One-Equation Turbulence Model for Aerodynamic Flows, AIAA Paper 92-0439, Reno, NV, 1992.

[5] D.C. Wilcox, Turbulence Modeling for CFD, DCW Industries, Inc., La Cañada, 1993.

[6] A.Wick and F.Thiele, Dynamic mesh generation with grid quality preserving methods, Proc. Second International Symposium on Finite Volumes for Complex Applications, Duisburg, Hermes Science Publications, 1999.

[7] J. Yan, L. Xue and F. Thiele, Performance of Multigrid Method on Biharmonic Grid Generation. In: Proceedings of the 6th International Conference on Numerical Grid Generation in Computational Field Simulations. University of Greenwich, July 6-9, 1998.

Notes on Numerical Fluid Mechanics

Available volumes

Volume 75: E.H. Hirschel (ed.): Numerical Flow Simulation II. CNRS-DFG Collaborative Research Programme, Results 1998–2000. ISBN 3-540-41608-0

Volume 74: D. Henry, A. Bergeon (eds.): Continuation Methods in Fluid Dynamics. Contributions to the ERCOFTAC/EUROMECH Colloquium 383, Aussois, France, 6.–9. September 1998. ISBN 3-540-41556-4

Volume 73: M. Griebel, S. Margenov, P. Yalamov (eds.): Large-Scale Scientific Computations of Engineering and Environmental Problems II. Proceedings of the Second Workshop on "Large-Scale Scientific Computations" Sozopol, Bulgaria, June 2–6, 1999. ISBN 3-540-41551-3

Volume 72: W. Nitsche, H.-J. Heinemann, R. Hilbig (eds.): New Results in Numerical and Experimental Fluid Mechanics II. Contributions to the 11th AG STAB/DGLR Symposium Berlin, Germany 1998. ISBN 3-540-41552-1

Volume 70: W. Hackbusch, G. Wittum (eds.): Numerical Treatment of Multi-Scale Problems. Proceedings of the 13th GAMM-Seminar, Kiel, January 24–26, 1997. ISBN 3-540-41562-9

Volume 69: D. P. Hills, M. R. Morris, M. J. Marchant, P. Guillen (eds.): Computational Mesh Adaptation. ECARP – European Computational Aerodynamics Research Project. ISBN 3-540-41548-3

Volume 67: C. H. Sieverding, G. Cicatelli, J. M. Desse, M. Meinke, P. Zunino: Experimental and Numerical Investigation of Time Varying Wakes behind Turbine Blades. Results of the Industrial and Material Technology Aeronautics Research Project AER2-CT92-0048, 1992–1996. ISBN 3-540-41546-7

Volume 66: E. H. Hirschel (ed.): Numerical Flow Simulation I. CNRS-DFG Collaborative Research Programme. Results 1996–1998. ISBN 3-540-41540-8

Volume 64: R. Friedrich, P. Bontoux (eds.): Computation and Visualization of Three-Dimensional Vortical and Turbulent Flows. Proceedings of the Fifth CNRS-DFG Workshop on Numerical Flow Simulation, München, Germany, December 6 and 7, 1996. ISBN 3-540-41557-2

Volume 63: M. Fiebig, N. K. Mitra (eds.): Vortices and Heat Transfer. Results of a DFG-Supported Research Group. ISBN 3-540-41558-0

Volume 62: M. Griebel, O. Iliev, S. D. Margenov, P. S. Vassilevski (eds.): Large-Scale Scientific Computations of Engineering and Environmental Problems. Proceedings of the First Workshop on "Large-Scale Scientific Computations" Varna, Bulgaria, June 7–11, 1997. ISBN 3-540-41560-2

Volume 57: H. Deconinck, B. Koren (eds.): Euler and Navier-Stokes Solvers Using Multi-Dimensional Upwind Schemes and Multigrid Acceleration. Results of the BRITE/EURAM Projects AERO-CT89-0003 and AER2-CT92-00040, 1989–1995. ISBN 3-540-41535-1

Printing: Mercedesdruck, Berlin
Binding: Buchbinderei Lüderitz & Bauer, Berlin

GPSR Compliance

The European Union's (EU) General Product Safety Regulation (GPSR) is a set of rules that requires consumer products to be safe and our obligations to ensure this.

If you have any concerns about our products, you can contact us on ProductSafety@springernature.com

In case Publisher is established outside the EU, the EU authorized representative is:

Springer Nature Customer Service Center GmbH
Europaplatz 3
69115 Heidelberg, Germany

Batch number: 10755284

Printed by Printforce, the Netherlands